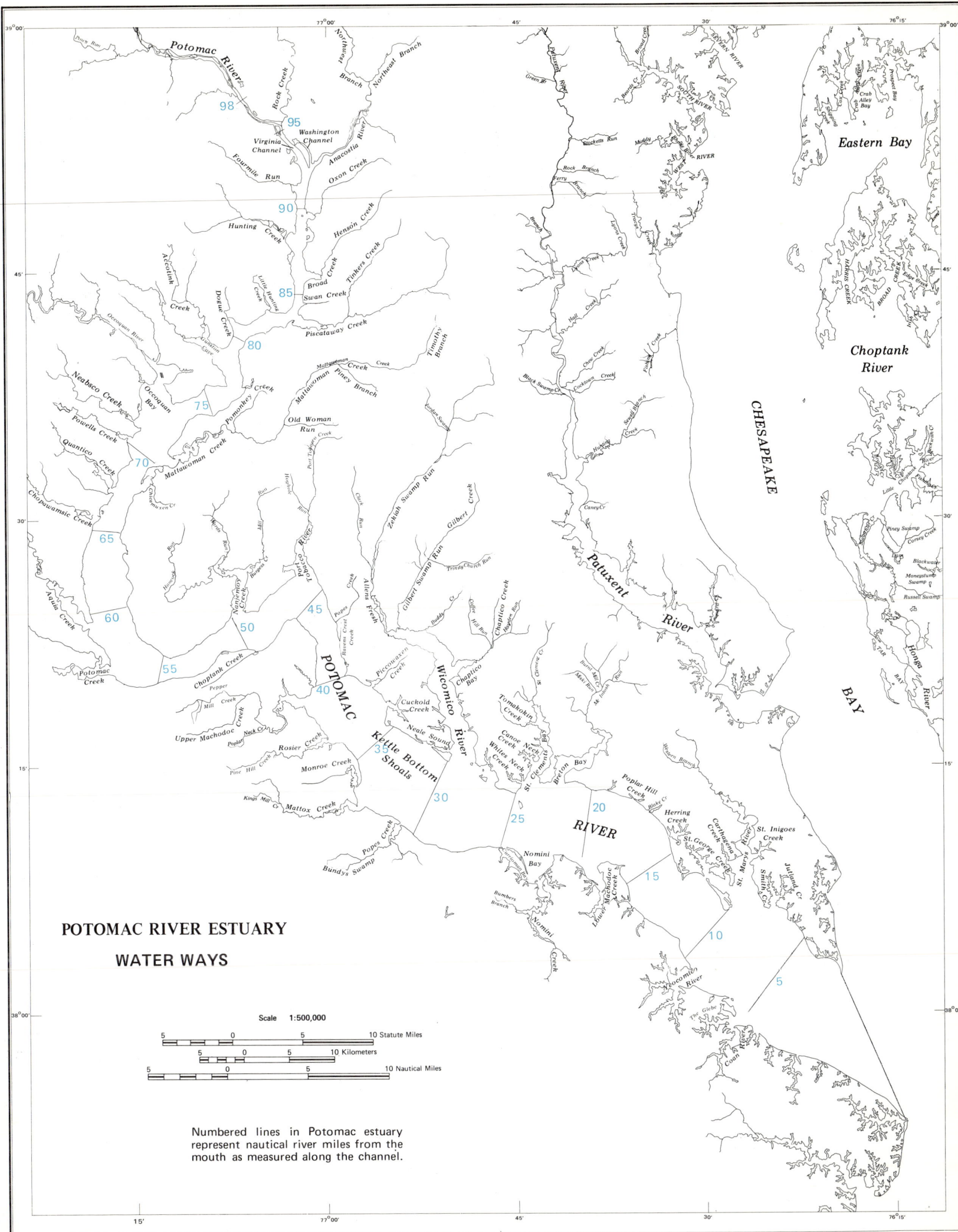

POTOMAC RIVER ESTUARY

WATER WAYS

Scale   1:500,000

5   0   5   10 Statute Miles
5   0   5   10 Kilometers
5   0   5   10 Nautical Miles

Numbered lines in Potomac estuary represent nautical river miles from the mouth as measured along the channel.

Potomac River

Rock Creek

98

95

Washington Channel

Virginia Channel

Fourmile Run

Anacostia River

Oxon Creek

Northwest Branch

Northeast Branch

90

Hunting Creek

Henson Creek

Accotink Creek

Dogue Creek

Broad Creek

Tinkers Creek

85

Swan Creek

Little Hunting Creek

Piscataway Creek

Occoquan River

80

Timothy Branch

Mattawoman Creek

Piney Branch

Neabsco Creek

Occoquan Bay

75

Pomonkey Creek

Old Woman Run

Powells Creek

Mattawoman Creek

70

Quantico Creek

Chopawamsic Creek

Zekiah Swamp Run

Gilbert Creek

65

Port Tobacco River

Gilbert Swamp Run

Trinity Church Run

Aquia Creek

60

POTOMAC

45

Nanjemoy Creek

50

Allens Fresh

Wicomico River

Chaptico Creek

Chaptico Bay

Potomac Creek

55

Choptank Creek

40

Piccowaxen Creek

Cuckold Creek

Tomakokin Creek

Upper Machodoc Creek

Rosier Creek

Kettle Bottom Shoals

Neale Sound

Canoe Neck Creek

Whites Neck Creek

St. Clements Bay

Monroe Creek

35

30

Breton Bay

Poplar Hill Creek

Mattox Creek

Popes Creek

25

20

Herring Creek

St. George Creek

Carthagena Creek

St. Inigoes Creek

Bundys Swamp

Nomini Bay

15

St. Marys River

Smiths Creek

Bumbers Branch

Lower Machodoc Creek

10

Jutland Creek

Nomini Creek

Coan River

5

Yeocomico River

The Globe

CHESAPEAKE BAY

Patuxent River

RIVER

Eastern Bay

Choptank River

South River

West River

Honga River

39°00'

45'

30'

15'

38°00'

77°00'

45'

30'

76°15'

15'

Environmental Atlas
of the
Potomac Estuary

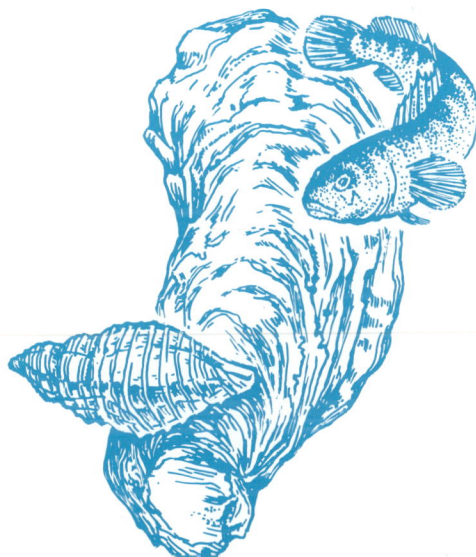

Environmental Atlas
of the
Potomac Estuary

Alice J. Lippson
Michael S. Haire
A. Frederick Holland
Fred Jacobs
Jorgen Jensen
R. Lynn Moran-Johnson
Tibor T. Polgar
William A. Richkus
of the
Environmental Center, Martin Marietta Corporation
Prepared for
Power Plant Siting Program, Maryland Department of Natural Resources

Copyrights

PHOTOGRAPHS

Cover — © 1979, Brenda J. Bodian
Chapter 1 — © 1978, John C. Mannone
Chapter 2 — © 1977, Robert deGast
Chapter 3 — © 1965, Robert deGast
Chapter 4 — © 1971, John V. Martin
Chapter 5 — © 1977, Robert deGast
Chapter 6 — © 1978, J. Gordon and T. Weeks/Environmental Photography
Chapter 7 — © 1978, J. Gordon and T. Weeks/Environmental Photography
Chapter 8 — © 1978, J. Gordon and T. Weeks/Environmental Photography
Chapter 9 — © 1974, M. E. Warren
Chapter 10 — © 1959, M. E. Warren

ILLUSTRATIONS

Chapter 4 and Appendix Table 1

Illustration of *Ceramium:*

Redrawn after Isabella A. Abbott and E. Yale Dawson: *How to Know the Seaweeds.* 2nd ed. Copyright © 1978 by the publisher, Wm. C. Brown Company, Dubuque, Iowa. Used with permission.

Illustrations of *Carteria cordiformis. Ceratium hirudinella. Chroomonas, Coelastrum reticulatum. Cryptomonas ovata. Dinobryon cylindricum. Eudorina. Exuviella marina. Glenodinium cinctum. Oocystis. Pediastrum duplex. Peridinium. Phacotus lenticularis. Pyramimonas tetraryhnchus. Skeletonema costatum. Sphaerocystis schroeteri. Spirogyra. Staurastrum. Thalassiosira. Tribonema:*

Redrawn after F. E. Fritsch: *The Structure and Reproduction of the Algae*, Vol. I. Copyright © 1956. Used with permission of the publisher, Cambridge University Press. New York, N.Y.

Illustrations of *Fragilaria. Microcoleus.* and *Rhizosolenia:*

Redrawn after G. W. Prescott: *The Algae: A Review.* Copyright © 1968 by the author, published by Houghton Mifflin Company, Boston, Mass. Used with permission.

Illustrations of *Cladophora gracilis. Ectocarpus tomentosus,* and *Enteromorpha prolifera:*

Redrawn after William Randolph Taylor: *Marine Algae of the Northeastern Coast of North America.* 2nd ed. Copyright © 1957 by the University of Michigan Press. Ann Arbor, Mich. Used with permission.

Chapter 6

Illustration of the barnacle nauplius in Fig. 6-2:

Redrawn after Charles C. Davis: *The Marine and Fresh-Water Plankton.* Copyright © 1955 by the Michigan State University Press. East Lansing, Mich. Used with permission.

Illustration of the cladoceran *Leptodora* sp. in Fig. 6-1:

Redrawn after Kenneth L. Gosner: *Guide to Identification of Marine and Estuarine Invertebrates.* Copyright © 1971 by John Wiley & Sons, New York, N.Y. Used with permission.

Illustration of the barnacle cypris in Fig. 6-2:

Redrawn after M. Grant Gross: *Oceanography — a View of the Earth.* Copyright © 1972 by Prentice-Hall, Inc., Engelwood Cliffs, N.J. Used with permission.

Illustrations of the cladoceran *Bosmina longirostris* and the true shrimp *Palaemonetes vulgaris* in Fig. 6-1:

Redrawn after Robert W. Pennak: *Fresh-Water Invertebrates of the United States.* Copyright © 1953 by The Ronald Press Company, New York, N.Y. Used with permission.

Chapter 7

Illustrations of *Corophium lacustre* and *Leptocheirus plumulosus:*

Adapted from E. L. Bousfield: *Shallow-Water Gammaridean Amphipoda of New England.* Copyright © 1973 by the National Museum of Canada. Used by permission of the publisher, Cornell University Press.

Illustrations of *Chirodotea almyra. Cyathura polita,* and *Leptochelia rapax:*

Redrawn after Kenneth L. Gosner: *Guide to Identification of Marine and Estuarine Invertebrates.* Copyright © 1971 by John Wiley & Sons, Inc., New York, N.Y. Used with permission.

Illustration of the oligochaete *Aeolosoma leidyi:*

Redrawn after Robert W. Pennak: *Fresh-Water Invertebrates of the United States.* Copyright © 1953 by The Ronald Press Company, New York, N.Y. Used with permission.

Illustration of *Cerebratulus lacteus:*

Redrawn after Ralph I. Smith: *Keys to Marine Invertebrates of the Woods Hole Region.* Contribution II, Systematics — Ecology Program. Copyright © 1964, by Marine Biological Laboratory. Woods Hole, Mass. Used with permission.

Chapter 8

Illustrations of alewife and Atlantic croaker:

From illustrations in Alice Jane Lippson: *The Chesapeake Bay in Maryland: An Atlas of Natural Resources.* Copyright © 1973 by The Johns Hopkins University Press, Baltimore, Md. Used with permission.

Illustrations of bluegill, tessellated darter, largemouth bass, spottail shiner, white sucker, and yellow perch:

From illustrations drawn by Mr. Anker Odum and Mr. Peter Buerschaper for W. B. Scott and E. J. Crossman: *Freshwater Fishes of Canada.* Copyright © 1973 by the authors. Used with permission.

Illustration of brook lamprey ammocoete:

Redrawn after illustration made by R. Lynn Moran-Johnson for Johnson C. S. Wang and Ronnie J. Kernehan: *Fishes of the Delaware Estuaries: A Guide to the Early Life Histories.* Copyright © 1979 by the authors. Published by EA Communications, Ecological Analysts, Inc., Towson, Md.

Chapter 9

Illustration of mallard and lesser scaup:

From illustrations in Alice Jane Lippson: *The Chesapeake Bay in Maryland: An Atlas of Natural Resources.* Copyright © 1973 by The Johns Hopkins University Press, Baltimore, Md. Used with permission.

PRODUCTION CREDITS

Design and layout — Schneider Design Associates, Inc., Baltimore, Md.
Typography — Service Composition Co., Baltimore, Maryland
Map production and book printing — Williams & Heintz Map Corporation. Washington, D.C.

Acknowledgements

We are indebted to Ms. Brenda J. Bodian[1] for her invaluable and substantial contributions as the editor responsible for managing production and technical editing of this Atlas. Her enthusiasm, diligence, and initiative were critical to the completion of the project.

The authors would like to thank three artists whose fine illustrations assist greatly in conveying information and contribute to the aesthetic quality of this work. Ms. Elaine Kasmer[2] spent many painstaking hours producing most of the graphic and illustrative art. We are also grateful to Ms. Martha H. Heigel[3] and Mr. Daniel A. Wilhide[1] for their efforts.

Special acknowledgements go to Ms. Lois A. Craig,[1] Ms. Nadine S. Johansen,[1] and Ms. Charlotte E. Clark[1] for their careful typing and retyping of the manuscript, for attending to the many details, and for maintaining consistency of style.

In the gathering of source material, librarians Mrs. Helen Lang,[4] Ms. Rosalind P. Cheslock,[1] and Ms. Judith G. Watts[1] provided an invaluable service. The excellent technical assistance of Messrs. Joseph Arlauskas, Jr.,[1] George Krainak,[1] Allart Kok,[1] and Robert P. LaBelle,[1] and Ms. Judith H. Marcus[1] contributed substantially to the synthesis and verification of information.

We appreciate the time taken by the following individuals to review all or part of the draft version of this Atlas: Drs. Melbourne R. Carriker,[5] L. Eugene Cronin,[6] John M. Dean,[7] and Saul B. Saila;[8] and Messrs. Elder A. Ghigiarelli, Jr., Thomas Hopkins, Jr., Duane Pursley, Joshua Sandt, William S. Sipple, and Gary J. Taylor, all of the Maryland Department of Natural Resources.[9]

And, finally, we thank Drs. Paul R. Massicot, Myron H. Miller, and Randy Roig, and Mr. Levio E. Zeni of the Maryland Power Plant Siting Program and Dr. Leonard H. Bongers, Messrs. Kenneth Jarmolow, Arthur E. Koski, and Donald R. Talbot of Martin Marietta Corporation for their personal interest in and administrative support of this project.

---

[1] Martin Marietta Corporation, Environmental Center and Martin Marietta Laboratories, Baltimore, Md.

[2] Freelance, Baltimore, Md. (contact through Martin Marietta Corporation, Environmental Center, Baltimore, Md.)

[3] University of Maryland, Chesapeake Biological Laboratory, Solomons, Md.

[4] U.S. Department of the Interior, National Marine Fisheries Service, Library, Oxford, Md.

[5] University of Delaware, College of Marine Sciences, Newark, Del.

[6] Chesapeake Research Consortium, Parole, Md.

[7] University of South Carolina, Belle W. Baruch Institute, Columbia, S.C.

[8] University of Rhode Island, Graduate School of Oceanography, Narragansett Bay Campus

[9] Maryland Department of Natural Resources, Annapolis, Md.

Credits

These credits are specially singled out because no reference citations are made for the materials provided. We are grateful to many of the following individuals and institutions for contributing information or data unavailable from publications or data banks.

Chapter 1

John C. Mannone, Westinghouse Electric, Idaho Falls, Idaho — photograph

Chapter 2

Robert deGast, Annapolis, Md. — photograph

Chapter 3

Robert deGast, Annapolis, Md. — photograph

Robert L. Lippson, University of Maryland, Chesapeake Biological Laboratory, Solomons, Md. — information from original field data sheets on bottom and surface salinities

William T. Mason, U.S. Department of the Interior, Fish and Wildlife Service (formerly with Interstate Commission on the Potomac River Basin) — information on sedimentation

V. James Rasin, Interstate Commission on the Potomac River Basin, Bethesda, Md. — up-to-date information on water quality status

United States Environmental Protection Agency, Annapolis Field Office, Md. (Donald Lear; Maria O'Malley; Susan Smith) — water quality data not available from the EPA STORET system

Chapter 4 and Appendix Table 1

Cambridge University Press — permission to use illustrations made from figures in *Structure and Reproduction of the Algae*, Vol. 1, by F. E. Fritsch, © 1956*

Judith L. Connor, University of Maryland, Center for Environmental and Estuarine Studies, Horn Point Laboratory, Horn Point, Md. — information on seaweed distributions in the lower Potomac estuary, and review of taxonomy in Appendix Table 1

Houghton Mifflin Company — permission to use illustrations made from figures in *The Algae: A Review* by G. W. Prescott, © 1968*

Michael E. Kachur, Academy of Natural Sciences of Philadelphia, Pa. — information on taxonomy of blue-green algae

John V. Martin, Ecological Analysts, Inc., Baltimore, Md. — photograph, and review of taxonomy in Appendix Table 1

The University of Michigan Press — permission to use illustrations made from figures in *Marine Algae of the Northeastern Coast of North America*, 2nd ed., by G. J. Taylor, © 1957*

Wm. C. Brown Publishers — permission to use an illustration made from a figure in *How to Know the Seaweeds*, 2nd ed., by I. A. Abbott and E. Y. Dawson, © 1978*

Chapter 5

Harold M. Castle, Maryland Department of Natural Resources, Annapolis, Md. — information from original field data on wetlands in three Maryland counties

Robert deGast, Annapolis, Md. — photograph

Chapter 6

Dale R. Calder, South Carolina Research Laboratory, Charleston, S. C. — permission to use illustrations of the anthomedusa *Sarsia tubulosa* and the leptomedusa *Lovenella gracilis* in Fig. 6-1, which were made from figures in *Hydroids and Hydromedusae of Southern Chesapeake Bay*. Special Papers in Marine Science 1, Virginia Institute of Marine Science

Julius Gordon and Townsend E. Weeks, Environmental Photography, Newark, Del. — photograph

Michigan State University Press — permission to use an illustration made from a figure in *The Marine and Fresh-Water Plankton*, by C. C. Davis, © 1955*

John Wiley & Sons, Inc. — permission to use an illustration made from a figure in *Guide to Identification of Marine and Estuarine Invertebrates*, by K. L. Gosner, © 1971*

Prentice-Hall, Inc. — permission to use an illustration made from a figure in *Oceanography: A View of the Earth*, by M. G. Gross, © 1972*

Paul A. Sandifer, South Carolina Marine Research Laboratory, Charleston, S.C. — permission to use the illustration of shrimp zoea in Fig. 6-2 made from a figure in his Ph.D. dissertation, *Morphology and Ecology of Chesapeake Bay Decapod Crustacean Larvae*, University of Virginia

The Ronald Press Company — permission to use illustrations made from figures in *Fresh-Water Invertebrates of the United States*, by R. W. Pennak, © 1953*

Chapter 7

Dale R. Calder, South Carolina Marine Research Laboratory, Charleston, S.C. — permission to use an illustration of *Bougainvillia rugosa* made from a figure in *Hydroids and Hydromedusae of Southern Chesapeake Bay*. Special Papers in Marine Science 1 Virginia Institute of Marine Science

Cornell University Press — permission to use illustrations made from figures in *Shallow-water Gammaridean Amphipoda of New England*, by E. L. Bousfield, © 1973*

Julius Gordon and Townsend E. Weeks, Environmental Photography, Newark, Del. — photograph

Frank L. Hamons, Jr., Maryland Department of Natural Resources, Water Resources Adm., Annapolis, Md. — information on soft-shell and brackish-water clams

George H. Harmon, Maryland Department of Natural Resources, Water Resources Adm., Annapolis, Md. — field data sheets on benthic organisms in the upper Potomac estuary

Martha H. Heigel, University of Maryland, Chesapeake Biological Laboratory, Solomons, Md. — permission to use her illustrations of *Pectinaria gouldii* and *Heteromastus filiformis* and Figure 7-1.

Robert L. Lippson, University of Maryland, Chesapeake Biological Laboratory, Solomons, Md. — information on the distribution of blue crabs and other benthic macroinvertebrates and species list of benthic macroinvertebrates in the Potomac estuary

Marine Biological Laboratory, Woods Hole, Mass. — permission to use an illustration made from a figure in *Keys to Marine Invertebrates of the Woods Hole Region*, edited by R. I. Smith, © 1964*

Robert M. Norris, Potomac River Fisheries Commission, Colonial Beach, Va. — data and distributional information on clam and oyster beds

Erik Rasmussen, The Isefjord Laboratory, Vellerup Vig, Denmark — permission to use illustrations made from figures in his book *Systematics and Ecology of the Isefjord Marine Fauna (Denmark)*, reprinted from *Ophelia*, Vol. II

The Ronald Press Company — permission to use an illustration made from a figure in *Fresh-Water Invertebrates of the United States*, by R. W. Pennak, © 1953*

John Wiley & Sons, Inc. — permission to use illustrations made from figures in *Guide to Identification of Marine and Estuarine Invertebrates*, by K. L. Gosner, © 1971*

Chapter 8

Julius Gordon and Townsend E. Weeks, Environmental Photography, Newark, Del. — photograph

The Johns Hopkins University Press — permission to use illustrations from *The Chesapeake Bay in Maryland: An Atlas of Natural Resources*, by A. J. Lippson, © 1973*

Robert L. Lippson, University of Maryland, Chesapeake Biological Laboratory, Solomons, Md. — information on fish distributions

Jay O'Dell, Maryland Department of Natural Resources, Fisheries Adm., Annapolis, Md. — information on finfish distributions and anadromous fish spawning migrations

W. B. Scott, Huntsman Marine Biological Laboratory, St. Andrews, New Brunswick, Canada, and E. J. Crossman, Royal Ontario Museum, Toronto, Ontario, Canada — permission to use figures from their book *Freshwater Fishes of Canada* published by the Fisheries Board of Canada, © 1973. Drawing of tessellated darter adapted from illustration of *Etheostoma niger* by Mr. Anker Odum. The other drawings used were made by Mr. Anker Odum and Mr. Peter Buerschaper.*

Johnson C. S. Wang, Ecological Analysts, Inc., Towson, Md. and Ronnie J. Kernehan, RMC, Middletown, Del. — permission to use illustration of brook lamprey ammocoete drawn by R. Lynn Moran-Johnson for their book *Fishes of the Delaware Estuaries: A Guide to the Early Life Histories* published by EA Communications, Ecological Analysts, Inc., Towson, Md. © 1979.

Chapter 9

Larry J. Hindman, Maryland Department of Natural Resources, Wildlife Adm., Annapolis Md. — field data sheets from 1960-76 waterfowl surveys

The Johns Hopkins University Press — permission to use illustrations from *The Chesapeake Bay in Maryland: An Atlas of Natural Resources*, by A. J. Lippson, © 1973*

Duane Pursley, Maryland Department of Natural Resources, Wildlife Adm., Annapolis, Md. — information on mammals in Maryland portion of Potomac drainage

Vernon D. Stotts, Maryland Department of Natural Resources, Wildlife Adm., Annapolis, Md. — information on waterfowl distributions

Gary J. Taylor, Maryland Department of Natural Resources, Wildlife Adm., Annapolis, Md. — information on eagle and osprey nest sites

Marion E. Warren, Annapolis, Md. — photograph

Chapter 10

Alexandria Drafting Company, Alexandria, Va. — information for producing Figure 10-1

Kirby A. Carpenter, Potomac River Fisheries Commission, Colonial Beach, Va. — information on the oyster and clam fishery and on the use of different commercial fishing gear in the Potomac estuary

Robert M. Norris, Potomac River Fisheries Commission, Colonial Beach, Va. — information on the oyster and clam fishery and on the use of different commercial fishing gear in the Potomac estuary

Marion E. Warren, Annapolis, Md. — photograph

---

*See copyright page for list of illustrations.

Table of Contents

Folio Maps:

Introduction

2 The growing awareness in recent years of the need to protect the Potomac estuary's resources has led planners to sponsor detailed and quantitative studies designed to integrate knowledge about the estuarine system's physical and biological features. This Atlas was conceived and sponsored by the Maryland Power Plant Siting Program as a management aid for evaluating the regional implications of environmental changes caused by the operation of existing power plants and for evaluating the suitability of sites for future facilities along the estuary. This compendium of information can also be productively used in regional planning for industrial, agricultural, and urban development activities.

The portion of the Potomac from Washington, D.C., to the Chesapeake Bay is under the influence of the tides and salt water intruding from the Bay, and is thus considered an estuary. Estuaries, including the Potomac, are complex environments that have a variety of habitats and biological components. They are also greatly influenced by dynamic physical and chemical processes. Suitable planning for management and conservation in heterogeneous environments such as the Potomac can proceed only from knowledge of the behavior, distribution, and interdependence of the biotic and abiotic elements of the system. Such planning would allow for continued growth in human activities that could have impacts on the estuary. The Atlas is intended as an information base from which planners and scientists can identify problem areas requiring further investigation, especially in cases where various uses, value judgments, and concerns about the estuarine environment could conflict.

Mapping was chosen as the most effective way of depicting and summarizing the features of the biological zones and the interactions of physical and biological components in the estuary. Mapping also allows delineation of regions that may be vulnerable to human activities by superimposing concurrent uses of various habitats by different biota. These uses are interdependent and are also controlled by prevailing physical and chemical conditions.

The maps, with the accompanying graphs and tables, represent the synthesis of most of the published and much of the unpublished material pertaining to the Potomac. The existing information was gathered and distilled at various levels of detail to bring forth the essential and typical characteristics of the system. Some of the sources used were raw data gathered in occasional and unrelated scientific studies and monitoring programs. However, a great deal of the detailed and synoptic information was derived from the results of systematic, large-scale investigations such as those sponsored by the Maryland Power Plant Siting Program and those carried out by consultants to industry. The larger programs are briefly described in Appendix Table 10.

Because of the great disparities in the quality, nature, and detail of the data used, each of the maps in the Atlas has been coded to characterize its informational content and to clarify the intent of each map for the user. These codes do not indicate judgments about the basic quality of the data used to construct the maps; instead, since only data deemed of adequate technical reliability were mapped, the coding reflects the degree of generalization made from the data and the limitations implicit in data synthesis. Each Atlas map is coded in the upper right hand corner with one of the following categorizations:

D represents a mapping of *descriptive* information; *qualitative* patterns are derived from descriptive sources.

I represents a mapping that is intended to be *interpretive; qualitative* patterns are shown which represent well-known distributions that are either derived from selected and incomplete data, or that are expected to be associated with patterns derived from quantitative, published information describing Potomac populations and/or populations in other similar systems.

R represents a mapping that is intended to be *representative; quantitative* patterns are presented that are typical and that are not expected to exhibit much variation over time.

M represents a mapping that displays *monitored* conditions found during specific periods; *quantitative* patterns are shown from monitoring of conditions that may change in time.

The text on the back of each of the nine folded folio maps explains the map's intended use and lists the sources of information presented. These explanatory notes replace the coding used for the maps in the text since many of the folio maps contain both qualitative and quantitative data. Folio maps 1, 2, and 3 present basic geographical information, and Folio Map 9 depicts the locations and sizes of power plant cooling water flows and wastewater discharges. The five remaining maps are guides for relating species of interest to their habitats and to associated organisms, and are intended to be used with the more detailed maps, tables, and discussions in the Atlas.

The text of the Atlas is organized into three general subject areas — the physical and chemical characteristics of the estuary, the biota of the estuary, and finally, human exploitation of the Potomac's renewable resources. The first three chapters describe the physical and chemical characteristics of the Potomac and the properties and processes that establish the estuarine habitat zones which delimit biotic distributions. The biota of the estuary are discussed in Chapters 4 through 9 in a logical progression from one biological group to another. Plant forms — phytoplankton and other algae in the water column, submerged plants rooted to the bottom, and wetland vegetation along the shores — are presented first because they are the crucial links in the food chain between the nutrients in the water and animals, and because they respond most directly to changes in physical conditions and nutrient levels. Estuarine invertebrates, fish, birds, and mammals are then discussed in relation to the physical and chemical characteristics of the environment. Chapters 7 and 8 discuss benthic species and fish, and are the most detailed since they describe the natural resources of the estuary that are of most direct importance to people. Chapter 10 discusses the spatial and temporal patterns of the fisheries harvest from the estuary and the species composition of the catch.

Ten appendix tables follow the chapters. Appendix Tables 1 through 8 are associated with the chapters on biota, and inventory the species recorded in the Potomac. In addition, the endangered species that have been or may be found in the Potomac are listed in Appendix Table 9, and Appendix Table 10 briefly describes the major scientific monitoring and impact assessment programs that have contributed data to this Atlas.

The glossary defines specialized and technical terms, and a conversion table is provided for converting metric and English units.

Reference numbers in the text and on the accompanying figures and tables relate to an abbreviated list of references at the end of each chapter. For readers who wish to investigate a topic further, an alphabetical list of complete reference citations follows the conversion table. The detailed subject index should enable the reader to easily locate information in the text, figures, appendix tables, and folio maps.

The inside covers of the Atlas provide additional aids to the reader. The maps on the front inside cover include the common place names, waterway names, and nautical-river-mile designations used in the Atlas. Because many nautical-river-mile systems have been used by different organizations and government agencies working on the Potomac, one system was arbitrarily chosen for use in the Atlas — the one that has been employed by the Chesapeake Bay Institute for many years. The matrix on the back inside cover of the Atlas summarizes the physical, chemical, biological, resource use, and cultural features of the system at different nautical-river-mile locations. This matrix can serve as a guide for selecting components of concern for more detailed review.

A wide spectrum of interdependent elements of the ecosystem is covered in the Atlas, but all aspects of the system are not treated equally, either because information was lacking in some areas or because certain elements would have a limited effect on management decisions. Physical and chemical processes are presented only as they relate to biological distributions and functions. Although insects and the bacteria, fungi, protozoans, and other microbiota in the Potomac play important roles in the ecosystem, they are mentioned only briefly, and amphibians and reptiles are not discussed.

The Atlas deals with a specific estuary, but the distributions of the animals and plants described and the ecological role that each plays within the estuarine system are controlled by physical and biological processes that are generally common to all temperate-zone estuaries. Thus, we hope that the information and its presentation will serve as a prototype that will be useful in the management of other coastal plain estuaries.

Estuaries

Estuaries

Most definitions of the word estuary do not reflect the uniqueness of these bodies of water as habitats for living things. By scientific definition, an estuary is "a semi-enclosed body of water that has a free connection with the open sea and within which sea water is measurably diluted with fresh water derived from land drainage."[1] Even though this physical description is precise, it fails to convey the dynamic nature of the physical processes operating in estuaries, nor does it explain the roles that these processes play in shaping the character of aquatic and terrestrial life in and around an estuary.

Salty waters intruding from the seaward ends of estuaries are gradually mixed by tides with fresh waters from upstream, creating continuous salinity gradients along their lengths. Vertical and cross-stream salinity gradients are also maintained by tidally-induced mixing and by other processes such as wind-induced turbulence. Nutrients brought in by river waters, as well as pollutants introduced by man, are distributed in estuaries by the same processes that control salinity distribution. Circulation and transport processes also influence temperature and sediment distributions, and play major roles in determining the geometry of the basin by influencing erosion and deposition patterns.

The dynamic processes occurring in estuaries create several fairly predictable, but heterogeneous, environments. Assemblages of organisms found in estuaries range from fresh to marine in character, and the physical environment may encompass vigorously flushed to somewhat stagnant waters. The circulation, storage, and recycling of materials in estuaries, together with occasional large imports of nutrients from drainage areas, assure a lasting supply of ingredients to maintain high levels of biological productivity in all estuarine habitats. The high primary productivity of plants is reflected by the presence of numerous and diverse groups of organisms that consume plant material. Together, the plants and animals form a complex estuarine food web in which there are many interdependencies that are normally in balance.

The linkage between the physical and chemical factors in an estuary and its biological composition is indeed close. Even under normal conditions, physical and chemical variations from one location to another may be large. However, through evolution, many organisms found in estuaries have adapted to tolerate extreme conditions and to form viable associations. To appreciate the biological, physical, and chemical characteristics of estuaries and their interdependencies, it is instructive to delve into the general nature of biotic and abiotic processes. It is the intent of this chapter to characterize general estuarine processes.

PHYSICAL AND CHEMICAL PROPERTIES
Estuarine Circulation and Salinity Distribution

To the most casual observer, the only movement of water in estuaries seems to be the ebbing and flooding of the tide. However, other flows that are masked by the oscillating tidal currents can be detected by current meters and by the displacement of floating buoys. These flows

result from fresh water flushing through estuarine basins and setting up complex, nontidal circulation patterns in the process. It is the particular character of the nontidal circulation pattern that physically distinguishes one type of coastal plain estuary from another. The major coastal plain estuarine types can be described either as highly stratified (i.e., unmixed), partially mixed, or fully mixed (i.e., vertically homogeneous). This nomenclature refers to the stratification of water density, which in turn is mostly determined by the salt content of water and, to a lesser extent, by temperature. The degree of mixing between salt and fresh waters and the maintenance of a particular salinity pattern in an estuary are closely tied to the character of its nontidal flows.

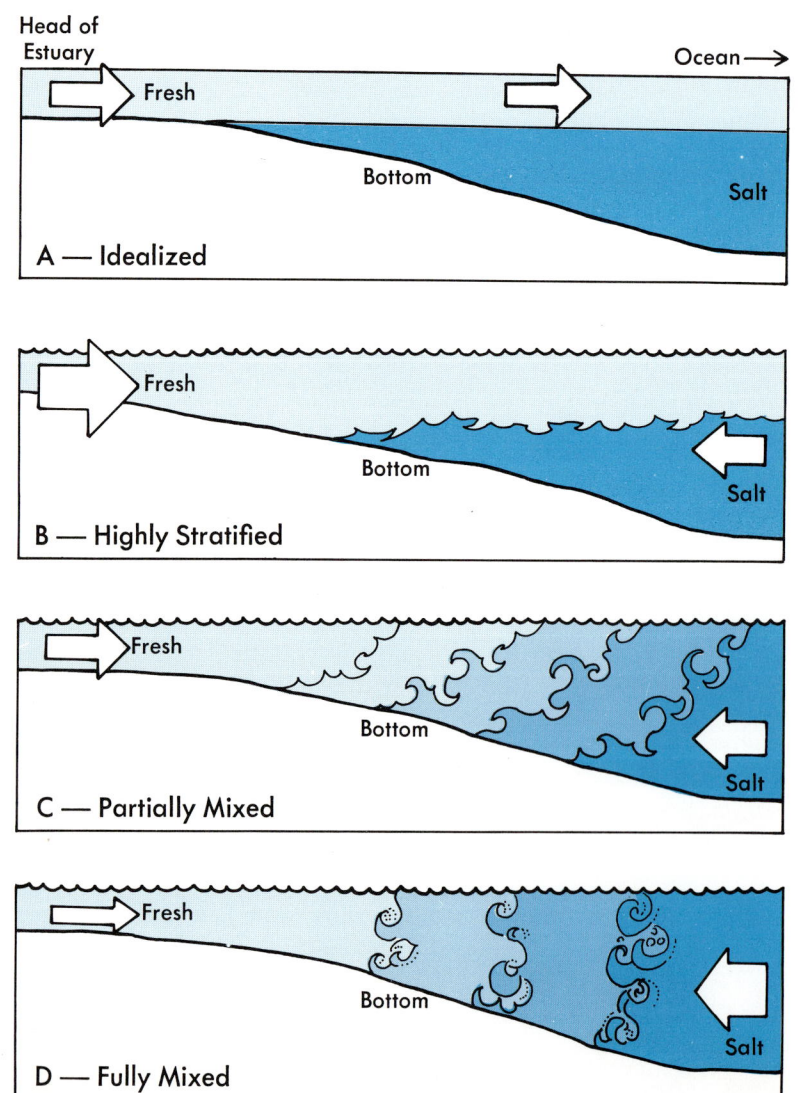

Figure 1-1. *Schematic diagram of coastal plain estuary types, indicating direction and degree of mixing. Arrows show direction of net mass transports of water, and the arrow size indicates the relative magnitudes of the transports.*

Estuaries

Estuaries

Most definitions of the word estuary do not reflect the uniqueness of these bodies of water as habitats for living things. By scientific definition, an estuary is "a semi-enclosed body of water that has a free connection with the open sea and within which sea water is measurably diluted with fresh water derived from land drainage."[1] Even though this physical description is precise, it fails to convey the dynamic nature of the physical processes operating in estuaries, nor does it explain the roles that these processes play in shaping the character of aquatic and terrestrial life in and around an estuary.

Salty waters intruding from the seaward ends of estuaries are gradually mixed by tides with fresh waters from upstream, creating continuous salinity gradients along their lengths. Vertical and cross-stream salinity gradients are also maintained by tidally-induced mixing and by other processes such as wind-induced turbulence. Nutrients brought in by river waters, as well as pollutants introduced by man, are distributed in estuaries by the same processes that control salinity distribution. Circulation and transport processes also influence temperature and sediment distributions, and play major roles in determining the geometry of the basin by influencing erosion and deposition patterns.

The dynamic processes occurring in estuaries create several fairly predictable, but heterogeneous, environments. Assemblages of organisms found in estuaries range from fresh to marine in character, and the physical environment may encompass vigorously flushed to somewhat stagnant waters. The circulation, storage, and recycling of materials in estuaries, together with occasional large imports of nutrients from drainage areas, assure a lasting supply of ingredients to maintain high levels of biological productivity in all estuarine habitats. The high primary productivity of plants is reflected by the presence of numerous and diverse groups of organisms that consume plant material. Together, the plants and animals form a complex estuarine food web in which there are many interdependencies that are normally in balance.

The linkage between the physical and chemical factors in an estuary and its biological composition is indeed close. Even under normal conditions, physical and chemical variations from one location to another may be large. However, through evolution, many organisms found in estuaries have adapted to tolerate extreme conditions and to form viable associations. To appreciate the biological, physical, and chemical characteristics of estuaries and their interdependencies, it is instructive to delve into the general nature of biotic and abiotic processes. It is the intent of this chapter to characterize general estuarine processes.

PHYSICAL AND CHEMICAL PROPERTIES
Estuarine Circulation and Salinity Distribution

To the most casual observer, the only movement of water in estuaries seems to be the ebbing and flooding of the tide. However, other flows that are masked by the oscillating tidal currents can be detected by current meters and by the displacement of floating buoys. These flows

result from fresh water flushing through estuarine basins and setting up complex, nontidal circulation patterns in the process. It is the particular character of the nontidal circulation pattern that physically distinguishes one type of coastal plain estuary from another. The major coastal plain estuarine types can be described either as highly stratified (i.e., unmixed), partially mixed, or fully mixed (i.e., vertically homogeneous). This nomenclature refers to the stratification of water density, which in turn is mostly determined by the salt content of water and, to a lesser extent, by temperature. The degree of mixing between salt and fresh waters and the maintenance of a particular salinity pattern in an estuary are closely tied to the character of its nontidal flows.

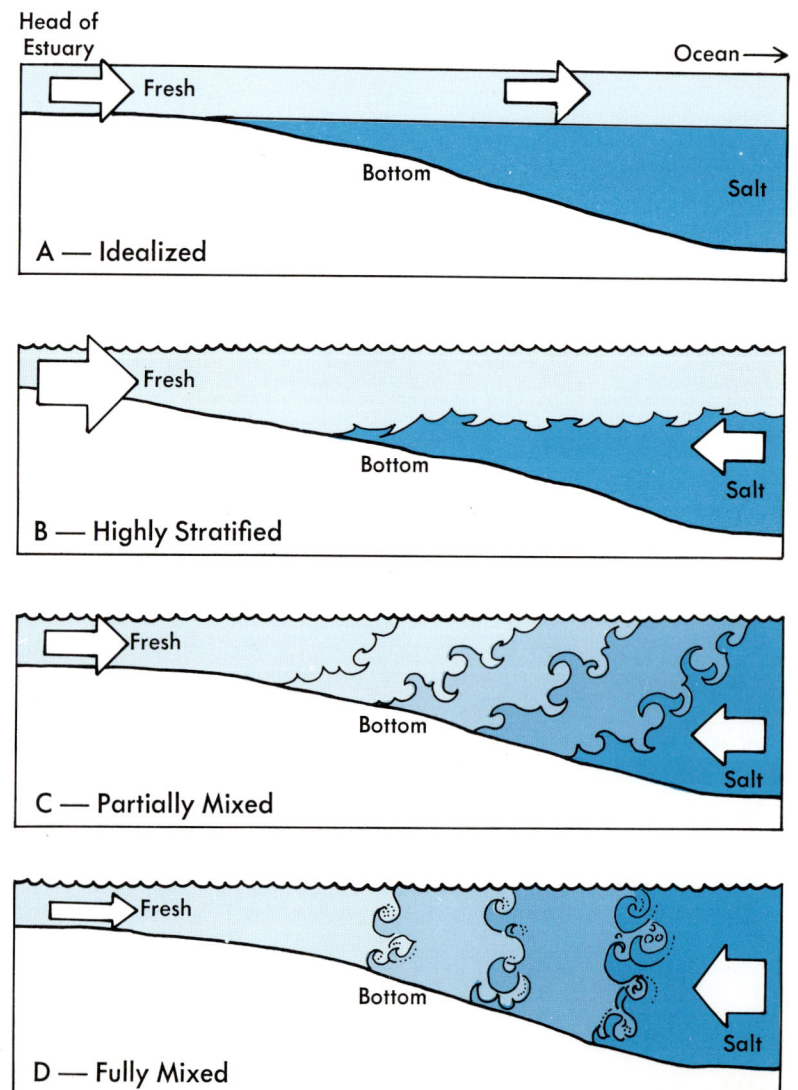

Figure 1-1. *Schematic diagram of coastal plain estuary types, indicating direction and degree of mixing. Arrows show direction of net mass transports of water, and the arrow size indicates the relative magnitudes of the transports.*

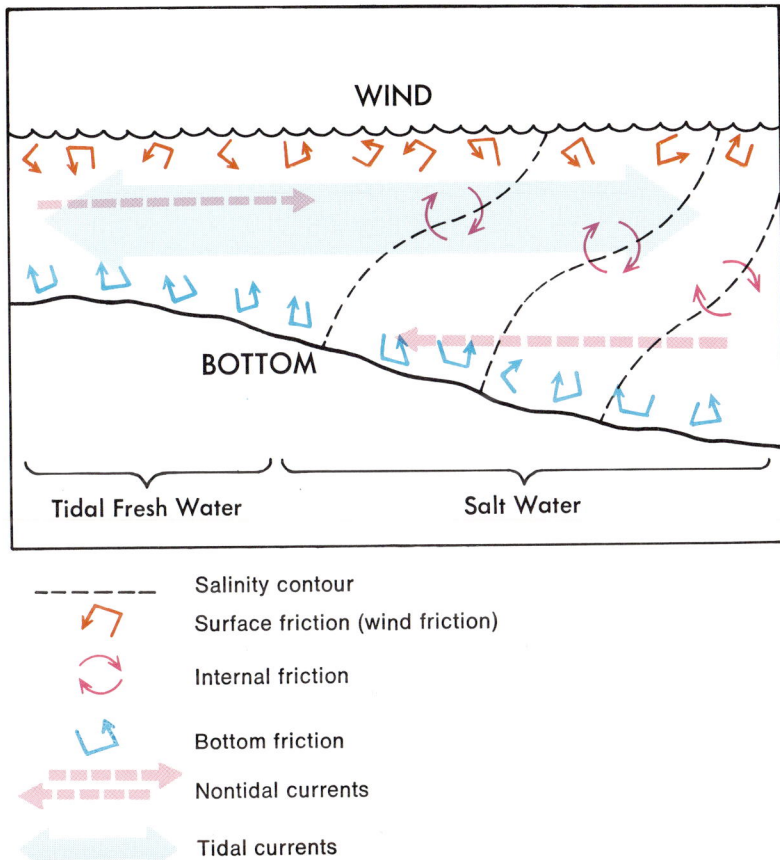

WIND

BOTTOM

Tidal Fresh Water | Salt Water

- - - - - Salinity contour

Surface friction (wind friction)

Internal friction

Bottom friction

Nontidal currents

Tidal currents

Figure 1-2. *Mixing forces and flows in estuaries.*

In an idealized situation, without tides and wind and with little friction, fresh water would enter the upstream end of the estuary and, because it is less dense, would float out to sea on top of a layer of salt water (Fig. 1-1A). The interface between saline and fresh water would extend upstream approximately to the point where the mean sea level intersects the bottom of the basin. There would be a small amount of mixing between the two discrete density layers, and practically all motion would occur in the downstream direction in the fresher surface layer alone. Because the basin would generally widen towards the mouth, the velocity would gradually decrease downstream. Since the basin would be deeper towards the mouth, a lengthwise profile of it would show that the saline layer was wedge-shaped, with its top forming an interface between the fresh and saline layers that was horizontal and parallel to the water surface.

In real estuaries, the tides, wind, and bottom friction, as well as internal friction, interact to produce mixing (Fig. 1-2). At the interface between salty and fresh water, the friction between the overriding freshwater layer and the denser saltwater layer will drag the leading edge of the salt front somewhat downstream.

Together, the amount of fresh water, the degree of tidal influence, and the geometry of the estuarine basin determine the physical

character of a coastal plain estuary, which strongly influences the nature of its biological communities.

HIGHLY STRATIFIED ESTUARIES

In highly stratified estuaries (Fig. 1-1B), which are also referred to as salt wedge estuaries, the density stratification between fresh and saline waters is relatively undisturbed by frictional forces and turbulence. These estuaries have steep vertical salinity gradients and small nontidal flows directed upstream in the bottom layers. The mass transport out of a highly stratified estuary in the upper layer is not appreciably greater than freshwater input at the head because frictional forces sweep only a small portion of the underlying saline layer seawards with the fresh water. The amount of water transported downstream in the surface layer usually keeps increasing because this layer constantly picks up saline waters from below in its passage. Since the salt water lost to the outgoing flow can only be replaced by inflow at the mouth, an upstream flow is induced in the saltier bottom layers.

Although this process of the surface layer picking up water from the more saline bottom waters is not dramatic in this type of estuary, it is a general characteristic of estuaries with some amount of vertical mixing. Thus, when the tidal currents are averaged out in any cross section of a highly stratified estuary, the upper layer tends to have a downstream nontidal transport that exceeds freshwater flow into the estuary. The difference between this transport and the transport directed upstream in the lower layer always equals the cumulative freshwater input to the point in the estuary where these transports are measured (Fig. 1-3). This situation applies even when there is no freshwater flow coming into the estuary from tributaries. When there are no tributaries along the estuarine portion of a river, the net downstream transport of water always equals the river input of fresh water at the head of the estuary.

For the water column to remain highly stratified in a salt wedge estuary, the ratio of the river flow to tidal flow must be large, and the ratio of width to depth must be small. These characteristics assure that relatively little vertical and longitudinal mixing will take place and

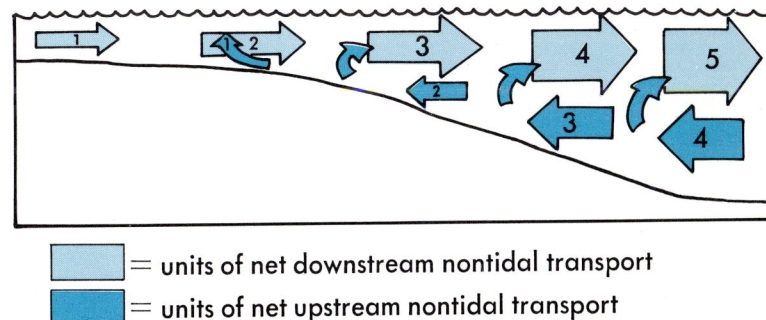

☐ = units of net downstream nontidal transport

☐ = units of net upstream nontidal transport

Figure 1-3. *Net transports in estuaries resulting from estuarine flows and mixing. At any one point along an estuary, the difference between upstream- and downstream-directed transports is equal to the freshwater input to that point. In this example with no tributaries, the difference is equal to the input at the head of the estuary. (Source: Ref. 2)*

that a large amount of fresh water will be available to maintain density stratification.

PARTIALLY MIXED ESTUARIES

In shallower and wider estuaries, tidal influence increases. In the resulting partially mixed estuaries, the nontidal transports in both surface and bottom layers can amount to more than the river input of fresh water. The Potomac below Washington, D.C., is an estuary of this type except during periods of low flow when it approaches being a stratified type. As a result of the relatively greater mixing between fresh and saline waters, the salinity distribution in any partially mixed estuary differs markedly from that in highly stratified estuaries (Fig. 1-1C). Whereas salinity differences occur predominantly in the vertical direction in the latter, salinity increases steadily down the estuary at all depths in a partially mixed estuary. In the middle portions of such estuaries the longitudinal salinity gradients can be almost linear (Fig. 1-4), with the isohalines (lines of equal salinity) equally spaced. Undiluted tidal fresh water* is likely to occur only near its head. Vertical salinity contours are very similar in shape along the length of a partially mixed estuary. The greatest mixing occurs at mid-depth, where the greatest vertical changes in salinity occur. Turbulence created by bottom friction and strong winds also plays an important role in shaping the contours.[3][4][5] Figure 1-2 depicts some of the major mechanisms and resulting flow and salinity patterns within such coastal estuaries.

Since water is displaced by nontidal flows and is dispersed by tidal and wind-induced turbulence, a parcel of water (if one could be marked) would travel downstream in the surface layer and upstream in the bottom layer, and parts of it would also be continuously dispersed into the surrounding water. If discrete particles were released and could maintain themselves in particular density layers, the resulting displacements would depend on the nontidal flows. The nontidal flows can be measured, in fact, by such displacements over many tidal cycles. Figure 1-5 shows schematically how tidal oscillations are averaged out over several tidal cycles so that the net displacement is due only to nontidal flows. Theoretically, there is a mid-depth position at which a particle would fail to undergo any net displacement in either direction because nontidal flow is zero.

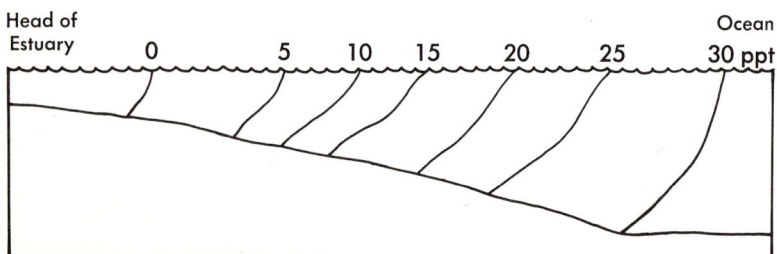

Figure 1-4. *Salinity gradients in a partially mixed estuary. Lines designate equal salinity contours (isohalines). Numbers designate the concentration of salt as ppt (parts salt per thousand parts of water by weight).*

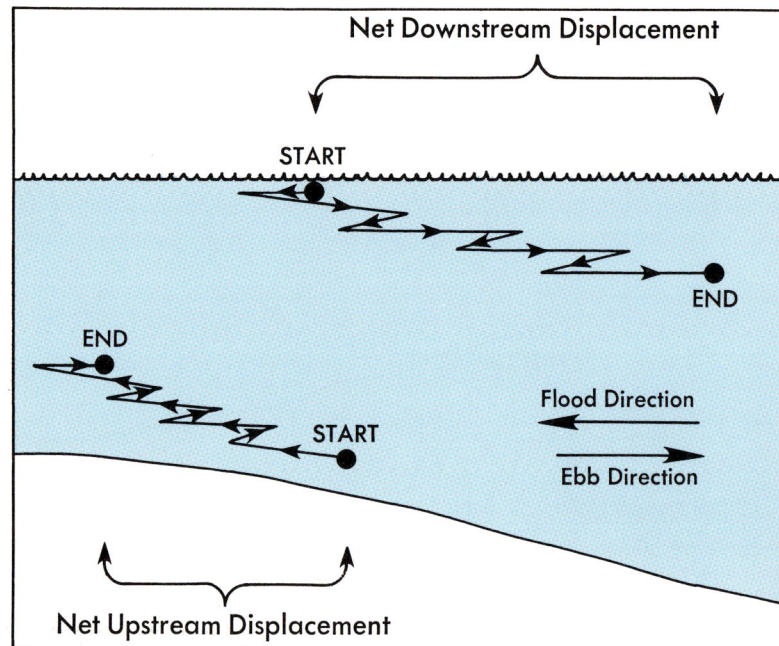

Figure 1-5. *Net movement of a particle in each layer of a two-layered flow system. (Source: Ref. 2)*

FULLY MIXED ESTUARIES

In a fully mixed, vertically homogeneous estuary (Fig. 1-1D), there is no significant vertical stratification, mixing occurs predominantly in the longitudinal direction, and the salt balance is maintained by longitudinal mixing alone. This condition is created in estuaries that have large tidal fluxes compared to river input and small cross-sectional areas, which together amplify the effects of bottom frictional forces. If the basin is wide enough, the Coriolis force, which is due to the earth's rotation** will distribute the net seaward flow of estuaries in the Northern Hemisphere in a manner that will create greater seaward transport on the right-hand side of the estuary (looking downstream). The compensating upstream return flow will tend to be along the opposite side. The Coriolis effect can also be inferred from the measurable cross-stream bending of isohalines in a highly stratified to partially mixed estuary such as the Potomac (see surface salinities in Fig. 3-10, Chapter 3). While salinity would increase towards the mouth along both sides of the estuary under high river discharge conditions (as in spring), the right-hand side of the estuary would be fresher than the left.

If the basin is narrow, lateral frictional forces will produce homogeneity of density in the cross-stream as well as the vertical direction. Under these conditions, nontidal flow is seaward everywhere and is relatively constant across any cross section. The salt intrusion,

---

* This is the stretch of fresh water in the river above the upper limit of salt intrusion which is still subject to the oscillations of tidal currents.

** In the Northern Hemisphere, the Coriolis force bends flows to the right as viewed in the direction of their motion.

however, extends only a few tidal excursions (characteristic distances associated with the back and forth particle displacements shown in Fig. 1-5), because salt balance can be maintained only by a larger tidal mass transport into the estuary on the flood stage than out of it during the ebb stage. The difference is obviously made up by the nontidal transport out of the estuary.

The estuarine types described above are not discrete, but are merely the three most characteristic types in a continuum generated by nontidal circulation patterns that result in predictably distinct salinity distributions. The topography of the basin and freshwater river input have the greatest influence on the circulation and resulting salinity patterns in an estuary. Because river input may fluctuate radically even over short periods of time, an estuary may change from one type into another, even within the same season. As fresh flow increases, measurable salinity is also pushed further down the estuary. In temperate climate estuaries, lowest salinities generally occur in spring when high freshwater flows occur due to melting snows and high spring precipitation. In the Potomac, the estuary resembles a highly stratified type during low flow conditions in summer when the water column is also stabilized by steep temperature gradients due to solar heating. During other parts of the year, especially when freshwater flows into the basin are high, much more vertical mixing occurs, and the Potomac estuary takes on the characteristics of a partially or well-mixed system.

Nutrient Distribution

Estuarine circulation distributes not only salt but all other components dissolved in the water. The constituents most vital to the biota of the estuary are certain dissolved organic and inorganic chemicals, or nutrients. To photosynthesize, plants need the sun's energy, water, and carbon dioxide. Plant biomass increases by synthesis of new protoplasm for which nutrients and trace elements are necessary. The principal nutrients are nitrogen and phosphorus compounds, calcium, and silicon. Trace elements such as iron and magnesium are not usually limiting to plant growth in estuaries.[6] However, a reduced supply of nitrogen or phosphorus can limit plant growth, and supplies of these nutrients fluctuate radically.

Nitrogen and phosphorus are the key chemicals utilized by plants in the synthesis of new biomass. The original sources of these nutrients are the atmosphere and the land. The uptake of inorganic chemical forms by plants, the subsequent utilization of plant material by animals, and the decay of plant biomass, are all parts of the biogeochemical cycle. Processes within this cycle allow complex organic compounds containing nitrogen and phosphorus that are stored in the tissues of living organisms to be broken down into the inorganic forms available for use by plants. Some of the nutrients may be permanently lost through abiotic processes — they may be flushed out of the system by currents, or they may react with the constituents of bottom sediments.

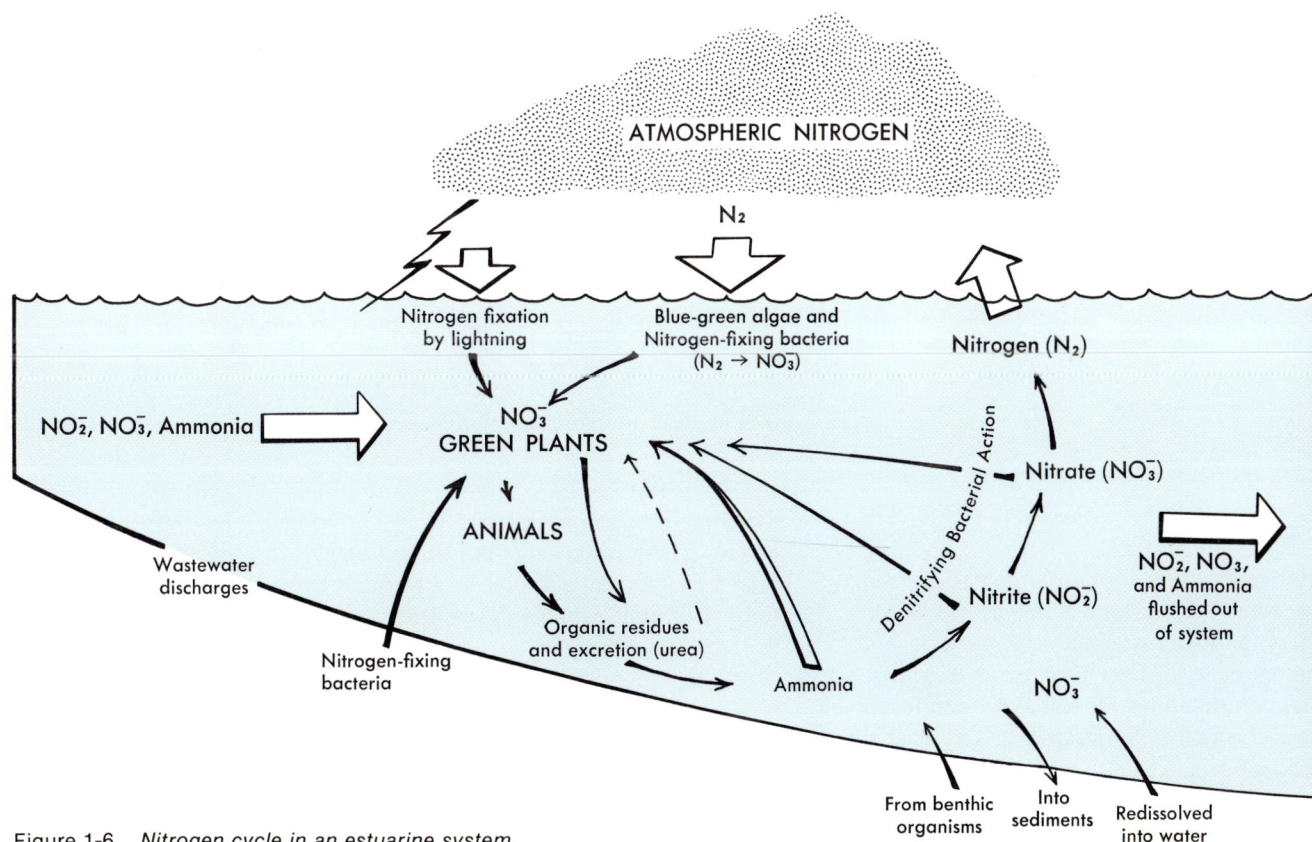

Figure 1-6. *Nitrogen cycle in an estuarine system.*

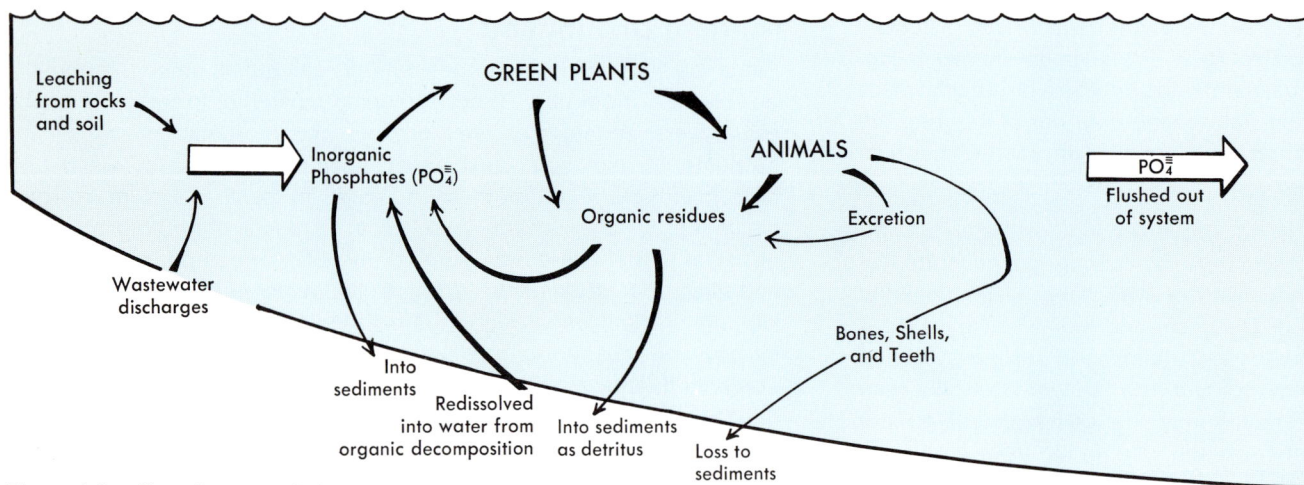

Figure 1-7. *Phosphorus cycle in an estuarine system.*

However, most nutrients are stored in sediments temporarily and are eventually re-released into the water in soluble forms.

The nitrogen cycle (Fig. 1-6) is more complex than the phosphorus cycle (Fig. 1-7). Most of the nitrogen in estuaries comes from forest and agricultural drainage in the upper freshwater basins, from discharges of sewage treatment plants, and from storm runoff associated with urbanization. Nitrogen from the atmosphere may also be directly fixed (changed into a form usable by plants) by lightning or by some blue-green algae. Nitrogen fixation by bacteria in the sediments may provide an additional input.

Nitrogen enters estuaries in organic and inorganic states. The most common inorganic forms of nitrogen are nitrate (NO_3^-) and nitrite (NO_2^-) compounds and ammonia. The most common organic forms are amino compounds originating from animal and plant proteins. Almost all algae can use nitrites and nitrates, although they generally perfer ammonia and organic nitrogenous forms such as urea.[7] Organic nitrogen passes into the water via wastes from the metabolic processes of animals (mostly as urea which quickly decomposes into ammonia) or via the organic residue of dead and decaying plant and animal biomass (detritus). Whereas ammonia can be oxidized directly to nitrite and nitrate, denitrifying bacteria are needed to break down animal protein compounds, first to nitrites and then to nitrates. Inorganic nitrogen compounds may be released again to the atmosphere, lost to the bottom sediments, or flushed out of the system.

Phosphorus is also found in water in both inorganic and organic forms. Its cycling within the estuarine system is relatively simple (Fig. 1-7). Phosphates (PO_4^-) enter rivers and streams as they are leached out of rock formations and soil, or enter from wastewater discharges. Organic phosphate compounds that are released from living matter through decomposition or excretion are reconverted by microbial action into inorganic phosphorus.[8] Much of the phosphorus entering the aquatic system becomes tied up in bottom sediments or is lost through flushing. Since bones, teeth, and shells do not decompose readily, the phosphorus bound in them is essentially lost to the system. Normally, the amount of phosphorus in flowing bodies of water is low, but, as is discussed in Chapter 3, the large quantities of human and animal wastes entering estuaries greatly increase the levels of available phosphorus.

Temperature

In estuaries, seasonal temperature cycles are far more important to species distributions than the relatively small longitudinal or vertical temperature differences that may be seen at any one time. Water temperature and day length are the principal factors that initiate migration and reproduction of estuarine organisms and that substantially affect behavior.

Water temperatures in relatively shallow bodies of water closely follow the seasonal changes in air temperatures. At any one time, water temperatures along the lengths of most estuaries vary only a few degrees, in contrast to much larger relative differences in salinity. Average monthly water temperatures are usually at most a few degrees higher than the average monthly air temperatures (see, for example, Fig. 3-17).

Temperature has a secondary effect on the stratification of estuarine waters, as it does in the Potomac (Fig. 1-8). In summer, denser saline waters moving up the estuary on the bottom are colder than surface waters, which are warmed by high ambient air temperatures. In winter, while surface water temperatures are dropping rapidly in response to colder air temperatures, deeper waters respond more slowly to atmospheric conditions. The result is an inverted temperature stratification with warmer waters at the bottom. Salinity generally has a greater effect on the density of water than temperature. In late spring and fall, however, salinity stratification can be created or destroyed by rapid heating or cooling of the surface and by consequent convection. Nonetheless, higher salinity waters always tend to be near the bottom.

SUMMER

Strong Sun

high ambient air temperature

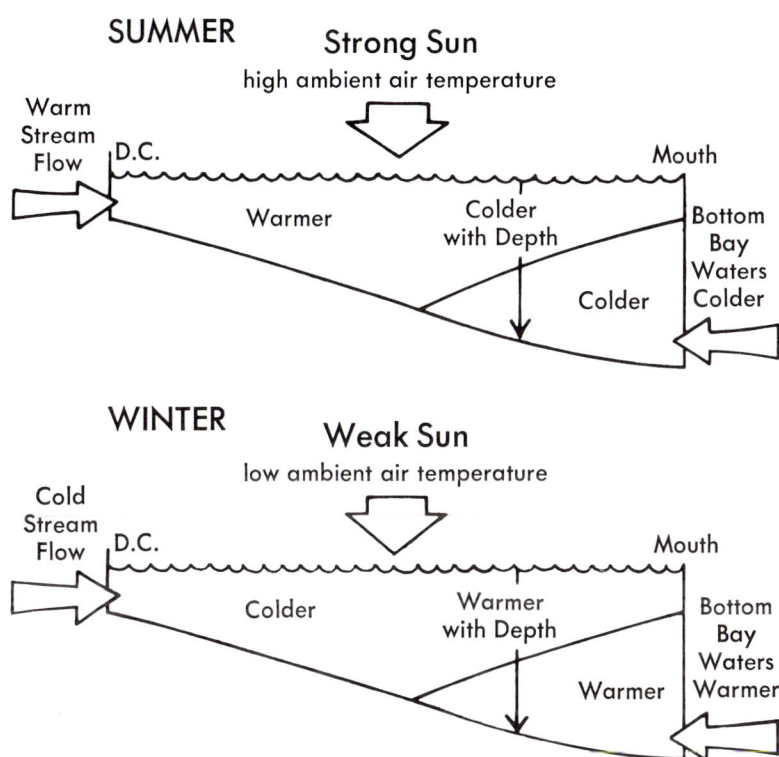

Warm Stream Flow · D.C. · Mouth · Warmer · Colder with Depth · Colder · Bottom Bay Waters Colder

WINTER

Weak Sun

low ambient air temperature

Cold Stream Flow · D.C. · Mouth · Colder · Warmer with Depth · Warmer · Bottom Bay Waters Warmer

Figure 1-8. *Schematic diagram of seasonal temperature patterns in the Potomac estuary.*

Sedimentation

Next to salinity, sediment type is the most important parameter in determining the structure of benthic communities (see Chapter 7). Bottom sediment characteristics also influence the cycling of some nutrients and pollutants in estuaries. The bottoms of estuaries are covered with sedimentary deposits composed of unconsolidated particles originally eroded from terrestrial rocks and soil. In various places, sediments may also contain particles that are insoluble remains of animals, such as bits of shell, bones, and teeth; the siliceous remains of the microscopic algal forms called diatoms (see Chapter 4); and particles precipitated by chemical reactions in the water. Suspended particles are continually carried into estuaries from the upper riverine systems and from shore erosion. Sediments are also transported into estuaries from offshore sources; however, the greatest amount of sediment is contributed by river transport from upstream sources. The manner in which suspended particles of varying size and composition are sorted, deposited, and reworked by currents, waves, and turbulence determines the distribution of various bottom sediment types along an estuary.

In estuaries, sediments go through many cycles of suspension, deposition, and resuspension before being buried by other sediments.[9] Some sediments are permanently flushed out of the system. The size of a particle or grain will determine whether it settles out for long periods,

whether it is flushed out of an estuary, or whether it will continually participate in resuspension and deposition cycles. [9] [10] As shown by the classical Hjulstrom curve[11] in Fig. 1-9, water velocities in relation to particle size will determine whether erosion, transport, or deposition occurs. Consequently, sediment distributions are strongly influenced by the water motions that can put particles into suspension.

Tidal currents, combined with wave action induced by winds, are usually strong enough to suspend considerable amounts of material and keep them in motion. The greatest concentration of suspended material often occurs at the interface region where fresh and saline waters meet and where density-dependent estuarine circulation patterns may keep particles in suspension. This phenomenon, usually referred to as "the turbidity maximum," is not due to resuspension by water motions; rather, finer particles brought downstream by river flow aggregate into larger particles in response to electric charges induced on them as they encounter salt water. This process is called flocculation. The turbidity maximum usually extends a distance of 10 to 20 nautical river miles downstream from the point where salinity first occurs (Fig. 1-10).[3] Turbidity affects the penetration of light and therefore primary production.

The sediment loads from upstream are deposited along estuaries in a fairly predictable pattern. At the head of some estuaries, such as the Potomac, heavier loads of bedrock and coarser particles often mark the fall line. Sediment types normally range from softer, loosely compacted mixtures of silts and clays near the heads of estuaries to harder, more compact mixtures of mostly clays and some silts near the mouths. Local physical processes may alter these expected patterns. Laterally,

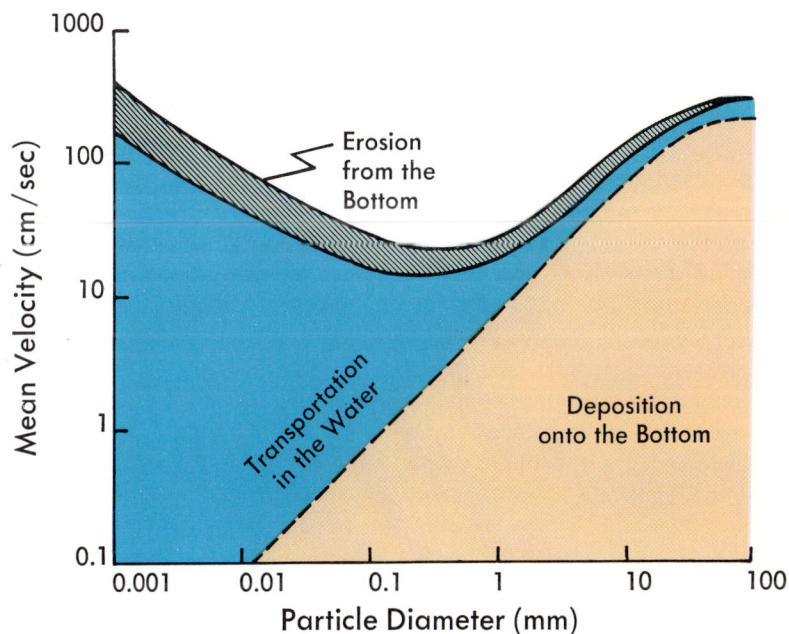

Figure 1-9. *Current velocities necessary for erosion, transportation, and deposition of unconsolidated sediments with various particle diameters. (Source: Ref. 11)*

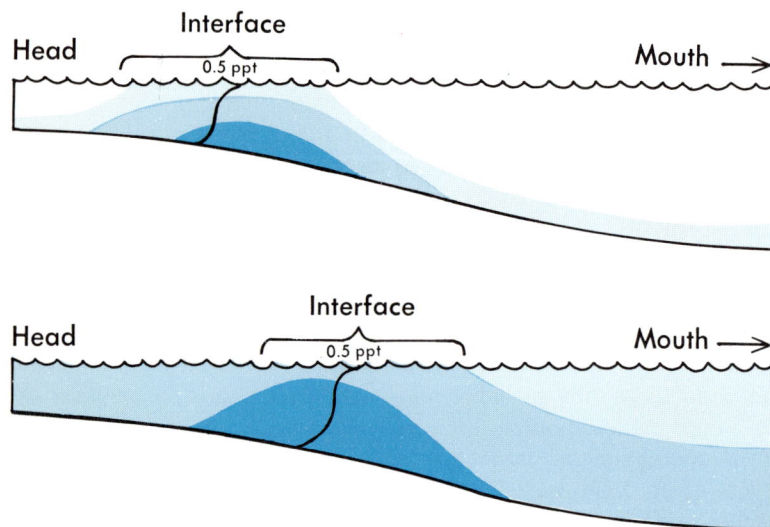

Figure 1-10. *Sediment suspension under low flow conditions (top) and high flow conditions (bottom), showing turbid regions (the "turbidity maxima") under each condition.*

across the width of estuaries, there is a general gradient from soft, fine, loosely compacted sediments in midchannel to larger-grained and firmer sediments near the shores.[12] Where tidal currents are strong, scouring occurs, leaving larger-grained deposits behind. Where current and wave action is minimal (usually in protected embayments and coves), mud flats of finer particles are created.

Water Depth Regimes

Variation in water depth is an important determinant of biological distribution patterns. However, organisms do not react to water depth changes directly, but to differences in current velocity and bottom type, and to local temperature fluctuations that are associated with changes in bottom depth. Most estuaries along the Atlantic Coast are relatively shallow. The average depth of the Chesapeake Bay is only 21 feet, even though the depth of the channel may exceed 100 feet in some places. Surprisingly, the relatively small depth variations in estuaries seem very significant to species distributions, apparently because these small variations are usually associated with considerably larger changes in currents and sedimentation patterns. Certain fish seek deep waters to spawn because the often swifter channel currents keep their light, floating eggs suspended; other spawning fish seek quiet shallows where their eggs, which are often adhesive, will be less disturbed. The benthic species associations in estuaries vary due to different sediment distributions that are usually associated with particular depths. As tides rise and fall over broad, low-profile shores, ecologically distinct habitats of intertidal mud flats are formed. Communities of animals found in these habitats are more numerous and diverse than those found in the higher velocity, deeper estuarine environments.

BIOLOGICAL ASSOCIATIONS IN ESTUARIES

The biological organization of aquatic systems may be described as a pyramid of populations that converts inorganic chemicals and the sun's energy into living matter through levels of plants, herbivores, and carnivores. At each higher trophic (feeding) level, from the smallest protozoan to the largest carnivorous fish, the total biomass is successively less. In addition, the decaying and unutilized biomass from all trophic levels is reconverted to inorganic components by microbial action to replenish the nutrient supply of the system (Fig. 1-11).

The primary producers in water are microscopic floating algae (phytoplankton), some photosynthetic bacteria, macroalgae, and rooted aquatic plants, which use the sun's energy to convert inorganic nutrients into organic compounds and oxygen. The aquatic primary consumers include free-floating microscopic invertebrate animals (zooplankton) that graze on phytoplankton, and benthic (bottom-dwelling) invertebrates that filter phytoplankton from the water. These animals in turn provide food for a wide spectrum of larger animals and thus serve as secondary producers. Some organisms that are considered zooplankton during some stage of their life cycle, such as the larval stages of fish and the young planktonic nauplioid stage of barnacles, feed on phytoplankton and on true zooplankton. Other, larger organisms, such as hydroids, anemones, and sea squirts, also depend on zooplankton as their major food source. From this level, the ecological pathways to higher and larger forms go through steps of "big fish eating little fish." Yet, a direct step-to-step chain is not always maintained, and a complex food web relationship usually evolves, with many interdependencies among forage and predator organisms. Of course, the recycling of particulate and dissolved materials, including the regeneration of inorganic nutrients and the maintenance of a decomposer community of organisms, is all important in sustaining and stabilizing trophic structure in the estuarine food web.

Phytoplankton

Phytoplankton are the most important primary producers in estuaries that have relatively small areas of wetlands. In estuaries that have large areas of wetlands, the decaying vegetation plays a more important role in providing energy to higher trophic levels.

Most phytoplankton biomass is produced in the euphotic zone near the water surface where light energy is sufficient for photosynthetic activity. Estuarine circulation and flow have an important influence on phytoplankton species distributions. For example, net surface downstream movement brings freshwater algal communities from upstream into salinities where they cannot survive. Thus, their downstream distributions are truncated. Estuarine circulation carries many algal cells from near the surface to lower waters where photosynthetic activity is reduced or ceases, depending on depth and turbidity. However, recirculation of bottom waters towards the surface returns many algal cells to regions where photosynthesis and reproduction are renewed. The seasonal pattern in water column stability, which makes it

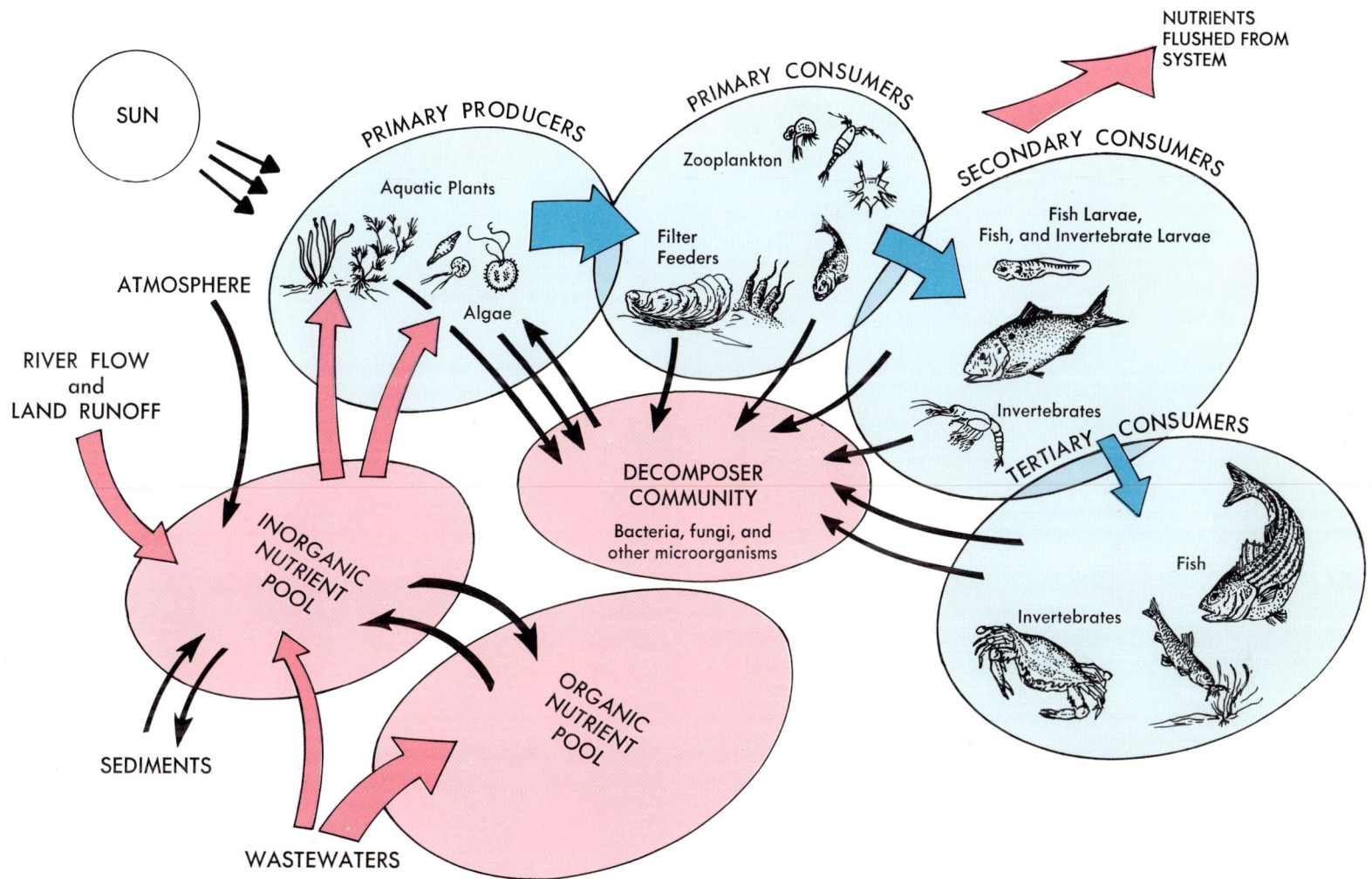

Figure 1-11. *The aquatic food web, showing the relationship of various trophic levels.*

easier or harder for vegetative cells (cells that are not in their reproductive stage) to reach surface waters, probably plays a significant role in regulating community composition.

Zooplankton

In aquatic systems, the most abundant forms of zooplankton are three groups of small invertebrates: rotifers, cladocerans, and copepods. Rotifers and cladocerans dominate in fresh and low salinity waters; copepods are usually the most abundant group in high salinity estuarine and ocean waters. Members of all three groups are important primary consumer organisms in estuaries.

Estuarine circulation is important for regulating the distribution of zooplankton. Although organisms do have limited powers of locomotion, their horizontal displacement is most strongly affected by prevailing currents. Zooplankton are, however, capable of effective vertical movement. As the patch migrates to deeper water, it is captured in the

net bottom upstream flow. Complex interactions of this vertical behavioral movement with nontidal and tidal currents may provide mechanisms by which zooplankton communities keep from being flushed out of an estuary.

The generally productive phytoplankton populations in estuaries support large zooplankton populations. Often, there is a related succession of phytoplankton and zooplankton populations, which is especially evident when zooplankton numbers increase dramatically a few weeks after a period of peak phytoplankton productivity.

Filter Feeders

Of considerable importance as primary consumers, along with zooplankton, larval fish, and filter-feeding adult fishes such as menhaden, are the myriad forms of filter-feeding benthic animals. Some of the larger filter-feeding benthos include clams, oysters, and worms, which consume not only phytoplankton but detritus (decaying organic material), bacteria, and protozoans. The filtering mechanisms of each species select specific size ranges of food from the water. The essentially nonmigratory benthic invertebrates (in particular, those that draw in water and siphon off food particles) are dependent on estuarine circulation to carry food to them. Estuarine circulation also serves as the means of dispersing these species by distributing their planktonic larvae. Many benthic organisms are basically locked to bottom substrates and are at the mercy of every environmental hazard that may come their way. They cannot escape from heavy sedimentation, harmful chemical additives, or other pollutants. Some benthic filter feeders, however, actually clean estuarine waters by removing large quantities of sediments, detritus, and associated bacteria without harming themselves.

Upper Trophic Levels

The higher trophic levels of feeding are often not as clearly definable as the lower levels. The most important predators on zooplankton are the larvae of fish; larval stages of the larger invertebrates such as shrimp, crabs, barnacles, and jellyfish; small pelagic and benthic invertebrates; and plankton-feeding fish. These animals, in turn, are preyed on by larger fish and invertebrates.

Species from different phylogenetic groups can be in the same trophic level, and species from the same group can be primary, secondary, or tertiary consumers. For example, many fish species are herbivorous, carnivorous, and omnivorous during different developmental stages. Plankton-feeding species (fish or jellyfish) graze on phytoplankton and zooplankton. Scavengers such as crabs, worms, and some molluscs are omnivores, but also serve as primary decomposers by breaking down dead tissue into smaller particles that can be more readily decomposed by bacteria and fungi. Terrestrial consumers are also multi-level feeders. Shorebirds and waterfowl feed on seaweeds, aquatic plants, filter-feeding molluscs, and fish; mammals and man feed primarily on molluscs, crustaceans, and fish from estuaries. Thus, the estuarine food web has only a few direct linkages at lower trophic levels, but the linkages become much more numerous at higher levels.

CHARACTERISTIC ASSOCIATIONS WITHIN SALINITY ZONES

Salinity probably influences the distribution and functioning of biota in an estuary more than any other natural environmental factor. Various ranges of salt concentration can be associated with characteristic zones that have distinctive biotic communities.

Salinity Zones

The Venice System[13] is a well-accepted method of characterizing salinity zones and covers the salinity ranges from riverine regions to the ocean. The freshwater category in the Venice System has been modified in this Atlas to account for the tidal and nontidal regions found in rivers with estuarine portions (Fig. 1-12):

Type of System	Zones	Salinity
Riverine	Nontidal Fresh	0 ppt
Estuarine	Tidal Fresh	0 - 0.5 ppt
	Oligohaline	0.5 - 5.0 ppt
	Mesohaline	5.0 - 18.0 ppt
	Polyhaline	18.0 - 30.0 ppt
Marine	Euhaline	> 30.0 ppt

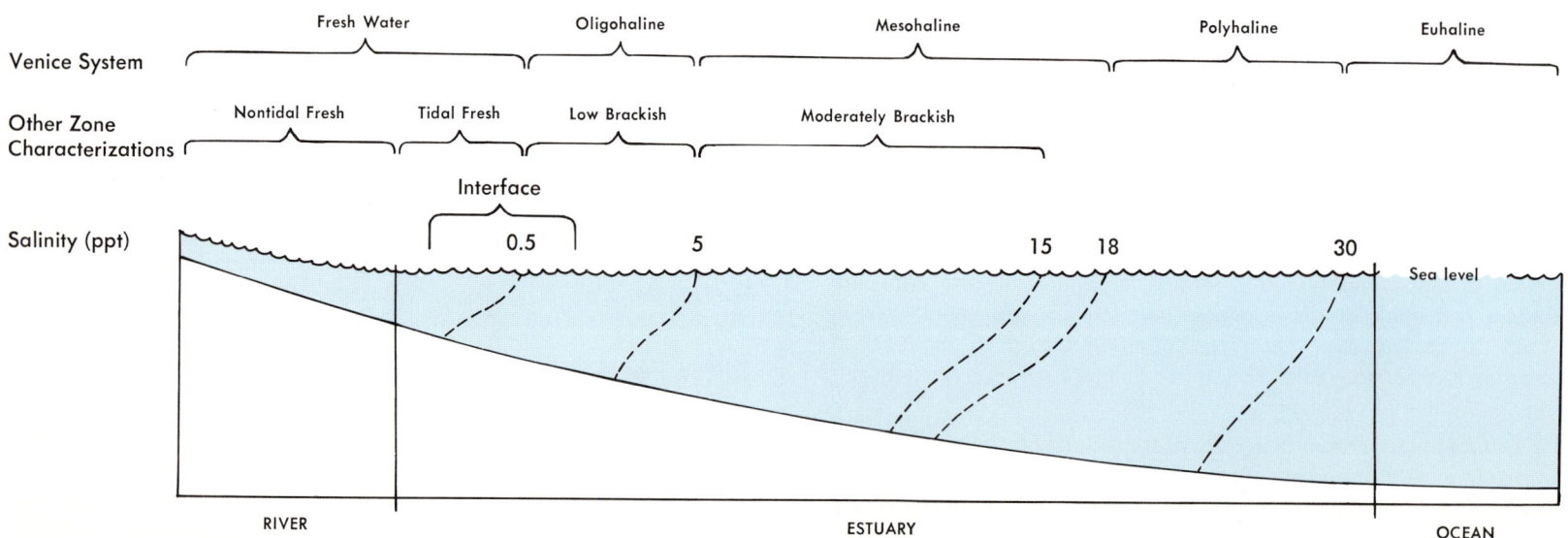

Figure 1-12. *Terminology used to characterize the various salinity zones in estuaries.*

By common definition, the upper end of an estuary is marked by the first point where tidal effects are no longer present, that is, the boundary between tidal and nontidal freshwater regions.*

There are other, less specific terms that are commonly used to describe estuarine regions by salinity. The following are used in this Atlas:

Low brackish and moderately brackish are somewhat imprecise terms that are often used to describe salinity habitats. A low brackish area roughly corresponds to the oligohaline zone, and moderately brackish roughly corresponds to the mesohaline zone (up to 15 ppt).

The interface region, where fresh water first meets the salt front, is the transition zone between tidal fresh and oligohaline waters. The tidal fresh and interface region is where anadromous fish mainly spawn (see below).

The zones move upstream or downstream seasonally with changing freshwater flows. Their boundaries, as well as their linear extents, may respond to changes in river input that last from a few days to several years and range from the short-term heavy rainfalls associated with hurricanes to long-term droughts. Although the distribution of species with wide salinity tolerances may not respond to changes in the position or extent of salinity zones, the distribution of species entering the estuary during spawning or feeding migrations might vary considerably from year to year.

Associations Within Salinity Zones

The three major categories of organisms found in estuaries, as defined by the salinity habitats in which they normally occur, are the estuarine, marine, and freshwater forms.

Estuarine animals or plants are the resident organisms that have physiologically adapted through evolutionary processes to live and thrive through all their life stages within estuarine salinities. Most are euryhaline (able to live over a wide range of salinities). They may also exist, although not necessarily optimally, in totally fresh water or in the ocean. The number of truly estuarine species is relatively small compared to the variety of marine forms, but an estuarine species is often more abundant within an estuary than are its relatives living in the ocean. Estuarine species are most abundant in oligohaline and mesohaline regions.

Many **marine** species are also found within estuaries. Their numbers decrease upstream as salinities drop. The more euryhaline a marine species is, the greater its penetration into an estuary. Stenohaline marine organisms (those able to live only within a limited salinity range) are found only at the mouths of estuaries.

Freshwater organisms are at the other end of the spectrum. Many can live in low brackish waters, but usually, relatively few freshwater species are found in the mesohaline regions of estuaries.

Most fish species fall into one of these three categories. However, the terms **anadromous** and **catadromous** are used to denote those fishes that migrate into and out of estuaries, respectively, to spawn. During migrations, these fishes cover the whole range of salinity from fresh to salt or vice versa. When they are not spawning, anadromous fishes are found in high salinity estuarine and marine environments, whereas catadromous species are mostly confined to regions encompassing the fresh to mesohaline zones. **Semianadromous** fishes spend most of their lives in estuarine waters, but seek fresh water to spawn.

All elements of the aquatic food web, from primary producers to top level predators, are found in every salinity habitat zone in estuaries. However, the relative importance and species composition of each group changes from one salinity regime to the next, depending on the salinity tolerances of individual species; and as the species change, so do the uses and ecological roles of each zone.

Most mid-Atlantic coastal plain estuaries have characteristic biotic associations in each salinity zone. The **tidal fresh regions** of these estuaries are often subjected to high nutrient levels from wastewater discharges and large sedimentary loadings from the upstream watershed and from urban construction activities, which frequently occur at their heads. Primary production flourishes due to the high nutrient levels. Freshwater algal species (greens and blue-greens) proliferate in spring and summer and constitute a ready food source for large populations of freshwater rotifers, cladocerans, and copepods. In winter, diatoms, which are the most ubiquitous algae in estuaries, provide the major source of primary production. Benthic invertebrate species are relatively sparse. Tidal fresh water and the interface region at its lower limits are particularly important to commercial fisheries since it is there that major spawning activity by anadromous and semianadromous fishes takes place and that their hatched young are nurtured. These species include striped bass, river herrings, American shad, and white perch. The large phytoplankton and zooplankton populations characteristic of this zone feed the early developmental stages of all these fish. The number of fish species in tidal fresh zones may be relatively high, due primarily to the many freshwater species infiltrating from upriver and the anadromous and semianadromous species entering from downstream. But, when the spawning anadromous and semianadromous fish have left, the total population of fish is generally less than in higher salinity zones.

The low brackish **oligohaline areas** in mid-Atlantic coastal plain estuaries are the transition zones between populations of freshwater and estuarine organisms. Nutrient levels are usually adequate for maintaining normal phytoplankton biomass without the high production that is so prevalent upstream. In oligohaline regions, the species structure of zooplankton populations changes, because numbers of cladocerans and rotifers decline radically with a slight increase in salinity. Above 1 ppt salinity, rotifers and cladocerans begin to decline in numbers, and copepods become more abundant as salinity

---

* A recent classification system[14] proposes that the limit of an estuary be set at the point where salt content is less than 0.5 ppt (i.e., at the oligohaline-tidal freshwater line). Such a limit is not geographically fixed because of fluctuations of salinity regimes. The extent of tide is a more logical point of reference, particularly in relatively long estuaries such as the Potomac where the oligohaline zone may shift a distance of 20 nautical miles or so seasonally (see Chapter 3).

increases. Although copepod numbers may fluctuate in this zone, the total biomass remains high due to their relatively large sizes. Copepods provide food for larval and juvenile fish migrating downstream from the tidal fresh spawning grounds. These regions are also used to a limited extent as spawning areas by anadromous species. Oligohaline zones serve as nursery grounds for commercially harvested ocean-spawning fish — such as Atlantic menhaden, spot, and Atlantic croaker — whose bottom-oriented larvae are transported far from their spawning grounds at the mouths of estuaries by the net upstream movement of bottom waters. Resident estuarine fish in this zone are not as numerous as they are downstream.

In the **mesohaline zones** of coastal plain estuaries, estuarine species dominate all trophic levels. Copepods dominate the zooplankton populations but are often less numerous than they are upstream. These regions support most of the estuarine forage fish and thus are important feeding grounds for larger predator fish (such as striped bass and bluefish) that are valuable to commercial and sportfishermen. They serve as nursery areas for many marine fish species. Mesohaline regions are also the prime areas for commercial oyster and clam production. In the summer, commercially harvestable numbers of blue crabs are generally found in the mesohaline regions.

The higher salinity **polyhaline and euhaline zones** approach the marine habitat in their characteristics. (The Potomac is too far from the sea to have a significant polyhaline zone.) Primary production is moderate compared to that in the upstream estuarine zones. The composition of the zooplankton populations is extremely variable — both estuarine and marine forms occur, but copepods still dominate. Commercial shellfish species are abundant, including in many instances, oysters, soft-shell clams, and blue crabs. Blue crabs, whose larval stages require high salinities to survive, spawn in these regions. The fish fauna is varied, with increasing numbers of marine species present towards the mouths. Tidal and nontidal transports are greater in polyhaline regions than in the other regions of estuaries. Therefore, the rate of flushing may also be greater, and pollutants are dispersed faster.

Each coastal estuary has its own individual set of environmental characteristics that determine species distributions within its boundaries. The extent and location of zones within each estuary depend on its riverine discharge (flow), length, and basin geometry. The extent of each zone varies significantly among estuaries. Some short estuaries with little watershed area and low river discharge have steep salinity gradients and, thus, poorly defined habitat differences; one salinity zone grades quickly into the other. Long estuaries with extensive watersheds and relatively high freshwater flow, such as the Potomac, the Chesapeake Bay, and Delaware Bay, have relatively well-defined zones. The same zones in each of these systems support similar species types, display the same seasonal succession, and perform the same ecological functions.

The remaining chapters characterize the Potomac estuary by the distributions of its natural resources, by the seasonal patterns of its physical and biological components, and by the ecological functions in its habitats.

Chapter 1 — References

1. Pritchard, 1967a
2. Cronin, L. E., and A. J. Mansueti, 1971
3. Schubel and Meade, 1977
4. Pritchard, 1967b
5. Boyce Thompson Institute for Plant Research, 1977
6. Jaworski, Lear, and Villa, 1971
7. McCarthy, Taylor, and Taft, 1977
8. Odum, 1959
9. Postma, 1967
10. Gross, 1972
11. Hjulstrom, 1939
12. Scherk, 1973
13. Carriker, 1967
14. Cowardin, Golet, and LaRoe, 1977

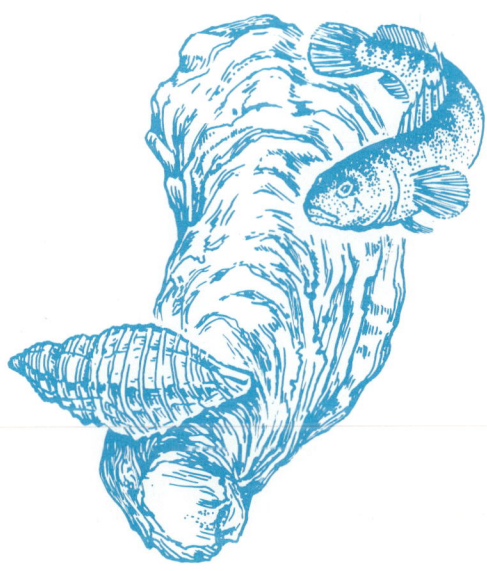

The
Visible
Boundaries

The Visible Boundaries

Before the Pleistocene glacial age, many great river systems drained the eastern slopes of the long mountain ridges along the North American continent. The greatest of these was the Susquehanna, which had a watershed of thousands of square miles with boundaries extending as far north as upstate New York and as far west as western Pennsylvania. As the Susquehanna meandered southward to the Atlantic, cutting through the Piedmont foothills, it was joined by waters from hundreds of streams, large and small. The largest of these tributaries was the Potomac River, which drained the southwestern slopes of the system. The Potomac, along with other southern tributaries (the York, James, and Rappahannock rivers of Virginia), cut deep channels across the ancient coastal plain ledges.[1] At the end of the last Pleistocene glacial age, from about 15,000 to about 9,000 years ago, sea levels rose with the melting retreat of the glaciers. Water inundated the valleys of the coastal plain rivers[2] and eventually reached the base of the Piedmont hills at what is now called the fall line. These tidal waters flooded the lower portions of the Susquehanna River basin, drowning the valleys inland for almost 180 statute miles [290 kilometers (km)].[3] Thus, as seawater intruded into the lower reaches of the Susquehanna basin, the Chesapeake Bay and the estuarine portions of all its tributaries, including the Potomac River, came into being.

GEOGRAPHY

Boundaries of the Upper Potomac River and the Estuary

Today, the Potomac River flows through three physiographic provinces (Fig. 2-1). It arises in the Allegheny Plateau division of the Appalachian Province, flows through the Greater Appalachian Valley and the Blue Ridge Mountain divisions, and crosses the Piedmont Province and the Coastal Plain Province before it enters the Chesapeake Bay.[4] The fresh headwaters originate high in the Appalachian mountain system, close to the southwest corner of Maryland. Here, the North Branch flows northeastward for 100 statute miles (161 km), forming a part of the Maryland and West Virginia border. As it passes eastward from Cumberland, many other large rivers, such as the South Branch, the Shenandoah, and the Monocacy, contribute substantially to the combined flow. These tributaries and many other smaller streams flow north through Virginia and West Virginia and south through Pennsylvania and Maryland to meet the Potomac. The riverine segment terminates near the fall line, which is demarked by a series of rapids at Washington, D.C. At Little Falls, 300 statute miles (483 km) from its source, the river comes under the influence of tides, and the estuary begins. The estuary runs another 113 statute miles (182 km) before it meets the Chesapeake Bay. It is the estuarine portion that is discussed in this Atlas, including both the main stem of the Potomac (the main trunk of the estuary, excluding the tributaries) and the tidewater portions of all tributaries that enter the river between Little Falls (Fig. 2-2) and the mouth (defined by a straight line from Point Lookout, Maryland, to Smith Point, Virginia).

The Potomac is part of the State of Maryland except for the section flowing through Washington, D.C. Along the estuary, the Maryland state line follows the Virginia shore at the mean low water line, crossing the mouths of Virginia tributaries at designated points (see Folio Map 1 for state boundaries).

Watershed

The Potomac River watershed has a total drainage area of about 9.4 million acres[5] [38,040 million square meters (m^2)] (see Table 2-1). Of all East Coast rivers, its watershed ranks fourth in area;[6] of all rivers in the United States, its watershed ranks twenty-first.[5] The extensive watershed area of the upper Potomac basin (unshaded area on Fig. 2-2)

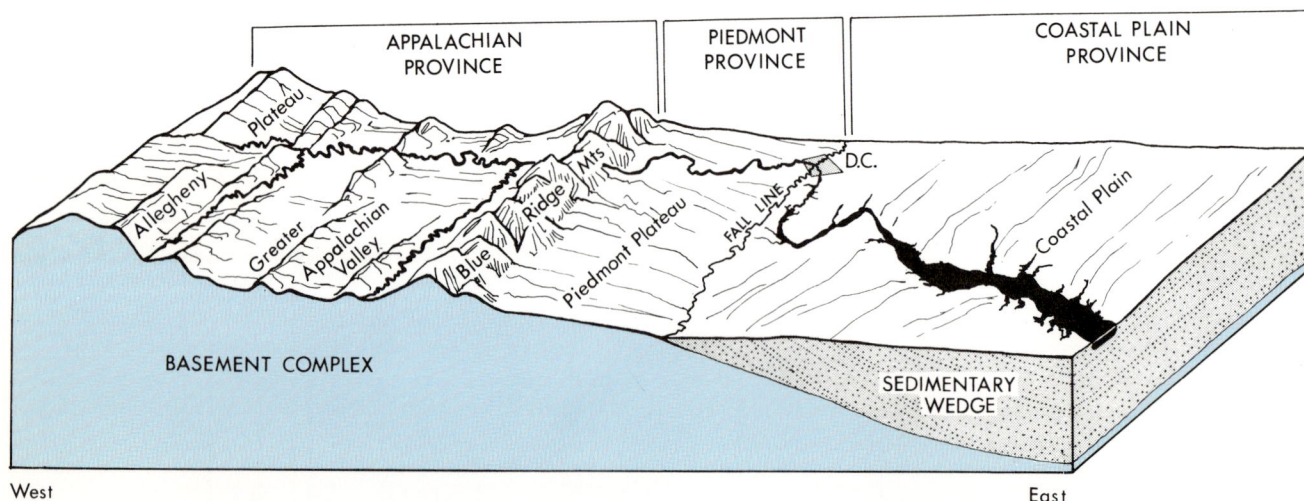

Figure 2-1. *Physiographic provinces traversed by the Potomac River. The Appalachian Province is divided into three divisions. Province designations from Ref. 4.*

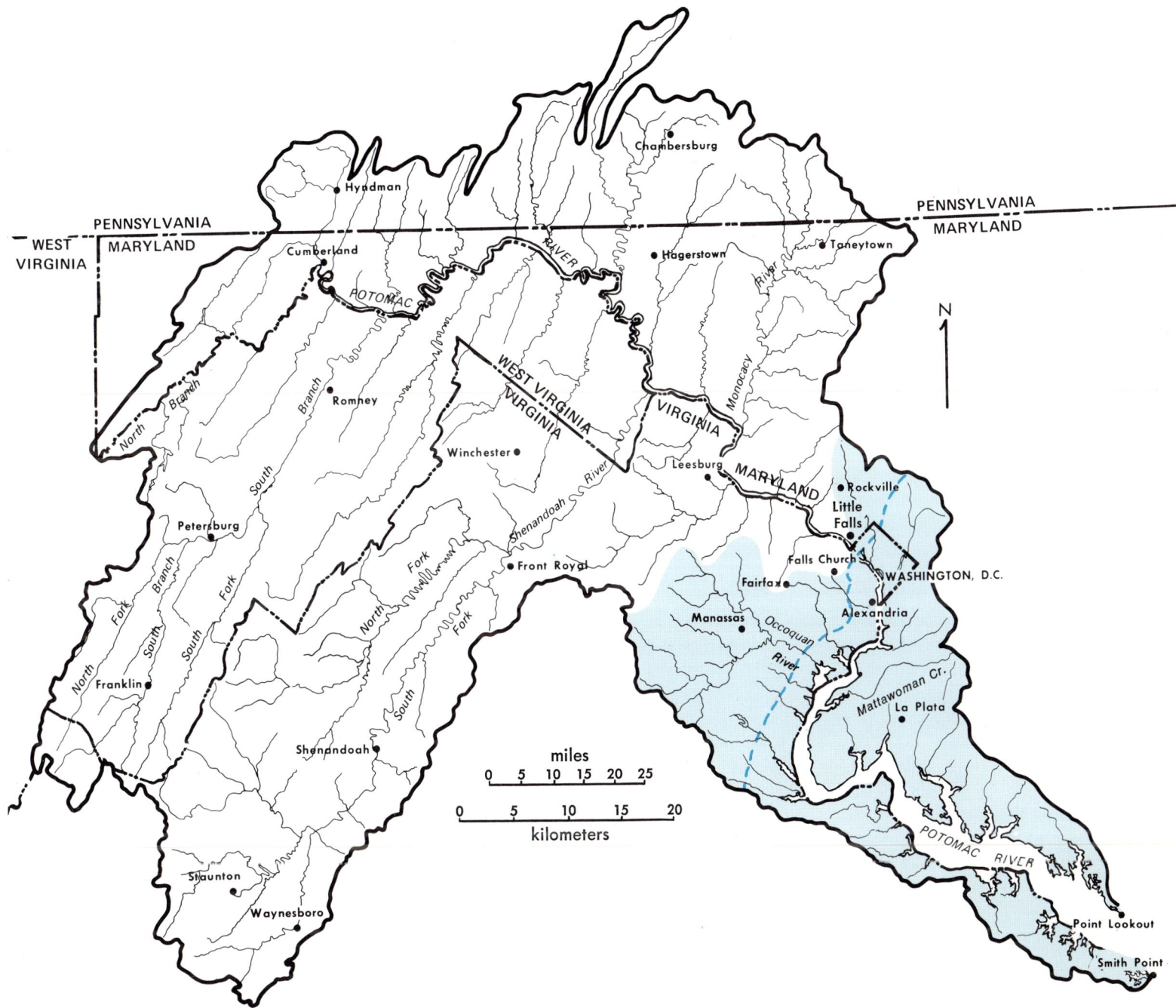

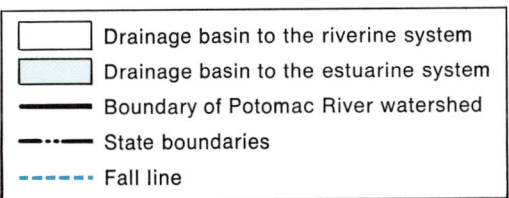

Figure 2-2. *Watersheds of the Potomac River. Base map courtesy of the Interstate Commission on the Potomac River Basin.*

stretches 30 miles (48 km) into Pennsylvania to the north and encompasses most of the eastern part of West Virginia, the Appalachian Mountains region of northern Virginia, and most of western Maryland. The watershed collects, on the average, a yearly precipitation of approximately 45 inches [1.1 meter (m)], which results in a yearly average freshwater flow of 11,190 cubic feet [313 cubic meters (m³)] per second at the head of the estuary.[7]

The watershed of the estuarine portion (shaded area) begins at the headwaters of the uppermost tributaries draining into the estuary, and includes their nontidal and tidal waters. On the Virginia side, the estuarine watershed covers 947,700 acres (3,800 million m²).[5] The total Maryland estuarine watershed is appreciably smaller — 676,350 acres (2,700 million m²).[5] However, Maryland has a greater area of tidal wetlands than Virginia.

Table 2-1 — Basin Drainages to the Potomac Estuary [(a)]

VIRGINIA		
Watershed Name	Drainage Area (in acres)	Total for County
Cities of Arlington and Alexandria		
Potomac estuary tributaries	15,520	15,520
Fairfax County		
Bull Run (tributary to Occoquan River)	51,620	
Accotink Creek (tributary to Gunston Cove)	22,690	
Occoquan River (lower segment, N side)	44,580	118,890
Fauquier County		
Cedar Run ⎫	68,140	
Kettle Run ⎪ Headwaters to Occoquan River	6,220	
Broad Run ⎬	30,780	
Bull Run ⎭	820	
Headwaters to Aquia Creek	4,820	110,780
Prince William County		
Cedar Run and Occoquan River (S side)	62,590	
Kettle Run (tributary to Broad Run)	10,190	
Broad Run (tributary to Occoquan River)	36,080	
Bull Run (S side, tributary to Occoquan River)	49,680	
Quantico Creek and Neabsco Creek	53,400	
Chopawamsic Creek (N side)	9,010	220,950
Loudoun County		
Bull Run (headwaters to Occoquan River)	16,250	16,250
Stafford County		
Chopawamsic Creek and Aquia Creek	96,410	
Potomac Creek	32,160	128,570
King George County		
Pepper Mill Creek and Upper Machodoc Creek	54,520	
Pine Hill Creek and Rosier Creek (N side)	12,950	67,470
Westmoreland County		
Potomac estuary tributaries from Rosier Creek (S side) to Nomini Creek	58,600	
Potomac estuary tributaries from Nomini Creek to Yeocomico River (N side)	50,670	109,270
Northumberland County		
Potomac estuary tributaries	160,000	160,000
TOTAL VIRGINIA ESTUARINE WATERSHED DRAINAGE		947,700

MARYLAND		
Watershed Name	Drainage Area (in acres)	Total for County
Montgomery County (includes part of Washington, D.C.)		
Rock Creek	52,140	
Anacostia River	38,110	90,250
Prince Georges County (includes part of Washington, D.C.)		
Anacostia River	65,430	
Cabin Branch and Beaverdam Creek (tributaries to Anacostia River)	3,910	
Potomac estuary tributaries from Anacostia River to Piscataway Creek	44,820	
Piscataway Creek	40,380	
Mattawoman Creek (N side)	16,010	
Zekiah Swamp Run (headwaters)	4,900	175,450
Charles County		
Potomac estuary tributaries from Pomonkey Creek to Wicomico River	93,500	
Mattawoman Creek (S side)	34,460	
Port Tobacco River	18,810	
Zekiah Swamp Run (tributary to Wicomico River)	67,080	
Gilbert Swamp Run (tributary to Wicomico River)	26,120	239,970
St. Marys County		
Potomac estuary tributaries from Wicomico River to Point Lookout	93,500	
Gilbert Swamp Run (headwaters)	2,500	
Chaptico Creek (tributary to Wicomico River)	22,480	
McIntosh Run	21,900	
St. Marys River	27,300	
St. Inigoes Creek (tributary to St. St. Marys River)	3,000	170,680
TOTAL MARYLAND ESTUARINE WATERSHED DRAINAGE		676,350 [(b)]

TOTAL MARYLAND AND VIRGINIA WATERSHED DRAINAGE TO ESTUARY 1,624,050 ACRES

(a) English units used as in original source (Table 1 in Ref. 5). Multiply acres by 4,046.8564 to get square meters.

(b) Does not include Maryland's Potomac estuary surface area of 278,050 acres.

MORPHOLOGY OF THE POTOMAC ESTUARY BASIN

The Potomac estuary generally meanders in a south, southeastward direction, except for a sharp bend about halfway downstream — its most distinctive morphological feature. It is a comparatively broad estuary, generally increasing in width downstream from just a little over 200 feet (61 m) across near its beginning at Chain Bridge, Washington, D.C., to nearly 10 statute miles (16.1 km) across at its confluence with the Chesapeake Bay (see Folio Map 1). It is the longest and broadest estuary entering the Bay. With its many tributary rivers, creeks, and bays, the Potomac estuarine shoreline length of 1,121 statute miles (1,804 km; Table 2-2) far exceeds its midchannel length of 113 statute (98 nautical) miles (182 km).

Table 2-2 — **Shoreline Length Along the Potomac Estuary and Its Tributaries (in statute miles)** [a]

Maryland		Virginia	
Charles County	182	Fairfax County (with the	
Prince Georges County	43	City of Arlington)	37
St. Marys County	297	Prince William County	26
Total	522	Stafford County	42
		King George County	61
		Westmoreland County	136
		Northumberland County	297
		Total	599

TOTAL POTOMAC RIVER ESTUARINE SHORELINE	1,121
TOTAL CHESAPEAKE BAY SYSTEM SHORELINE	7,325

(a) English units used as in original source (pages B-V-43 through B-V-45 in Ref. 1). Multiply statute miles by 1.60934 to get kilometers.

Topography

The topography of the Potomac watershed grades from the high mountains and Piedmont foothills of the upper riverine portion to the low coastal plain elevations of the estuarine region. Folio Map 2 shows the topography bordering the estuary. The land along the Potomac below the Rt. 301 bridge is generally low lying and gently slopes on both sides towards the river from elevations of no more than 150 feet (46 m), except between Popes Creek (Virginia) and Nomini Bay where cliffs fringe the Virginia shore. Above the Rt. 301 bridge, the land on either side of the estuary rises more abruptly. Low bluffs meet the water's edge at many points along the upper estuary. They are the dominant shoreline feature of both shores around the bend of the estuary at Mathias Point and along the Maryland shore from Maryland Point upstream to the cliffs at Indian Head. The fall line cuts relatively close to the mouth of all the Virginia tributaries, so that the tidewater portions of these streams are smaller, and the elevations rise more sharply than do those along the tributaries on the Maryland side.

Bathymetry

The Potomac is a relatively shallow estuary, with depths ranging from 119 feet (36.3 m) off Mathias Point, to the shallow tidal marshes and mud flats that are exposed at low tide (bathymetry shown on Folio Map 2). Twenty selected cross-stream transects on the map depict the changes in depths along the estuary. At the mouth, these bottom profiles are broad and gently sloping. At the upper end of the estuary, they become narrower and more sharply angled. The overall average depth, calculated from Table 2-3, is 19.7 feet (6 m), which is somewhat less than the average for the whole Chesapeake Bay system of 21.2 feet (6.5 m).[3] The average and maximum depths along each nautical mile segment of the mainstem channel are shown in Table 2-3, and Table 2-4 lists the average depths of the tributaries.

The Potomac estuary is navigable by small, shallow-draft boats from its mouth upstream to Little Falls. For larger vessels, the U.S. Army Corps of Engineers maintains a navigational channel to a depth of approximately 24 feet (7.3 m) up to Washington, D.C.[10] From the mouth upstream, the major stretches of the dredged channel (Folio Map 2) are: the channel segment through Kettle Bottom Shoals offshore from Neale Sound, the channel segment along the eastern side of the estuary between Maryland Point and Douglas Point, one across the mouth of Mattawoman Creek, and another between Dogue and Hunting creeks. The lower reaches of the Anacostia River are also kept clear. Many other shorter channel segments are cut to 6- or 7-foot (1.8- or 2.1-m) depths to allow passage of smaller boats into tributaries, creeks, or sheltered harbors.

Surface Areas and Water Volumes

There is a general downstream increase in both surface area and water volume per 1-mile segment along the estuary (Table 2-3 and Fig. 2-3), although the uneven geometry of the basin makes the changes from one segment to the next irregular. Surface areas range from a little over 500,000 square meters for the nautical mile segment at the uppermost end of the estuary, to over 60 million square meters for the triangular segment at the mouth (Fig. 2-3).[9] The main stem of the estuary can be divided into three general regions in which each mile segment has a similar mean low water (MLW) surface area:

- a narrow reach downstream from Washington, D.C., to Indian Head
- a stretch of moderate and more varied proportions from Indian Head downstream to Morgantown
- a broad lower portion from Morgantown to the mouth.

The volume of the 1-mile segments ranges from slightly less than 2.5 million cubic meters in the upper part of the estuary to a little more than 250 million cubic meters at the mouth. The average total volume (Table 2-3) of the tidal Potomac at MLW is 7,059 million cubic meters[8][9] and accounts for almost 8.7 percent of the average total MLW volume of the Chesapeake Bay and its tributaries. Lowest values of volume for any 1-nautical-mile mainstem segment occur in the area of Hains Point. Water volumes and surface areas of the tributaries are given in Fig. 2-3 and Table 2-4.

Table 2-3 — **Average and Maximum Depths, Surface Areas, and Water Volumes for Nautical River Mile Segments of the Potomac Estuary Main Stem**

Nautical River Mile	Average Depth [a] (meters)	Maximum Depth [b] (meters)	Surface Area [c] (millions of square meters)	Water Volume [d] (millions of cubic meters)	Nautical River Mile	Average Depth [a] (meters)	Maximum Depth [b] (meters)	Surface Area [c] (millions of square meters)	Water Volume [d] (millions of cubic meters)
0-1 [e]	14.0	19.5	60.92	852.88	50-51	5.9	13.2	5.40	32.64
1-2	10.3	20.0	19.87	203.86	51-52	7.2	13.2	4.12	29.84
2-3	9.1	20.5	23.43	213.31	52-53	4.9	14.0	5.66	27.54
3-4	8.9	20.5	22.71	201.66	53-54	5.3	14.0	5.03	26.75
4-5	9.9	20.0	20.57	202.01	54-55	6.6	14.0	4.64	30.77
5-6	9.5	20.0	21.47	202.01	55-56	4.4	13.0	8.43	36.87
6-7	10.2	18.2	19.91	203.12	56-57	3.5	8.0	11.41	39.06
7-8	11.4	18.2	17.13	194.90	57-58	3.6	8.0	10.38	36.56
8-9	8.5	18.3	21.04	180.30	58-59	2.9	8.0	11.66	33.96
9-10	10.7	18.3	14.92	161.17	59-60	4.4	8.0	8.41	37.12
10-11	11.9	21.0	12.65	150.99	60-61	3.8	8.0	9.98	38.48
11-12	12.5	26.0	12.85	161.06	61-62	3.8	9.0	9.60	37.38
12-13	11.2	26.0	14.08	157.73	62-63	4.4	9.0	8.59	38.43
13-14	11.4	25.0	12.24	140.00	63-64	4.5	12.2	8.26	37.41
14-15	8.3	24.4	14.63	121.78	64-65	4.6	12.2	7.18	33.20
15-16	8.9	21.0	12.78	113.19	65-66	4.7	9.6	5.62	26.33
16-17	9.8	18.2	11.69	114.05	66-67	5.6	10.0	4.54	25.48
17-18	8.0	17.0	14.10	112.67	67-68	5.8	10.0	4.75	27.84
18-19	8.2	12.2	13.75	111.51	68-69	7.0	10.0	4.06	28.93
19-20	7.6	12.0	14.50	109.34	69-70	5.4	11.0	6.67	36.51
20-21	7.7	10.0	13.65	106.34	70-71	4.6	11.0	7.63	35.10
21-22	7.0	10.2	15.22	106.92	71-72	4.6	11.0	6.22	28.60
22-23	7.2	11.4	14.71	106.56	72-73	4.4	13.0	5.32	23.92
23-24	7.5	12.2	14.40	108.67	73-74	4.5	13.0	4.41	20.23
24-25	7.2	15.2	13.38	96.47	74-75	4.6	23.0	5.19	24.19
25-26	6.4	15.2	12.16	78.53	75-76	6.0	23.0	4.02	24.46
26-27	6.2	13.0	13.26	82.23	76-77	6.8	18.0	2.99	20.39
27-28	5.3	11.2	18.22	97.01	77-78	4.5	11.0	3.57	15.61
28-29	5.8	10.0	18.84	108.54	78-79	4.5	11.0	2.77	12.31
29-30	6.0	11.0	18.29	107.98	79-80	5.3	11.0	2.37	13.23
30-31	5.7	11.0	17.27	97.55	80-81	3.7	11.0	2.36	12.40
31-32	5.7	18.0	16.26	91.00	81-82	3.4	8.0	2.94	10.42
32-33	5.4	8.0	16.98	91.95	82-83	3.5	8.0	3.17	10.45
33-34	5.1	9.0	17.75	90.90	83-84	4.1	20.0	2.21	8.13
34-35	4.4	9.0	15.14	67.95	84-85	3.9	20.0	1.55	6.81
35-36	6.1	18.2	8.55	52.39	85-86	3.5	15.0	1.95	7.55
36-37	7.1	21.2	10.70	75.65	86-87	3.5	12.0	2.75	9.53
37-38	6.2	21.2	14.23	87.95	87-88	3.1	13.0	2.58	8.16
38-39	5.4	21.0	13.73	73.23	88-89	3.2	13.0	2.71	8.87
39-40	5.0	23.5	11.30	55.03	89-90	2.6	10.0	2.77	7.67
40-41	6.2	23.5	6.86	41.78	90-91	1.7	9.0	2.09	3.97
41-42	7.2	26.5	5.52	39.25	91-92	1.8	9.0	2.44	4.18
42-43	9.3	26.5	4.75	43.50	92-93	2.9	8.5	2.54	6.75
43-44	8.9	24.0	4.59	40.62	93-94	5.1	7.0	1.24	6.17
44-45	5.9	23.0	5.89	34.47	94-95	1.4	6.5	1.31	2.39
45-46	9.2	20.4	3.34	30.85	95-96	2.0	6.6	1.18	2.36
46-47	8.6	20.4	3.60	30.85	96-97	1.8	6.6	1.36	2.34
47-48	5.1	13.0	6.86	34.40	97-98	5.2	20.0	0.55	2.87
48-49	4.1	13.0	9.28	37.53	Average for Entire Estuary	6.0			
49-50	5.3	9.0	6.73	35.61	**TOTAL**			953.77	7059.45

(a) Average depths from Ref. 8.

(b) Maximum depths from Ref. 8. Deepest part at Mathias Point is not represented in the tabulations.

(c) Surface areas are calculated mean low water (MLW) values in Ref. 9. Total Potomac mainstem MLW surface area = 1,251.48 million square meters (Ref. 9).

(d) Water volumes are calculated MLW values in Ref. 8.

(e) For this segment, average and maximum depths were estimated from U.S. Geological Survey Chart No. 557. The surface area is from Ref. 9. Volume was calculated from the average depth and surface area.

R

POTOMAC RIVER ESTUARY
WATER VOLUME
MEAN LOW WATER VOLUME
BY NAUTICAL MILE SEGMENT

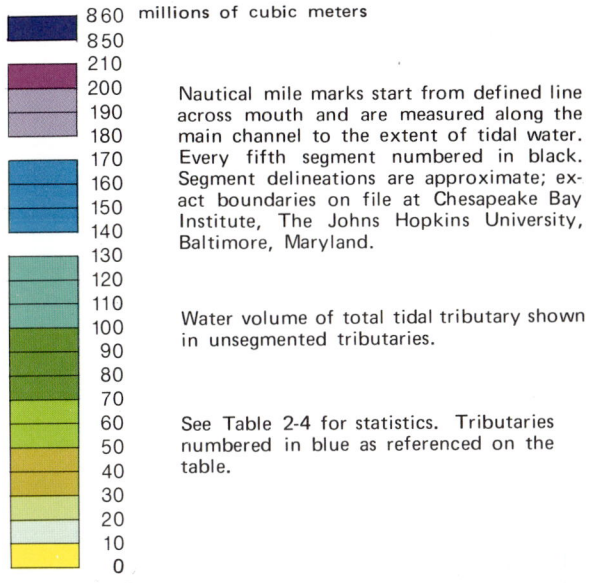

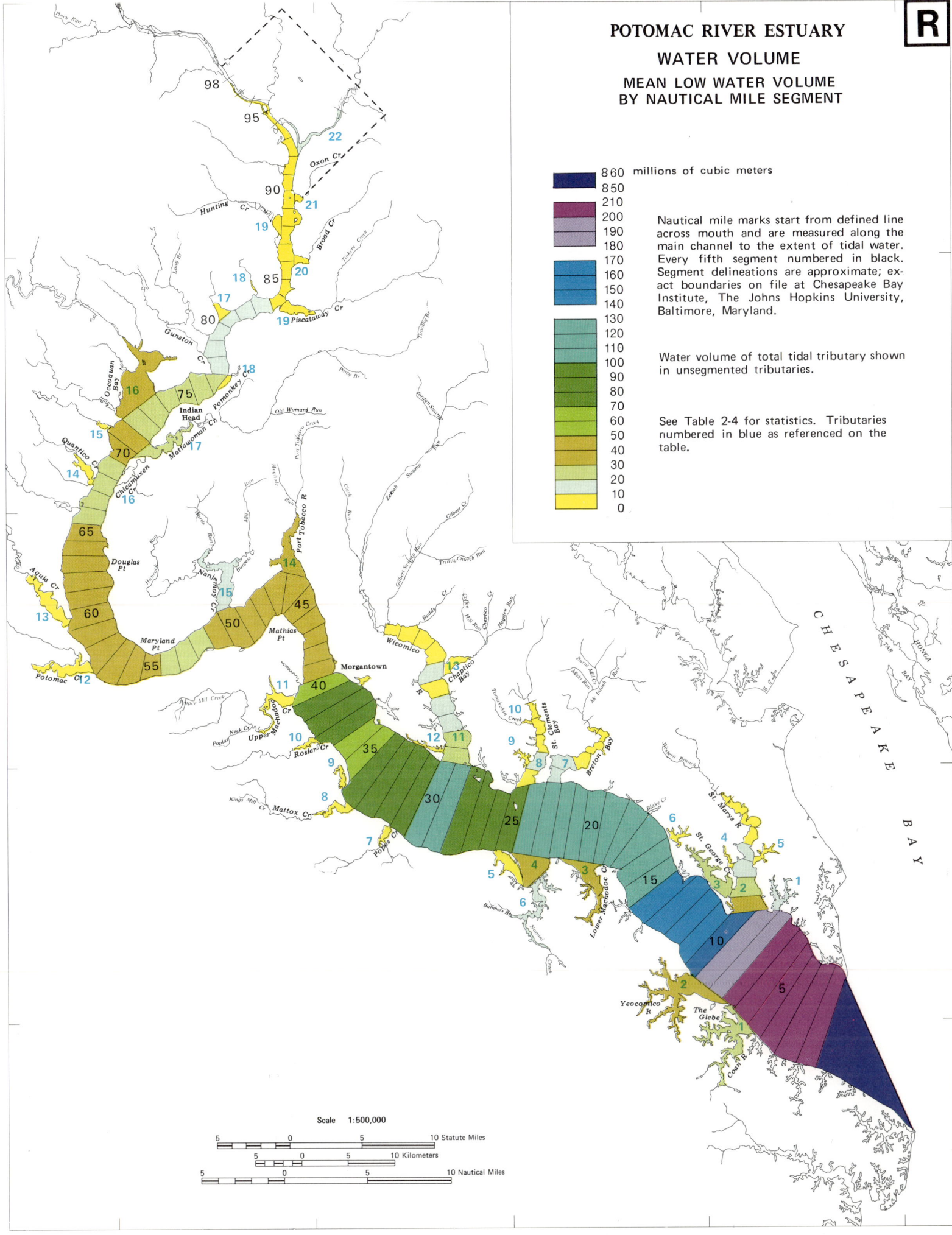

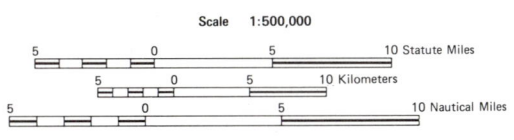

860 millions of cubic meters
850
210
200
190
180
170
160
150
140
130
120
110
100
90
80
70
60
50
40
30
20
10
0

Nautical mile marks start from defined line across mouth and are measured along the main channel to the extent of tidal water. Every fifth segment numbered in black. Segment delineations are approximate; exact boundaries on file at Chesapeake Bay Institute, The Johns Hopkins University, Baltimore, Maryland.

Water volume of total tidal tributary shown in unsegmented tributaries.

See Table 2-4 for statistics. Tributaries numbered in blue as referenced on the table.

CHESAPEAKE BAY

Scale 1:500,000

5 0 5 10 Statute Miles
5 0 5 10 Kilometers
5 0 5 10 Nautical Miles

Figure 2-3. *Relative water volumes of nautical mile segments of the Potomac estuary from the mouth to Washington, D.C., derived from statistics in Refs. 8 and 9.*

Depth Habitats

As discussed in Chapter 1, salinity, temperature, and substrate type are primary environmental factors influencing the distribution of biota; however, depth habitat is a major indirect determinant of the biological composition in any area. This Atlas and the folio maps use a particular nomenclature to define four distinct depth habitats: channel, mid-depth (shoal), shallow, and intertidal habitats.

Channel habitats* are regions where waters are 30 feet (9.1 m) or more in depth. These habitats are discontinuous along the estuary. In the lower estuary, there is a broad channel region that generally extends from the mouth upstream to the Wicomico River (see Folio Map 2). Depths range from 60 to 70 feet (18.3 to 21.4 m) towards the mouth, and

* "Channel" is used here in the sense of a habitat zone designation rather than in the sense of a navigation channel.

Table 2-4 — Average Depths, Surface Areas, and Water Volumes of Maryland and Virginia Tributaries to the Potomac Estuary

Tributary No. on Figures 2-3 and 2-4	Tributary Name or Nautical Mile Segment (a)	Average Depth (meters)	Surface Area (millions of square meters)	Water Volume (millions of cubic meters)
VIRGINIA				
1	Coan River and the Glebe	2.3	11.99	27.23
2	Yeocomico River	2.6	14.04	36.36
3	Lower Machodoc Creek	2.9	10.32	30.13
4	Nomini Bay	3.1	9.95	30.84
5	Currioman Bay	1.7	5.22	8.87
6	Nomini Creek	2.3	6.60	15.05
7	Popes Creek	0.6	1.72	1.05
8	Mattox Creek	1.5	2.82	4.15
9	Monroe Creek	0.9	1.72	1.62
10	Rosier Creek	1.2	1.92	2.36
11	Upper Machodoc Creek	2.3	1.55	3.55
12	Potomac Creek	0.6	5.84	3.56
13	Aquia Creek	1.5	5.56	8.34
14	Quantico Creek	1.0	3.25	3.28
15	Powells Creek	0.8	1.55	1.18
16	Occoquan Bay	1.6	22.25	35.38
17	Dogue Creek	1.1	1.75	1.99
18	Little Hunting Creek	0.9	0.56	0.51
19	Hunting Creek	1.0	2.26	2.10
MARYLAND				
1	Smith Creek	2.8	5.31	14.76
2	St. Marys River	3.6	27.77	111.94
	0-1	5.2	6.55	33.91
	1-2	3.9	7.23	28.17
	2-3	4.6	3.57	16.40
	3-4	5.1	2.00	10.29
	4-5	3.6	2.60	9.45
	5-6	2.6	2.36	6.21
	6-7	3.3	1.29	4.26
	7-8	1.5	2.17	3.24
3	St. George Creek	2.4	8.85	20.95
4	Carthagena Creek	3.0	1.86	5.51
5	St. Inigoes Creek	2.8	1.81	5.10
6	Herring Creek	1.4	2.09	2.97
7	Breton Bay	3.0	12.45	36.88
	0-1	4.3	2.48	10.75
	1-2	3.5	3.72	12.88
	2-3	3.1	2.79	8.52
	3-4	1.6	1.68	2.78
	4-5	1.0	1.37	1.39
	5-6	1.4	0.41	0.56
8	St. Clements Bay	3.3	10.37	35.95
	0-1	3.2	2.81	9.08
	1-2	4.4	2.53	11.03
	2-3	4.4	1.96	8.71
	3-4	2.5	1.96	4.82
	4-5	2.1	1.11	2.32
9	Canoe Neck Creek	2.5	0.96	2.36
10	Tomakokin Creek	1.7	0.31	0.54
11	Wicomico River	2.3	42.56	103.32
	0-1	5.0	5.25	26.04
	1-2	4.2	4.95	20.57
	2-3	2.7	4.80	13.16
	3-4	2.5	4.44	11.03
	4-5	2.3	4.17	9.54
	5-6	2.6	4.04	10.56
	6-7	1.8	3.64	6.49
	7-8	0.7	3.10	2.13
	8-9	0.6	3.38	2.13
	9-10	0.4	4.79	1.67
12	Neale Sound	1.2	1.57	1.85
13	Chaptico Bay	0.8	2.62	2.06
14	Port Tobacco River	2.0	22.23	44.46
15	Nanjemoy Creek	1.5	10.39	15.48
16	Chicamuxen Creek	1.4	2.27	3.13
17	Mattawoman Creek	1.2	22.25	25.85
18	Pomonkey Creek	1.3	1.19	1.57
19	Piscataway Creek	1.0	3.65	3.72
20	Broad Creek	1.1	1.50	1.62
21	Oxon Creek	1.9	0.69	1.30
22	Anacostia River	4.3	3.25	14.11

(a) Nautical river mile segments are measured along the channel and start at the line across the mouth of the Potomac River from Point Lookout to Smith Point.

Source: Ref. 9.

become shallower [to around 50 feet (15.2 m)] between Breton Bay and the Wicomico River. The part of the main channel lying off Colonial Beach is interrupted by fingers of sedimentary deposits, which create shoals where waters are only 10 to 15 feet (3.0 to 4.6 m) deep. The deepest spot in the main channel (119 feet or 36.3 m) is approximately midestuary, where the river bends around Mathias Point. In general, however, depths in this channel range from 50 to 80 feet (15.2 to 24.4 m). The next natural channel area upstream starts at Indian Head. From there, the channel continues intermittently to Washington, D.C., alternately hugging the Maryland and Virginia shores.

Mid-depth (shoal) habitats are regions where waters are 3 to 30 feet (0.9 to 9.1 m) deep. These habitats are found in many areas of the estuary, but there are two locations where they are extensive: Kettle Bottom Shoals and a region of the estuary from Nanjemoy Creek upstream to Indian Head. These two major shoal areas completely cross the estuary in most places and have somewhat deeper waters towards the Maryland side.

Shallow habitats are regions where waters are less than 3 feet (0.9 m) deep at mean low water. They exist throughout the estuary, but shallows that extend from shore to any significant distance out into the estuary are found primarily along the Virginia side of the lower half of the estuary and in the tributaries.

Intertidal habitats are shoreline areas that are alternately covered and uncovered with tidal waters and are sometimes called tidal flats. The intertidal system is defined as the area from the shoreward limit of waters at highest tides to the extreme low waterline of spring tides. The folio maps that delineate intertidal zones (maps 2, 3, 5, and 7) do so only to the mean low waterline. Intertidal habitats are mostly found in the lower half of the estuary, although shoaling around Washington, D.C., Gunston Cove, and Occoquan Bay has resulted in significant stretches of mud flats in the upper estuary. There are few intertidal flats in the midestuary segment.

In any area of the Potomac estuary, the biotic makeup will be strongly influenced by these depth habitats. In winter, the channel regions of the lower estuary provide a haven for overwintering fish and blue crabs. During the colder months, the channel habitat provides a warmer and more stable environment than other habitats in the estuary — the effects of winter storms and low ambient air temperatures are milder than in the shoals or shallows. In spring and summer, the many shallow coves and creeks and the broad shoal regions of the upper estuary are particularly suitable as fish spawning areas. Below the Rt. 301 bridge, there are also plentiful shallow habitats in the main stem and the tributaries, and broad stretches of shoals less than 18 feet (5.5 m) deep. These regions are particularly well suited for the proliferation of rich and varied benthic communities, some of which include the economically important oysters and clams.

Bottom Sediments

The Potomac estuary is carved out of the sediments of the Coastal Plain. These sediments are unconsolidated deposits of alternating layers of sand, silt, clay, diatomaceous earth, and gravel that form an eastwardly thickening sedimentary wedge (Fig. 2-1). This wedge lies over an ancient mass of hard, crystalline, metamorphic and igneous rock of the basement complex, which underlies the Piedmont Plateau and the mountain ranges to the west. The western beveled edge of the softer sedimentary wedge feathers out and forms the fall line where it comes in contact with the harder rocks of the Piedmont[1] (Fig. 2-1). The Coastal Plain sediment layer forms the "floor" of the Potomac and ranges in thickness from a few feet along the fall line to about 3,000 feet (914 m) at the mouth.[1] The sedimentary layers were formed over many epochs; some originated from sediments brought down from the western slopes; others are of marine origin.[1] On top of these ancient layers are the various sediments that have been brought into the estuary in more recent times (called surface bottom sediments) from the upper river and from the estuarine drainage system. The distributions of the different sediment types along the estuary depend on their grain size, the geometry of the estuary, water velocities, and current patterns.

Folio Map 3 shows general distributions of sediment types in the Potomac estuary. Since no detailed surveys of sediments have been undertaken in the Potomac, the major bottom sediment types can be grouped only into general categories of soft mud, hard mud, sand, shell, or rocks (gravel).[10] Soft sediments occur mostly at the head of the estuary, and there are some sandy areas towards the mouth bordering the lower shores. The uppermost regions of the estuary near Little Falls and Washington, D.C., are covered with rock and gravel beds. Throughout the estuary, less compacted sediments are generally found in the channels, and firmer muds and clays toward shore. The latter are often locally mixed with substantial amounts of sand, although they appear primarily as hard muds on the map. Shell areas are mostly in the lower estuary and are predominantly the remains of ancient oyster beds. One can readily see by comparing Folio Maps 3 and 6, that not all of the currently productive oyster beds are located on these ancient beds. Many are on firmer silts and clays or on sand.

Chapter 2 — References

1. *U.S. Department of the Army, 1973*
2. *Schubel and Meade, 1977*
3. *Cronin, L. E., 1973*
4. *Vokes and Edwards, 1968*
5. *Mason, 1974*
6. *Todd, 1970*
7. *U.S. Department of the Interior, 1965-1977*
8. *Cronin, W. B., and Pritchard, 1975*
9. *Cronin, W. B., 1971*
10. *U.S. Department of Commerce, 1974-1975*

The
Invisible
Boundaries

The Invisible Boundaries

The physical and chemical properties of Potomac estuarine waters set geographical limits that are as real to aquatic life as the tangible confines of the basin. Distinct habitats and biotic associations within them are circumscribed by invisible boundaries of salinity, water temperature, nutrients, dissolved oxygen, and toxic compounds, whose distributions are affected by river flow, tidal currents, and other physical processes. As a result, each habitat type contains a largely predictable association of biota that is adapted to prevailing conditions. To understand the relationship between biotic distributions and the factors controlling them, it is necessary to know something about the estuary's physical and chemical characteristics. These characteristics also affect water quality, a general term denoting the suitability of the waters for supporting biota.

PHYSICAL CHARACTERISTICS

The Potomac estuary is a tributary of the Chesapeake Bay and, together with the Bay and its other tributaries, is part of a vast coastal plain estuary system. The main physical characteristics of the Potomac estuary can be summarized as follows:

- It is under tidal influence that extends all the way to the fall line.
- For most of the year, it has only three of five possible salinity zones because the waters entering it from the Chesapeake Bay are only half as saline as ocean waters. Because of the considerable length of the estuary, these zones cover extensive stretches of water. In fall or during extremely low flow conditions, a fourth (polyhaline) zone occurs periodically near the mouth.
- The longitudinal salinity distributions and nontidal flows respond strongly to variations in river input.
- Downstream from the tidal fresh region, the estuary develops a two-layered net nontidal flow pattern.
- The flow patterns affect not only salinity and temperature, but also the distributions of nutrients and pollutants — variables that affect the general quality of the water.

Freshwater Discharge and Estuarine Circulation

Fresh water entering the Potomac estuary from the watershed of the upper basin is measured at stream gages placed at Little Falls Dam, 1 statute mile upstream from Washington, D.C. For the years of record, the annual average freshwater flow at Little Falls was 11,190 cubic feet per second (cfs), eighth highest of all U.S. Atlantic seaboard rivers (Table 3-1). And, of all U.S. rivers, the Potomac ranks twenty-sixth in flow.[4]

Flow rates can change dramatically from day to day, from season to season, and from year to year (Figs. 3-1, 3-2; Table 3-2). High flows normally occur in early spring as a result of runoff from melting snow and ice and high precipitation. Low flows characteristically occur in late summer and early fall when rainfall diminishes. The normal pattern may

Table 3-1 — Average Annual Freshwater Discharges of Major Rivers Along the Atlantic Coast of the United States (in cubic feet per second)

Susquehanna River	35,000
Connecticut River	16,470
Hudson River	14,400
Santee River System (South Carolina)	12,680
Altamaha River (Georgia)	12,420
Penobscot River (Maine)	12,020
Delaware River	11,850
Potomac River	11,190
Savannah River	10,830

Sources: Refs. 1, 2, 3.

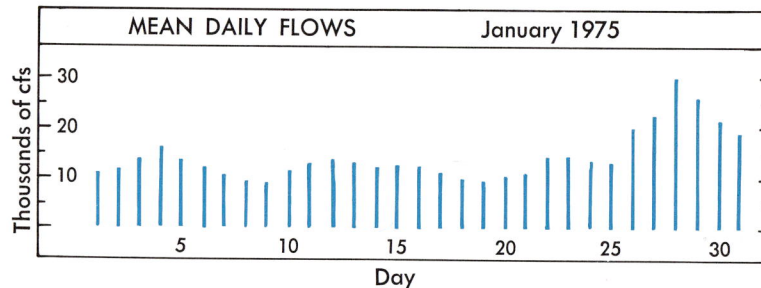

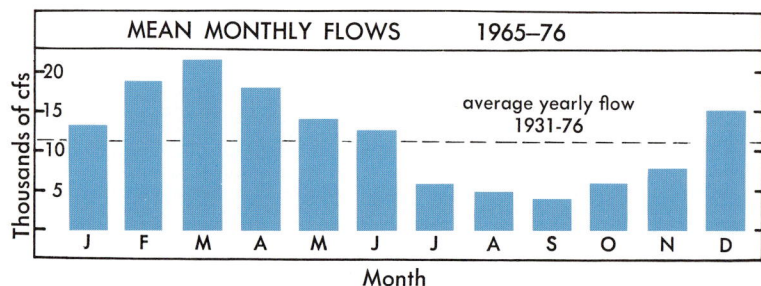

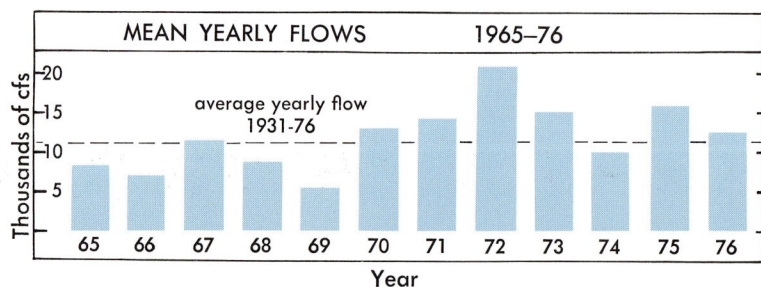

Figure 3-1. An example of mean daily, monthly, and yearly flows measured near Little Falls. Dashed line indicates the long-term mean yearly flow from 1931 to 1976. (Sources: Refs. 1,5)

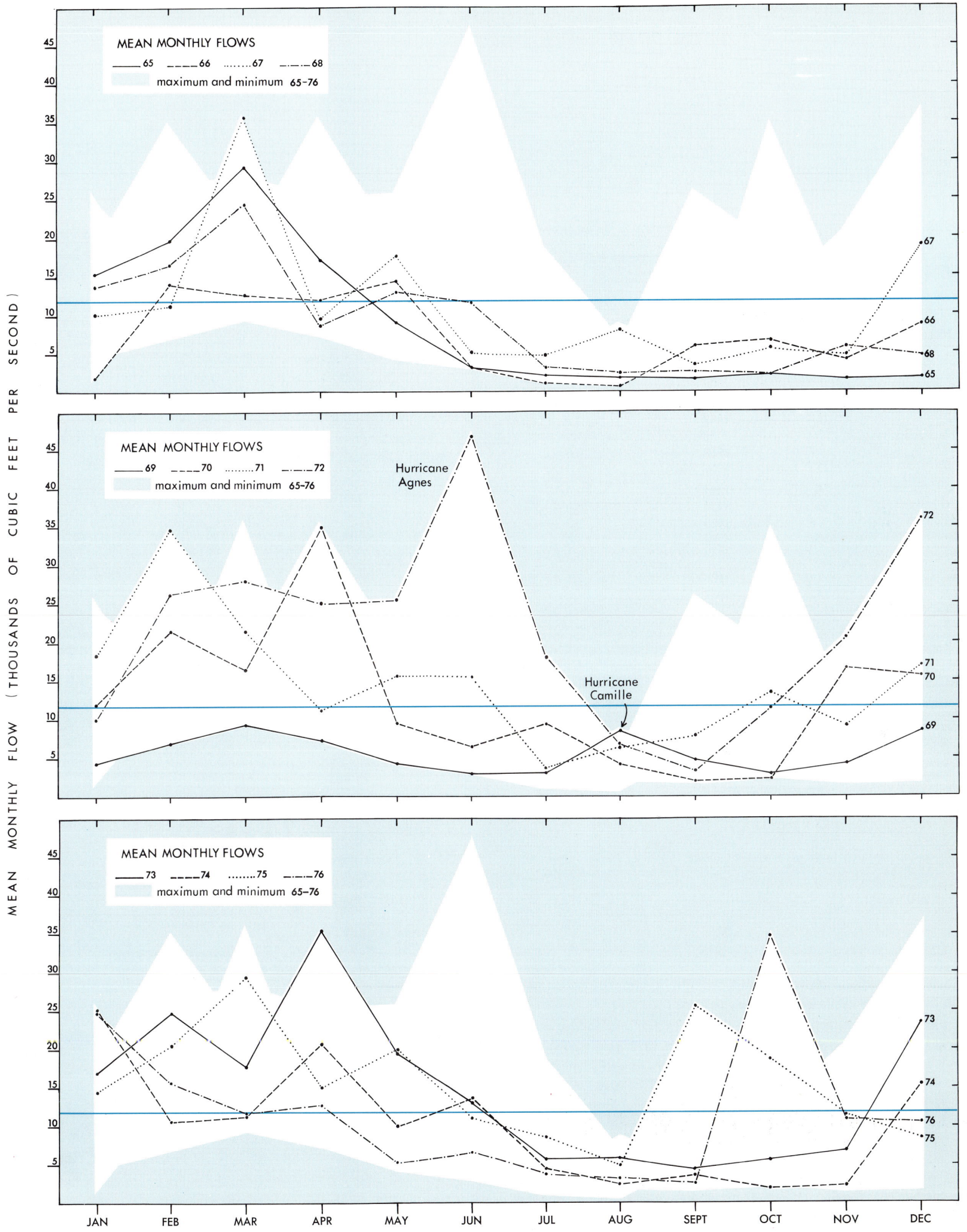

Figure 3-2. *Mean monthly flows, 1965-76, measured near Little Falls and adjusted for diversions for municipal water use. Dark blue lines indicate the long-term mean yearly flow from 1931 to 1976. (Source: Ref. 1)*

Table 3-2 — **Mean Yearly Discharges**[a] **Measured at the Head of the Potomac Estuary, Calendar Years 1931-76**

Year	Mean Yearly Discharge (in thousands of cfs)	Year	Mean Yearly Discharge (in thousands of cfs)
1931	6.0	1956	10.3
1932	10.9	1957	9.0
1933	13.4	1958	11.9
1934	8.0	1959	6.8
1935	11.8	1960	11.1
1936	16.0	1961	12.4
1937	18.3	1962	10.9
1938	7.2	1963	8.3
1939	10.7	1964	9.8
1940	12.5	1965	8.4
1941	7.0	1966	7.1
1942	14.3	1967	11.3
1943	10.9	1968	9.1
1944	9.6	1969	5.5
1945	12.1	1970	12.6
1946	9.1	1971	14.5
1947	7.0	1972	21.4
1948	13.0	1973	15.0
1949	13.7	1974	10.6
1950	13.4	1975	16.0
1951	12.8	1976	12.3
1952	15.7		
1953	11.7		
1954	7.6		
1955	11.5		

(a) All mean discharges were measured at Station 01646500, Potomac River near Little Falls, and were adjusted by adding amounts diverted for municipal use.
Sources: Refs. 1, 5.

be disrupted periodically by drastic changes in weather conditions, such as large storms or extended dry periods.

For the 46 years from 1931 to 1977 (Table 3-2), the highest mean yearly flow on record (21,400 cfs) occurred in 1972, primarily due to Hurricane Agnes — which created a mean monthly flow in June of more than 45,000 cfs — but also due to consistently high flows throughout the spring months before the hurricane, and an unseasonably wet period in November and December. The lowest mean annual flow during the past 47 years was 5,500 cfs, recorded in 1969 (Table 3-2). The maximum instantaneous peak flow measured at Little Falls since 1931 was 484,000 cfs (March 19, 1936); the minimum was 601 cfs (September 10, 1966).[1] Peak instantaneous flow during Hurricane Agnes was 359,000 cfs on June 24, 1972.[6]

Freshwater flow from the upper river drainage is augmented by runoff from all the tributaries below Little Falls (Fig. 3-3). By the time the

Potomac pours into the Chesapeake Bay, tributary runoff has added an average of 3,110 cfs to the mean freshwater flow of 11,190 cfs,[1] making an average total freshwater flow of 14,300 cfs.[7] The Potomac's freshwater contribution to the total Bay discharge is almost equal to the amount contributed by all other Bay tributaries except the Susquehanna, which supplies 52 percent of the 67,300 cfs total.[2]

As described in Chapter 1, water transport in and out of estuaries is complex, with both salty and fresh nontidal flows taking part in flushing water through the estuarine basin. At any point in the tidal freshwater region of the Potomac, the entire nontidal flow is directed downstream and equals the cumulative freshwater input to that point. As fresh water meets denser salty water, a two-layered flow pattern develops, with the net nontidal flow in the upper layer directed downstream and the net nontidal flow in the lower layer directed upstream (see Chapter 1).[8] In salty regions, the volume of water transport in both of these layers may at any point exceed the freshwater river input, but the difference between them is still equal to the cumulative freshwater input to that point.

An example of a two-layered flow pattern is illustrated by data from a survey in the vicinity of Morgantown, located in the mesohaline region (Fig. 3-4). In contrast to the nontidal velocities shown in the figure, the peaks of typical tidal flow measured at that time were 40 centimeters per

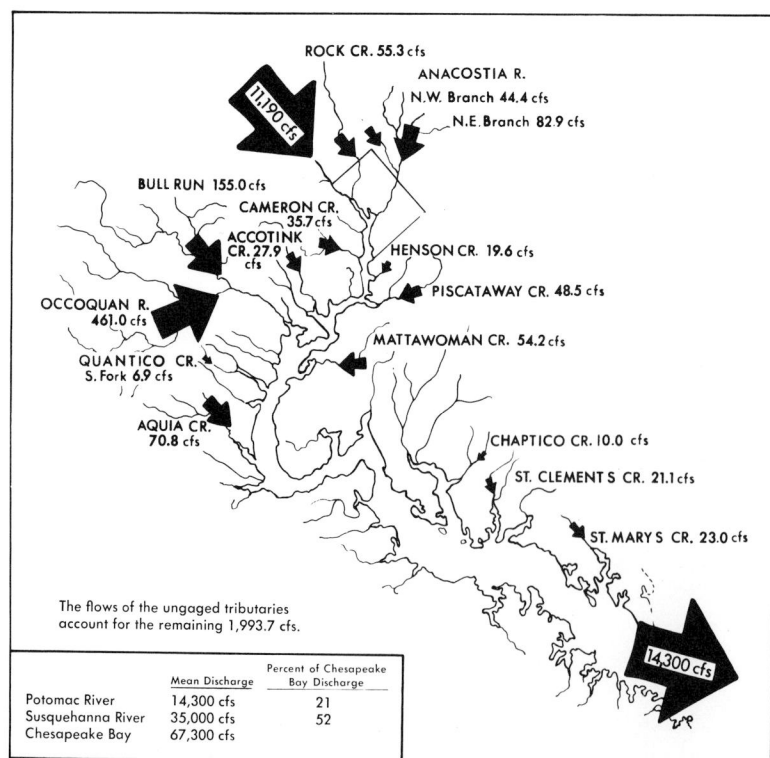

Figure 3-3. *Mean yearly discharges into the Potomac estuary from the upper river drainage and tributaries, measured in cubic feet per second. (Source: Ref. 1). Chesapeake Bay discharge and percent calculations derived from Ref. 2.*

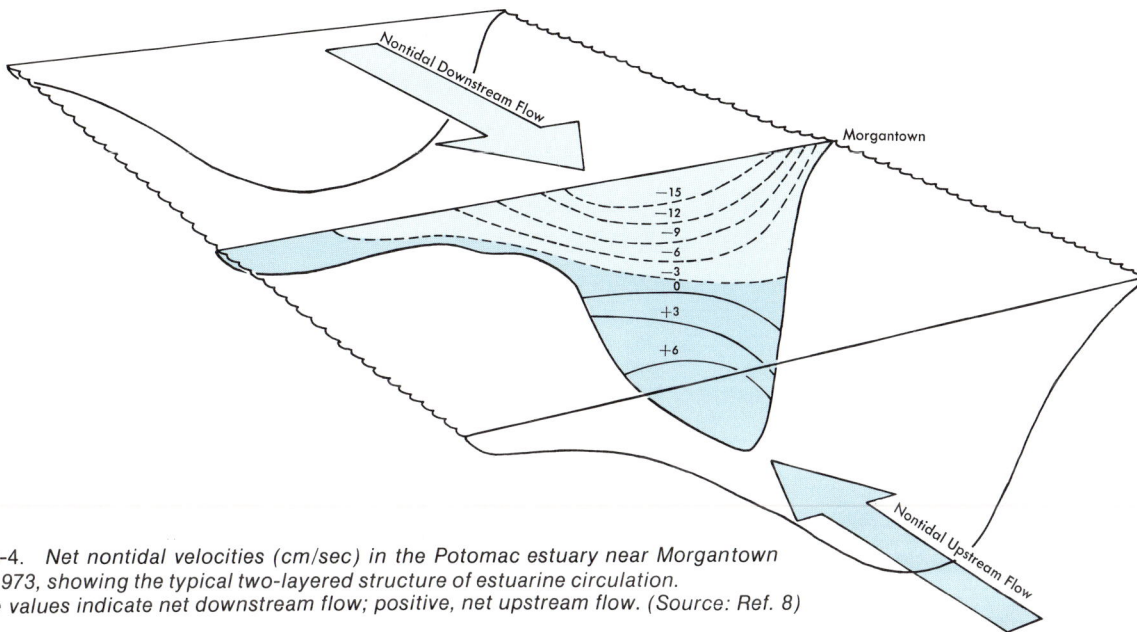

Figure 3-4. *Net nontidal velocities (cm/sec) in the Potomac estuary near Morgantown in May 1973, showing the typical two-layered structure of estuarine circulation. Negative values indicate net downstream flow; positive, net upstream flow. (Source: Ref. 8)*

second (cm/sec). Based on data from this survey, the nontidal transports were computed as:

- − 27,600 cfs directed downstream
- +   9,100 cfs directed upstream
- − 18,500 cfs net computed cumulative freshwater input from upstream

A negative value indicates net downstream transport; a positive one indicates net upstream transport.

Tides

Tides in the Potomac estuary are semidiurnal, with a period of 12.4 hours. Tidal heights and currents bear a complex relationship to each other along the length of the estuary. The tidal motion is greatly affected by the shape of the entire basin, as well as by local features. Since the tidal "wave" is partially reflected at the head of the estuary, the shapes of tidal height variations over time exhibit both progressive and standing wave characteristics. The wave fronts in progressive waves travel in time; thus, tidal crests seem to move up and down the estuary. For standing waves, minimum and maximum heights occur in the same locations. Nodal points are where no height changes occur, and their locations are stationary. These two types of patterns are mixed in the Potomac, and the nodal point of the standing wave occurs near Maryland Point (nautical river mile 55), where tidal range is minimal (Fig. 3-5).

The progression of the tidal wave in the Potomac is clearly seen when tidal heights along the estuary are compared to the rise and fall of the tide at Washington, D.C. (Fig. 3-6). When the tide is highest at Washington, D.C., the lowest tide occurs midway up the estuary near Morgantown (nautical river mile 40). Conversely, during low tide at

Washington, D.C., the highest tide occurs near St. Catherine Island (nautical river mile 28). Calculated mean tidal heights at 3-hour intervals from high and low tides at Washington, D.C., are listed in Table 3-3.

The mean tidal range, or the difference in heights between mean high water and mean low water, varies along the length of the estuary (Fig. 3-5). [9] [10] At Washington, D.C., the mean tidal range is significantly greater [0.88 meters (m)] than it is at the mouth (0.37 m). Near Maryland Point (nautical river mile 55), the mean tidal range is only 0.30 m.

Tidal range varies seasonally and with the phase of the moon. Some of the highest tides occur during periods of the full or new moon. The average "spring range" (the term used to denote the larger

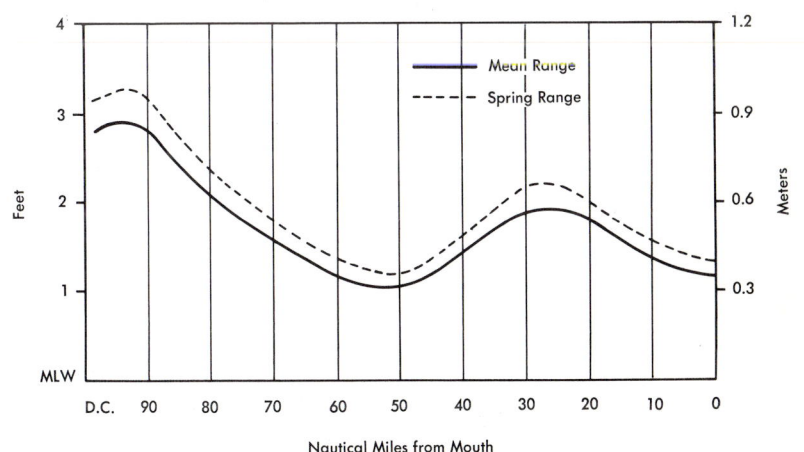

Figure 3-5. *Mean and spring ranges of tides along the Potomac estuary. (Source: Ref. 9)*

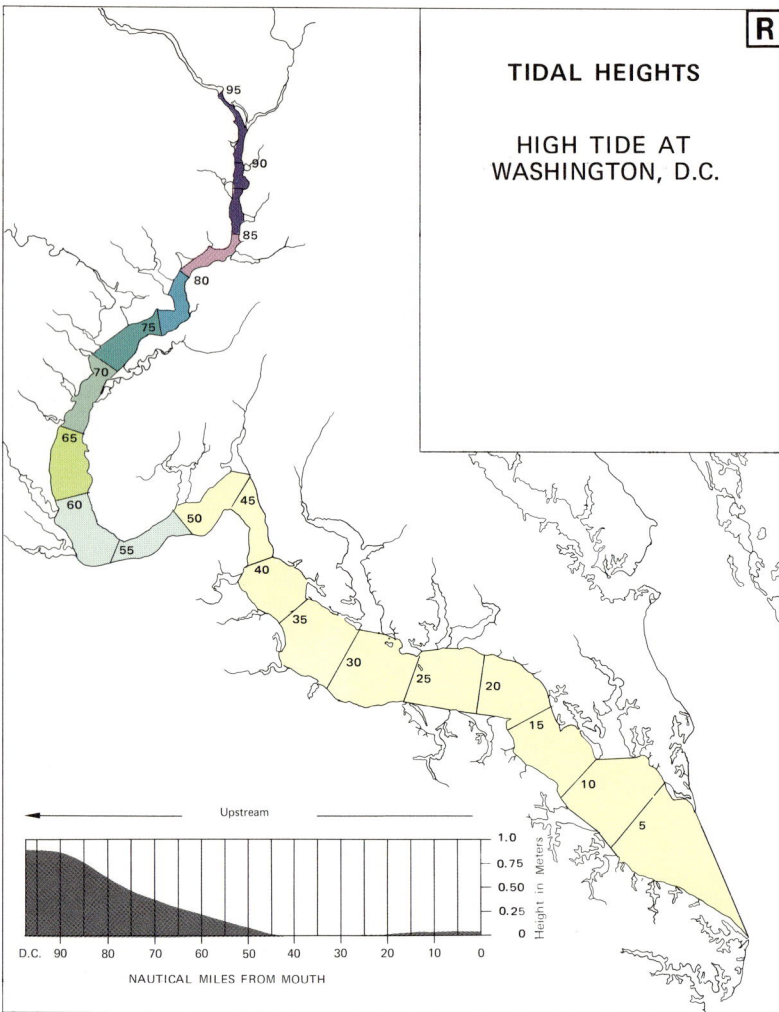

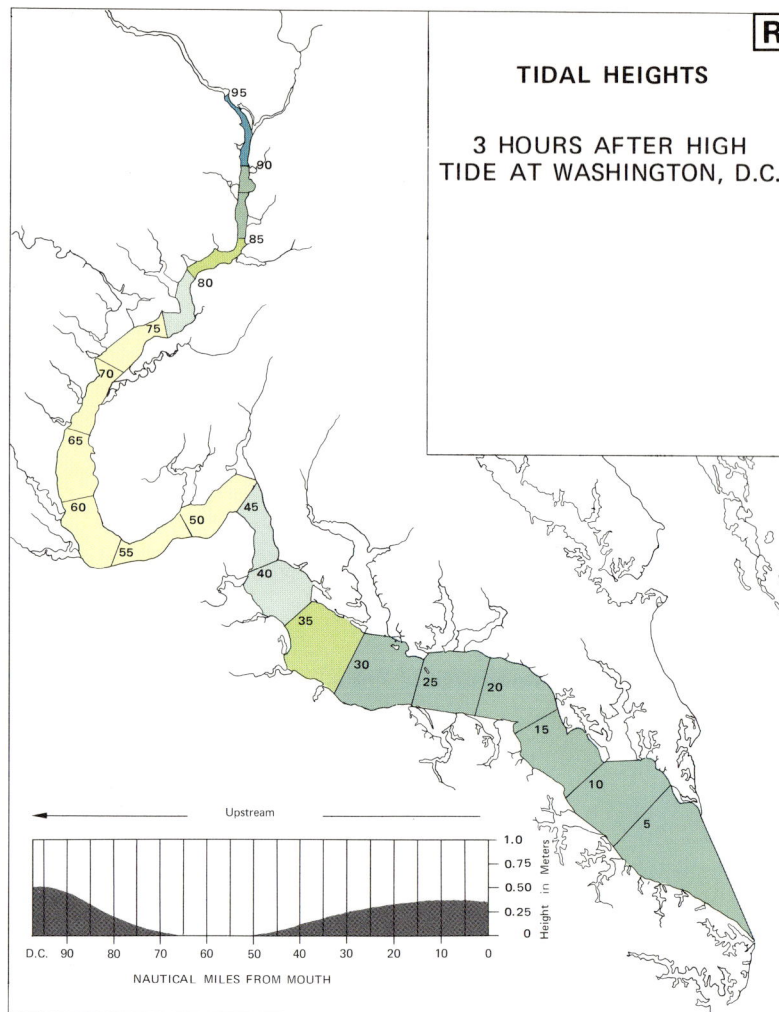

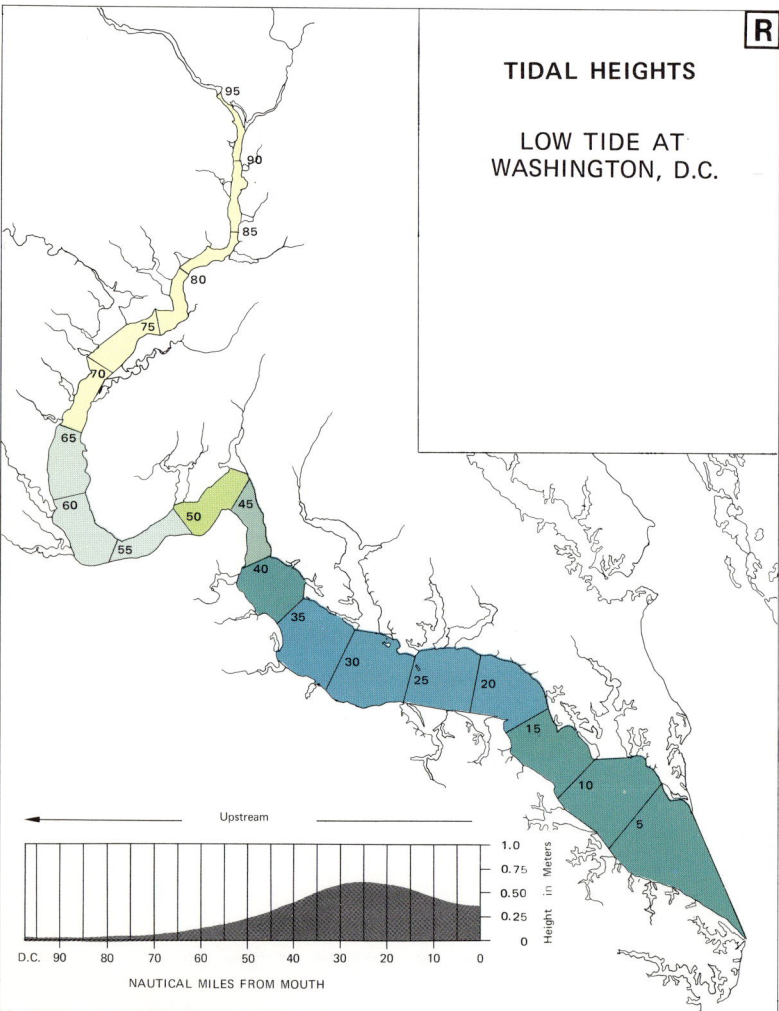

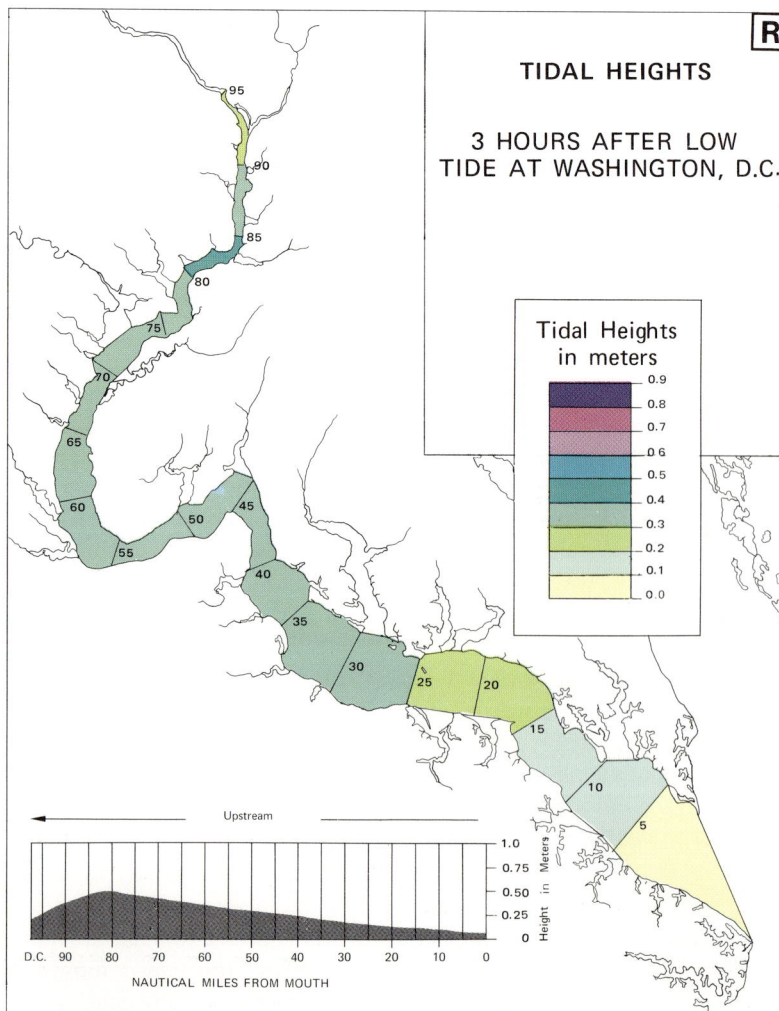

Figure 3-6. *Tidal heights for 5-nautical-mile segments of the Potomac estuary in relation to tidal phases at Washington, D.C. (Source: Ref. 10)*

Table 3-3 — **Mean Calculated Tidal Heights (in meters) in Relation to Tidal Phase at Washington, D.C.**

Nautical River Miles from Mouth	95	90	85	80	75	70	65	60	55	50	45	40	35	30	25	20	15	10	5	0
High Tide	0.90	0.81	0.68	0.57	0.44	0.33	0.26	0.17	0.11	0.04	0.01	0.00	0.02	0.02	0.04	0.04	0.04	0.07	0.07	
3 Hours after High Tide	0.50	0.37	0.23	0.14	0.07	0.03	0.01	0.01	0.02	0.06	0.12	0.18	0.25	0.31	0.34	0.34	0.32	0.32	0.34	
Low Tide	0.01	0.01	0.04	0.06	0.07	0.09	0.13	0.16	0.19	0.28	0.36	0.44	0.51	0.59	0.57	0.54	0.48	0.45	0.42	
3 Hours after Low Tide	0.27	0.39	0.40	0.39	0.35	0.38	0.36	0.33	0.34	0.34	0.34	0.32	0.32	0.31	0.27	0.23	0.16	0.13	0.09	

Source: Ref. 10.

semidiurnal ranges occurring at the times of full and new moons) is 1 m at Washington, D.C. (Fig. 3-5). Ranges during maximum tides, which occur in the spring and fall, may be even greater. For example, predicted tidal heights at Washington, D.C., during the spring full moons in May and June 1977 were 1.16 m.[9]

Weather has a marked effect on local tidal heights. Winds from the north or west that push water out of the estuary cause lower tides than normal. Winds from a southerly or easterly direction act as a barrier to water leaving the estuary, so that water builds up, creating higher than normal tides. Water levels may also increase or decrease a foot or more above or below average levels under extreme high-flow or drought conditions.

Tidal currents flow along the longitudinal axis, although in the broader sections, considerable cross-stream flows may develop (e.g., as they do in the shallows between Indian Head and Maryland Point).[11] Current velocities vary with tidal phase and bottom configuration, but in general, tend to be highest in constricted areas of the estuary (Fig. 3-7). Peak currents throughout the Potomac estuary average below 0.5 knots, as they do elsewhere in the Chesapeake Bay. Ebb currents generally have greater velocity than flood currents. Peak velocities of flood and ebb currents at various points along the estuary are listed in Table 3-4.

Peak velocities occur only for a short period during each tidal phase, and they do not generally coincide with either the time of high or low water. Maximum flood velocities develop from 0.5 to 1.5 hours before high tide and usually occur towards the bottom where they are augmented by the net upstream movement of denser bottom waters. Maximum ebb velocities usually occur at the surface of the water where they get an extra push downstream by the net downstream surface flow. Tidal currents, responding to changes in the elevation of the surface, vary with the lunar phase and are greatest in the periods that have the highest tidal range, during the full or new moon. The lunar cycle of maximum and minimum current velocities repeats every lunar month. Maximum current velocities can vary significantly between different locations in the same tidal system. Times of maximum and minimum velocities follow a predictable pattern — between any two points in a system, there is a constant difference in phase. For example, there is a

Table 3-4 — **Calculated Peak Tidal Velocities Along the Potomac Estuary**

Location	Flood knots	Flood cm/sec	Ebb knots	Ebb cm/sec
1 mile south of Cornfield Point	0.5	26	0.5	26
Midchannel off Cornfield Point	0.5	26	0.6	31
3.8 miles south of Cornfield Point	0.7	36	0.6	31
St. Marys River at Fort Point[a]				
Yeocomico River entrance[a]				
0.2 mile south of Piney Point	1.3	67	0.6	31
Midchannel off Piney Point	0.4	21	0.6	31
2.2 miles south of Piney Point	0.5	26	0.5	26
Lower Machodoc Creek entrance[a]				
Nomini Creek entrance at White Oak Point	1.2	62	1.2	62
Breton Bay entrance	0.6	31	0.4	21
St. Clements Bay entrance	0.4	21	0.9	46
1.8 miles southeast of St. Clements Island	0.4	21	0.9	46
1.1 miles southwest of St. Clements Island	0.6	31	0.8	41
Wicomico River entrance at Rock Point	0.5	26	0.6	31
Swan Point	0.3	15	0.8	41
Dahlgren Harbor Channel[a]				
Upper Machodoc Creek entrance	0.3	15	0.3	15
Just south of the Rt. 301 bridge	0.9	46	1.4	72
Persimmon Point	1.2	62	1.4	72
Port Tobacco River at Chapel Point[a]				
Maryland Point	1.1	57	1.4	72
Quantico	0.7	36	0.9	46
Quantico Creek entrance	0.5	26	0.5	26
2.3 miles east of Freestone Point	0.7	36	0.7	36
Hallowing Point	1.1	57	1.1	57
Jones Point	1.0	51	0.9	46
Hains Point	0.6	31	0.3	15
Anacostia River entrance[a]				
South Capitol Street Bridge, Washington, D.C.[a]				
Washington Channel, Washington, D.C.[a]				
Virginia Channel, Washington, D.C.	current seldom floods		0.6	31

(a) Currents are too weak and variable to predict.
Source: Ref. 12.

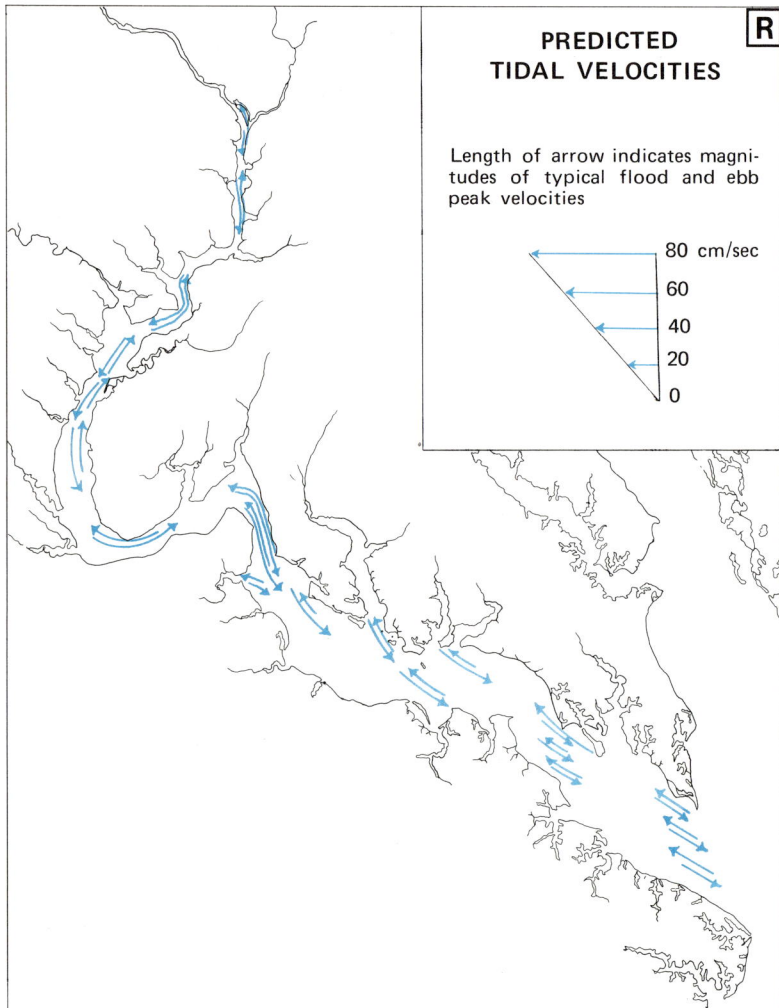

Figure 3-7. *Predicted tidal current velocities at various locations along the Potomac estuary, based on data in Table 3-4. (Source: Ref. 12)*

constant 4-hour lag between velocities appearing at the mouth of the Chesapeake Bay and those appearing at the mouth of the Potomac, as shown for part of a lunar cycle in Fig. 3-8. Between any two points in a system, there is also a constant magnitude ratio.

Tidal excursion, or the displacement of a particle of water during one-half tidal cycle from slack low water to slack high water, or vice versa, depends on velocity. In constricted areas, where velocities are highest, a tidal excursion may be 5 or more nautical miles. At Morgantown and Alexandria, two regions where the estuary narrows, tidal excursion has been calculated to be 5.9 nautical miles and 5.25 nautical miles, respectively. At Douglas Point, a region where the estuary is relatively wide and shallow, the tidal excursion has been estimated to be only 3.1 nautical miles.[13]

The interaction of nontidal and tidal currents and the vertical movements of some zooplankton and phytoplankton species allows these organisms to maintain themselves in their optimum habitats, even though they are passively transported by currents for the most part. The currents also bring marine and freshwater planktonic species into the estuary. For temporary plankton (eggs and larvae of benthic and fish species), currents largely determine whether these organisms will reach (and in the case of the benthos, settle out in) favorable habitats.

Salinity

During most of the year, the Potomac estuary has only three salinity zones — tidal fresh, oligohaline, and mesohaline (Fig. 3-9). The estuary is usually fresh above Indian Head and mesohaline from Colonial Beach downstream to the mouth, except during fall, when a short-lived tongue of polyhaline water may extend along the bottom near the mouth. During low flows, this zone may extend as far upstream as Piney Point. Maximum salinities at the mouth have been as high as 20 ppt (parts per thousand), but the long-term monthly averages reach only about 12 ppt. The region from Indian Head to Colonial Beach is the only stretch of the

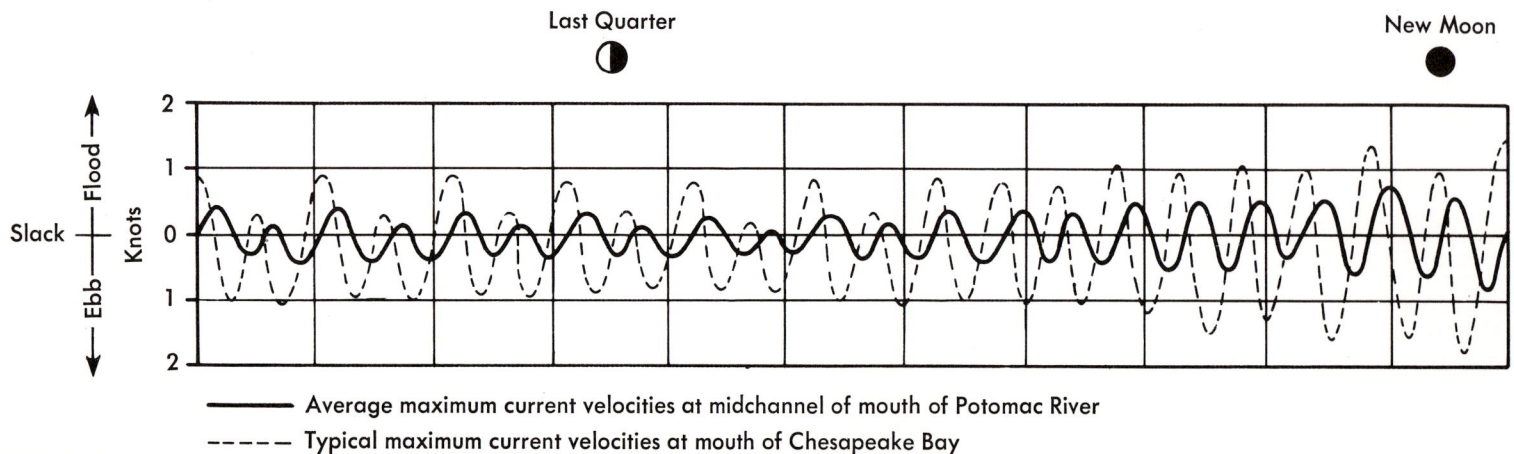

Figure 3-8. *Average maximum predicted current velocities near the mouth of the Potomac estuary and at the mouth of the Chesapeake Bay during part of a lunar cycle. Each segment is one day. (Source: Ref. 12)*

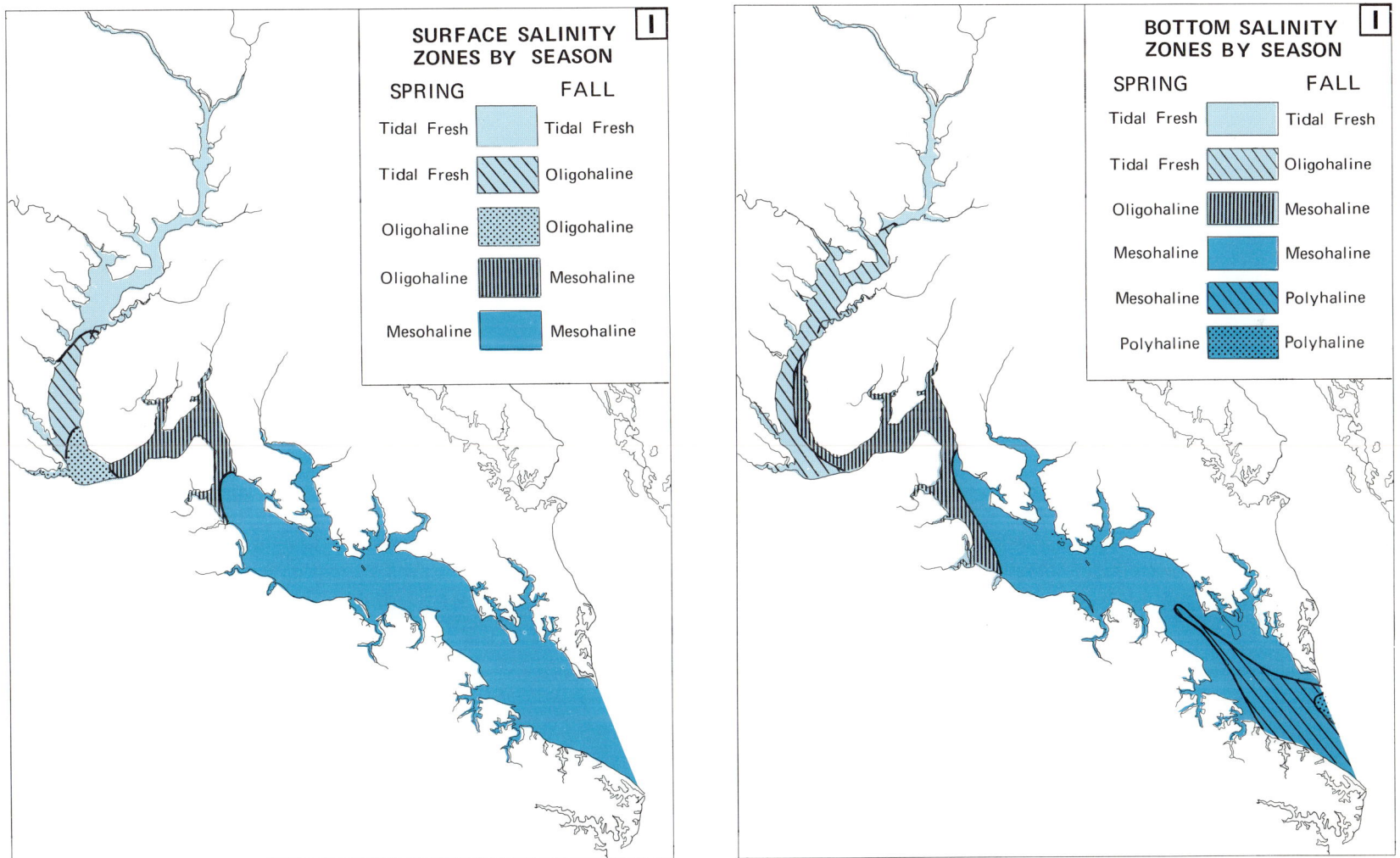

SURFACE SALINITY
ZONES BY SEASON

SPRING | FALL
Tidal Fresh | Tidal Fresh
Tidal Fresh | Oligohaline
Oligohaline | Oligohaline
Oligohaline | Mesohaline
Mesohaline | Mesohaline

BOTTOM SALINITY
ZONES BY SEASON

SPRING | FALL
Tidal Fresh | Tidal Fresh
Tidal Fresh | Oligohaline
Oligohaline | Mesohaline
Mesohaline | Mesohaline
Mesohaline | Polyhaline
Polyhaline | Polyhaline

Figure 3-9. *General locations of salinity zones at surface and bottom in the Potomac estuary during spring and fall, showing maximum extent of tidal fresh and polyhaline zones. (Sources: Refs. 14-17)*

estuary where the salinity zones will normally change seasonally (Fig. 3-10).

There are marked salinity differences across the estuary. Gravity is the strongest influence in the vertical distribution, but other forces (e.g., the Coriolis force due to the earth's rotation and centrifugal force due to the curves of the estuary) affect the lateral distribution. The Coriolis force is important near the surface in broader portions of the estuary, whereas the centrifugal force may become more important in the bends of the estuary. For example, these forces may cause the surface salinities along the Maryland shore to be 2 or 3 ppt greater than those along the opposite Virginia shore. In the lower estuary, the saltier waters are channelized along the bottom by basin topography. This leads to cross-sectional salinity patterns that are different from those upstream.

The salinity gradients in the tributaries of the Potomac reflect the amount of freshwater flow in each (established mostly by the extent of each watershed) and the level of salinity entering the tributary mouth from the main stem. The tributaries of the lower portion of the estuary, where salinities are highest, have compressed salinity zones. Midestuarine tributaries have broader, more distinctly separated salinity zones, while those of the upper estuary are completely tidal fresh.

Vertical salinity profiles (Fig. 3-11) show increasing salinities with depth. Typically, the most pronounced salinity stratification occurs in spring (May) and the least in fall (September and October).

Depending on location and season, the differences in salt content between surface and bottom or from one shore to the other can be less than 1 ppt or as great as 6 or 7 ppt. Average monthly values do not

AVERAGE SALINITY

JANUARY
Surface

SALINITY

0.5   5   10   15   ppt

| tidal fresh | oligo-haline | low mesohaline | high |

AVERAGE SALINITY

FEBRUARY
Surface

SALINITY

0.5   5   10   15   ppt

| tidal fresh | oligo-haline | low mesohaline | high |

JANUARY
Bottom

FEBRUARY
Bottom

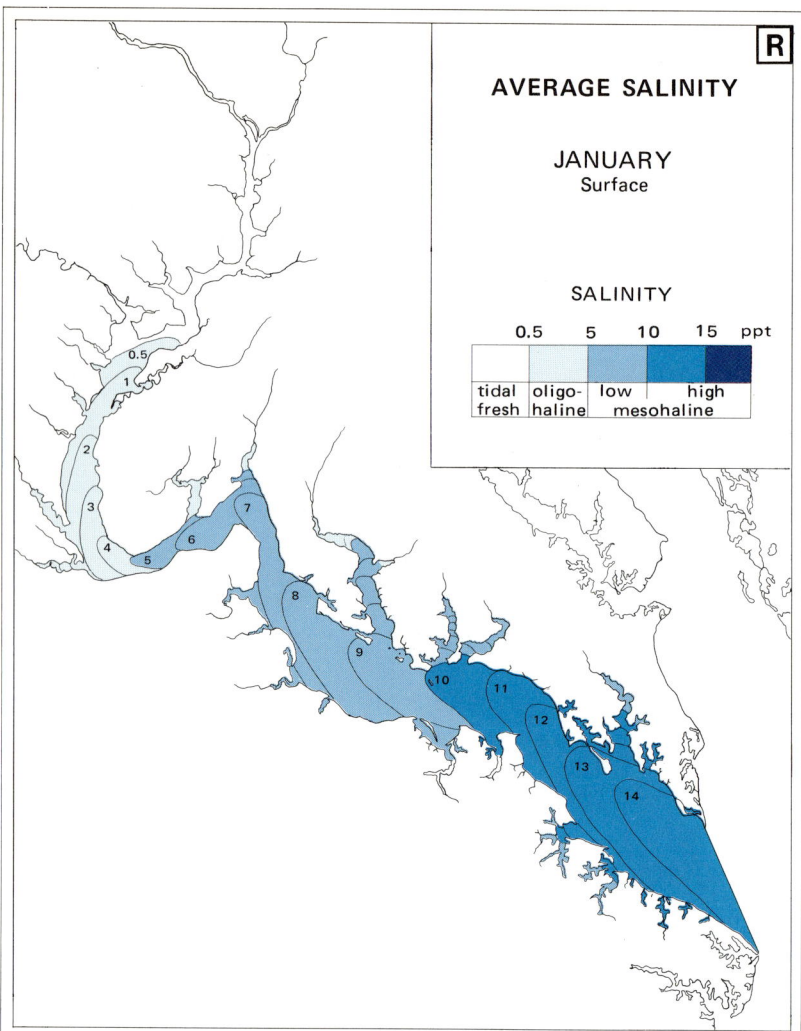

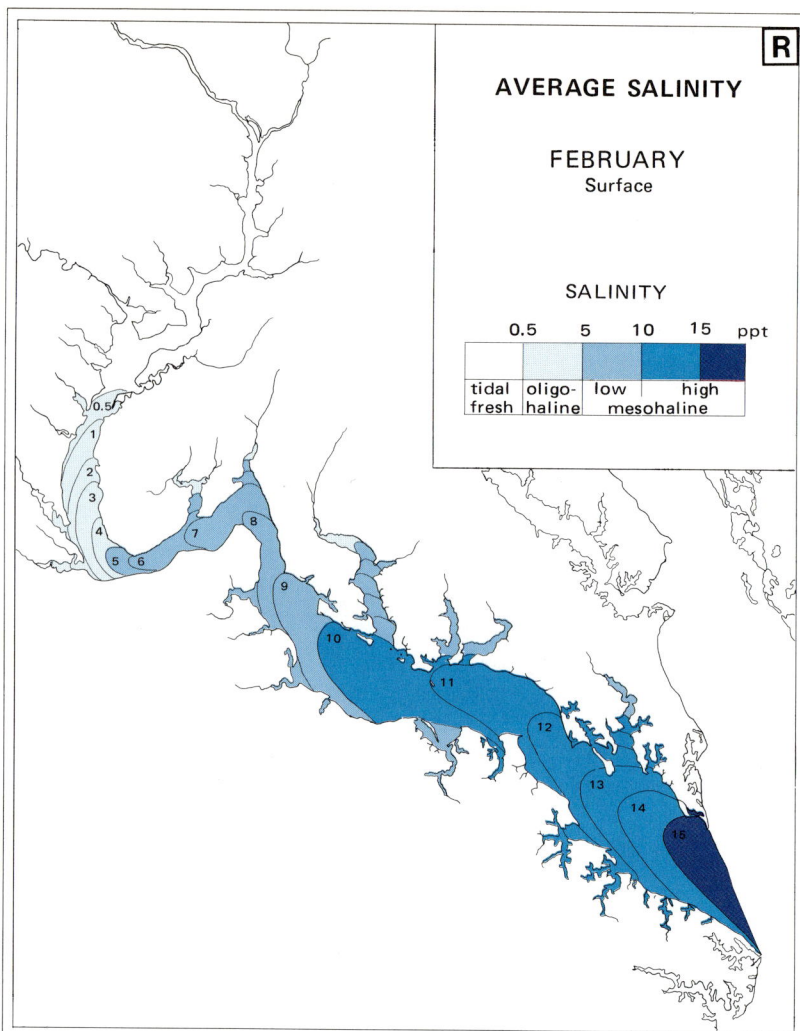

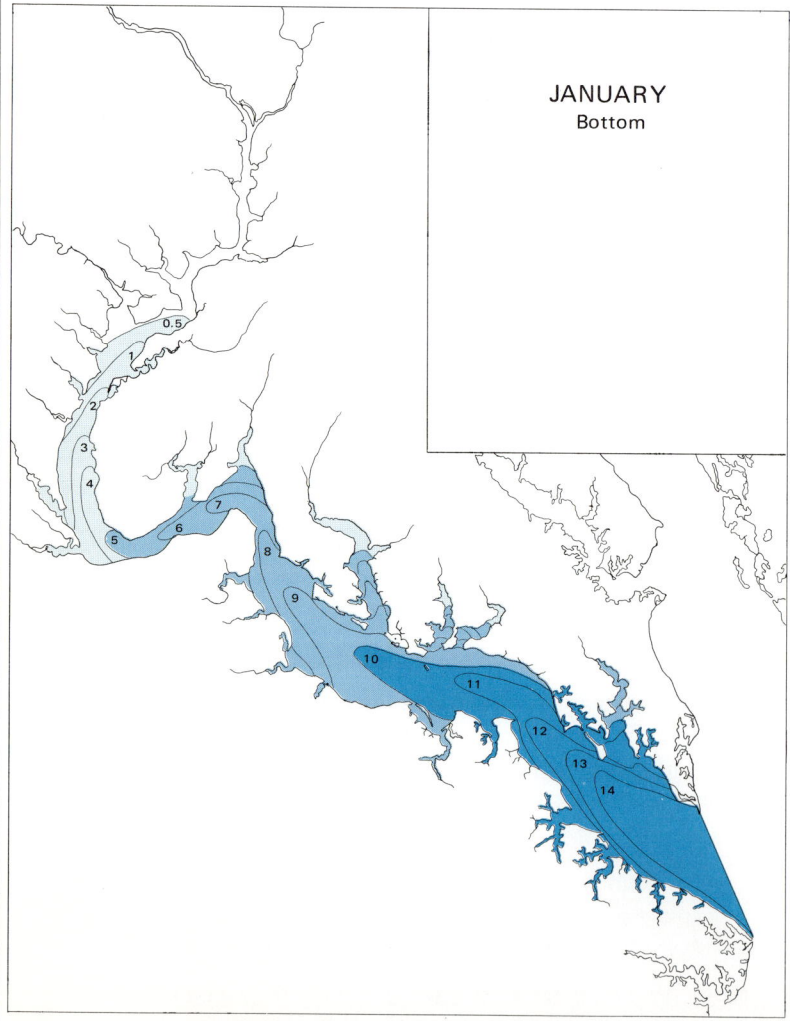

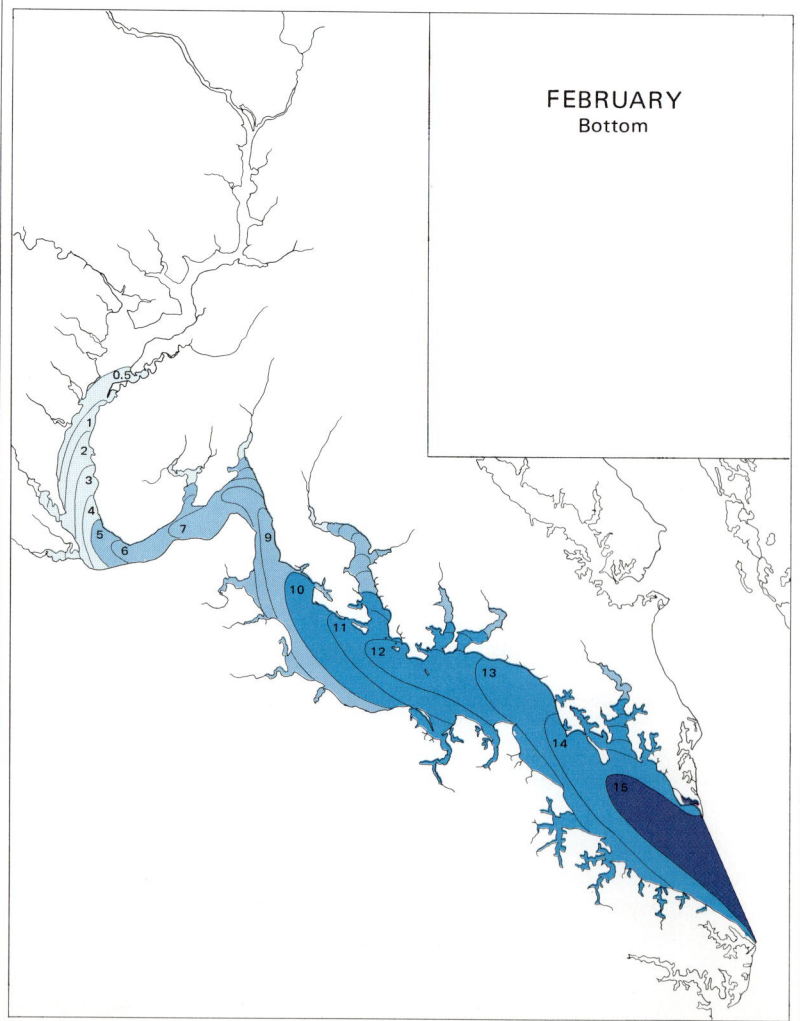

Figure 3-10.   Average monthly salinities at surface and bottom in the Potomac estuary for January through December. Salinity gradients in tributaries have been extrapolated from mainstem data. Bulk of data covers the period 1965-76. (Sources: Refs. 14, 15, 18)

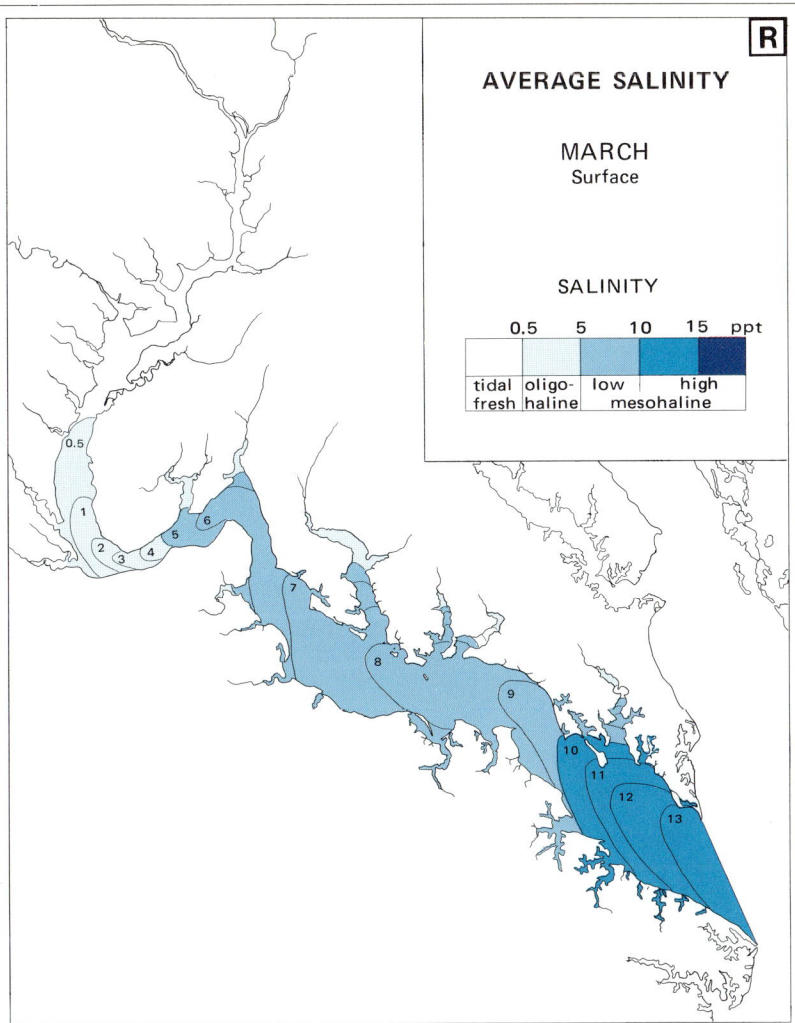

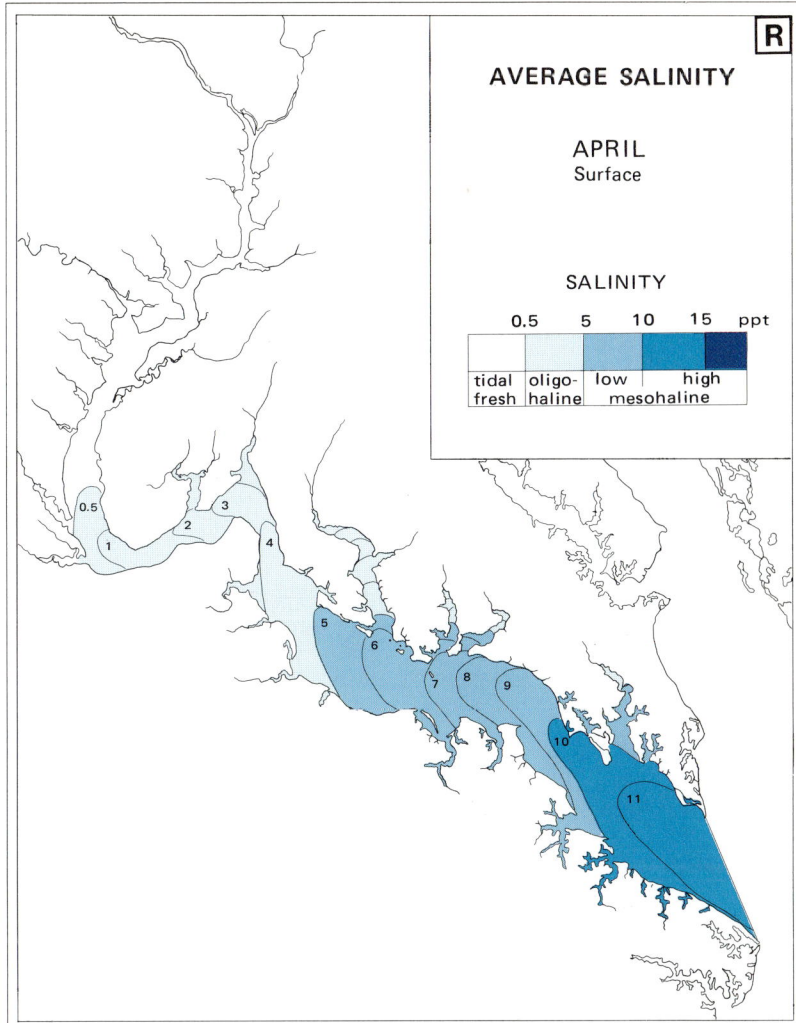

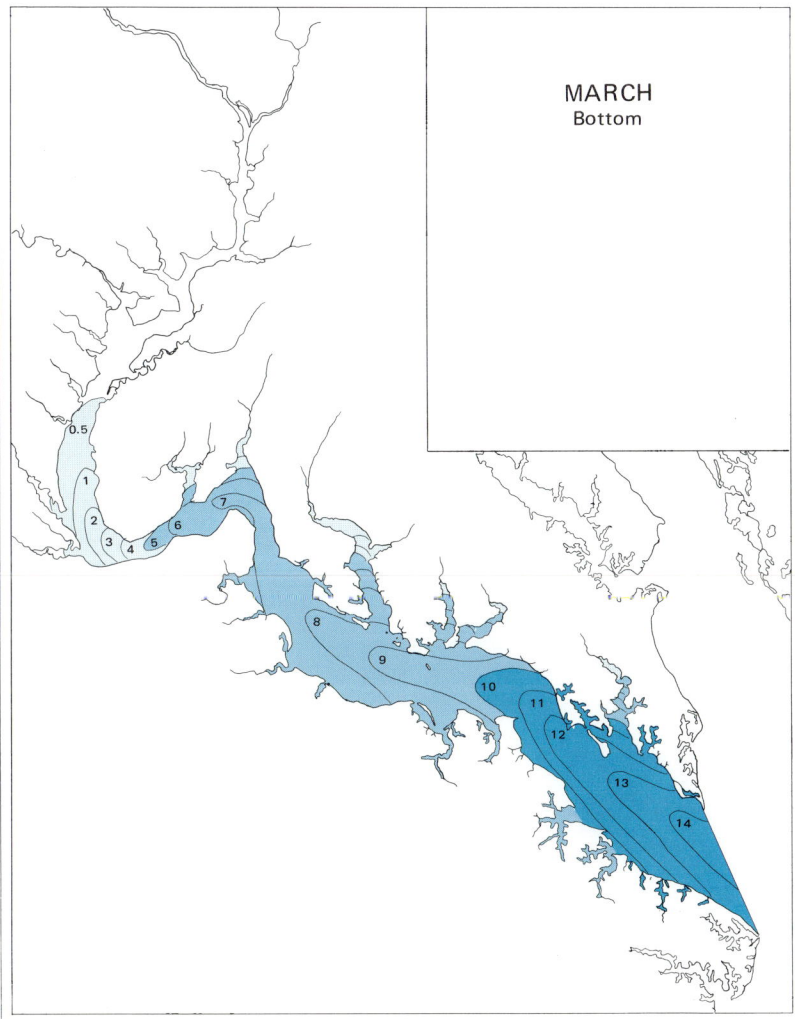

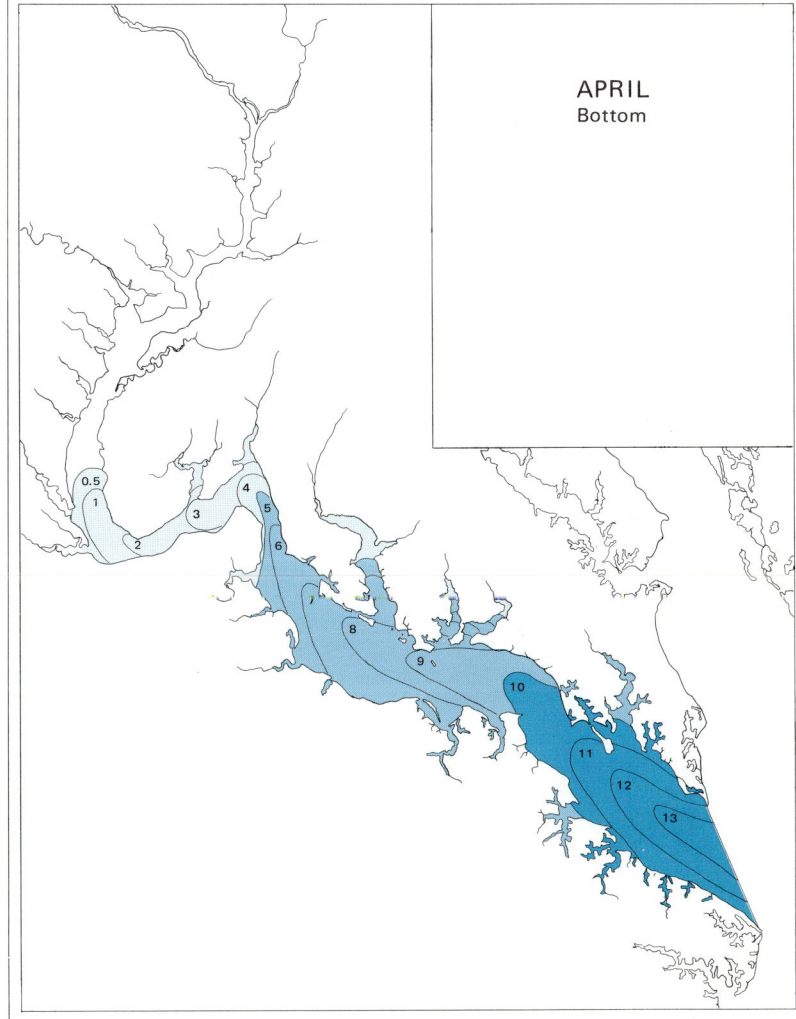

Figure 3-10. *Continued.*

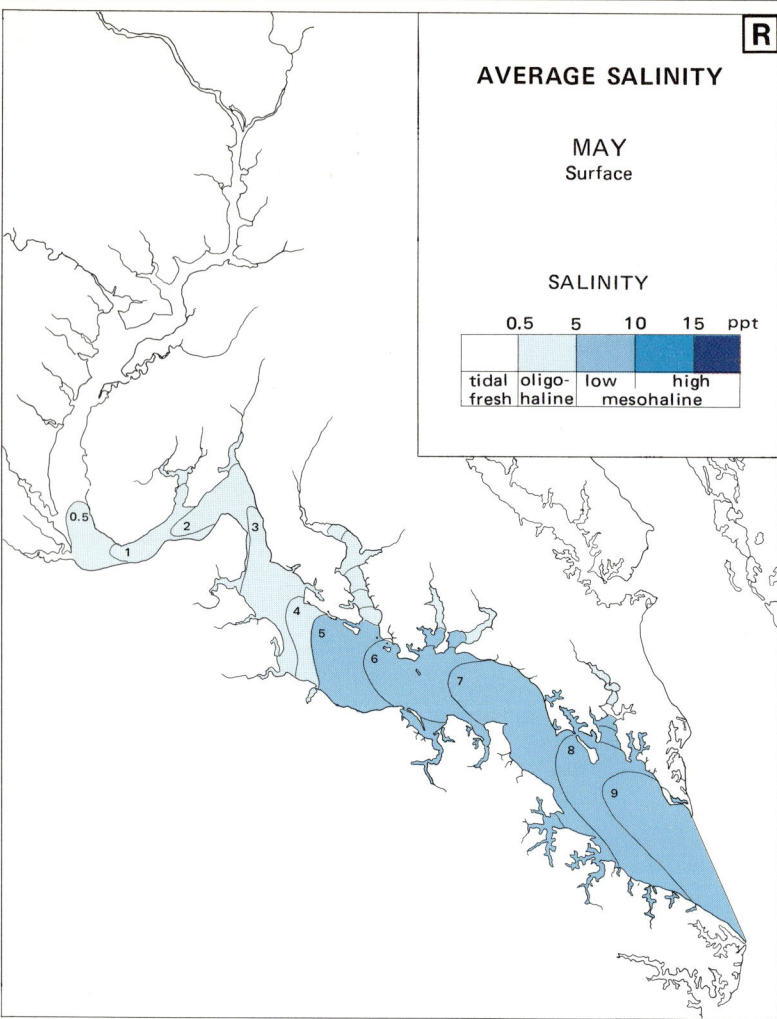

AVERAGE SALINITY

MAY
Surface

SALINITY

0.5 5 10 15 ppt

tidal | oligo- | low | high
fresh | haline | mesohaline

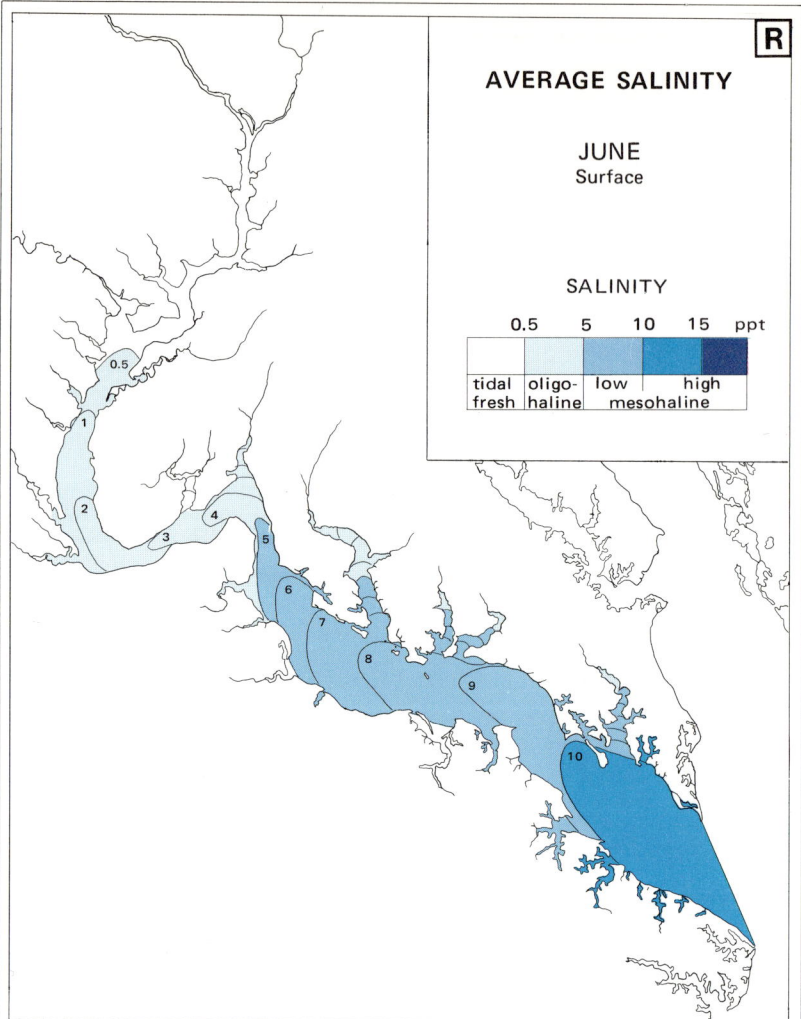

AVERAGE SALINITY

JUNE
Surface

SALINITY

0.5 5 10 15 ppt

tidal | oligo- | low | high
fresh | haline | mesohaline

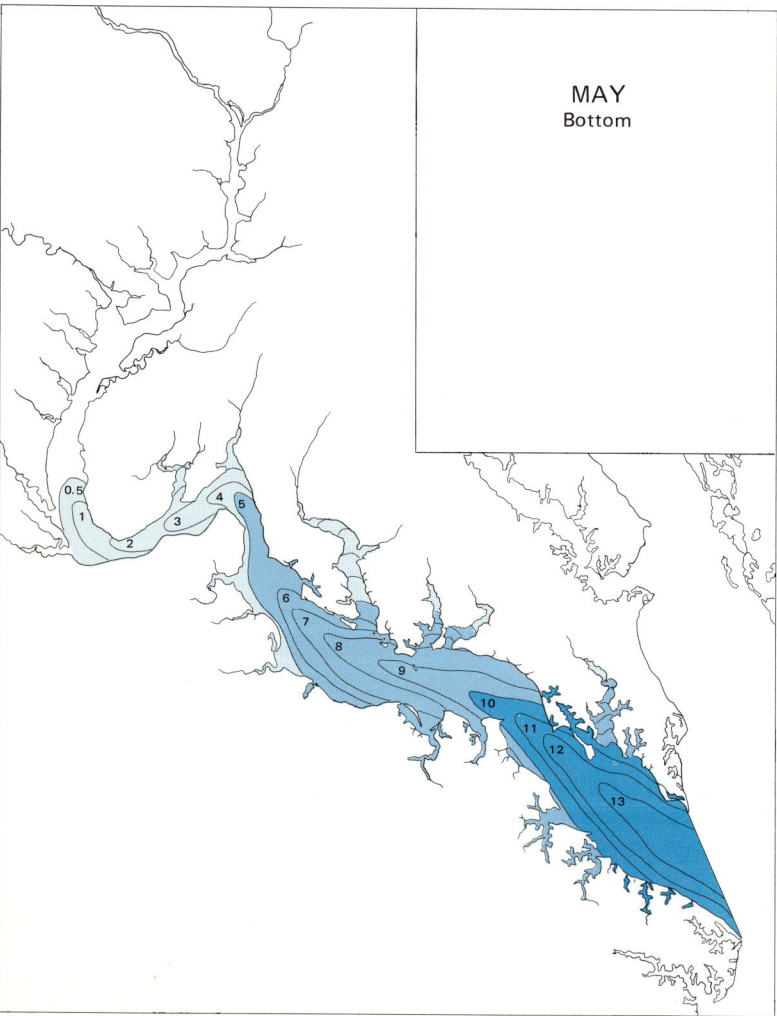

MAY
Bottom

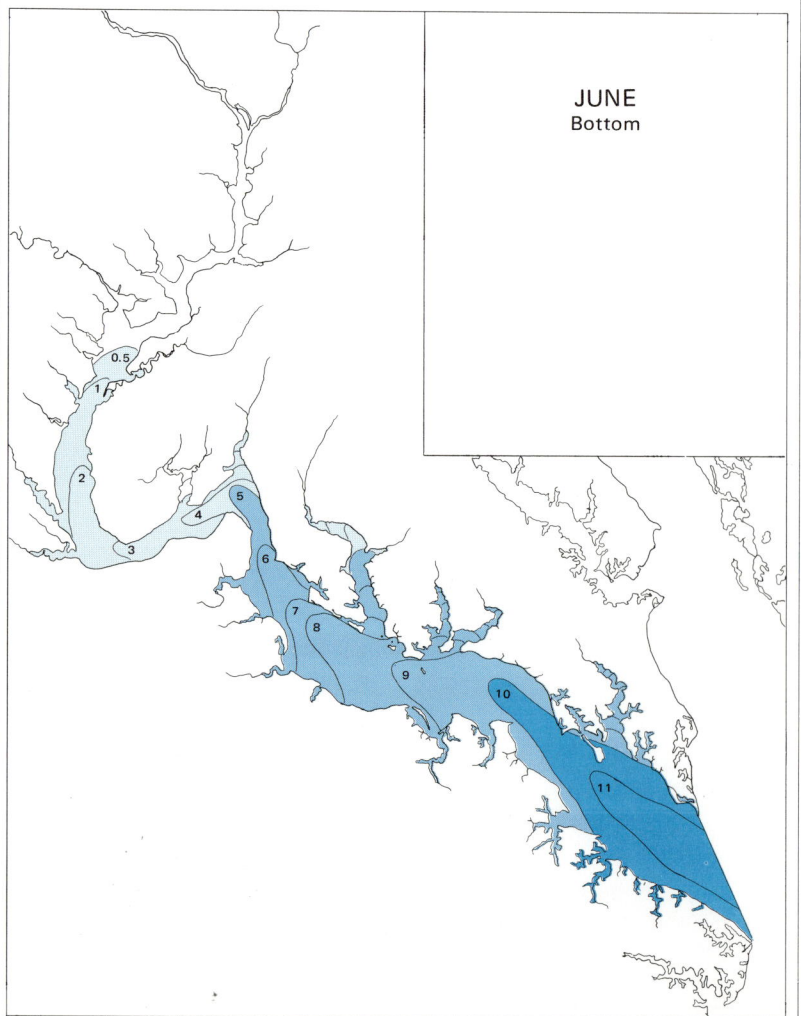

JUNE
Bottom

Figure 3-10. *Continued.*

40

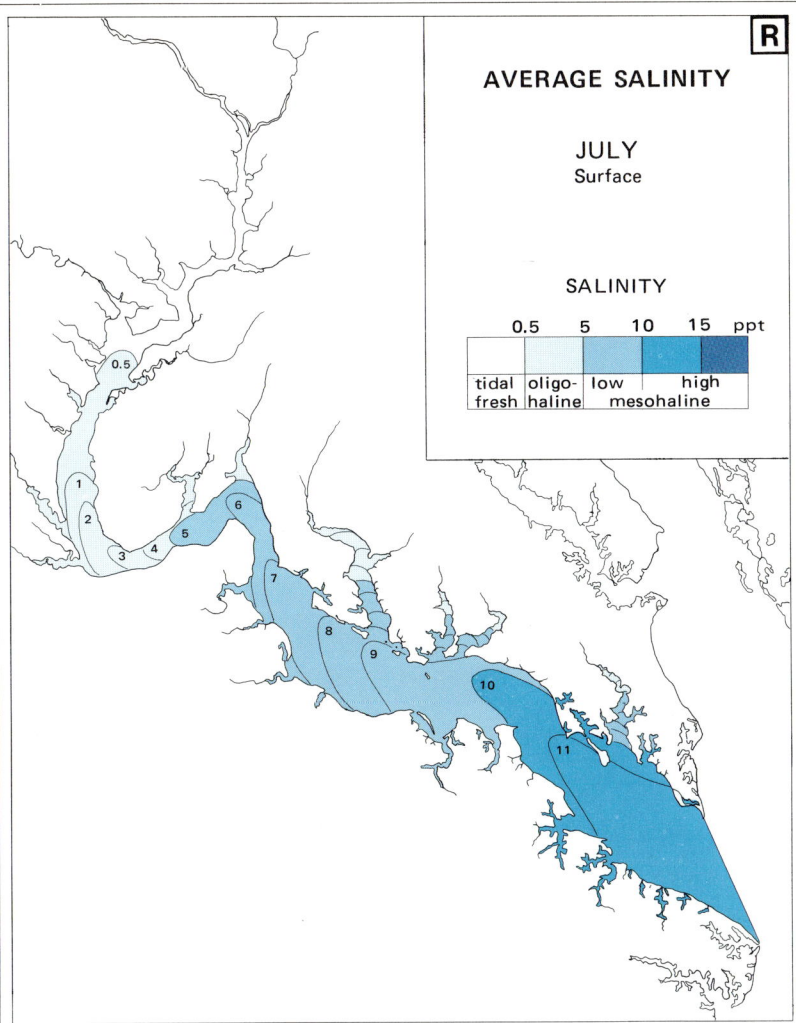

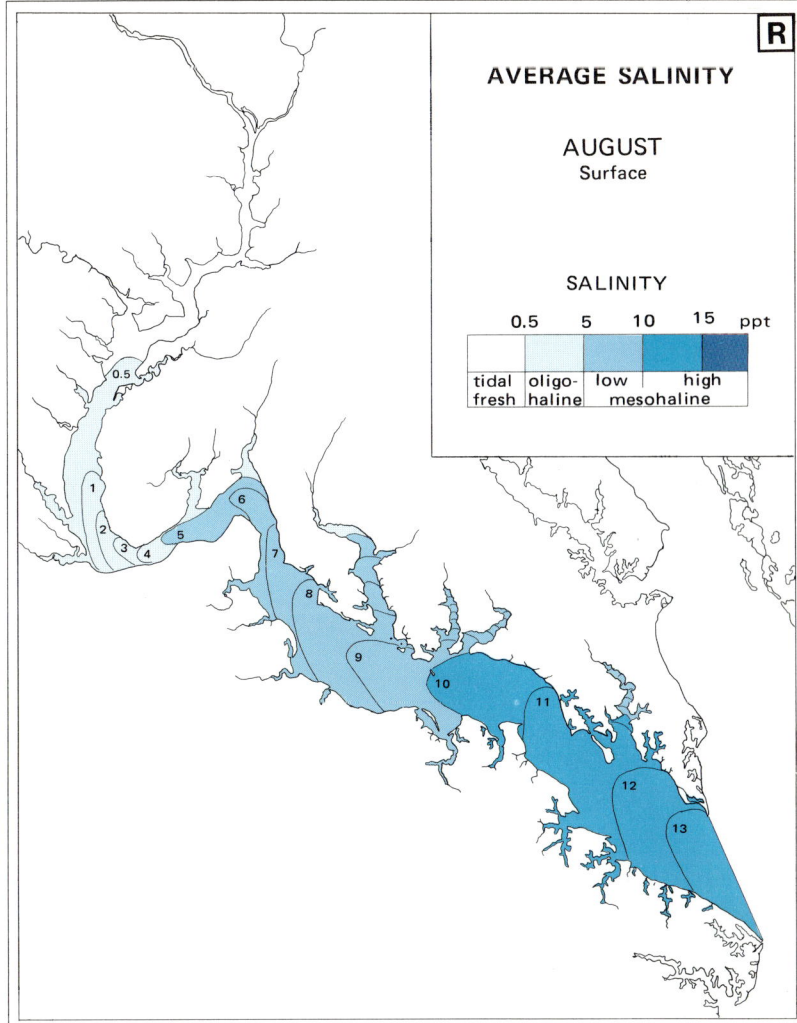

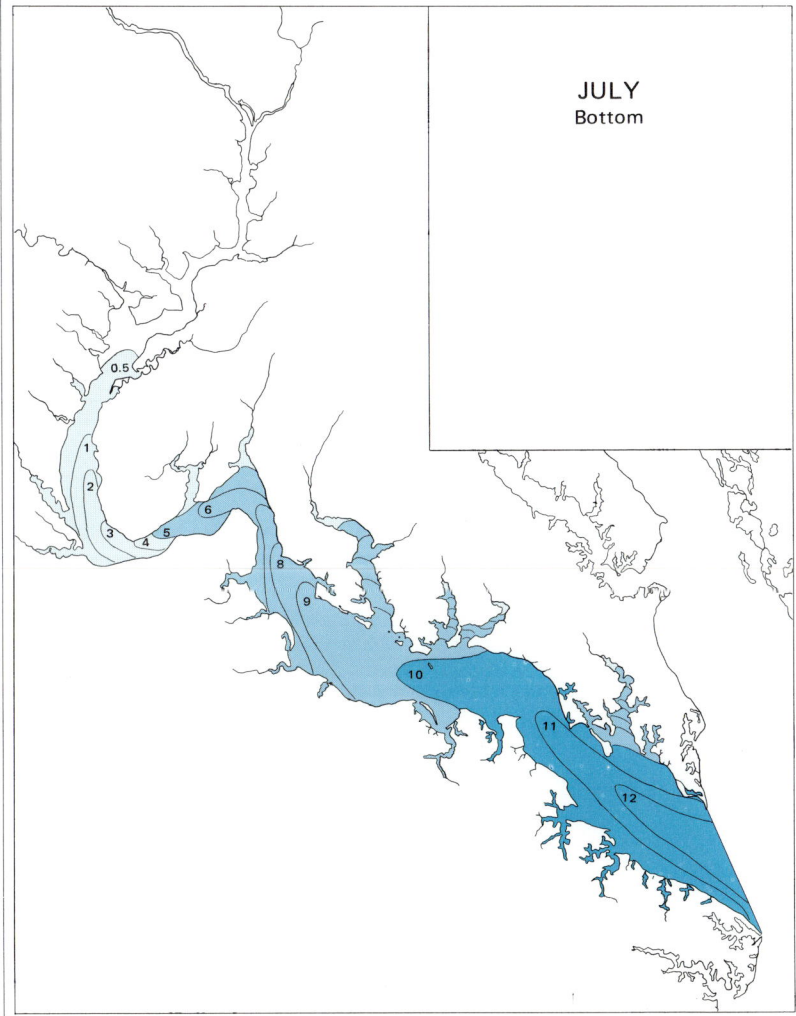

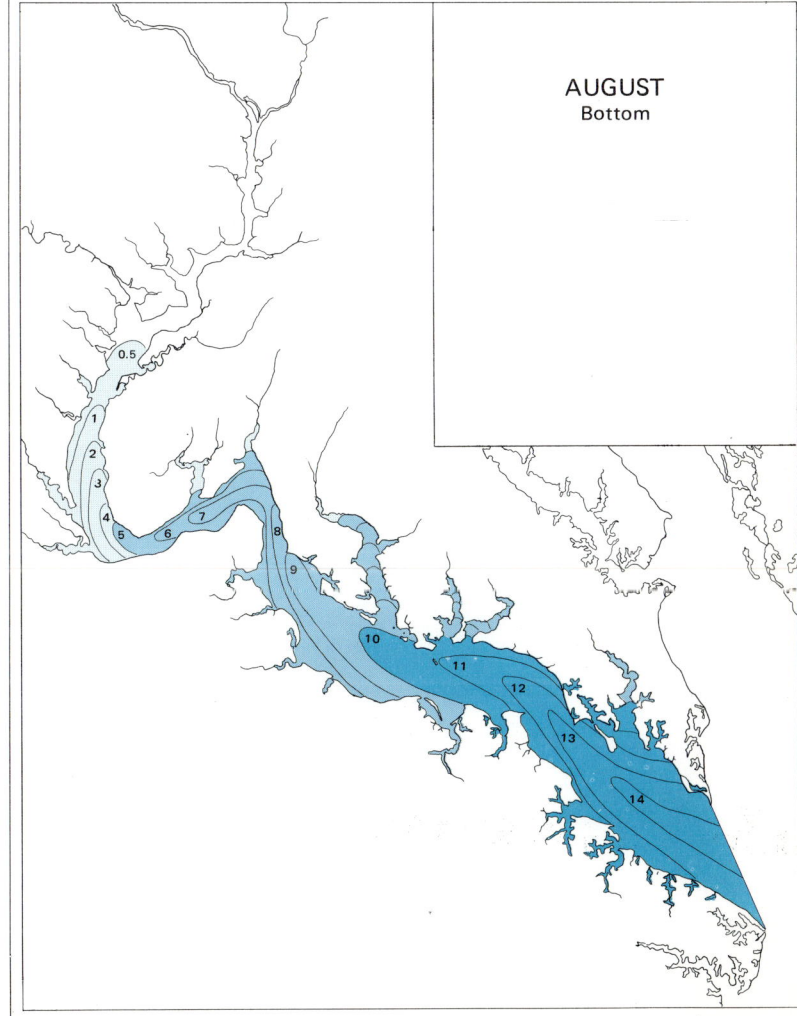

Figure 3-10. *Continued.*

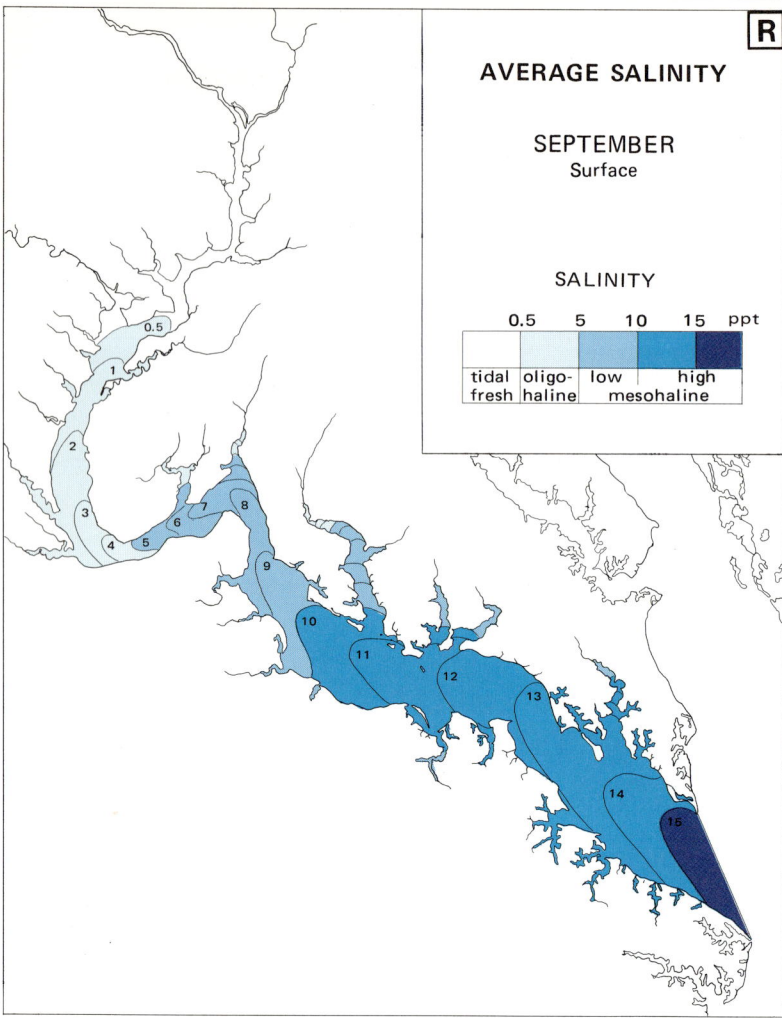

R

AVERAGE SALINITY

SEPTEMBER
Surface

SALINITY

0.5   5   10   15   ppt

tidal | oligo- | low | high
fresh | haline | mesohaline

R

AVERAGE SALINITY

OCTOBER
Surface

SALINITY

0.5   5   10   15   ppt

tidal | oligo- | low | high
fresh | haline | mesohaline

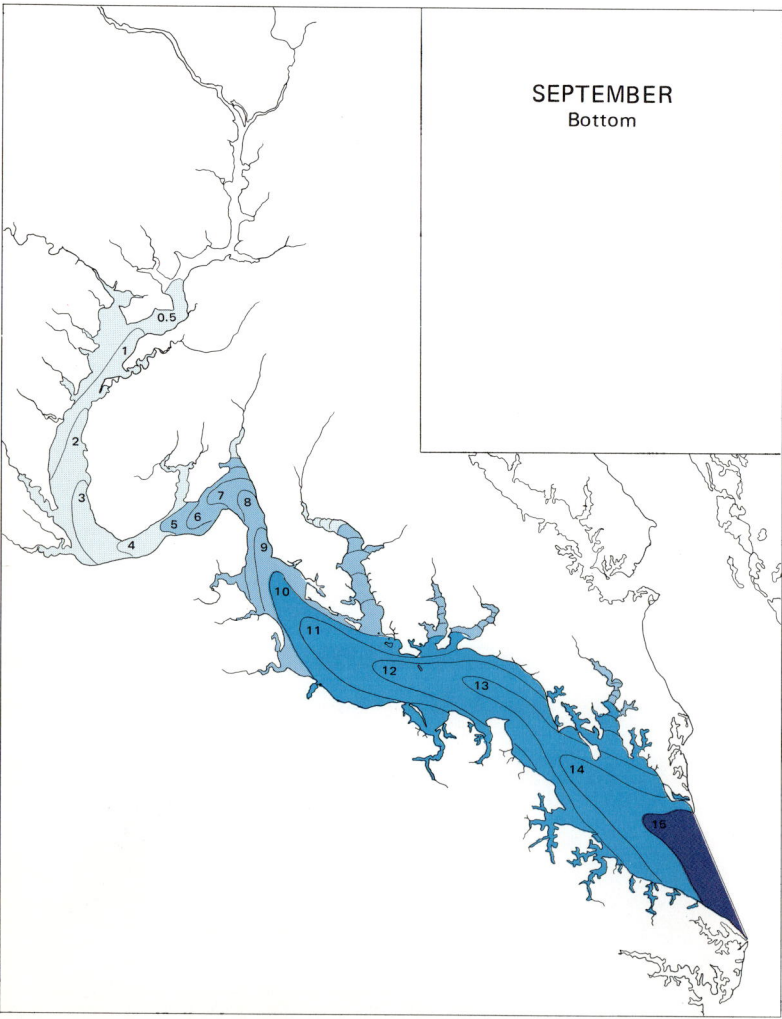

SEPTEMBER
Bottom

OCTOBER
Bottom

Figure 3-10. *Continued.*

AVERAGE SALINITY

NOVEMBER
Surface

SALINITY

0.5   5   10   15 ppt

tidal | oligo- | low | high
fresh | haline | mesohaline

AVERAGE SALINITY

DECEMBER
Surface

SALINITY

0.5   5   10   15 ppt

tidal | oligo- | low | high
fresh | haline | mesohaline

NOVEMBER
Bottom

DECEMBER
Bottom

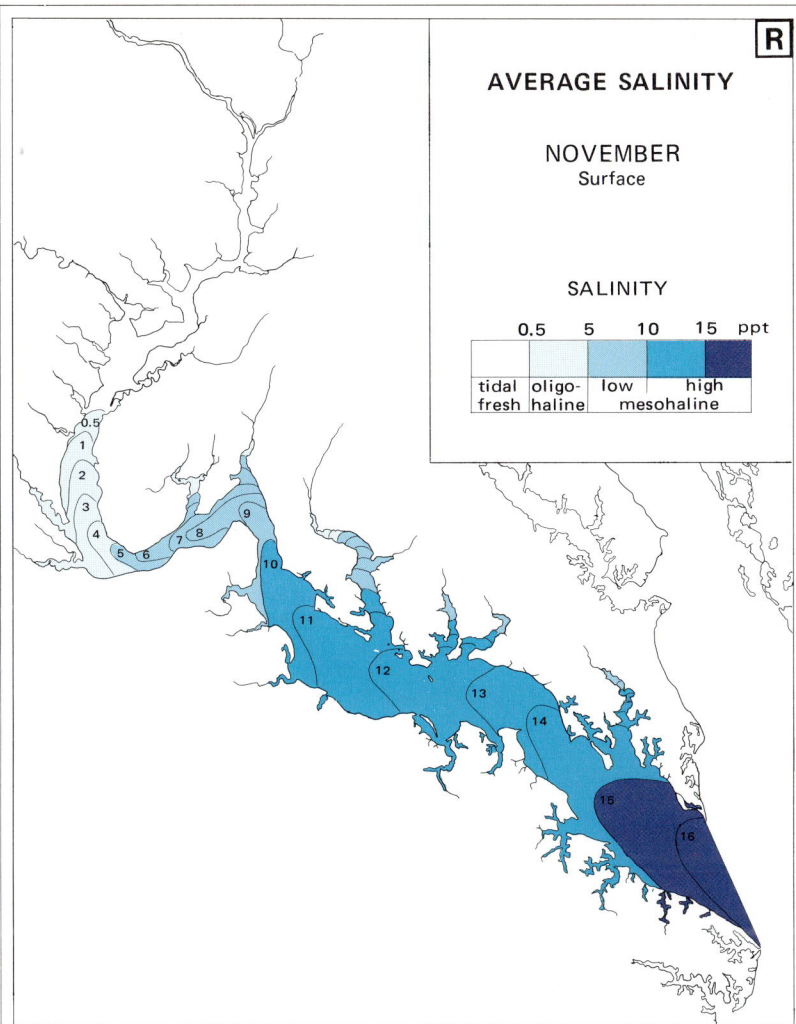

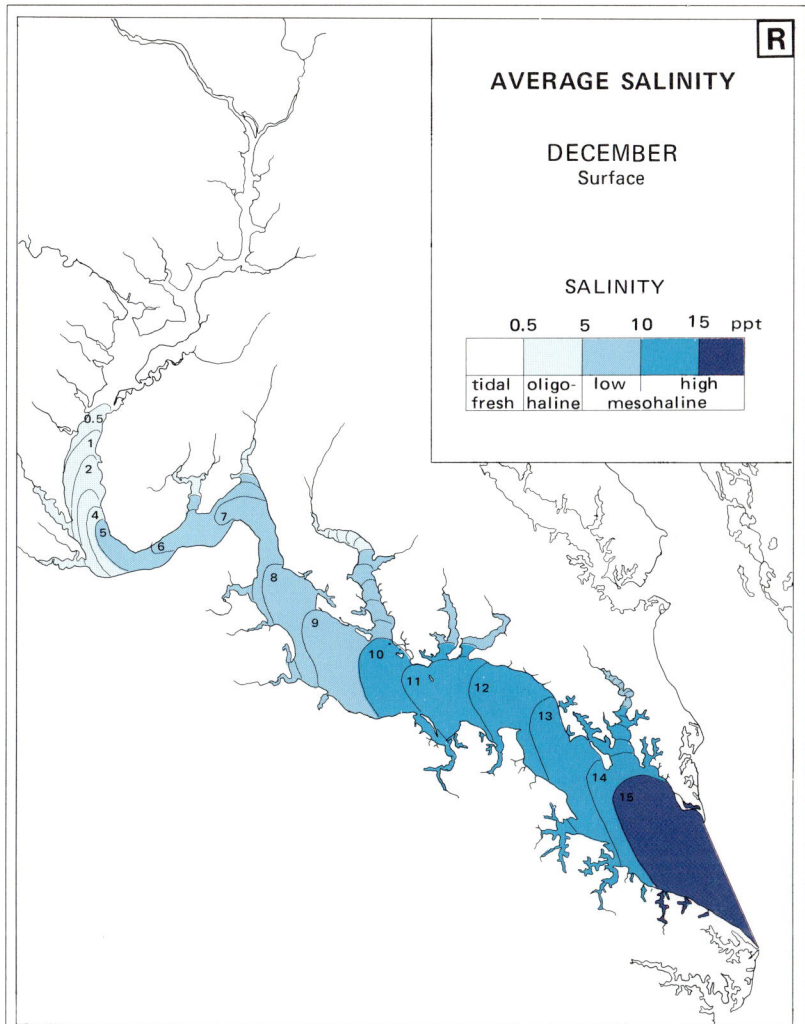

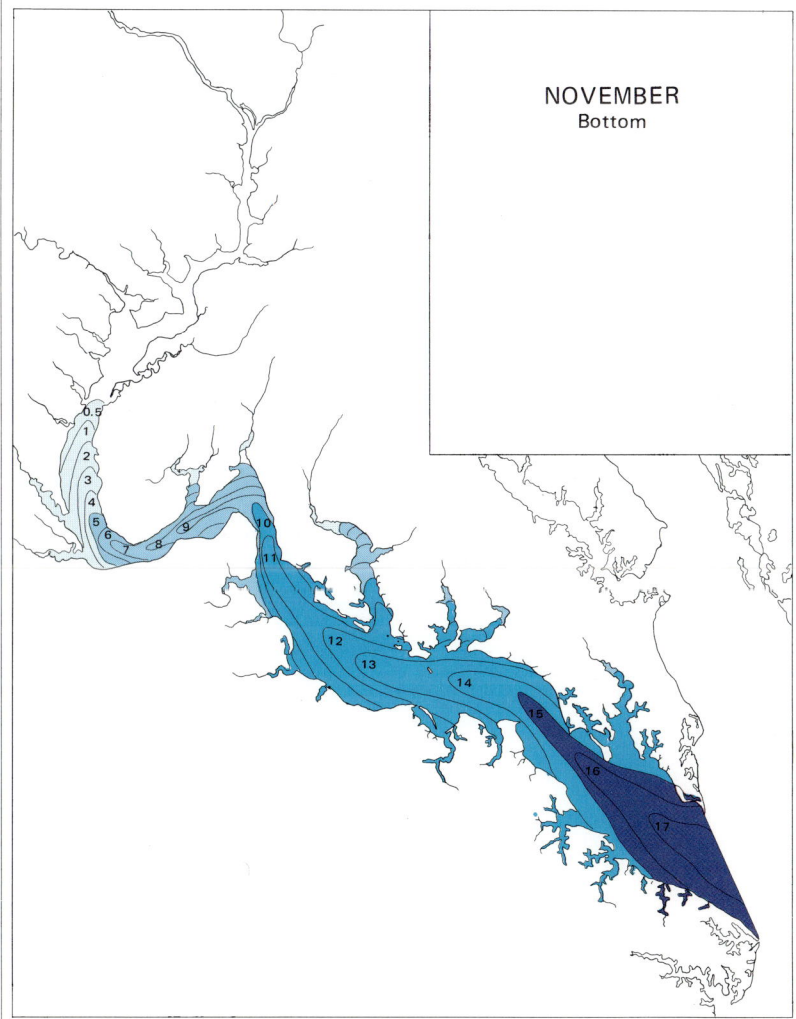

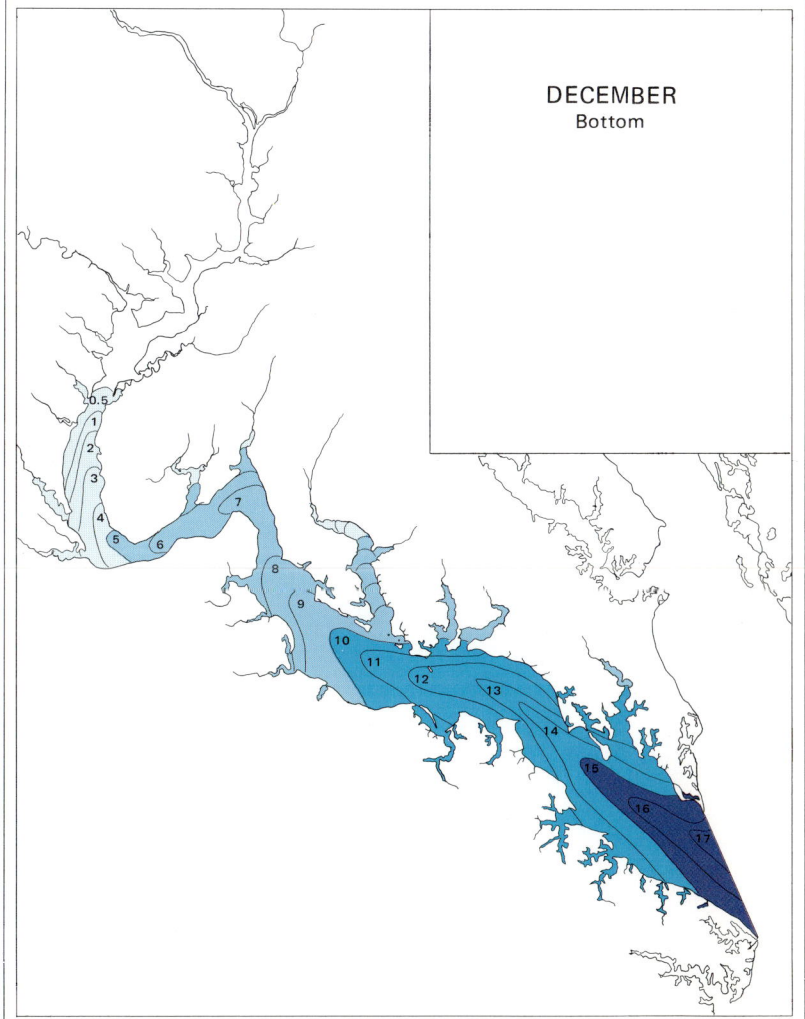

Figure 3-10. *Continued.*

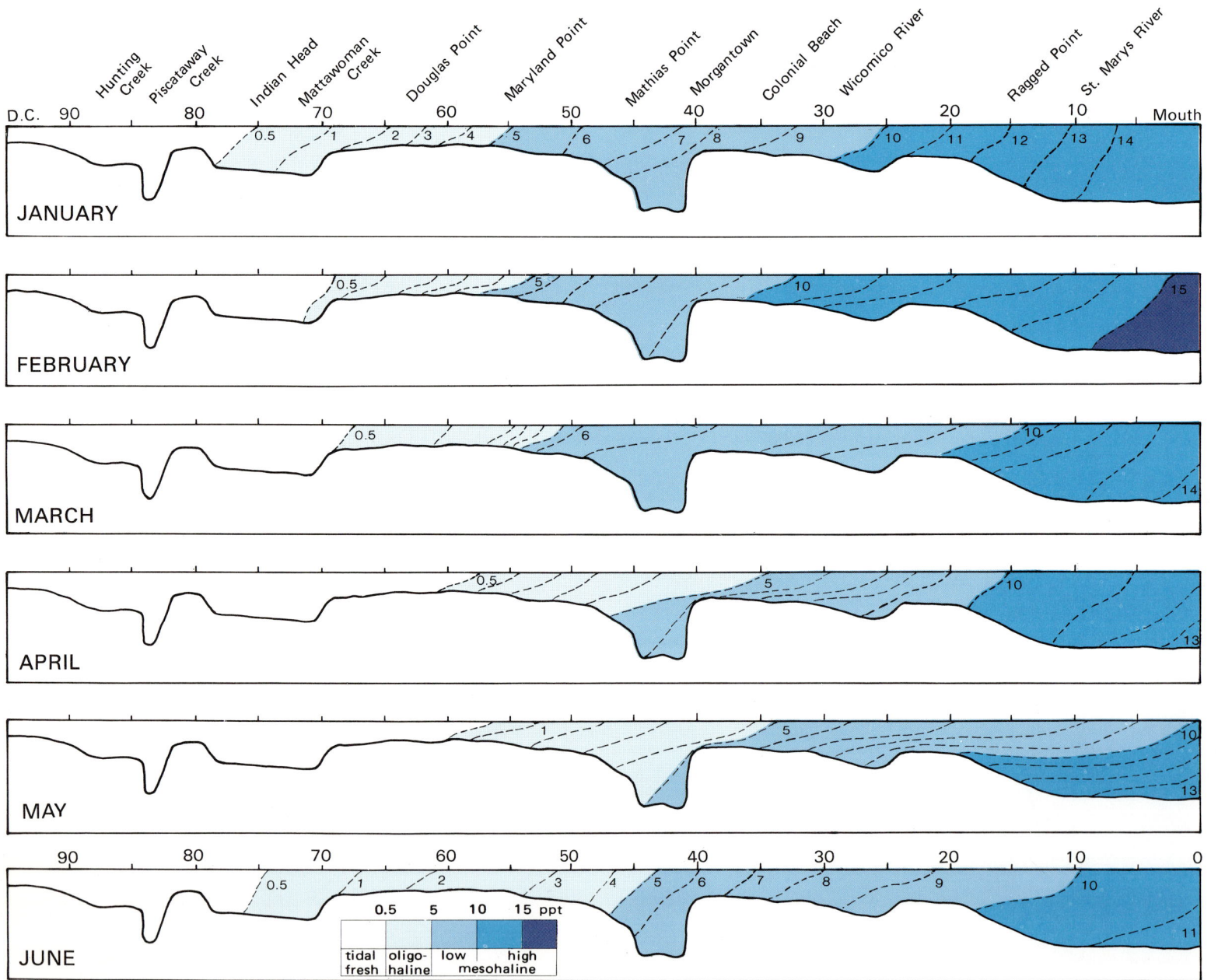

Figure 3-11. *Average monthly salinity profiles along the Potomac estuary, derived from surface and bottom salinities shown in Fig. 3-10. Numbers along water line are nautical river miles from the mouth. Bottom profiles show approximate water depths along the estuary at midchannel, with the deepest point at about 21 meters. (Sources: Refs. 14,15,18)*

AVERAGE MONTHLY
VERTICAL SALINITY PROFILES R

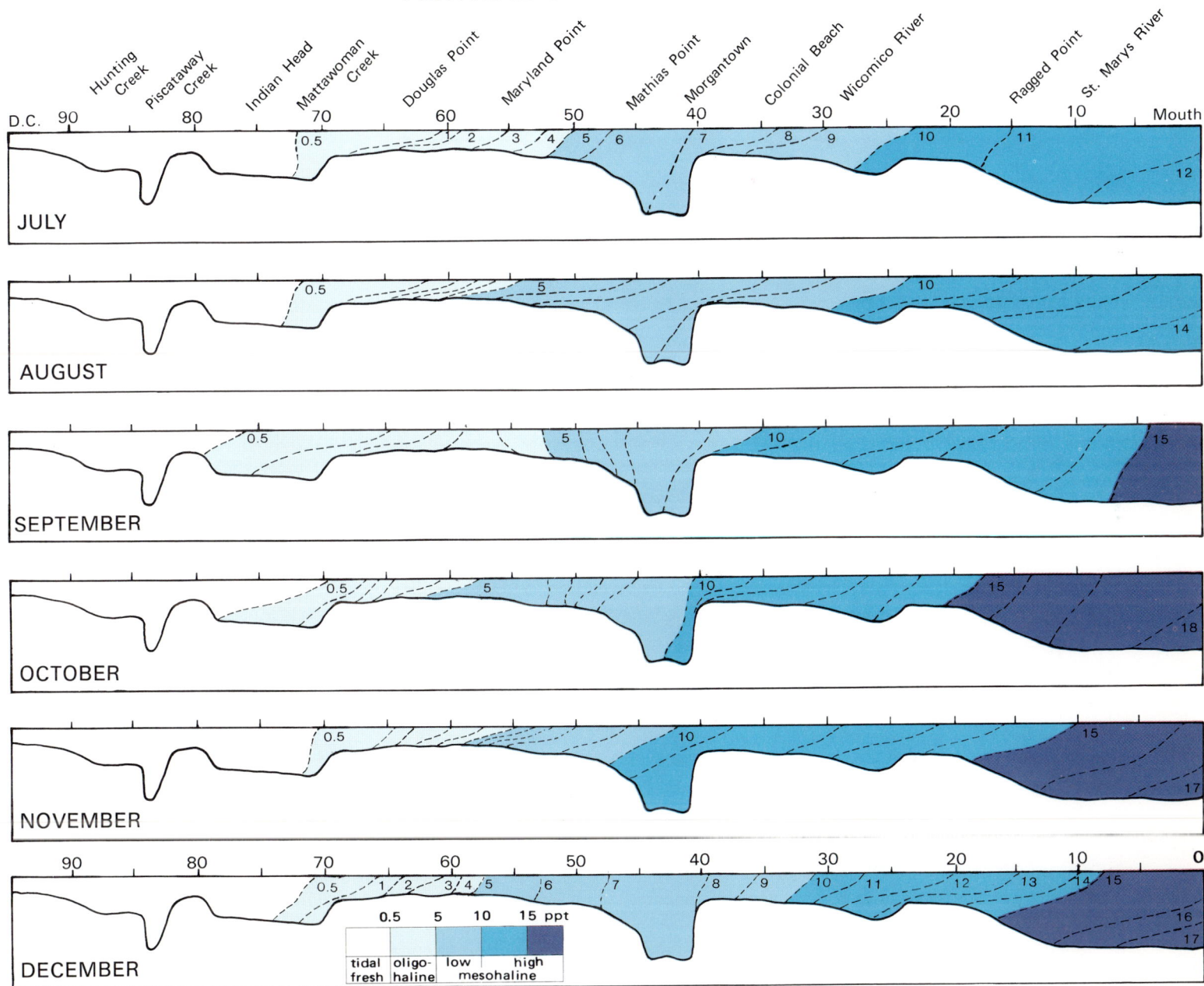

Figure 3-11. Continued.

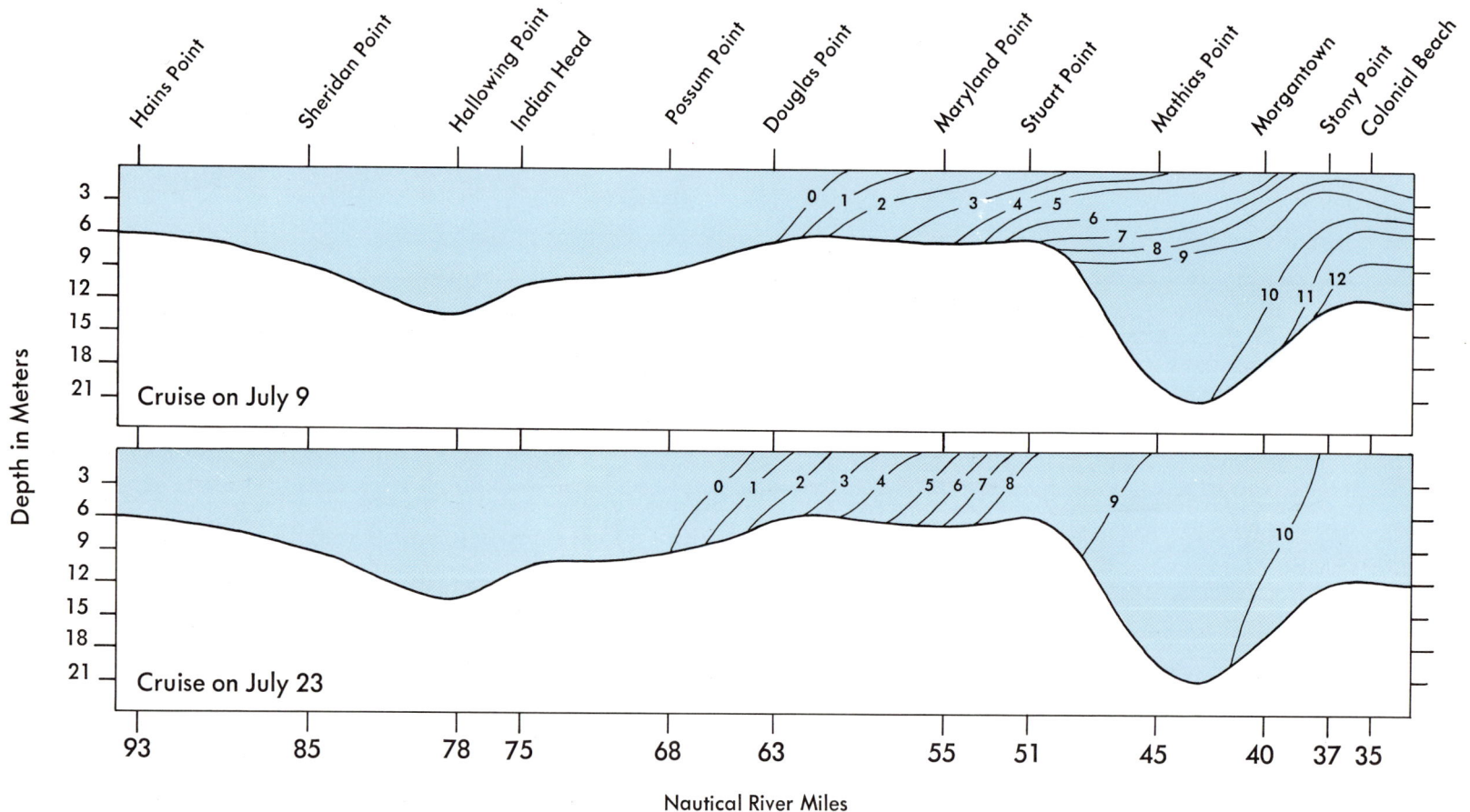

Figure 3-12. *Salinity profiles in the Potomac estuary from Hains Point downstream to Colonial Beach, from measurements taken two weeks apart in July 1974. Bottom profiles show approximate water depths along the estuary at midchannel. (Source: Ref. 18)*

necessarily reflect the magnitude of the difference at any one time. For example, Fig. 3-12 shows vertical salinity distributions along the upper half of the Potomac estuary from two consecutive sampling cruises conducted two weeks apart.[18] Within that short period, salinity differences of as much as 5 ppt between surface and bottom had dissipated so that near vertical homogeneity occurred.

The relationship of changing freshwater flows to longitudinal salinity distributions is clearly depicted in Fig. 3-13, where average monthly freshwater flows in 1969 and 1970 are shown in relation to observed salinities along the estuary. During a consistently low flow year (1969), salinities remained fairly evenly distributed along the estuary throughout all seasons. In contrast, in a normal flow year (1970), each peak of discharge is reflected in the downstream movement of fresher waters.

Heavy storms and droughts cause substantial changes in salinity. In the early summer of 1972, Hurricane Agnes dropped the salinities in the estuary so suddenly and dramatically that virtually fresh water (less than 1 ppt) was brought downstream to within 11 nautical miles of the mouth. A return to normal seasonal conditions was slow, and two months after the storm, the surface salinities at the mouth were 9 to 10 ppt, still below the seasonal norm of 13 to 15 ppt.[19]

The occurrence of such extreme conditions is significant, because the single most important factor affecting aquatic life in the estuary is salinity. Each of the Potomac's three major salinity zones (plus a small temporary extension of a fourth) contain characteristic communities. The upper tidal fresh region of the Potomac is characterized by anadromous fish spawning and nursery areas and by areas of high phytoplankton and zooplankton productivity. The oligohaline region

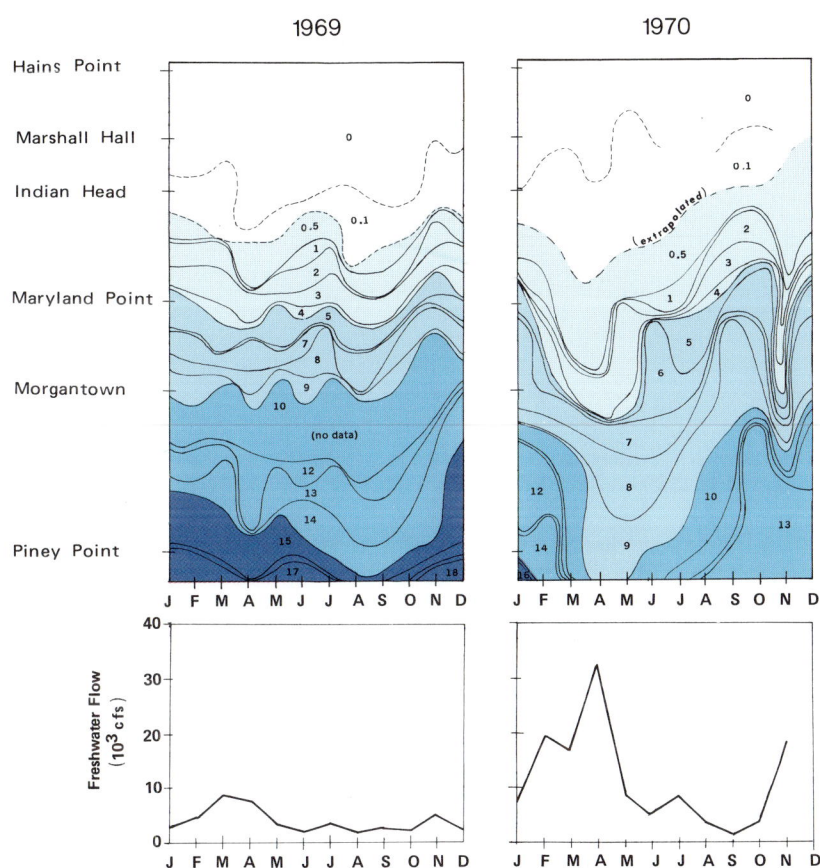

Figure 3-13. *Freshwater flow and its relationship to salinity distribution along the length of the Potomac estuary from Hains Point downstream to the mouth for a low-flow year (1969) and for a normal-flow year (1970). Isohalines are in parts per thousand. (Sources: Refs. 1,14)*

has major fish nursery areas, large forage fish populations, some anadromous fish spawning, and moderately large phytoplankton and zooplankton populations. The lower half, a typical mesohaline estuarine region, has large populations of forage fishes, predator fishes, and blue crabs and highly productive oyster and clam grounds. Any drastic changes in salinity, such as those caused by radical climatic events or large-scale water diversions, will affect the reproduction and growth of these populations.

Water Temperature

Average monthly surface water temperatures on the Potomac estuary range from a recorded low in January of 0°C to a high in July of 30.8°C.[15][20] Water along the upper estuary frequently freezes, and in severe winters, ice also forms in the saline waters of the lower estuary. At any one time, there is usually a temperature difference of only 2 or 3°C longitudinally along the estuary at constant depth. Even when

highly stratified conditions exist during the summer, the vertical temperature differences are rarely more than 5 to 6°C. Differences in temperatures generally follow the typical seasonal pattern found in temperate zone estuaries (see Chapter 1, Fig. 1-8). In the mesohaline portion of the Potomac, where moderately stratified conditions usually persist, waters are colder with depth in summer and warmer with depth in winter. In the upper, unstratified tidal fresh regions, a vertical seasonal shift is not so evident, and waters are generally colder with depth throughout the year except in the early spring months. Figures 3-14 and 3-15 show the longitudinal and vertical variation in water temperature relative to the monthly average river temperature. Blank segments in these figures indicate where little or no temperature monitoring has been conducted.

For any month in a given year, the average monthly temperatures at various locations along the estuary do not vary more than 5°C (see Fig. 3-14). The smallest monthly temperature extremes seem to occur in late summer and fall. However, temperature differences are extremely variable over short periods, as is shown in Fig. 3-16. On July 9, 1974,[18] large vertical temperature differences occurred, while just two weeks later, vertical stratification virtually disappeared.

The close relationship between water and air temperatures is apparent when average monthly air temperatures at Washington, D.C., are compared to average monthly water temperatures there (Fig. 3-17). In any 24-hour period, air temperatures may fluctuate over a broad range, changing under the sun's influence, while water temperatures change slowly, varying little from night to day. But daily average temperatures for the two are extremely close; monthly average water temperatures are only slightly higher than monthly average air temperatures. The only exception is in spring when high flows of colder mountain waters lower the water temperatures below the average ambient air temperatures.

Because the temperature changes in the estuary are moderate, organisms living there ordinarily do not have to adapt to rapid or radical changes. Their primary responses are to gradual seasonal shifts. As in all estuaries, life cycles in the Potomac follow patterns that are closely tied to seasonal temperature changes. In early spring, there is a general resurgence of biological activity throughout the estuary. Phytoplankton and zooplankton populations that were at low levels in winter suddenly increase their numbers. Fish and crabs begin to move towards warming shoreline waters from deep midestuary channels where they have overwintered. Anadromous fish begin to migrate to upstream spawning grounds. Some striped bass may arrive on the spawning grounds as early as March, but spawning activity is not triggered until temperatures exceed 10°C, usually in April. Meanwhile, in the lower estuary, just as predictably, oysters mature, but also wait for a thermal cue to spawn.[22] Seasonal sequences of abundance and distribution of each animal or plant group are the most obvious responses to temperature changes.

WATER TEMPERATURES

JANUARY

Surface Temperatures Only (°C)
Minimum   0.0
Maximum   6.2
Average   2.6

Deviation from January River Average
(2.8°C)

−2.5   −1.5   −0.5   +0.5   +1.5   +2.5 °C

0.3   1.3   2.3   3.3   4.3   5.3 °C

Actual Average Temperatures

WATER TEMPERATURES

FEBRUARY

Surface Temperatures Only (°C)
Minimum   2.0
Maximum   6.0
Average   3.2

Deviation from February River Average
(3.3°C)

−2.5   −1.5   −0.5   +0.5   +1.5   +2.5 °C

0.8   1.8   2.8   3.8   4.8   5.8 °C

Actual Average Temperatures

JANUARY

Bottom Temperatures Only (°C)
Minimum   0.8
Maximum   5.2
Average   2.9

FEBRUARY

Bottom Temperatures Only (°C)
Minimum   1.9
Maximum   6.0
Average   3.5

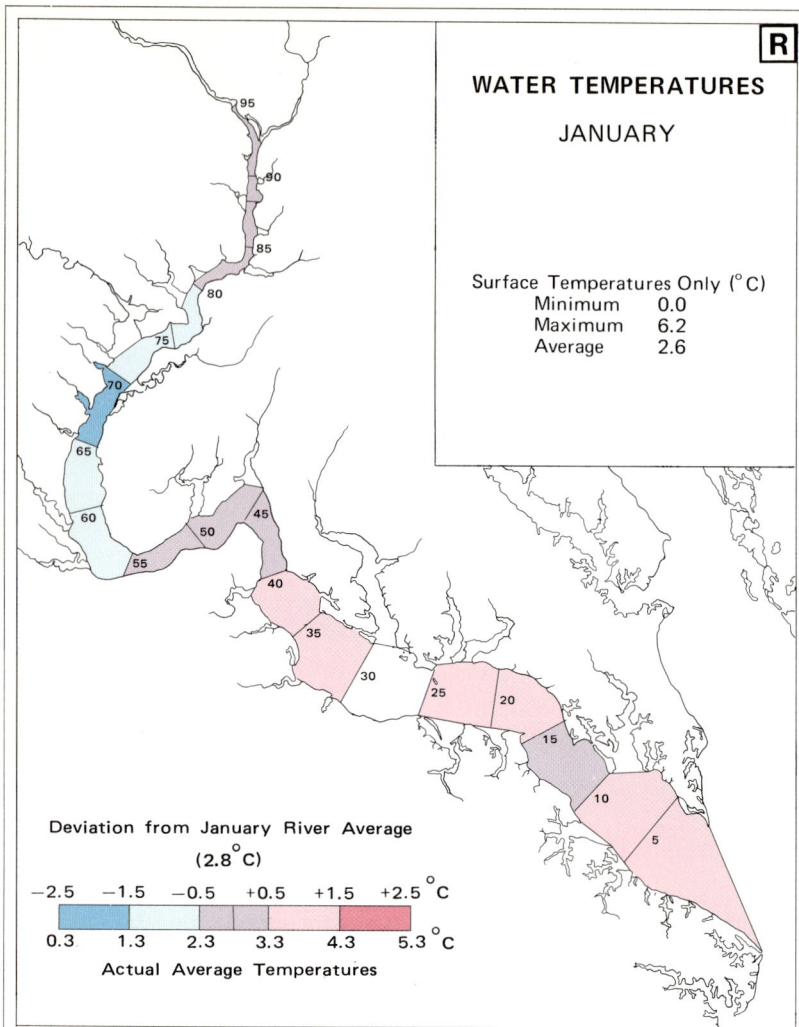

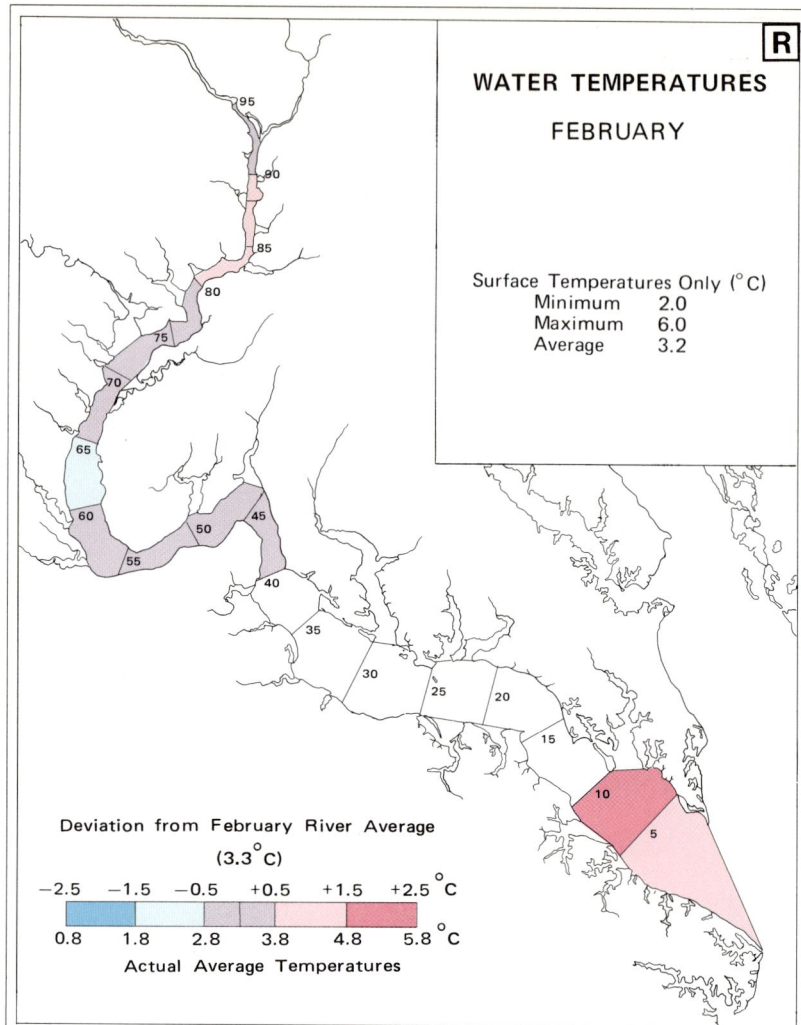

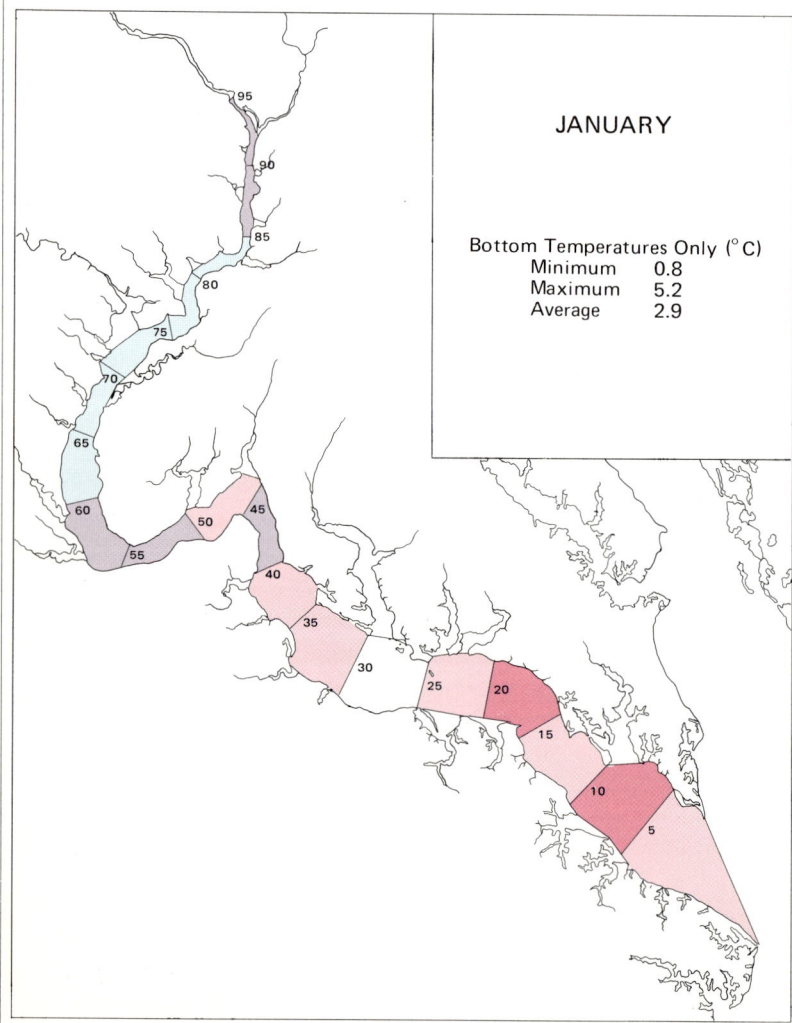

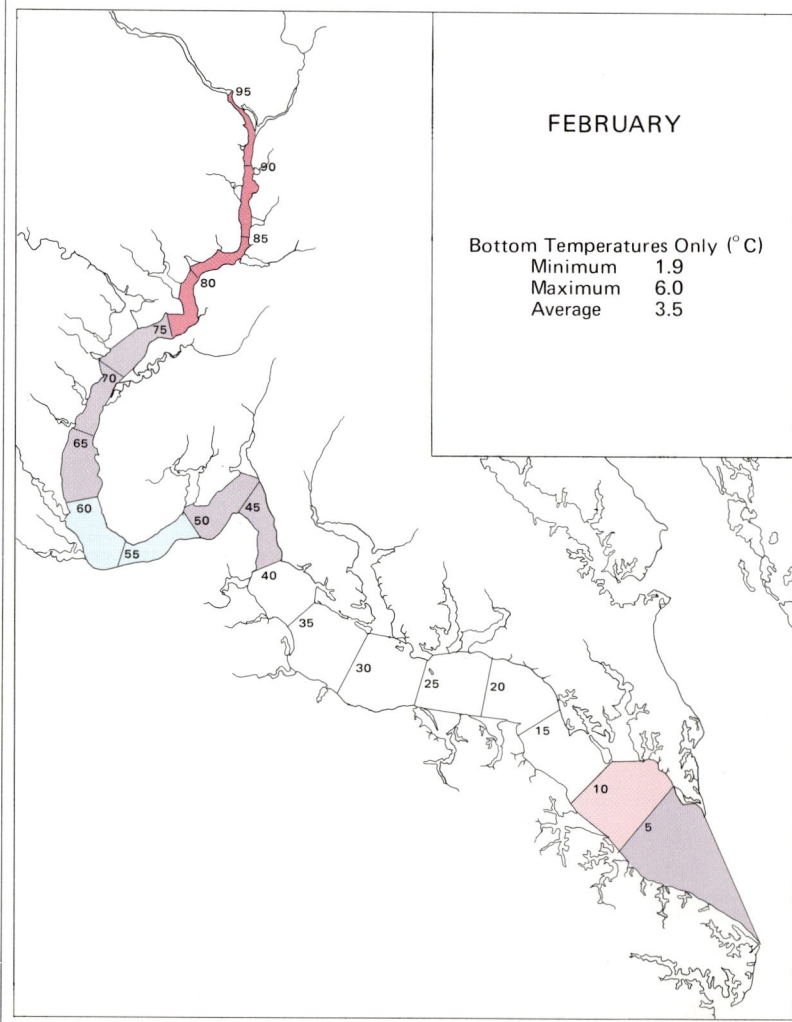

Figure 3-14.   *Deviation in surface and bottom water temperatures for each 5- nautical-mile segment of the Potomac estuary relative to the overall monthly river average for January through December. Monthly river averages include all surface and bottom temperatures. Segments are blank where little or no data were available. Bulk of data covers the period 1965-76. (Sources: Refs. 15,20)*

WATER TEMPERATURES

MARCH

Surface Temperatures Only (°C)
Minimum   4.5
Maximum   10.0
Average   6.6

Deviation from March River Average
(6.6 °C)

−2.5 −1.5 −0.5 +0.5 +1.5 +2.5 °C

4.1 5.1 6.1 7.1 8.1 9.1 °C
Actual Average Temperatures

WATER TEMPERATURES

APRIL

Surface Temperatures Only (°C)
Minimum   4.7
Maximum   17.4
Average   12.0

Deviation from April River Average
(11.6 °C)

−2.5 −1.5 −0.5 +0.5 +1.5 +2.5 °C

9.1 10.1 11.1 12.1 13.1 14.1 °C
Actual Average Temperatures

MARCH

Bottom Temperatures Only (°C)
Minimum   2.1
Maximum   9.5
Average   6.6

APRIL

Bottom Temperatures Only (°C)
Minimum   5.1
Maximum   18.6
Average   11.1

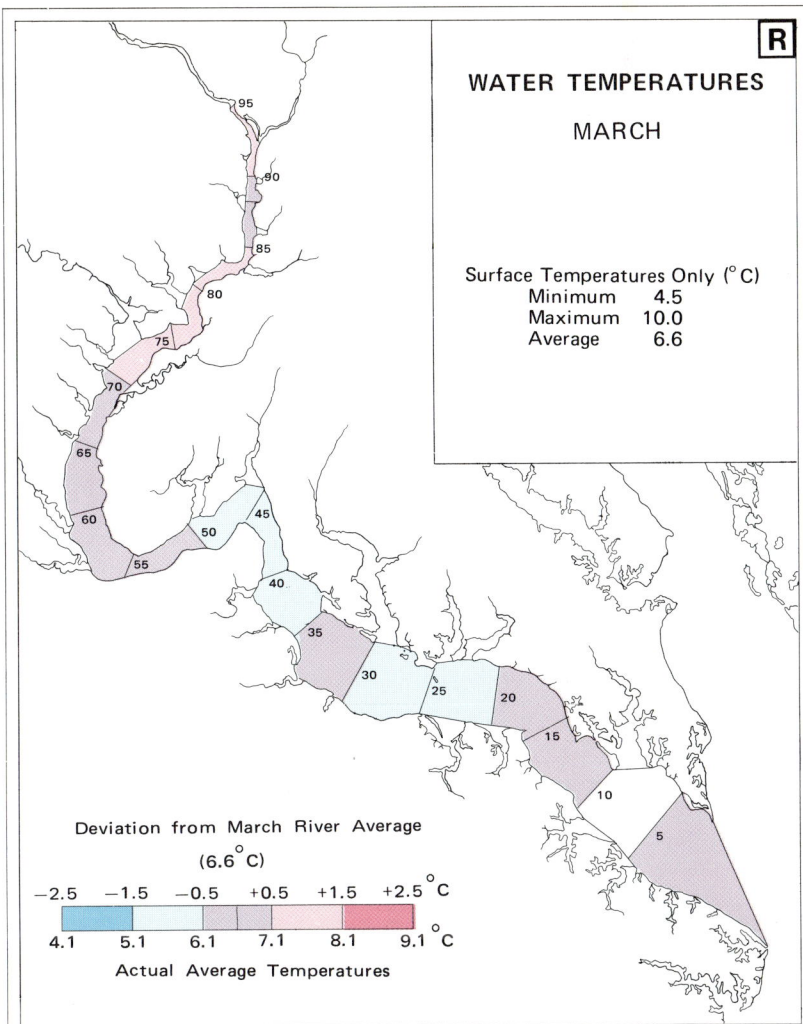

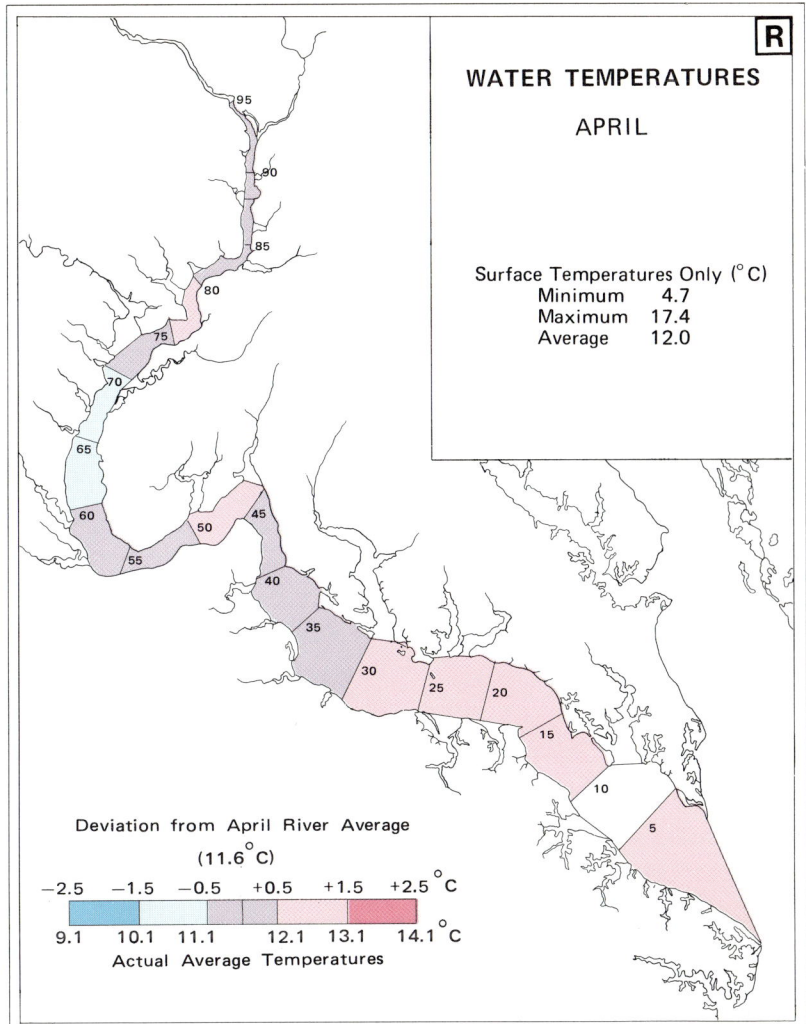

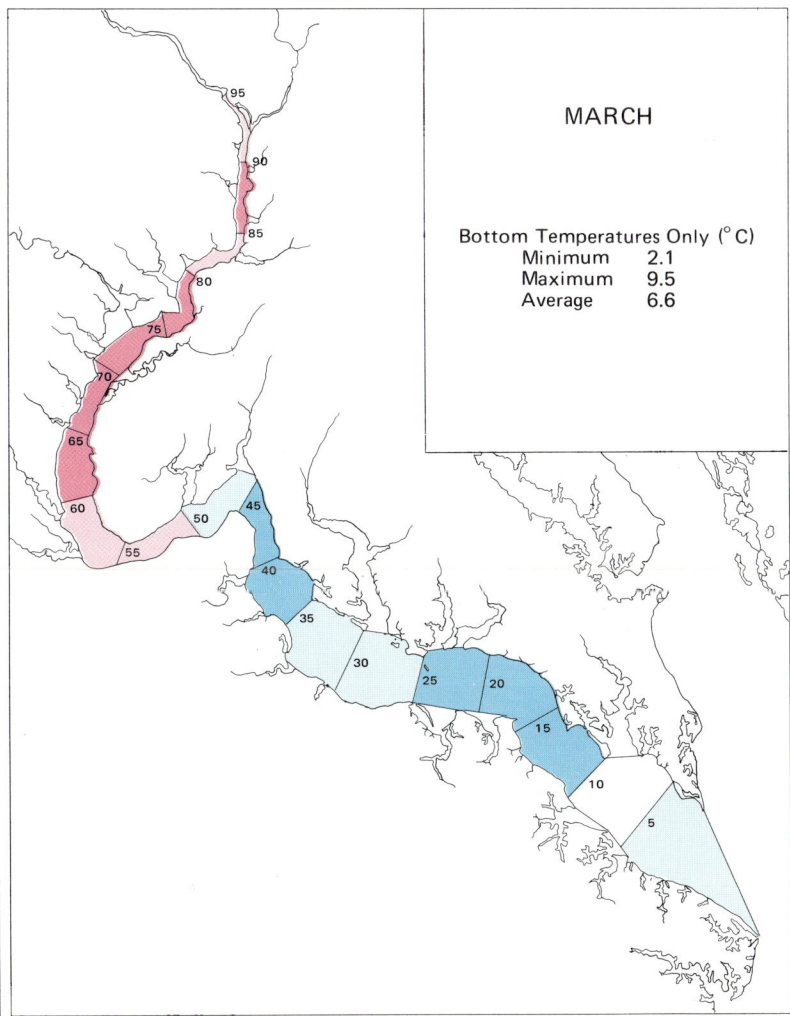

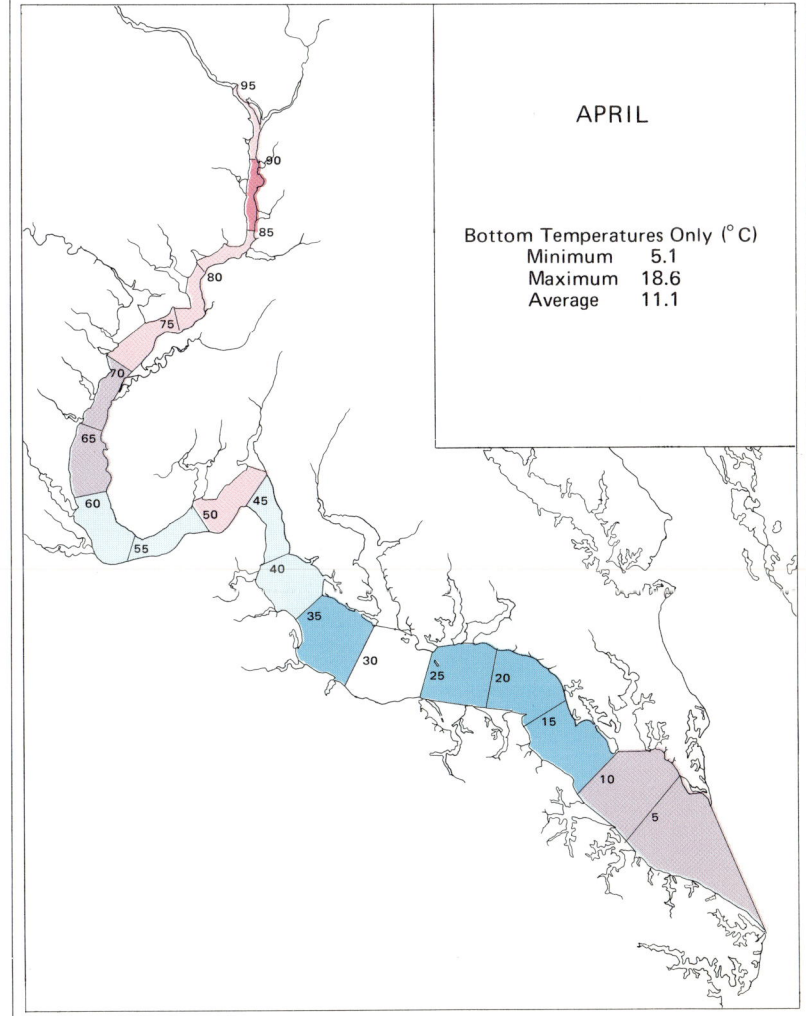

Figure 3-14. *Continued.*

WATER TEMPERATURES

MAY

Surface Temperatures Only (°C)
Minimum 14.1
Maximum 28.0
Average 18.5

Deviation from May River Average

(18.0°C)

-2.5 -1.5 -0.5 +0.5 +1.5 +2.5 °C

15.5 16.5 17.5 18.5 19.5 20.5 °C
Actual Average Temperatures

WATER TEMPERATURES

JUNE

Surface Temperatures Only (°C)
Minimum 17.4
Maximum 29.3
Average 23.8

Deviation from June River Average

(23.3°C)

-2.5 -1.5 -0.5 +0.5 +1.5 +2.5 °C

20.8 21.8 22.8 23.8 24.8 25.8 °C
Actual Average Temperatures

MAY

Bottom Temperatures Only (°C)
Minimum 12.0
Maximum 24.5
Average 17.5

JUNE

Bottom Temperatures Only (°C)
Minimum 16.0
Maximum 28.4
Average 22.7

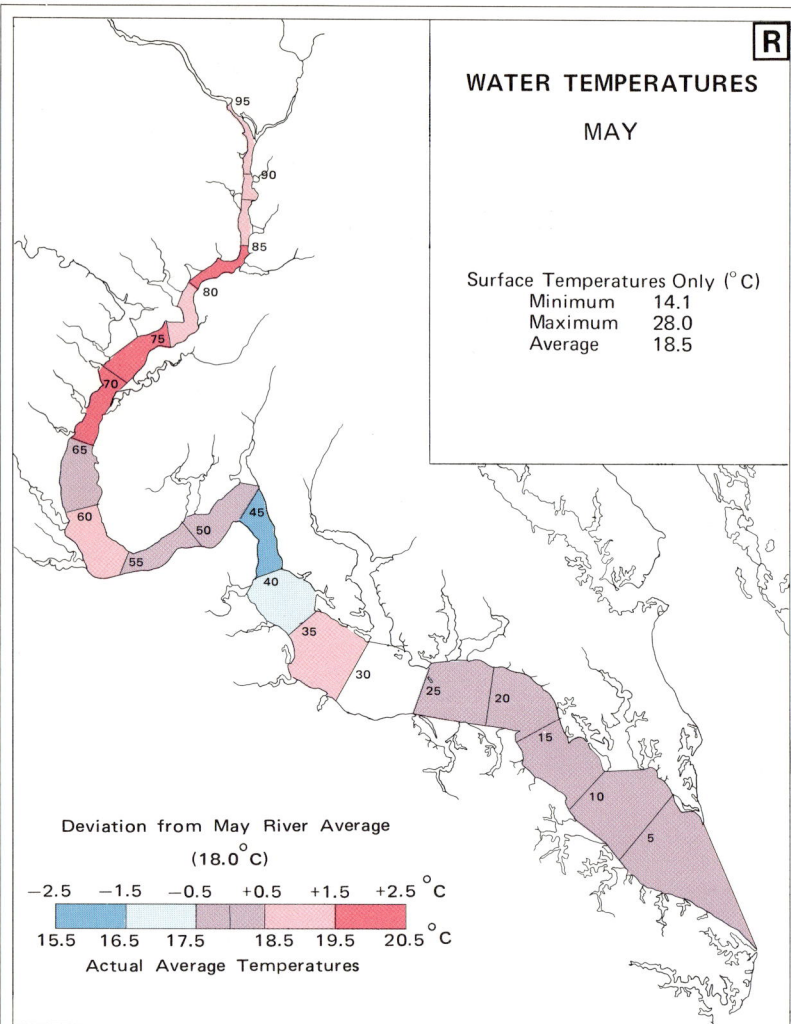

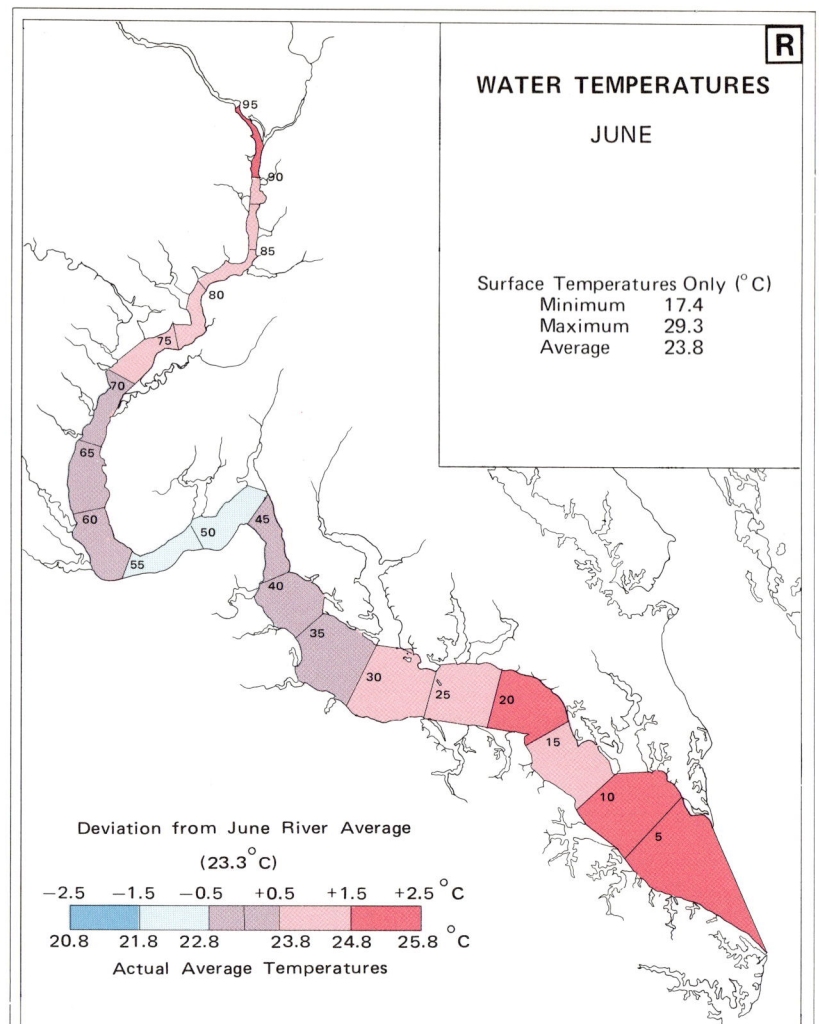

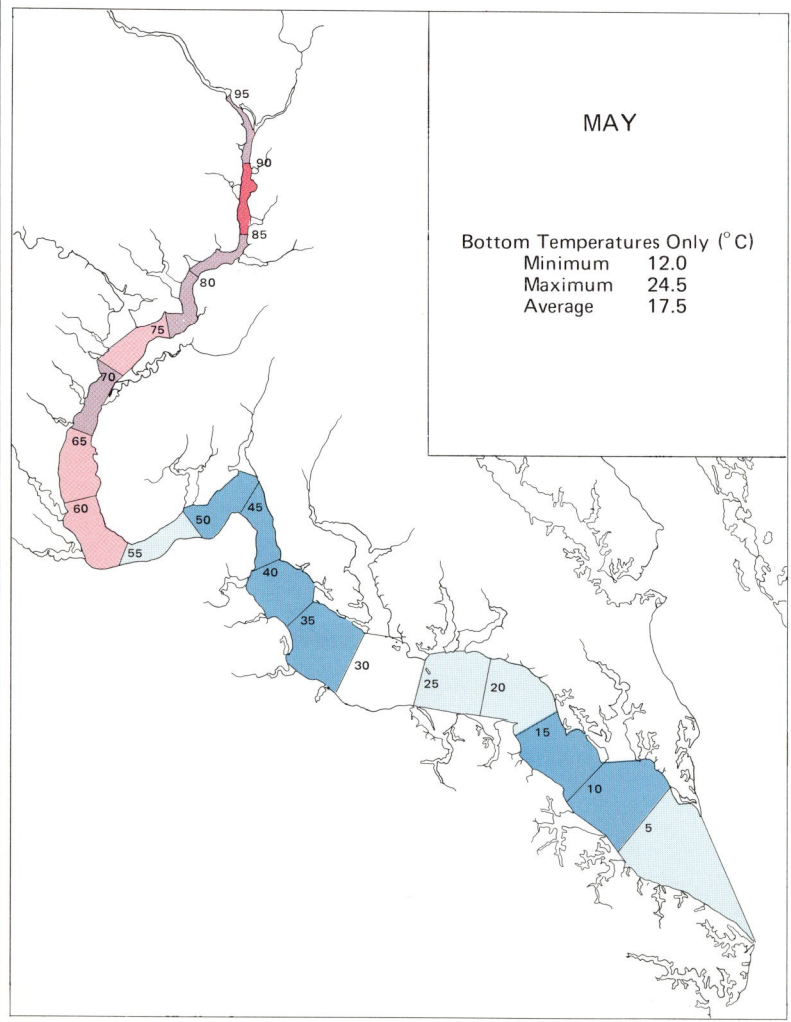

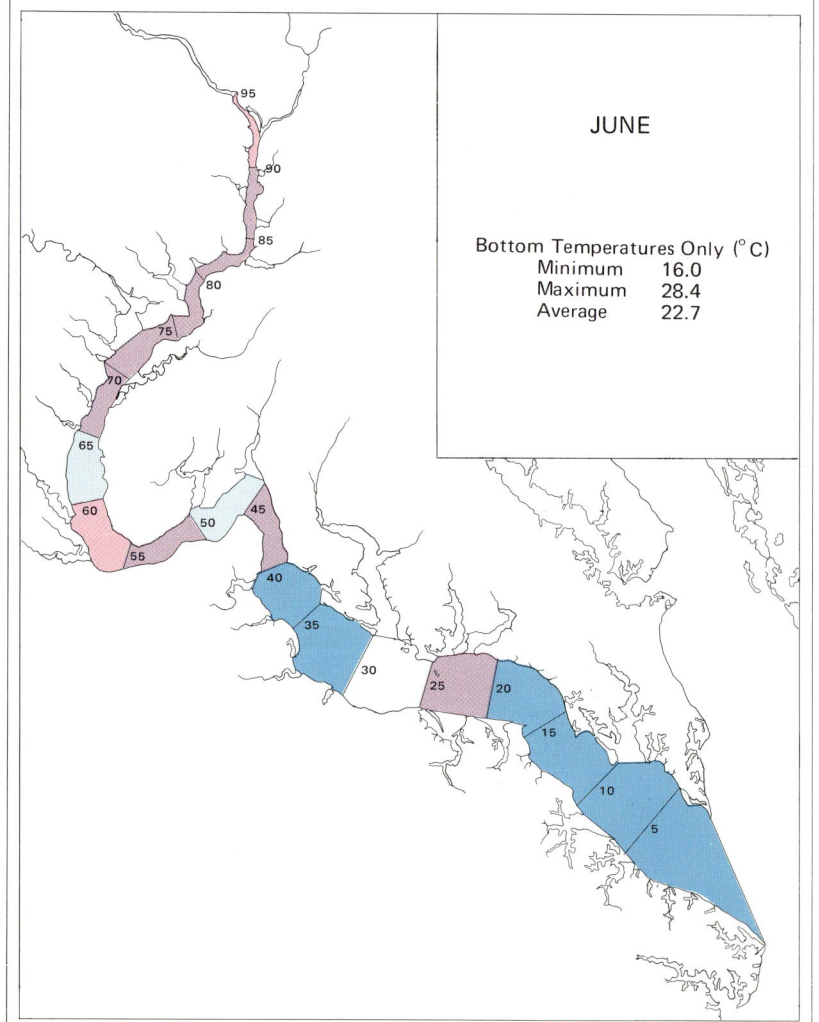

Figure 3-14. Continued.

WATER TEMPERATURES

JULY

Surface Temperatures Only (°C)
Minimum   17.8
Maximum   30.8
Average   26.2

Deviation from July River Average
(25.5 °C)

-2.5 -1.5 -0.5 +0.5 +1.5 +2.5 °C

23.0 24.0 25.0 26.0 27.0 28.0 °C
Actual Average Temperatures

WATER TEMPERATURES

AUGUST

Surface Temperatures Only (°C)
Minimum   22.2
Maximum   29.7
Average   26.0

Deviation from August River Average
(26.2 °C)

-2.5 -1.5 -0.5 +0.5 +1.5 +2.5 °C

23.7 24.7 25.7 26.7 27.7 28.7 °C
Actual Average Temperatures

51

JULY

Bottom Temperatures Only (°C)
Minimum   17.4
Maximum   28.8
Average   24.8

AUGUST

Bottom Temperatures Only (°C)
Minimum   23.0
Maximum   28.8
Average   26.3

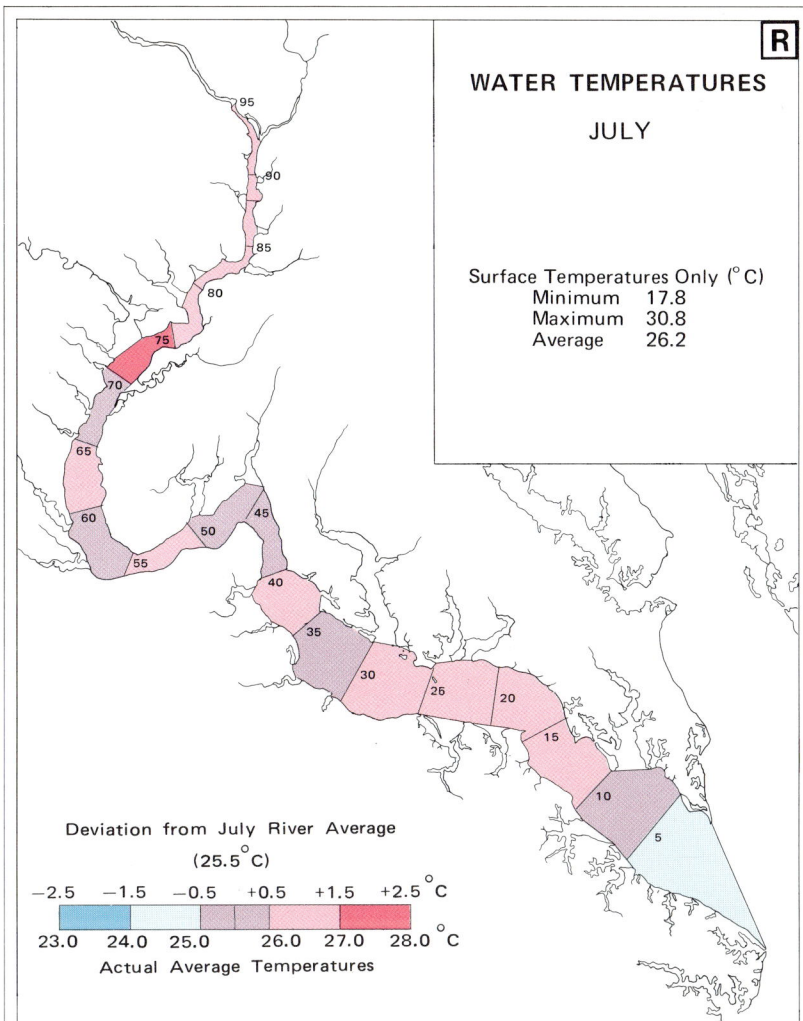

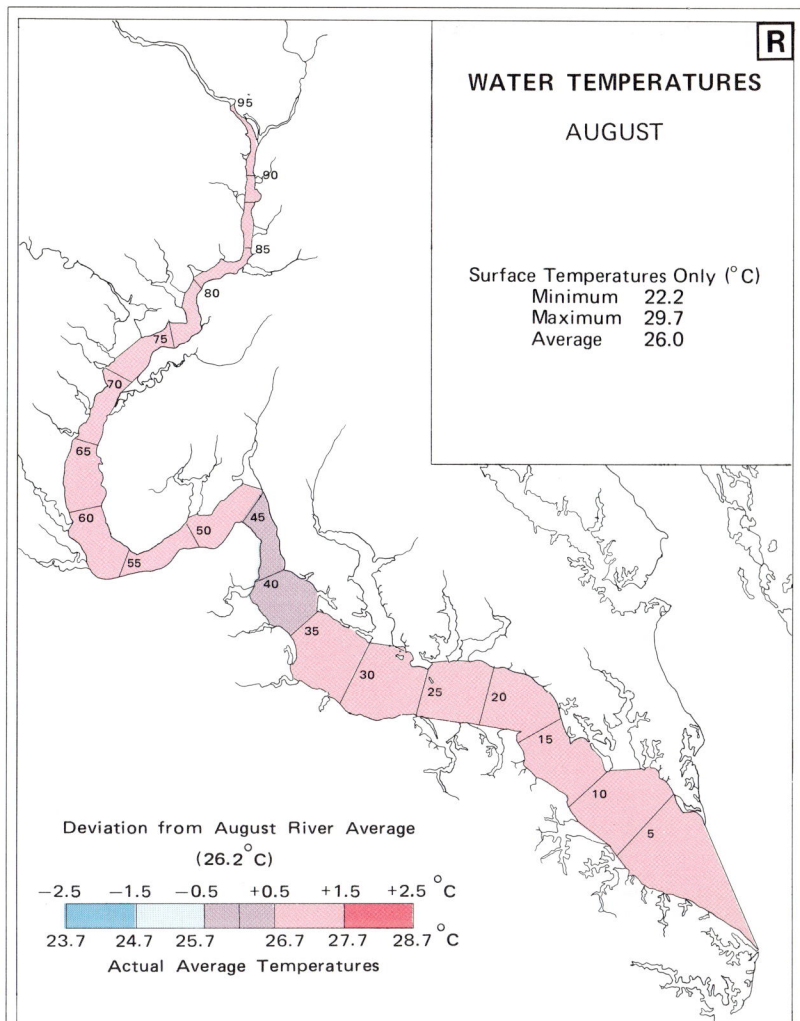

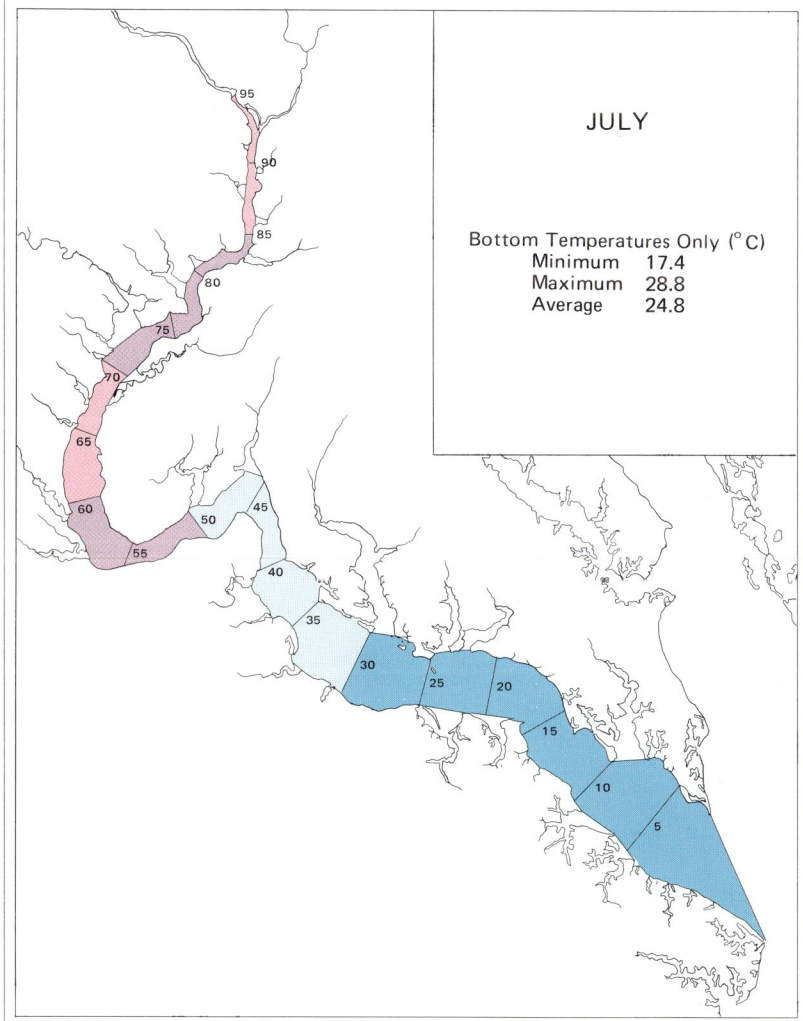

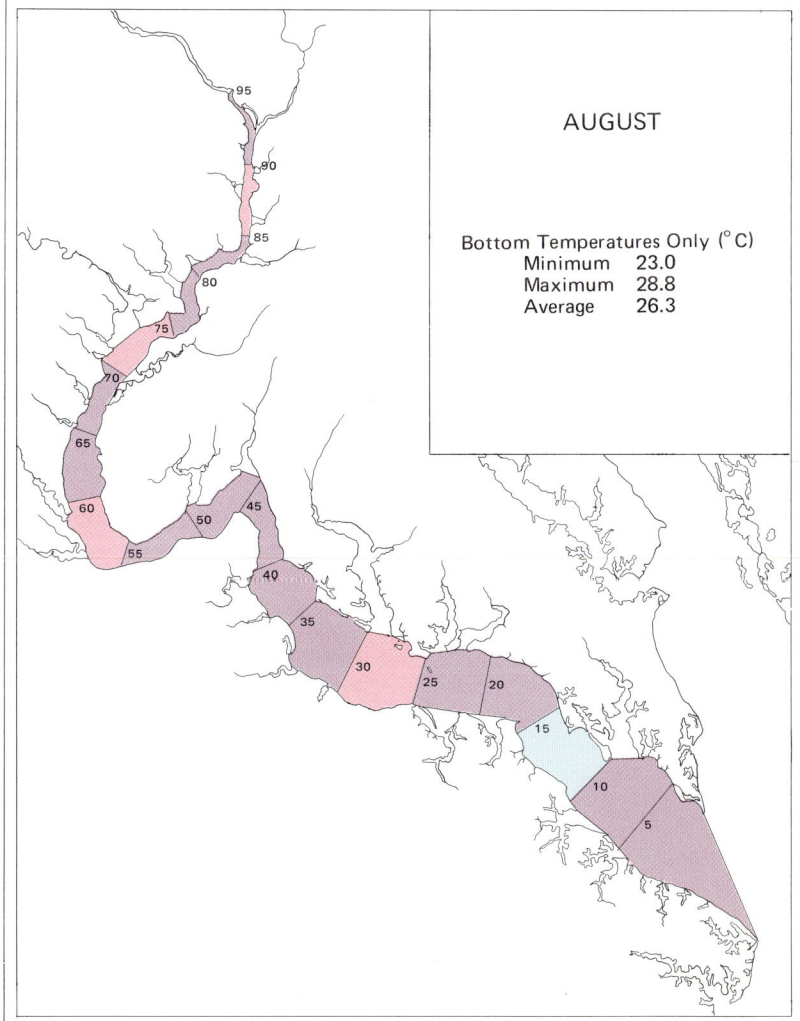

Figure 3-14. *Continued.*

WATER TEMPERATURES

SEPTEMBER

Surface Temperatures Only (°C)
Minimum   14.4
Maximum   31.1
Average   23.9

Deviation from September River Average
(23.8°C)

| −2.5 | −1.5 | −0.5 | +0.5 | +1.5 | +2.5 °C |

| 21.3 | 22.3 | 23.3 | 24.3 | 25.3 | 26.3 °C |

Actual Average Temperatures

WATER TEMPERATURES

OCTOBER

Surface Temperatures Only (°C)
Minimum   13.3
Maximum   25.4
Average   17.4

Deviation from October River Average
(17.2°C)

| −2.5 | −1.5 | −0.5 | +0.5 | +1.5 | +2.5 °C |

| 14.7 | 15.7 | 16.7 | 17.7 | 18.7 | 19.7 °C |

Actual Average Temperatures

SEPTEMBER

Bottom Temperatures Only (°C)
Minimum   15.0
Maximum   26.4
Average   23.8

OCTOBER

Bottom Temperatures Only (°C)
Minimum   13.2
Maximum   24.0
Average   16.9

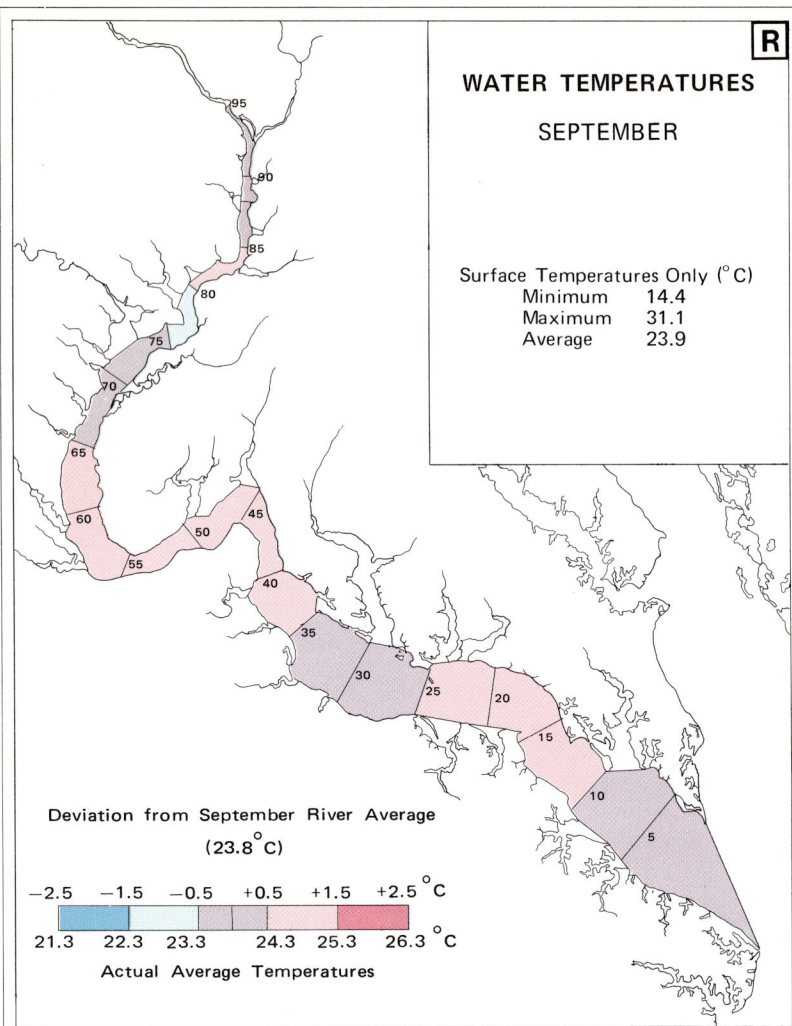

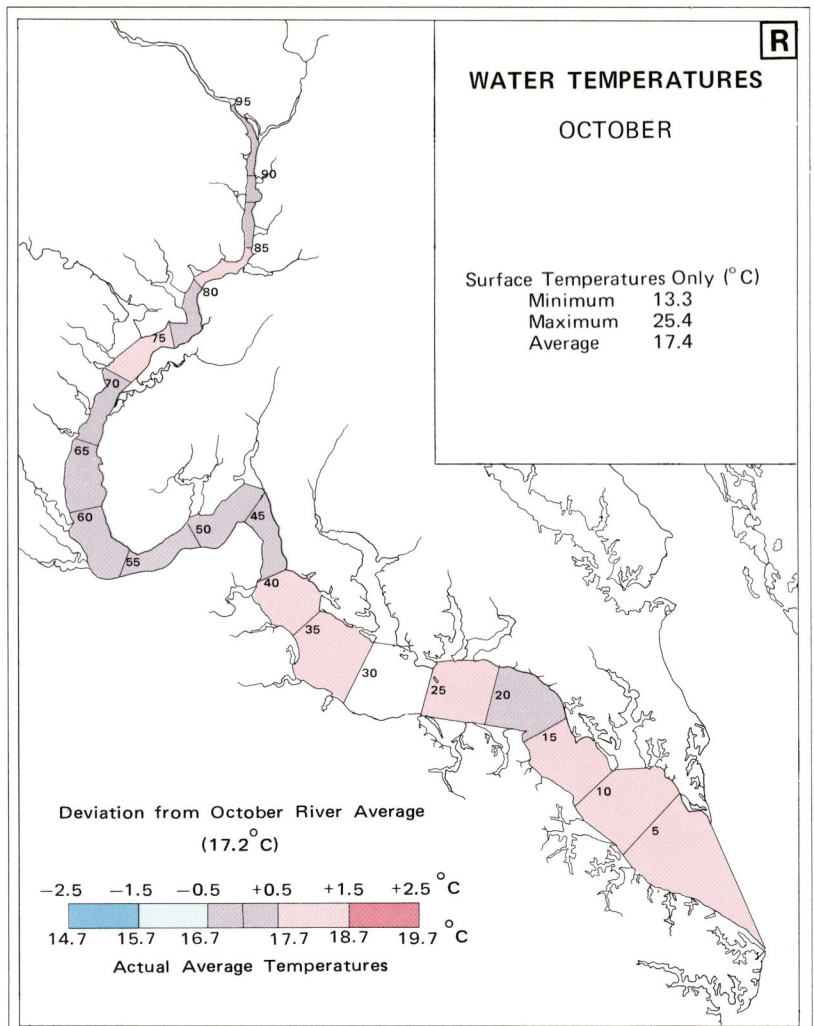

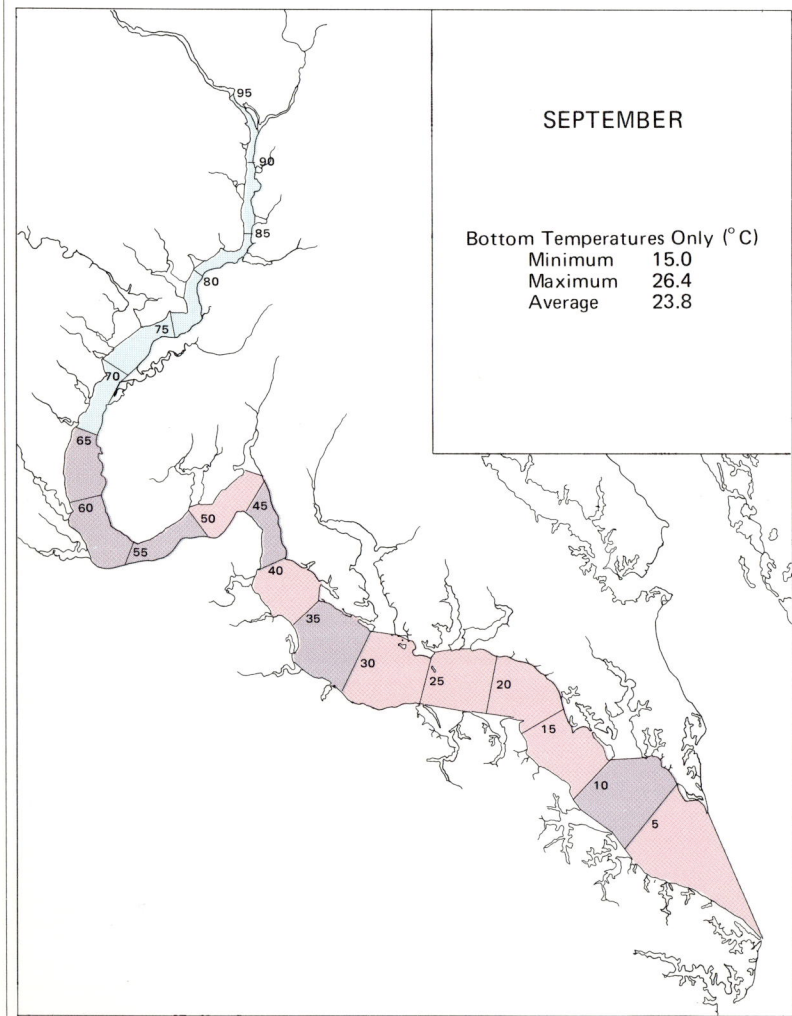

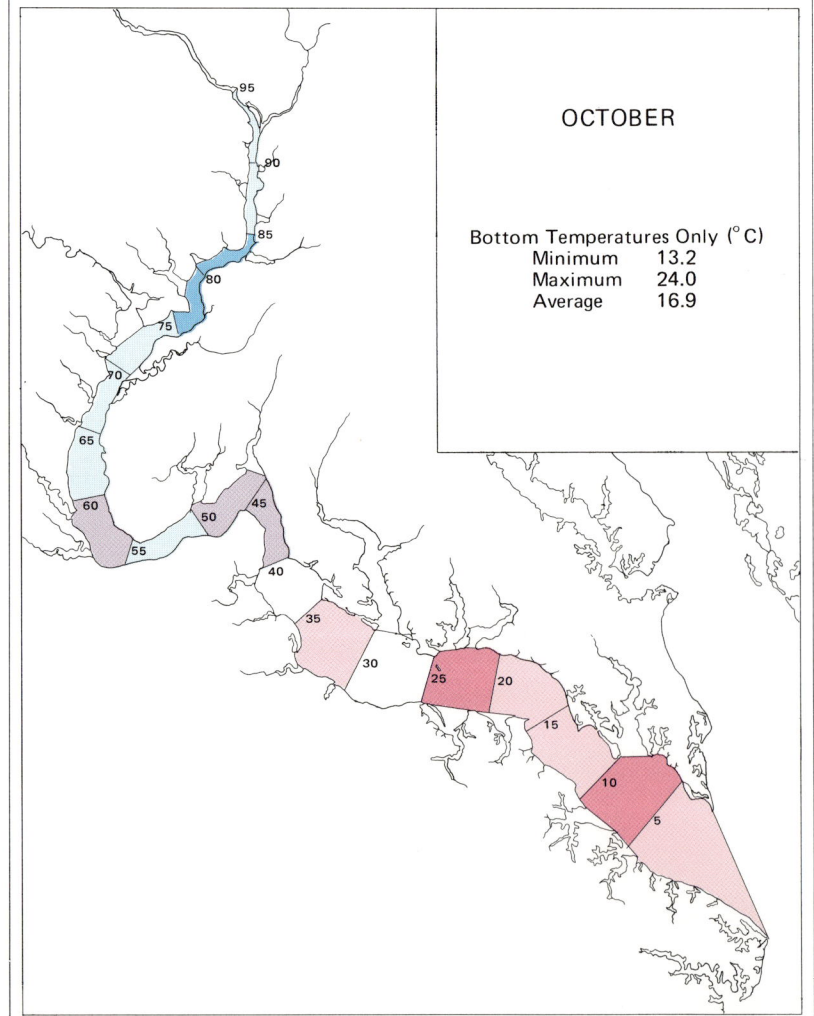

Figure 3-14. *Continued.*

R

WATER TEMPERATURES

NOVEMBER

Surface Temperatures Only (°C)
Minimum   6.4
Maximum   14.8
Average   10.9

Deviation from November River Average

(11.2°C)

| −2.5 | −1.5 | −0.5 | +0.5 | +1.5 | +2.5 °C |

| 8.7 | 9.7 | 10.7 | 11.7 | 12.7 | 13.7 °C |

Actual Average Temperatures

R

WATER TEMPERATURES

DECEMBER

Surface Temperatures Only (°C)
Minimum   1.3
Maximum   10.8
Average   6.4

Deviation from December River Average

(6.3°C)

| −2.5 | −1.5 | −0.5 | +0.5 | +1.5 | +2.5 °C |

| 3.8 | 4.8 | 5.8 | 6.8 | 7.8 | 8.8 °C |

Actual Average Temperatures

NOVEMBER

Bottom Temperatures Only (°C)
Minimum   8.0
Maximum   14.0
Average   11.1

DECEMBER

Bottom Temperatures Only (°C)
Minimum   2.3
Maximum   10.3
Average   6.3

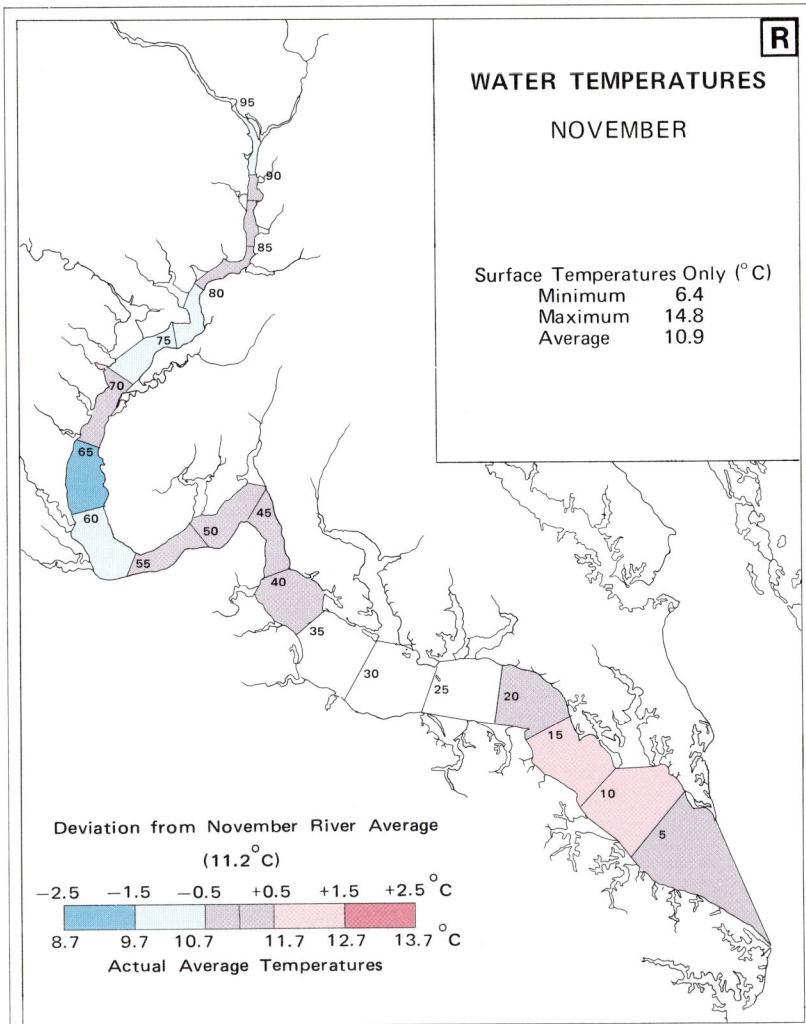

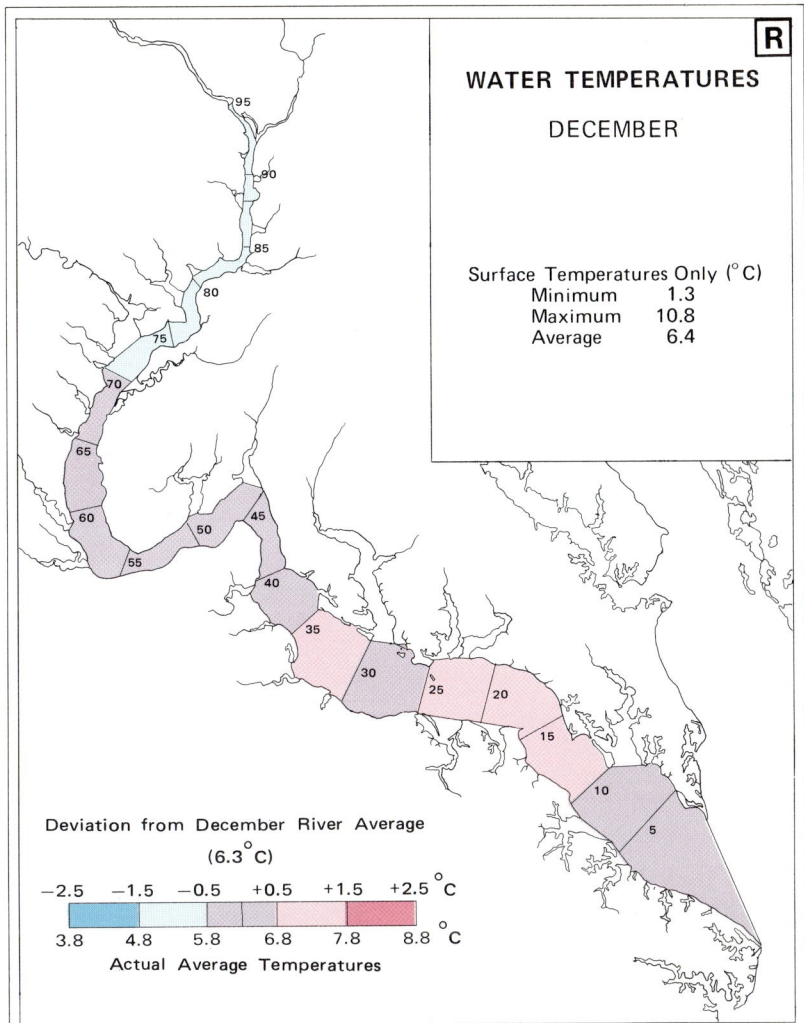

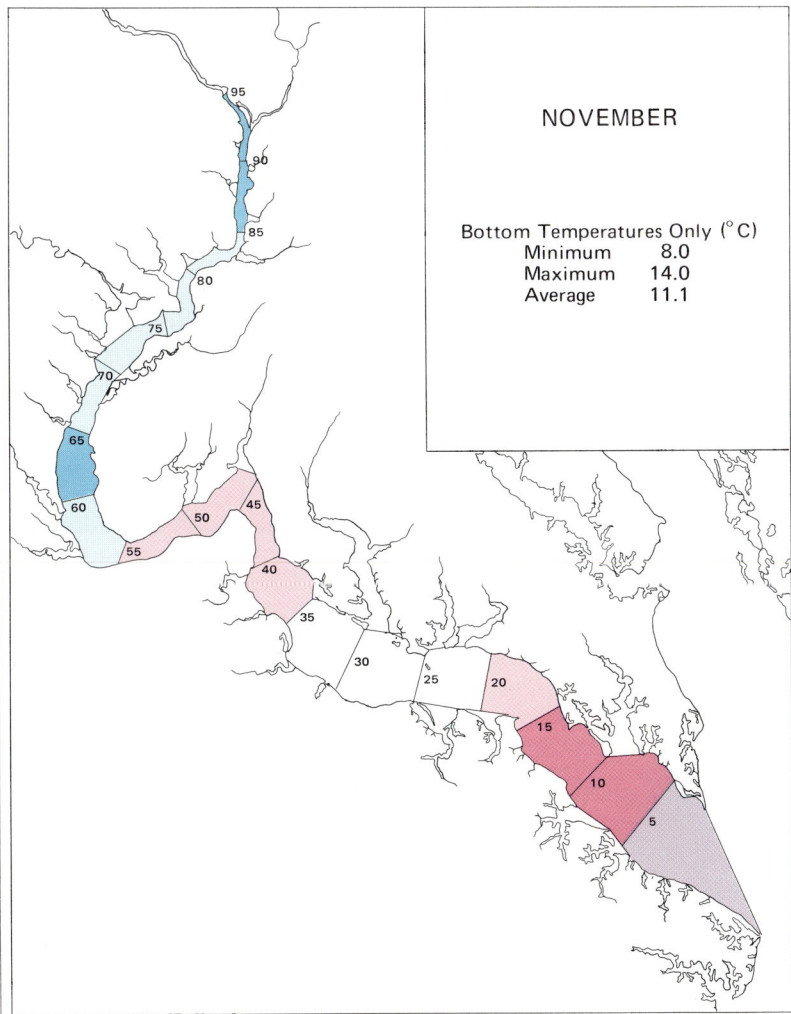

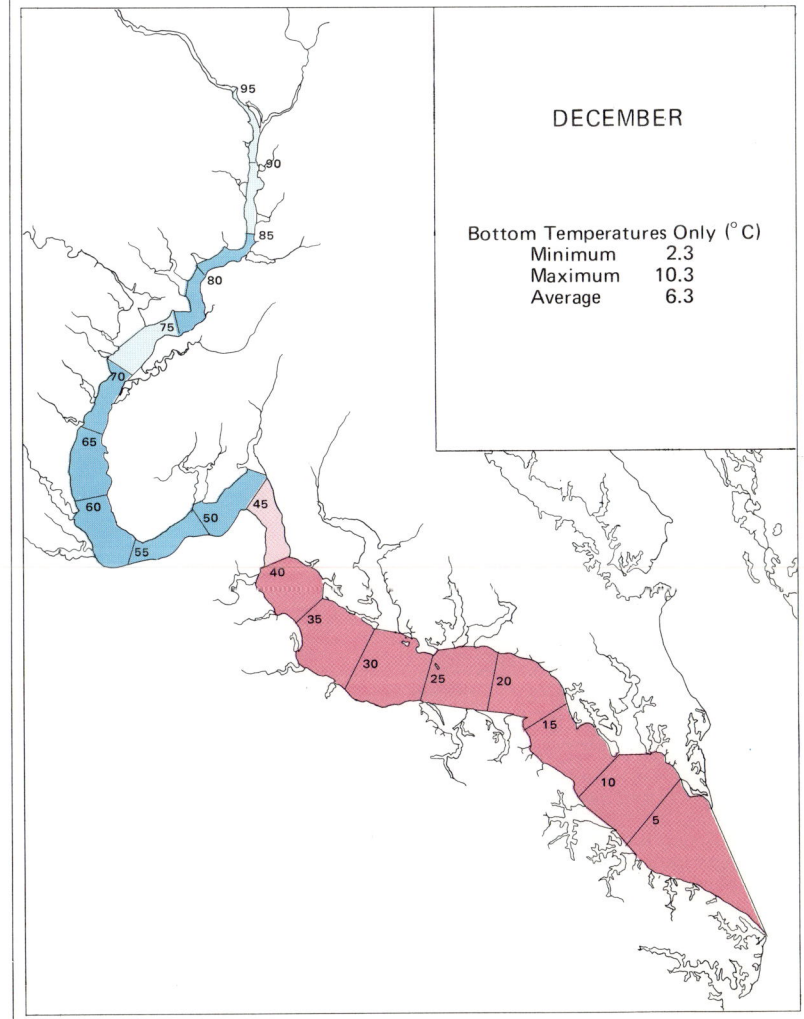

Figure 3-14. *Continued.*

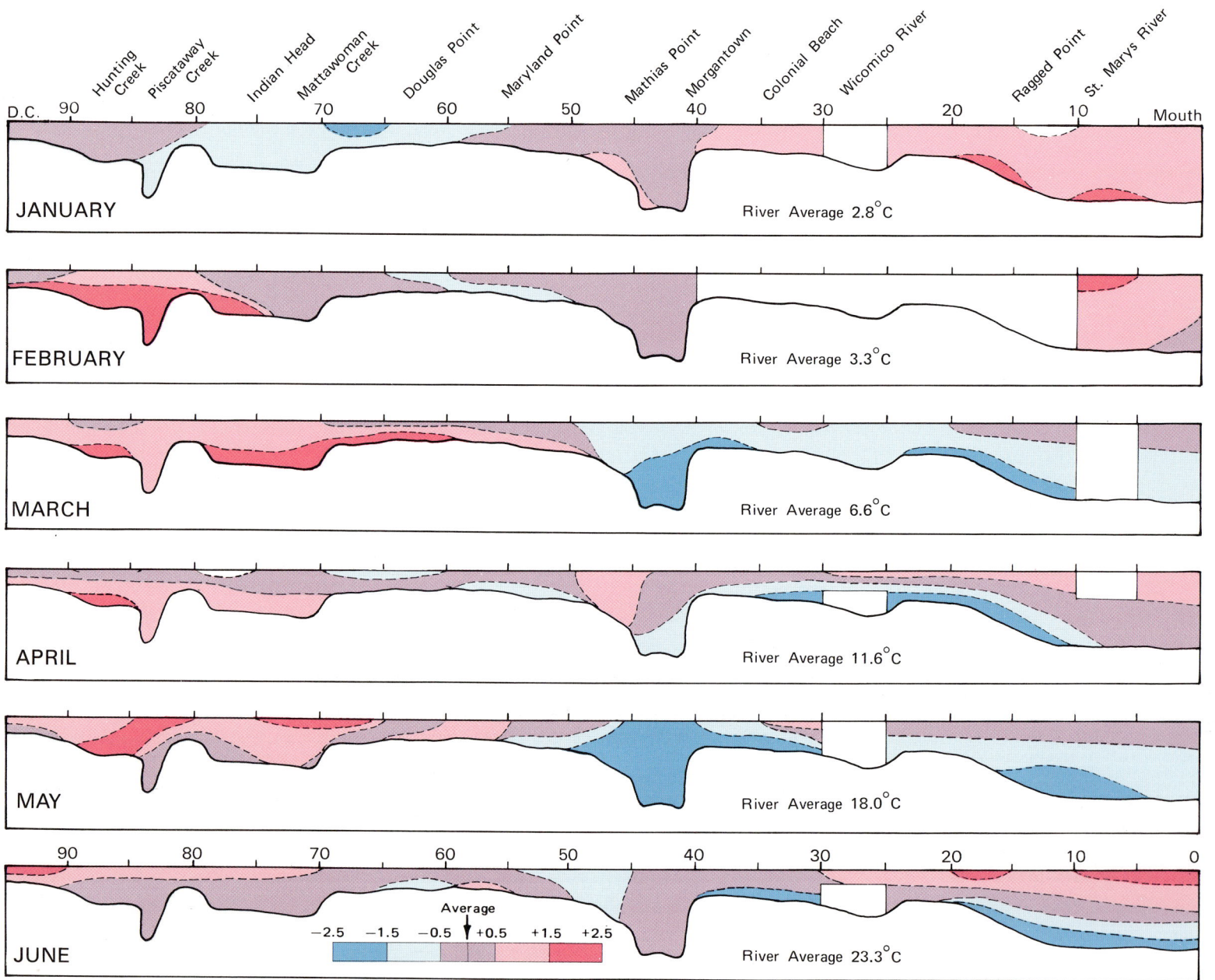

AVERAGE MONTHLY
VERTICAL TEMPERATURE PROFILES R

Figure 3-15. Monthly temperature profiles in relation to the overall river average of the Potomac estuary, derived from surface and bottom water temperatures in Fig. 3-14. Numbers along water surface line are nautical river miles from mouth. Bottom profiles show approximate water depths along the estuary at midchannel, with the deepest point at about 21 meters. Blank areas indicate where little or no data are available. (Sources: Refs. 15,20)

AVERAGE MONTHLY
VERTICAL TEMPERATURE PROFILES R

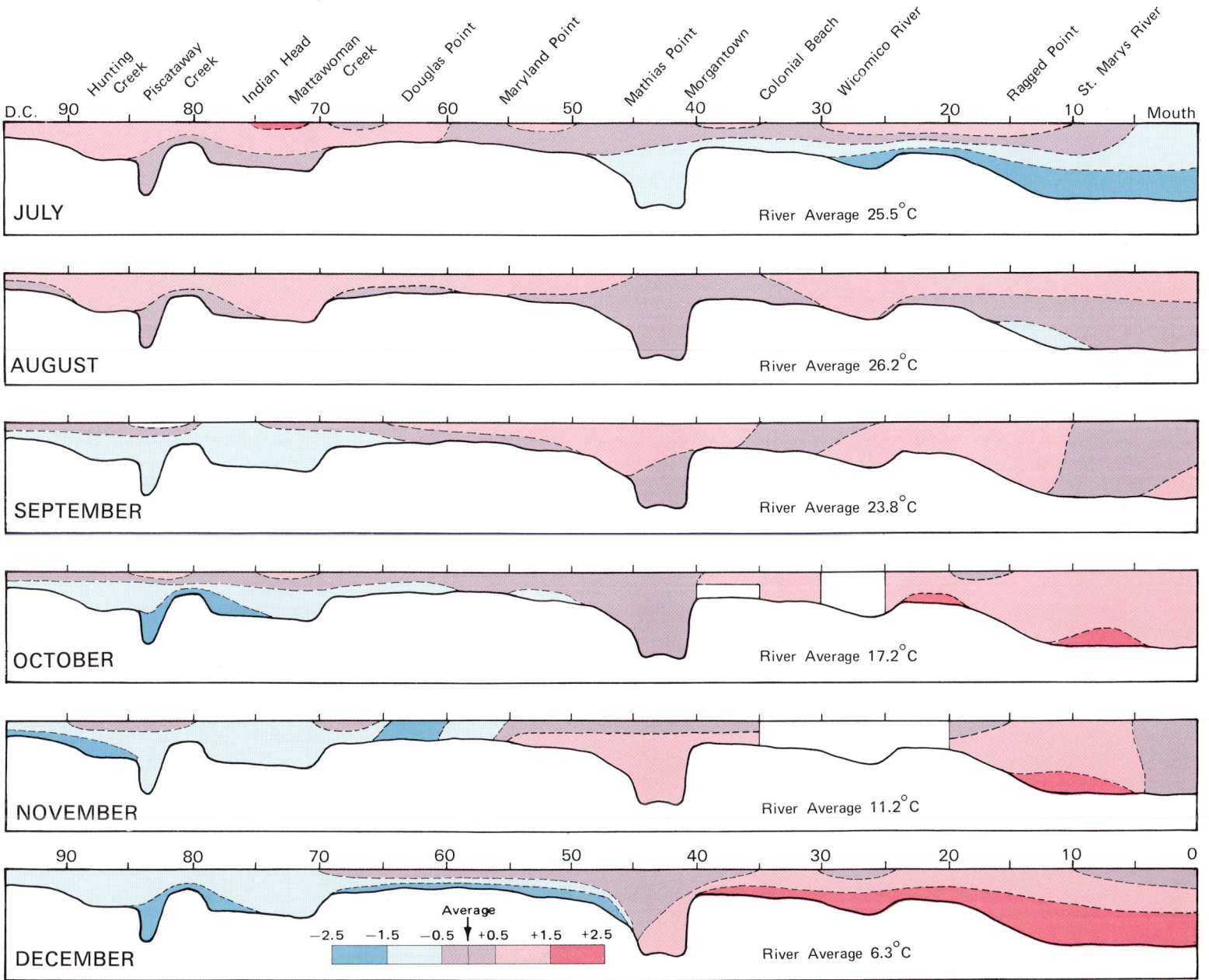

Figure 3-15. *Continued.*

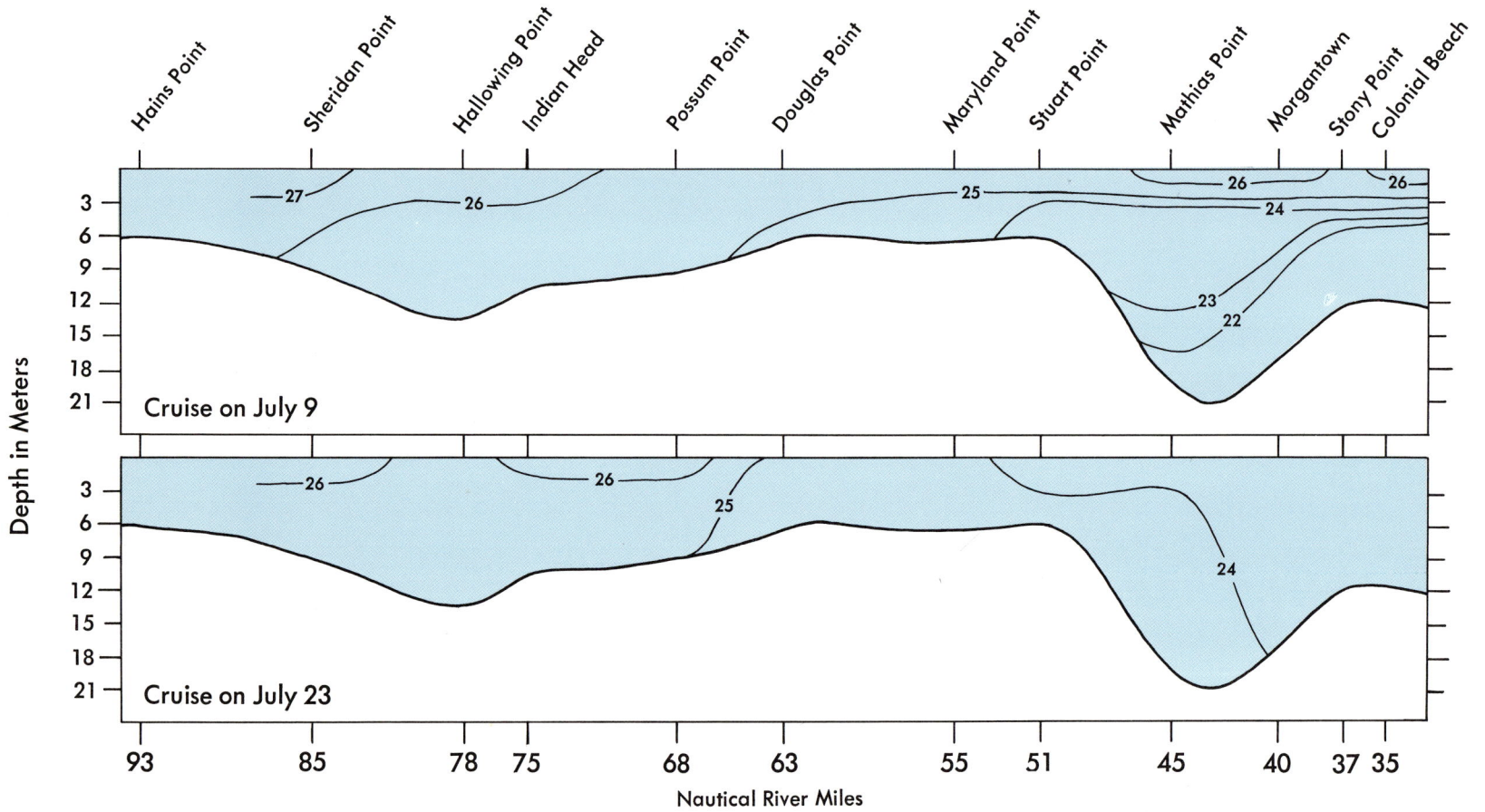

Figure 3-16. *Water temperature profiles in the Potomac estuary from Hains Point downstream to Colonial Beach, from measurements taken two weeks apart in July 1974. Bottom profiles show approximate water depths along the estuary at midchannel. (Source: Ref. 18)*

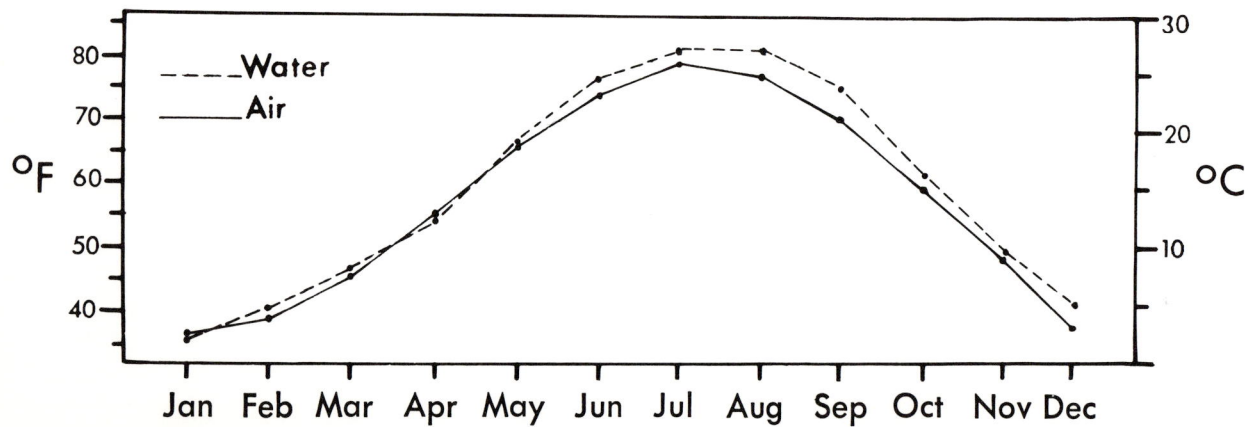

Figure 3-17. *Monthly average air temperatures (1951-60) and water temperatures (1965-74) at Washington, D.C. (Sources: Refs. 15,20,21)*

Sedimentation and Turbidity

Approximately 3.0 million tons of sediment enter the Potomac River basin annually, corresponding to an average yearly contribution of 170 tons of sediment per square mile of watershed area.[24] Direct erosion of the banks of the estuary is small (see, for example, Fig. 3-18), but the effects that upland erosion and the resultant sediment loading can have on water quality are great. Suspended sediments may have adverse effects on biota. Turbidity can reduce photosynthetic activity (and hence basic productivity) by inhibiting light penetration. Turbid waters can also degrade habitats for fish, and sudden sediment accumulation may smother fish eggs and benthic fauna.[24] Suspended sediments may also facilitate the dispersion and downstream transport of bacteria and toxic substances.[24]

As development increases along the Potomac River basin, the possibility of significant changes occurring in sediment loading and turbidity due to erosion becomes a matter of concern. The various land uses in the Potomac estuarine watershed (Table 3-5) have different effects on erosion rates. In urban areas, the removal of natural vegetation and the construction of impervious surface areas (e.g., commercial and residential buildings, parking lots, and highways) reduces natural ground absorptivity. This contributes to greater storm runoff which, in turn, increases peak flows and the possibility of flooding, the rate of erosion from surface lands and stream banks, and turbidity. It has been estimated that new urban development sites produce 50 to 200 tons of sediment per acre per year; as compared with 2 or 3 tons per acre per year from more established urban sites.[25] The construction of each mile of highway may contribute over 300 tons of sediment to nearby waterways.[23] In 1966, the urban areas around the District of Columbia contributed an estimated 682 tons per square mile (1.1 tons per acre) to the average annual sediment load of the Potomac estuary.[24] The current load is probably much lower because of reduced construction activity in the area.[26] In rural areas, erosion from storm waters can increase if land management and agricultural practices are poor. However, in the rural areas of the Potomac estuarine watershed, average annual sediment runoff is estimated to be only 48 tons per square mile (0.075 tons per acre).[24]

About 50 percent of the total sediment load entering the Potomac comes from the drainage basin above the Point of Rocks (Fig. 3-19), located 60 statute miles upstream of Washington, D.C., and 30 percent comes from the D.C. metropolitan area, which comprises only 2 percent of the entire drainage basin area.[23][24]

Sediment loading depends not only on land use, but also on the amount of freshwater flow through the system. Periods of high flow flush the greatest amounts of sediment through the estuary and usually

Table 3-5 — **Land and Water Areas of the Counties Bordering the Potomac Estuary, Percentage of Lands Applied to Various Uses, and Inhabitants per Acre** [a]

	Total Land Area [b] (acres)	Total Water Area (acres)	Total Land and Water Area (acres)	Percentage of Total Land Area Applied to Use:						Inhabitants Per Acre	
				Residential	Commercial/ Industrial	Agricultural	Woodlands	Open Lands	Wetlands	Total Land	Residential Land
District of Columbia	39,194	4,384	43,578	29	8	0	0	43	0	19.30	67.61
Maryland											
Prince Georges County	309,760	8,522	318,282	12	2	25	45	4	2	2.13	17.38
Charles County	293,760	109,298	403,058	2	<1	21	69	0	4	0.16	6.86
St. Marys County	238,720	293,801	532,521	4	<1	28	51	12	2	0.20	6.48
Virginia											
Alexandria	9,984	120	10,104	39	16	0	13	1	0	11.14	22.71
Arlington County	15,616	0	15,616	55	7	<1	0	9	<1	1.76	84.69
Fairfax County	259,200	7,040	266,240	2	2	9	43	32	1	0.50	24.69
Prince William County	222,080	4,480	226,560	2	1	24	49	3	1	0.14	5.14
Stafford County	173,440	3,840	177,280	3	1	14	56	8	2	0.07	4.92
King George County	113,920	3,200	117,120	1	<1	22	69	0	3	0.08	8.27
Westmoreland County	151,040	8,960	160,000	1	<1	31	61	1	5	0.10	4.88
Northumberland County	121,600	21,120	142,720	2	<1	28	62	0	8		

Residential — Includes all residential use, single to multi-family dwellings, mobile homes, etc.
Commercial/Industrial — Includes retail sales, offices, service activities, manufacturing plants, warehouses, refuse disposal, etc.
Agricultural — Includes croplands, pastures, and livestock and poultry raising areas.
Woodlands — Includes all forests (commercial and noncommercial) and woods on farmland.
Open Lands — Includes all vacant unused land not under any special category.
Wetlands — Includes all land regularly inundated with tidal water. (See Folio Map 4 for locations of wetlands.)

(a) Data for land areas from Table B-V-3 in Ref. 23. Inhabitants per acre from 1970 Census.
(b) Total area also includes acreage devoted to public areas, parks, highways, and railroads not separately tabulated here.

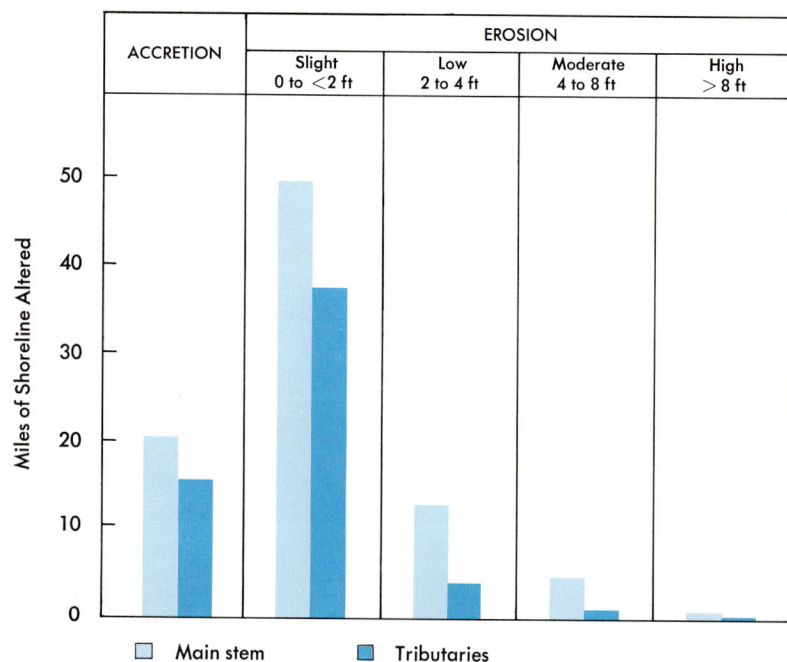

Figure 3-18. *Shore erosion and accretion on the Maryland side of the Potomac estuary from 1862 to 1971. (Source: Ref. 23)*

correspond to heavy rains. Storms increase suspended load levels by increasing runoff and resuspending bottom sediments.

Because of the constantly increasing problem of soil erosion, federal, state, and local governments have in recent years passed stringent controls on procedures for land clearing, soil movement, and other construction activities.[24] In urban areas, control measures include surface-stabilization treatments, sediment-trapping devices, and limitations on the extent of ground clearing.[25] There are few legal restrictions on agricultural procedures, but local soil conservation districts are working with farmers to promote improved conservation practices.[24][27]

CHEMICAL CHARACTERISTICS AND WATER QUALITY

Several chemical variables are used as gross indicators of water quality. Some of the more important ones are nitrogen and phosphorus compounds, whose chemical cycling and roles in primary productivity in the Potomac estuary will be briefly discussed here, along with the roles of dissolved oxygen (DO) and chlorophyll *a*.

Nitrogen and phosphorus compounds are vital nutrients for algae and other plants: nitrogen for synthesis and manufacture of plant protein, phosphorus for energy transfer in plant cells. Low concentrations of either of these limit plant growth; excess amounts can cause algal blooms or the proliferation of nuisance aquatic plants. Levels of

these two nutrients, by directly affecting phytoplankton biomass production, also influence dissolved oxygen concentrations, specific levels of which are necessary for the physiological functioning of aquatic animals. In a well-balanced system, plants help to maintain these required levels of dissolved oxygen. Reaeration created by turbulence at the air-water interface is the only other major mechanism for increasing the dissolved oxygen concentration. Dense animal and plant life in the presence of much decaying organic material increases the biochemical oxygen demand in the water, thereby depleting dissolved oxygen levels, sometimes to values intolerable to many organisms.

Chlorophyll *a*, a pigment present in almost all plants, can be used as a measure of phytoplankton biomass and, with other water quality indicators, may be used to establish cause-and-effect relationships between biotic and abiotic factors influencing water quality. Water quality in the upper Potomac estuary has been greatly affected by nutrient loading from municipal wastewaters (see Table 3-6). Folio Map 9 shows the locations of the major discharges into the Potomac estuary and lists the design-flow discharges of all the sewage treatment plants. The Blue Plains sewage treatment plant alone (at approximately nautical river mile 90) is designed to discharge over 280 million gallons per day (MGD).[28] An additional 110 MGD may be contributed by other local treatment plants down to Gunston Cove (nautical river mile 79). Table 3-7 lists the present flows and capacities and future planned capacities of the major sewage treatment plants.

The waters of the Potomac estuary and its tributary, the Anacostia, are also used to supply cooling water for five major electric power plants (Folio Map 9). Water is heated 10 to 17°F (5.5 to 9.4°C) as it passes through the cooling systems and is then returned to the estuary. Often, chlorine or some other biocide is added to the cooling water to prevent excessive buildup of algae and other fouling organisms in the system. The operations of these cooling systems may have localized effects on water quality.

The graphs in Figs. 3-20 through 3-25 show the averages and trends in concentrations of nitrogen and phosphorus compounds, dissolved oxygen, and chlorophyll *a* in the upper estuary for the years prior to and since 1972, and indicate improvement in overall water quality conditions over the last few years. The reason for dividing the data in this way is that the Federal Water Pollution Control Act Amendments of 1972 call for implementation of area-wide waste treatment to assure adequate control of sources of pollutants in each state. It seems reasonable to measure changes in water quality from the year when the act was implemented.*

---

* Since 1962, water quality in the Potomac has been monitored closely by federal, state, and municipal agencies. The data from these surveys are stored in the U.S. Environmental Protection Agency STORET System[20] and have been used as the basis for all major decisions on wastewater management in the upper Potomac.

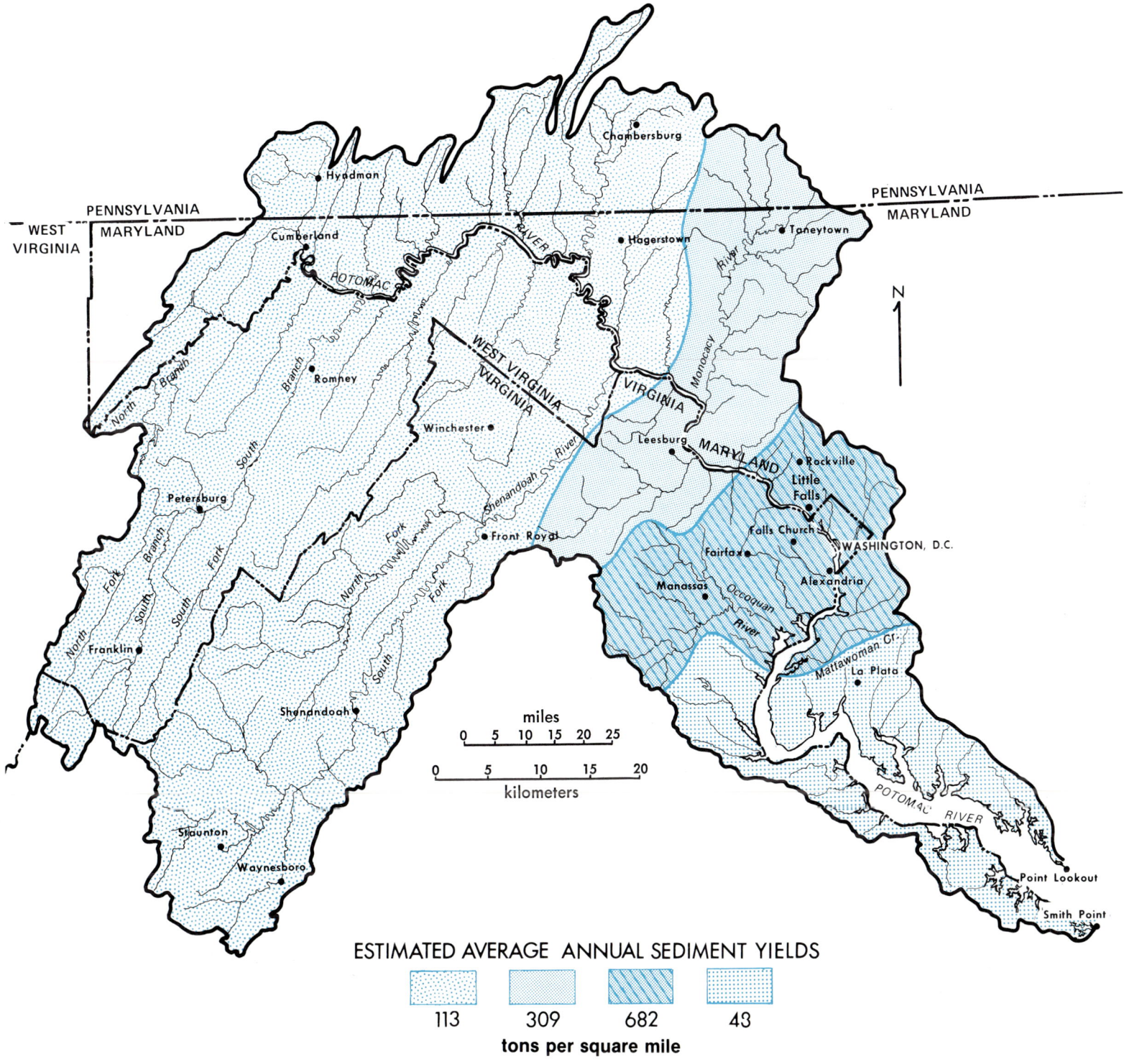

ESTIMATED AVERAGE ANNUAL SEDIMENT YIELDS

| 113 | 309 | 682 | 43 |

tons per square mile

Figure 3-19. *Estimated average annual sediment yields for the Potomac River drainage in 1966 (Ref. 24). The current load in urban areas around the District of Columbia is probably lower due to decreased construction activity since 1966 (Ref. 26).*

Table 3-6 — **Estimated Nutrient Loadings in the Potomac River Basin in 1974** [a]

	Area (acres) of Total Basin	Area (acres) Below Great Falls	Total Phosphate as Phosphorus (lb/day)		Inorganic Nitrogen as $NO_2^- + NO_3^-$ (lb/day)		Organic Nitrogen as TKN (lb/day)		Total Nitrogen (lb/day)	
			Total Basin	Below Great Falls	Total Basin	Below Great Falls	Total Basin	Below Great Falls	Total Basin	Below Great Falls
Municipal Discharge			28,465	24,786	4,344	3,784	55,672	47,499	60,016	51,283
Runoff										
Agricultural	3,737,600	1,740	2,477	813	35,040	10,540	3,796	1,136	38,836	11,676
Forest	5,184,000	1,300	1,322	212	16,200	2,600	3,240	520	19,440	3,120
Urban	467,200	630	261	225	2,190	1,890	475	410	2,665	2,300
TOTAL			32,525	26,036	57,744	18,814	63,183	49,565	120,957	68,379
Percentage from Municipal Discharge			87.5	95.2	7.5	20.1	88.1	95.8	49.6	75.0
Percentage from Runoff			12.5	4.8	92.5	79.9	11.9	4.2	50.4	25.0

(a) Adapted from Tables 13 and 15 in Ref. 24.

Table 3-7 — **Flow and Capacity of Major Sewage Treatment Plants on the Potomac Estuary in 1977 (Millions of Gallons Per Day)**

Plant	Present Flow	Capacity	Future Planned Capacity
Washington, D.C.			
Blue Plains [a]	276	309	309
Maryland			
Piscataway [b]	16	30	30
Virginia [c]			
Alexandria Sanitary Authority	21	27	54
Arlington County	21	24	30
Fairfax County			
Dogue Creek	2.5	5.0	—[d]
Little Hunting	4.25	6.6	—[d]
Lower Potomac	15	36	36
Westgate	11.3	13.7	—[e]
Upper Occoquan Sewage Authority Plant (UOSAP) [f]	5.9	10.9	39.3
Prince William County			
Mooney Plant (Potomac Regional)	—	—	12

(a) Values from Ref. 29.
(b) Values from Ref. 30.
(c) All values for Virginia from Ref. 31.
(d) Will be closed when the Mooney Plant becomes operative.
(e) Will be closed when expansion of the Alexandria Sanitary Authority Plant is completed.
(f) In operation since July 1978 and located on Bull Run, a tributary of Occoquan River.

Nitrogen

Three measurements have been used in the Potomac as indicators of nitrogen levels in the water: 1) the total amount of nitrites and nitrates (NO_2^- and NO_3^-), which are the major inorganic nitrogen compounds in water; 2) the total amount of organic nitrogen compounds, usually reported as total Kjeldahl nitrogen (abbreviated as TKN and referring to a specific testing method); and 3) ammonia levels.

Based on observations, nutrient levels have been designated in the Potomac estuary which, when exceeded, could produce nuisance algal blooms (increasing chlorophyll a levels to above 50 µg/liter). Analyses have shown that inorganic nitrogen levels must be less than 0.35 mg/liter to prevent blooms.[34] A smaller value of 0.3 mg/liter has been suggested as the upper level for the problem segment between Indian Head and Smith Point, and a larger value of 0.5 mg/liter for the estuary above Indian Head where poor light penetration limits algal growth to within 2 feet (0.6 m) of the surface.[24]

Municipal sewage discharges account for 75 percent of the annual total nitrogen loading into the Potomac estuary (Table 3-6). The nitrogen from sewage discharge is primarily composed of organic nitrogen (measured as TKN); that from runoff consists of nitrites and nitrates. The amount from runoff depends on freshwater flow, and the relative contribution of municipal nitrogen inputs varies accordingly. During low flow periods in summer and fall, up to 90 percent of the total nitrogen entering the Potomac comes from wastewater discharges and thus is mostly organic.[24] During high spring flows, however, an estimated 80 percent of the total nitrogen load into the estuary comes from the upper river basin and is primarily composed of NO_2^- and NO_3^-. These percentages do not take into account the biologically fixed nitrogen produced in the estuary.

The great amounts of organic nitrogen dumped into the estuary from the Blue Plains sewage treatment plant in summer are reflected in the peaking of ammonia (Fig. 3-20) and TKN values (Fig. 3-21) in the vicinity of the plant. The high levels of these nitrogenous compounds (primarily ammonia) are rapidly converted into living matter and are also oxidized into nitrites and nitrates within 9 nautical miles downstream of the discharge point. The distributional pattern of the peaks and depressions in average ammonia and TKN values are similar in the earlier (1965-71) and later (1972-75) periods, but more recently, there has been a noticeable reduction in total average levels. Inorganic nitrogen (Fig. 3-22) has consistently exceeded the critical level of 0.35 mg/liter from Washington, D.C., down to Maryland Point over both periods.

In the lower estuary, measurements made in 1965 and 1966 showed that total nitrogen was extremely low from Morgantown down to the Chesapeake Bay.[35] Inorganic nitrogen compounds were almost completely removed in midsummer by flushing and algal uptake.

Phosphorus

Phosphorus levels in water are determined by measuring phosphates, the only forms of phosphorus used by organisms. Of the estimated 32,525 pounds of phosphorus typically entering the Potomac basin each day, over 87.5 percent can be attributed to municipal sewage treatment plants (Table 3-6). The tremendous amount of phosphorus coming from municipal discharge is clearly shown by the peaking of phosphate levels just below the Blue Plains sewage treatment plant, which is at nautical river mile 90 (Fig. 3-23). Analyses of Potomac estuarine data have indicated that nuisance blooms can be reduced if total phosphorus as phosphate ($PO_4^{\equiv}$) is below 0.39 mg/liter; and/or if the inorganic phosphorus level is below 0.3 mg/liter.[34] In the period before 1972, average summer phosphate concentrations always exceeded that level, causing algal blooms above Morgantown (at approximately nautical river mile 40). However, concentrations declined somewhat downstream of the point of sewage effluent. Some improvement occurred from 1972 to 1975 when average concentrations dropped below the critical level downstream from Hallowing Point, approximately at nautical river mile 77. In these later years, during periods of low summer flow conditions, an even greater part (96 percent) of the phosphorus came from municipal discharges,[24] and phosphate levels in the upper estuary have at times reached 1.0 mg/liter or more.[20]

Due to the nature of the geochemical cycle of phosphorus in estuarine waters, these high levels do not persist along the estuary. Phosphates tend to precipitate and form granules that settle into the sediments. If dislodged by high flows, these granules become resuspended, but will only dissolve in saline waters. Therefore, for phosphates to become redissolved, the flow would have to be great enough to transport the granules into saline waters where, upon dissolution, they would either reprecipitate or be flushed out of the system. Measurements in 1965 and 1966[35] in the lower estuary showed

extremely low levels of both total and inorganic phosphorus, suggesting an almost total loss of phosphates from the water to the sediments. Measurements at a station just off the mouth of the Potomac in 1973 recorded a summer "high" of only about 0.002 mg/liter total phosphorus.[36]

Dissolved Oxygen

The concentration of dissolved oxygen in water depends on several physical and biological factors. Water temperature governs the solubility of oxygen in water, and turbulence increases the rate of aeration from the atmosphere. Biochemical demands created by the metabolic processes of aquatic organisms and by the decomposition of organic matter can also govern dissolved oxygen levels in water. The State of Maryland water quality standard for acceptable dissolved oxygen levels in the Potomac estuary is a minimum of 4.0 mg/liter and a minimum daily average of 5.0 mg/liter.[32] Minimum oxygen requirements differ greatly among animals, and regulatory standards are usually set at levels intended to ensure the survival of organisms with high oxygen requirements. Oxygen requirements of specific fish species have traditionally been used as criteria for establishing acceptable levels for aquatic systems, even though different fish may have very different requirements.

As with nutrients, dissolved oxygen levels follow a seasonal pattern. Winter oxygen levels are always substantially higher than summer levels, due to lower temperatures, greater turbulence, and reduced oxygen demand. The greater demand on oxygen levels in summer, which depletes dissolved oxygen, is primarily created by the decomposition of warm-weather algal blooms. In summer, the highly stratified water column and decreased river flow also result in diminished physical mixing and reoxygenation. Depressed summer oxygen conditions have existed between Blue Plains and Marshall Hall (at approximately nautical river miles 90 and 81, respectively) since 1965. However, the 1972-75 average values were always above 5.0 mg/liter; and the 1965-71 average levels during June, July, and August fell at or above the 4.0 mg/liter level (Fig. 3-24). In the middle reach of the estuary, the State of Maryland standard has generally been met.

Chlorophyll a and Bloom Conditions

Chlorophyll a concentration is an index of phytoplankton biomass and is therefore a particularly useful indicator of bloom conditions caused by nutrient enrichment. Extreme enrichment causes a condition commonly referred to as eutrophication of the water body, in which a few species of plankton or algae rapidly produce a large biomass, which then creates a high oxygen demand as it decays. Bloom conditions have been produced in the upper Potomac estuary by the great influx of nutrients from wastewater discharges and runoff from highly fertilized fields. When the total phosphorus and nitrogen concentrations exceed critical levels, nuisance algal blooms (defined as more than 50 μg/liter of chlorophyll a) are common.[37] The plants that grow most profusely

62

POTOMAC RIVER ESTUARY

M

AMMONIA
AVERAGE SUMMER SURFACE VALUES
JUNE, JULY, AUGUST, 1972–75

Milligrams per Liter

0.8 or more
0.6 – 0.8
0.4 – 0.6
0.2 – 0.4
0.0 – 0.2

Lines cross river close to locations of
sampling stations and correspond to red
tic marks on graph below.

Blue Plains

Oxon Cr.
Hunting Cr.
Broad Cr.
Tinkers Creek
Piscataway Cr.
Gunston Cr.
Pomonkey Cr.
Indian Head
Mattawoman Cr.
Old Womans Run
Quantico Cr.
Chicamuxen Cr.
Port Tobacco R.
Douglas Pt.
Aquia Cr.
Nanjemoy Cr.
Maryland Pt.
Mathias Pt.
Potomac Cr.
Morgantown
Wicomico R.
Chaptico Bay
St. Clements Bay
Breton Bay
St. Marys R.
St. George Cr.
Lower Machodoc Cr.
Yeocomico R.
The Glebe
Coan R.

CHESAPEAKE BAY

Scale   1:500,000

5   0   10 Statute Miles
5   0   5   10 Kilometers
5   0   5   10 Nautical Miles

Ammonia (mg./liter)

1.3
1.2
1.1
1.0
0.9
0.8
0.7
0.6
0.5
0.4
0.3
0.2
0.1

1965–71 Summer
Summer Average
1965–71 Winter
1972–75 Summer
1972–75 Winter

90   80   70   60   50   40
RIVER MILE

Red tic marks indicate locations
of sampling stations shown on map

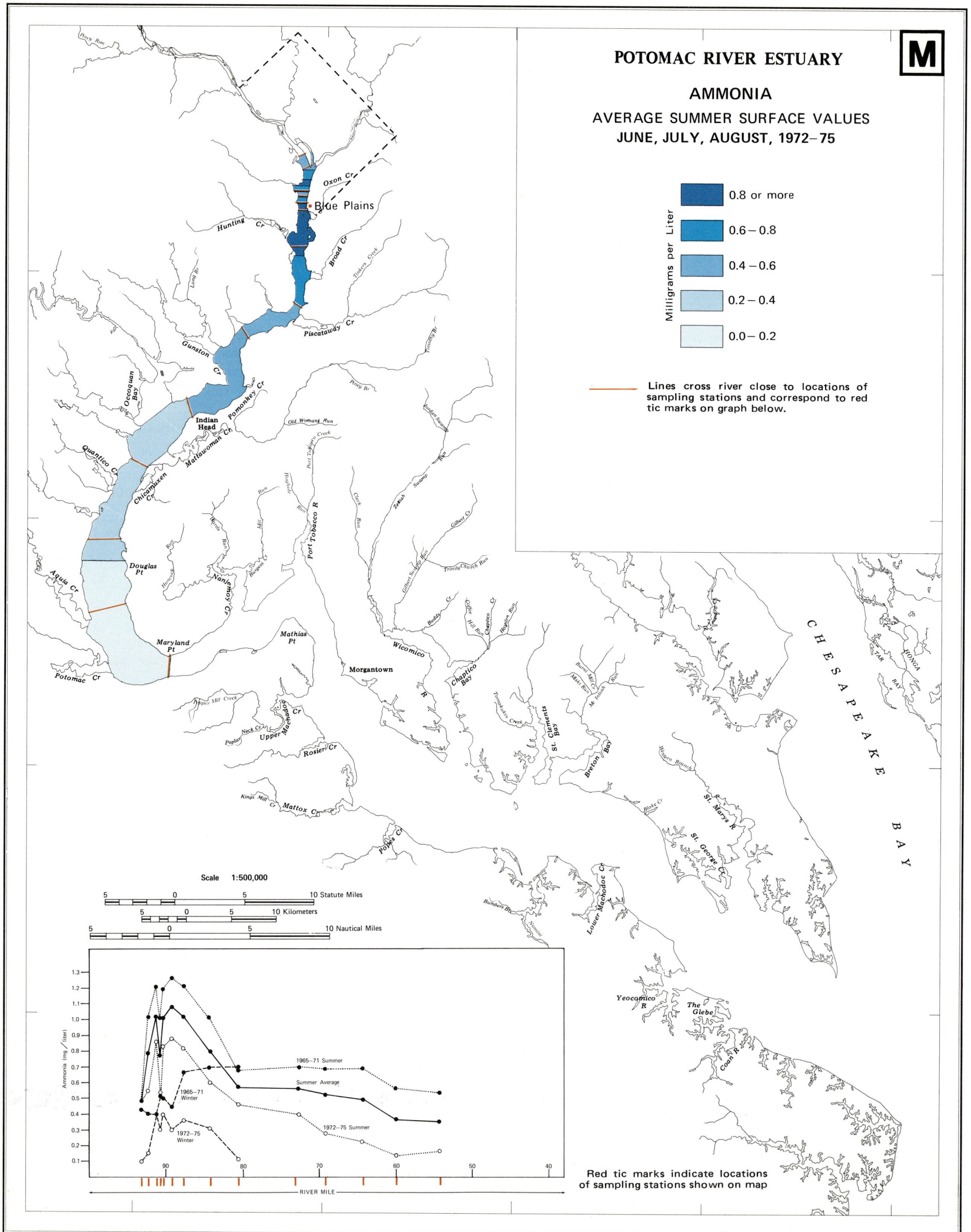

Figure 3-20.   *Average ammonia values in surface waters of the upper Potomac estuary during summer, 1972-75. Graph compares average summer and winter values for 1965-71 with those for 1972-75. (Source: Ref. 20)*

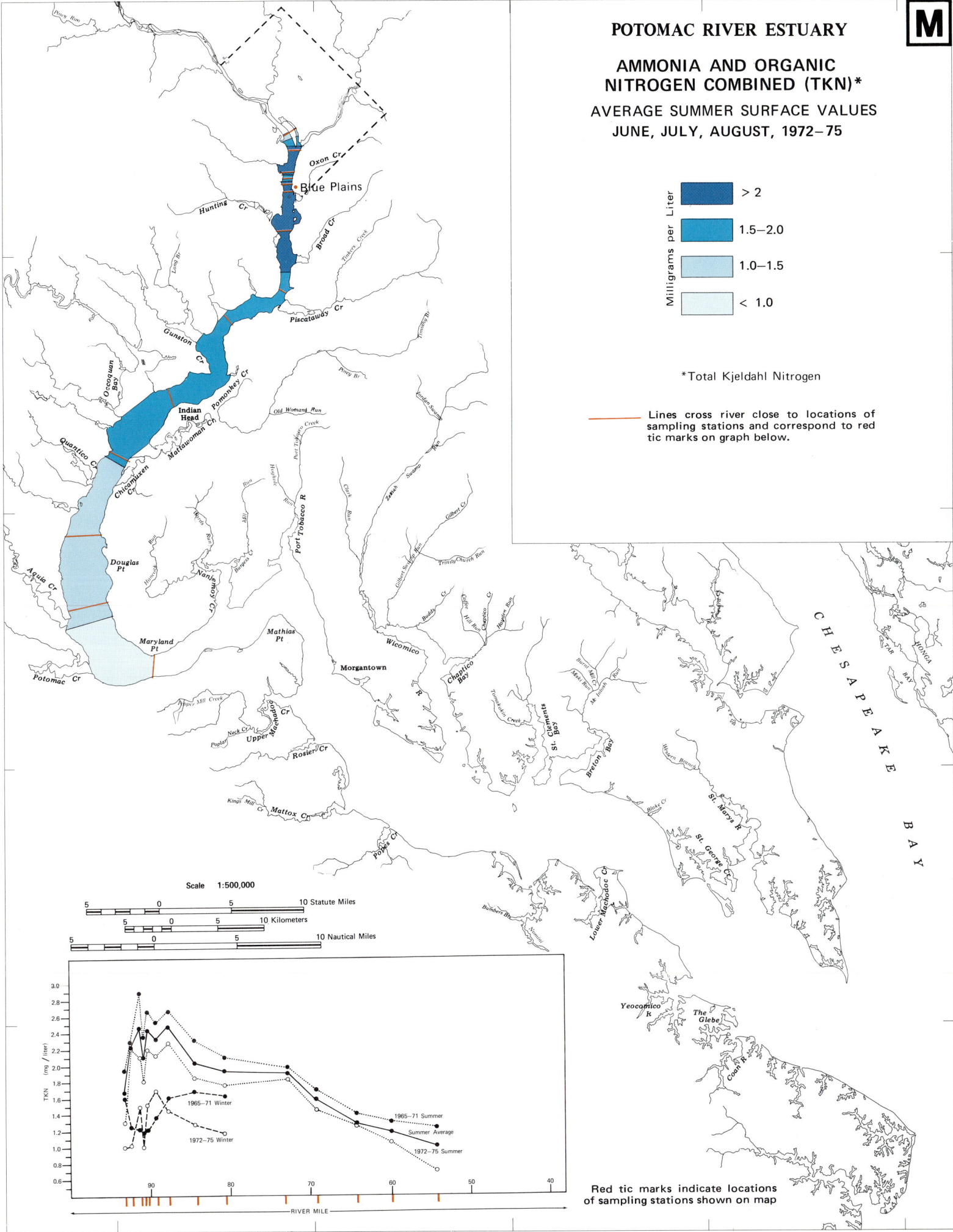

Figure 3-21. *Average combined ammonia and organic nitrogen values in surface waters of the upper Potomac estuary during summer, 1972-75. Graph compares average summer and winter values for 1965-71 with those for 1972-75. (Source: Ref. 20)*

POTOMAC RIVER ESTUARY

M

NITRITES PLUS NITRATES
(NO_2^- + NO_3^-)

AVERAGE SUMMER SURFACE VALUES
JUNE, JULY, AUGUST, 1972—75

Milligrams per Liter

> 0.75

< 0.75

Critical level of NO_2^- + NO_3^- at which algal blooms are induced is 0.35 mg/liter.

Lines cross river close to locations of sampling stations and correspond to red tic marks on graph below.

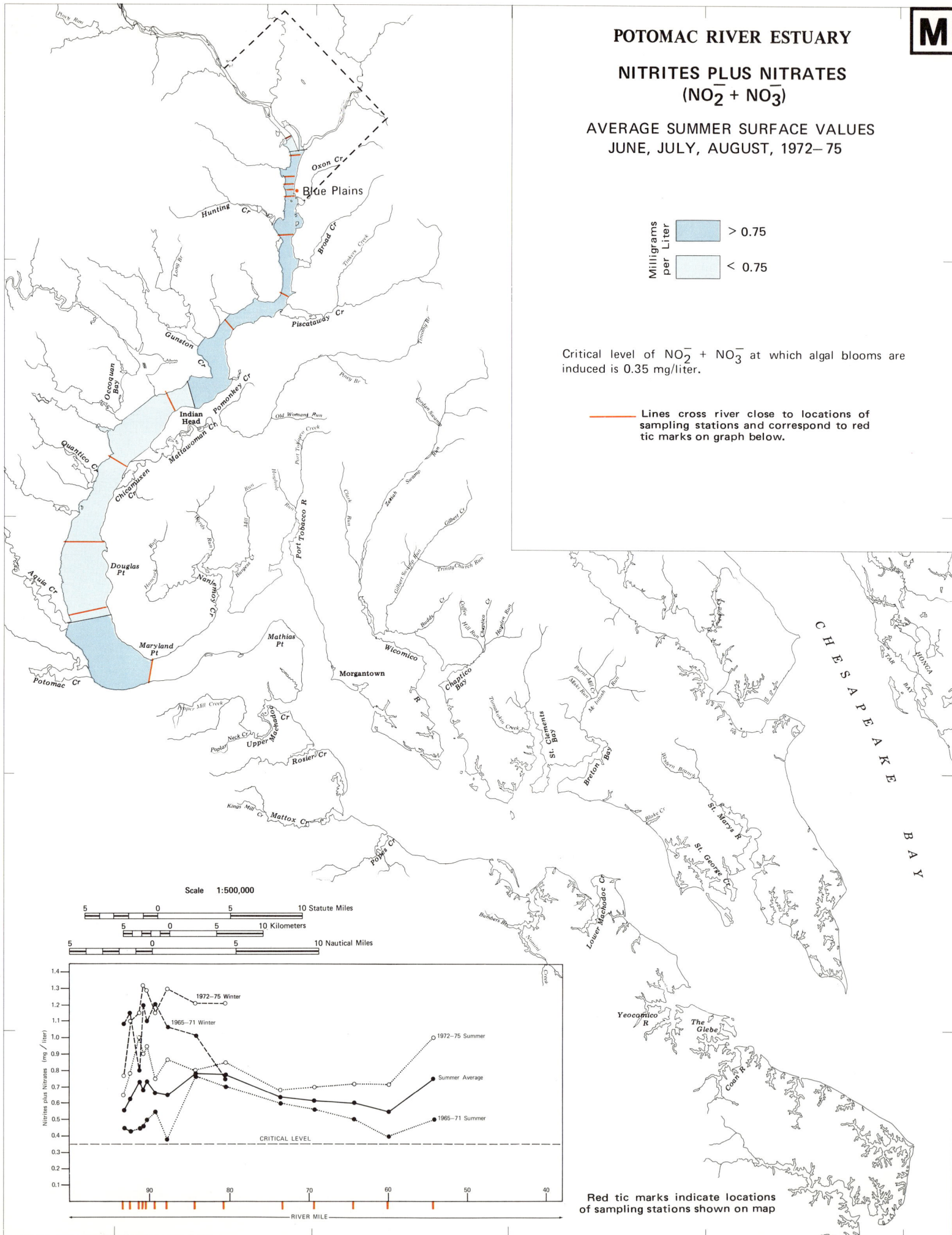

Scale   1:500,000

5   0   5   10 Statute Miles

5   0   5   10 Kilometers

5   0   5   10 Nautical Miles

Red tic marks indicate locations of sampling stations shown on map

1972–75 Winter

1965–71 Winter

1972–75 Summer

Summer Average

1965–71 Summer

CRITICAL LEVEL

RIVER MILE

Figure 3-22. *Average nitrite plus nitrate values in surface waters of the upper Potomac estuary during summer, 1972-75. Graph compares average summer and winter values for 1965-71 with those for 1972-75. (Source: Ref. 20; critical level from Ref. 24)*

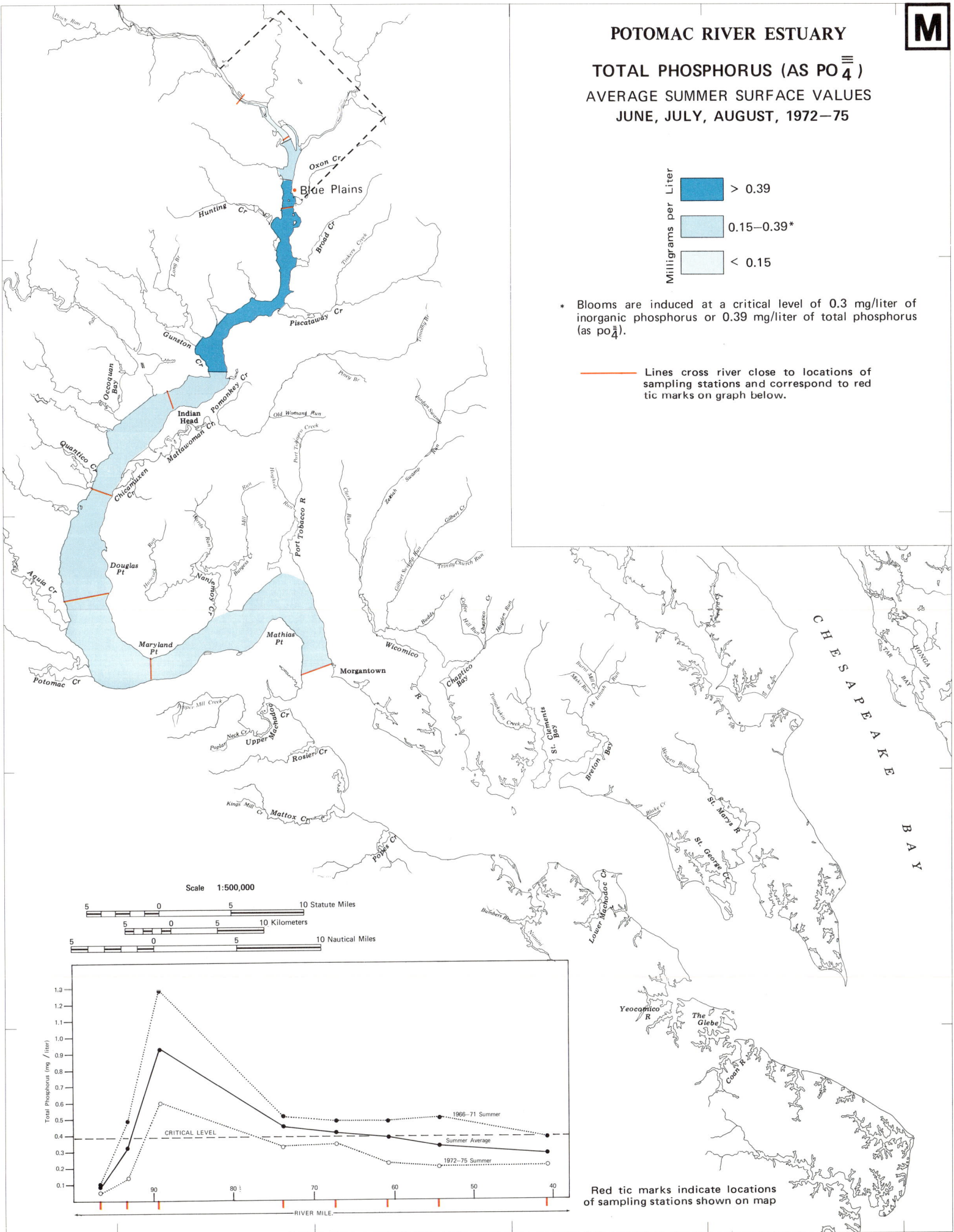

POTOMAC RIVER ESTUARY

TOTAL PHOSPHORUS (AS PO$_4^{\equiv}$)

AVERAGE SUMMER SURFACE VALUES
JUNE, JULY, AUGUST, 1972—75

Milligrams per Liter

> 0.39

0.15—0.39*

< 0.15

* Blooms are induced at a critical level of 0.3 mg/liter of inorganic phosphorus or 0.39 mg/liter of total phosphorus (as po$_4^{\equiv}$).

Lines cross river close to locations of sampling stations and correspond to red tic marks on graph below.

Red tic marks indicate locations of sampling stations shown on map

Figure 3-23. *Average total phosphorus values in surface waters of the upper Potomac estuary during summer, 1972-75. Graph compares average summer values for 1966-71 with those for 1972-75. (Source: Ref. 20; critical level from Ref. 24)*

POTOMAC RIVER ESTUARY

M

DISSOLVED OXYGEN
AVERAGE SUMMER SURFACE VALUES
JUNE, JULY, AUGUST, 1972–75

Milligrams per Liter

> 6.0

< 6.0

——— Lines cross river close to locations of sampling stations and correspond to red tic marks on graph below.

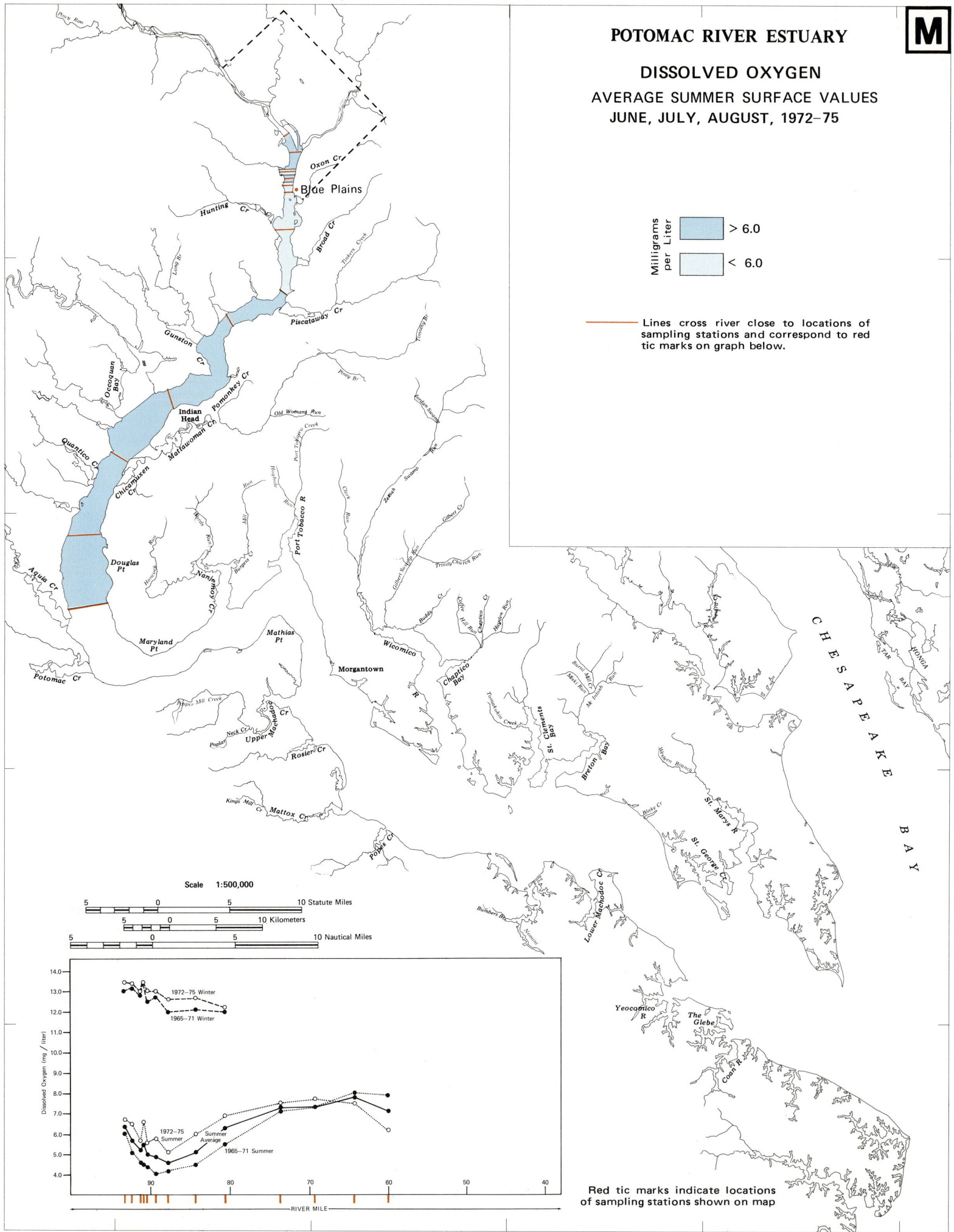

Blue Plains

Indian Head

Red tic marks indicate locations of sampling stations shown on map

Dissolved Oxygen (mg / liter)

14.0
13.0
12.0
11.0
10.0
9.0
8.0
7.0
6.0
5.0
4.0

1972–75 Winter
1965–71 Winter

1972–75 Summer
Summer Average
1965–71 Summer

90   80   70   60   50   40
RIVER MILE

Scale   1:500,000

5   0   5   10 Statute Miles
5   0   5   10 Kilometers
5   0   5   10 Nautical Miles

CHESAPEAKE BAY

Figure 3-24. *Average dissolved oxygen values in surface waters of the upper Potomac estuary during summer, 1972-75. Graph compares average summer and winter values for 1965-71 with those for 1972-75. (Source: Ref. 20)*

POTOMAC RIVER ESTUARY

M

CHLOROPHYLL *A*

AVERAGE SUMMER SURFACE VALUES
JUNE, JULY, AUGUST, 1972–75

Micrograms per Liter

> 25

< 25

Lines cross river close to locations of sampling stations and correspond to red tic marks on graph below.

Bloom conditions are considered to exist when chlorophyll *a* levels reach 50 μg per liter or greater.

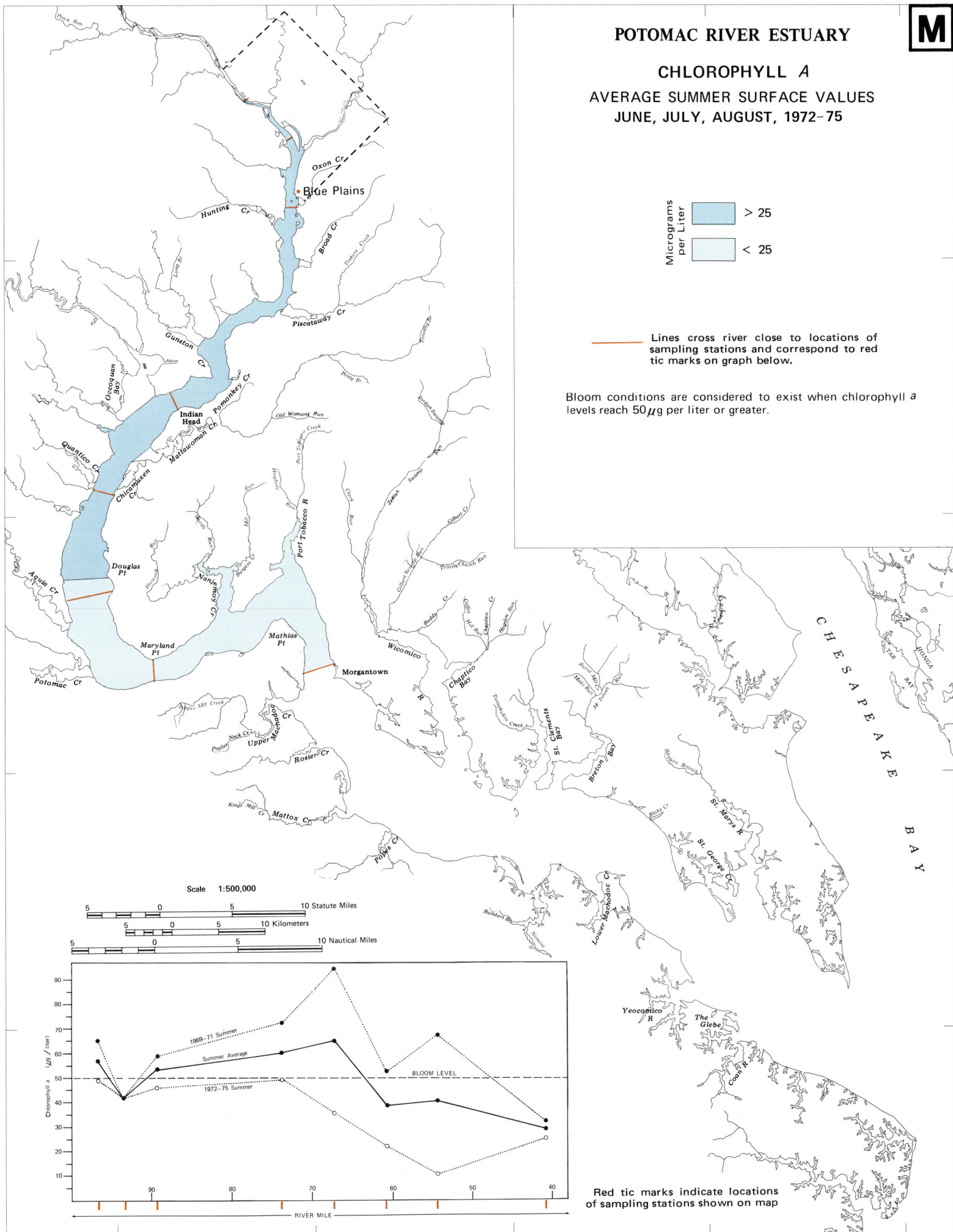

Scale   1:500,000

5   0   5   10 Statute Miles

5   0   5   10 Kilometers

5   0   5   10 Nautical Miles

1969–71 Summer

Summer Average

1972–75 Summer

BLOOM LEVEL

Chlorophyll *a* (μg / liter)

RIVER MILE

Red tic marks indicate locations of sampling stations shown on map

Figure 3-25. *Average chlorophyll a values in surface waters of the upper Potomac estuary during summer, 1972-75. Graph compares average summer values for 1969-71 with those for 1972-75. (Source: Ref. 20; bloom level from Ref. 24)*

under these enriched conditions are primarily freshwater blue-green and green algae and some submerged rooted aquatic plants.

The effects of enrichment in the Potomac became evident in the 1920s when a submerged rooted aquatic plant, water chestnut, proliferated to such an extent that mechanical removal was required to control its spreading (see Chapter 5). By the 1950s and 1960s, another rooted aquatic plant, Eurasian water milfoil, became dominant, filling many embayments from Indian Head to the mouth.[37] These nuisance species were generally not suitable foods for herbivores. Depleted oxygen levels also accompanied the overabundant growth and subsequent decay of these plants.

Around 1965, when water milfoil populations declined dramatically, possibly because of a virus, blue-green algal blooms (primarily of *Anacystis* species) started to appear. Gas-filled vacuoles in the bodies of these plants buoyed them to the surface where they formed massive rafts that were carried downstream by surface flows. Oxidation of dead and dying algae by microbial and chemical processes created a high biochemical oxygen demand. At the same time, oxygen replenishment from the atmosphere was obstructed by the thick surface growths. These growths have caused problems in the past, with persistent spring and summer blooms extending from Washington, D.C., downstream to Maryland Point and occasionally to Morgantown. Large masses of blue-green algae (most notably, species of *Anacystis*) have become less evident in the Potomac in the last few years, although localized concentrations are still seen. They are a nuisance and aesthetically objectionable, discouraging swimming and other recreational uses, and are possibly toxic to other organisms.[37]

The presence of *Anacystis* blooms is reflected in measurements of high levels of chlorophyll a (Fig. 3-25). The downstream extension of these blue-green freshwater species ceases in the transition zone between oligohaline and mesohaline waters, even though levels of nutrients there can support algal blooms. The decline in chlorophyll a levels in this region is caused not by the waning of nutrient concentrations as much as by the transition from fresh to brackish water algal species.

It would be expected that the lower nutrient concentrations in the lower estuary would be reflected by lower chlorophyll a values. In 1965 and 1966,[35] summer values there were only up to 30 μg/liter, while winter values remained below 10 μg/liter.

During the 1972-75 period, chlorophyll a averaged below the designated bloom level of 50 μg/liter along the entire upper estuary.[20] The estuary upstream from Indian Head (nautical river mile 75), however, continued to have periodic algal blooms and average chlorophyll a values close to problem levels. Inorganic nitrogen levels remained relatively high, and phosphorus and organic nitrogen levels, although somewhat reduced, were still present in the upper 25 miles of the estuary. High levels in the upper estuary remain a matter of concern.

Water Quality Standards

The State of Maryland has set several water quality standards for various classes of waters,[32] which are defined by their contemplated usage as follows:

Class I — Protected for use as water for contact recreation; for fish, other aquatic life, and wildlife

Class II — In addition to Class I safeguards, protected for shellfish harvesting

Class III — In addition to Class I safeguards, protected as natural trout waters

Class IV — In addition to Class I safeguards, protected as recreational trout waters

The major water quality parameters considered by the State are dissolved oxygen, bacterial content, pH, and temperature. Table 3-8 relates the water use classes to the desired values of these parameters. Regulations governing other surface water quality standards are published by federal and state agencies and should be consulted for details.

Table 3-8 — Maryland Water Quality Standards for Different Water Use Classes [a]

Dissolved Oxygen (mg/liter)

Class I	4.0 Minimum 5.0 Minimum daily average
Class II	4.0 Minimum 5.0 Minimum daily average
Class III	5.0 Minimum 6.0 Minimum daily average
Class IV	4.0 Minimum 5.0 Minimum daily average

Bacteria [b]

Class I	Fecal coliform: Max. log mean = 200 MPN/100 ml
Class II	Total coliform: Max. median = 70 MPN/100 ml Max. 10% > 230 MPN/100 ml (5 tubes) or > 330 MPN/100 ml (3 tubes)
Class III	Fecal coliform: Max. log mean = 200 MPN/100 ml
Class IV	Fecal coliform: Max. log mean = 200 MPN/100 ml

pH

All classes	6.5 - 8.5

Temperature [c]

Class I	90°F max. (32°C) [d]
Class II	90°F max. (32°C) [d]
Class III	68°F max. (20°C) [d]
Class IV	75°F max. (23.9°C) [d]

(a) Standards for dissolved oxygen, bacteria, and pH from Ref. 32. Temperature standards given are as amended in Ref. 33.

(b) **Fecal coliform** — bacteria that characteristically inhabit the intestines of warm-blooded animals; considered indicators of recent fecal pollution.
MPN — most probable number per 100 ml of sample.
Total coliform — bacteria found in soil and runoff, on plants and insects, and in old sewage; considered indicators of less recent fecal pollution or other pollution sources of nonfecal origin.

(c) Temperature standards concern discharge of heated water and apply to the receiving waters outside of an immediate mixing zone.

(d) Or the ambient temperature of the receiving waters, whichever is greater.

The waters of the Potomac estuary main stem are designated by the State as Class I above Simms Point (at the mouth of Aquia Creek) and as Class II below Simms Point. Within the jurisdiction of the District of Columbia, six classes for water use are designated:

A — Water contact recreation
B — Wading
C — Fish and wildlife propagation
D — Recreational boating
E — Maintenance of fish life
F — Industrial water supply

For example, at Hains Point and the Woodrow Wilson Bridge, the water use classes are D, E, and F.

The Interstate Commission on the Potomac River Basin (ICPRB) makes periodic water quality assessments based on measurements at specific points along the Potomac. Their assessments consider the water quality components shown in Table 3-9. Figure 3-26 characterizes the water quality along the entire length of the main stem by extrapolating from measurements taken at the six stations used by the ICPRB in their assessments.

POTOMAC RIVER ESTUARY

WATER QUALITY STATUS

M

Fair—Good

Good

Good—Excellent

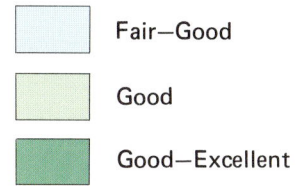

Unmarked Tributaries

Rock Creek: Poor—Fair
Hunting Creek: Poor
Remainder: No Data

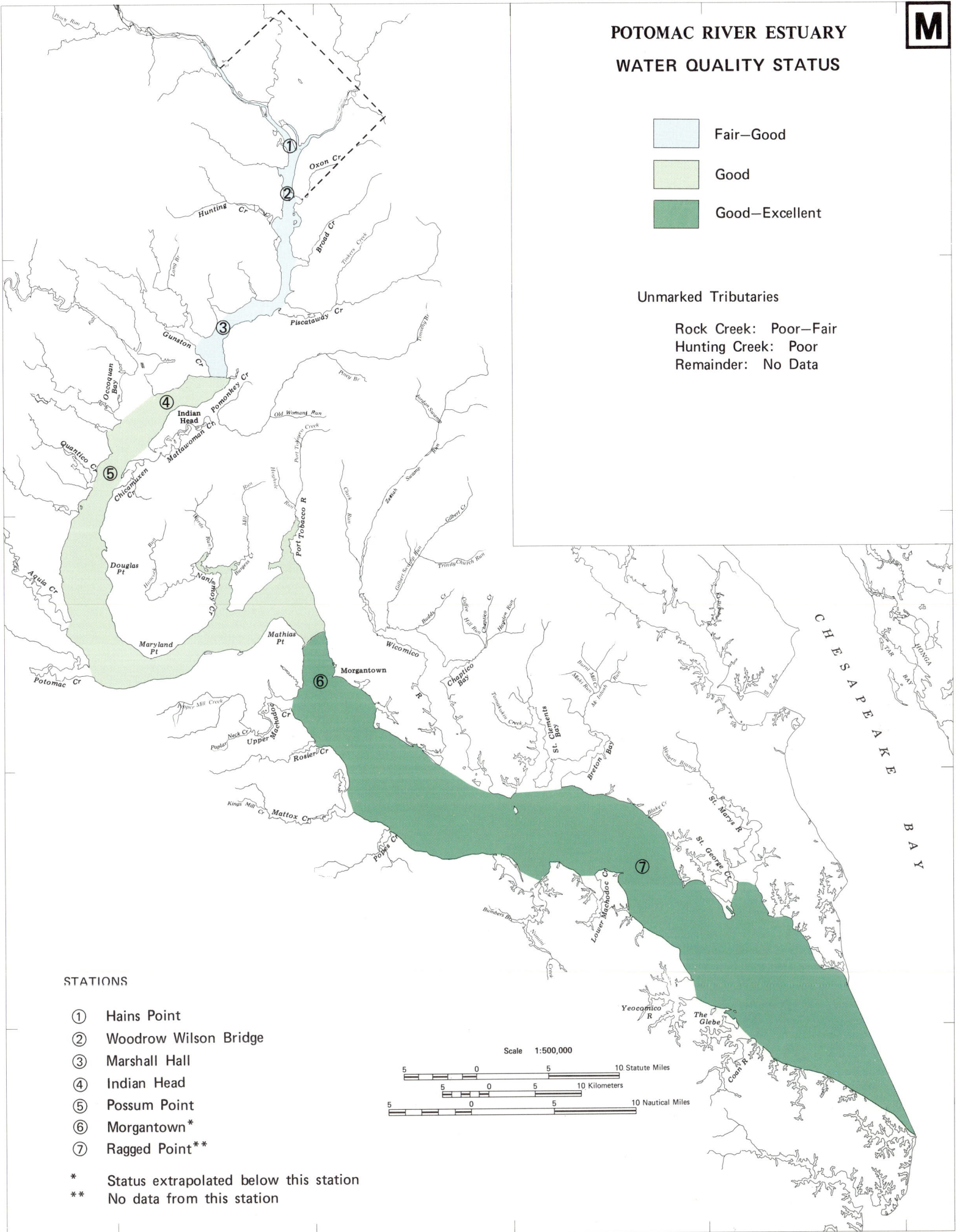

STATIONS

① Hains Point

② Woodrow Wilson Bridge

③ Marshall Hall

④ Indian Head

⑤ Possum Point

⑥ Morgantown*

⑦ Ragged Point**

* Status extrapolated below this station

** No data from this station

Scale 1:500,000

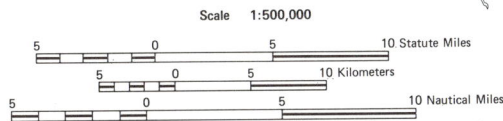

Figure 3-26. *General characterization of water quality along the Potomac main stem and some tributaries. Status designations are extrapolated from data collected at seven stations from 1975-76. (Source: Ref. 38)*

Table 3-9 — **Criteria for Classifying the Status of Water Quality In the Potomac** [a]

Parameter [b]	Status			
	Excellent	Good	Fair	Poor
pH [c]	6.9-8.0	8.0-8.5 or 6.5-6.9	8.5-9.5 or 5.5-6.5	>9.5 or <5.5
Dissolved oxygen (mg/liter) [d]	8.0-9.5	6.0-8.0	4.0-6.0	<4.0
Suspended solids (mg/liter)	<20	20-80	80-400	>400
Total organic carbon (mg/liter)	<5	5-20	20-35	>35
Nitrate-nitrogen (mg/liter) [e]	<0.2	0.2-0.6	0.6-2.0	>2.0
Total phosphorus as P (mg/liter) [e]				
Streams	<0.05	0.05-0.25	0.25-1.00	>1.0
Reservoirs, lakes, and ponds	<0.01	0.01-0.10	0.10-0.60	>0.6
Chlorophyll a (micrograms/liter)	<10	10-50	50-125	>125
Fecal coliform bacteria [f]	<200	200-1,000	1,000-5,000	>5,000

(a) Derived from Table 1 in Ref. 38 and Table 3 in Ref. 39.

(b) Parameters are based on fresh water at sea level and 25°C.

(c) pH values between 6.9 and 8.0 are considered optimal for aquatic life; pH values less than 5.5 or greater than 9.5 adversely affect fish spawning.

(d) DO concentrations between 8 and 9 milligrams per liter (mg/liter) at 25°C represent optimal conditions for aquatic life (zero mortality).

(e) Algal growth may be stimulated by inorganic nitrogen levels > 0.3 mg/liter in lakes and by total phosphorus levels of 0.01 to 0.03 mg/liter in lakes and > 0.1 mg/liter in streams.

(f) A most probable number of 200 bacteria per 100 milliliters of water is the fecal coliform standard for primary contact recreation (e.g., swimming, wading, waterskiing) at most stations.

Chapter 3 — References

1. *U.S. Department of the Interior, 1965-1977*
2. *Schubel, 1972*
3. *Todd, 1970*
4. *Mason, 1974*
5. *Mihursky et al., 1976*
6. *U.S. Department of the Army, 1974*
7. *Maryland Department of State Planning and the Chesapeake Bay Interagency Planning Committee, 1972*
8. *Bongers et al., 1975*
9. *U.S. Department of Commerce, 1977b*
10. *Cronin, W. B., 1971*
11. *Polgar, Ulanowicz, Pyne, and Krainak, 1975*
12. *U.S. Department of Commerce, 1977a*
13. *Mason and Flynn, 1976*
14. *U.S. Environmental Protection Agency, 1969-1972*
15. *Maryland Department of Natural Resources, 1969-1974*
16. *Lippson, R. L., 1969-1971*
17. *The Johns Hopkins University, 1949-1972*
18. *Maryland Department of Natural Resources, 1974-1976*
19. *Andersen et al., 1973*
20. *U.S. Environmental Protection Agency, 1965-1975*
21. *U.S. Department of the Army, 1973*
22. *Beaven, 1954*
23. *Maryland Geological Survey, 1975*
24. *Palmer, 1975*
25. *Allee, Kelso, and Rosenblatt, 1976*
26. *McCaw, 1978*
27. *Munkittrick, 1976*
28. *Washington Post, 1974*
29. *Strealy, 1978*
30. *Bratina, 1978*
31. *Moore, G.N., 1978*
32. *Mason, 1977*
33. *Maryland Department of Natural Resources, 1978a*
34. *Jaworski, Lear, and Aalto, 1969*
35. *Carpenter, Pritchard, and Whaley, 1969*
36. *Taft and Taylor, 1976*
37. *Jaworski, Lear, and Villa, 1971*
38. *Rasin, 1977*
39. *Mason, Rasin, McCaw, and Flynn, 1976*

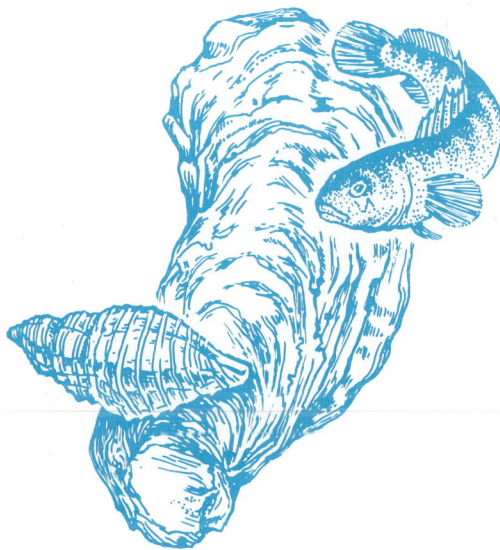

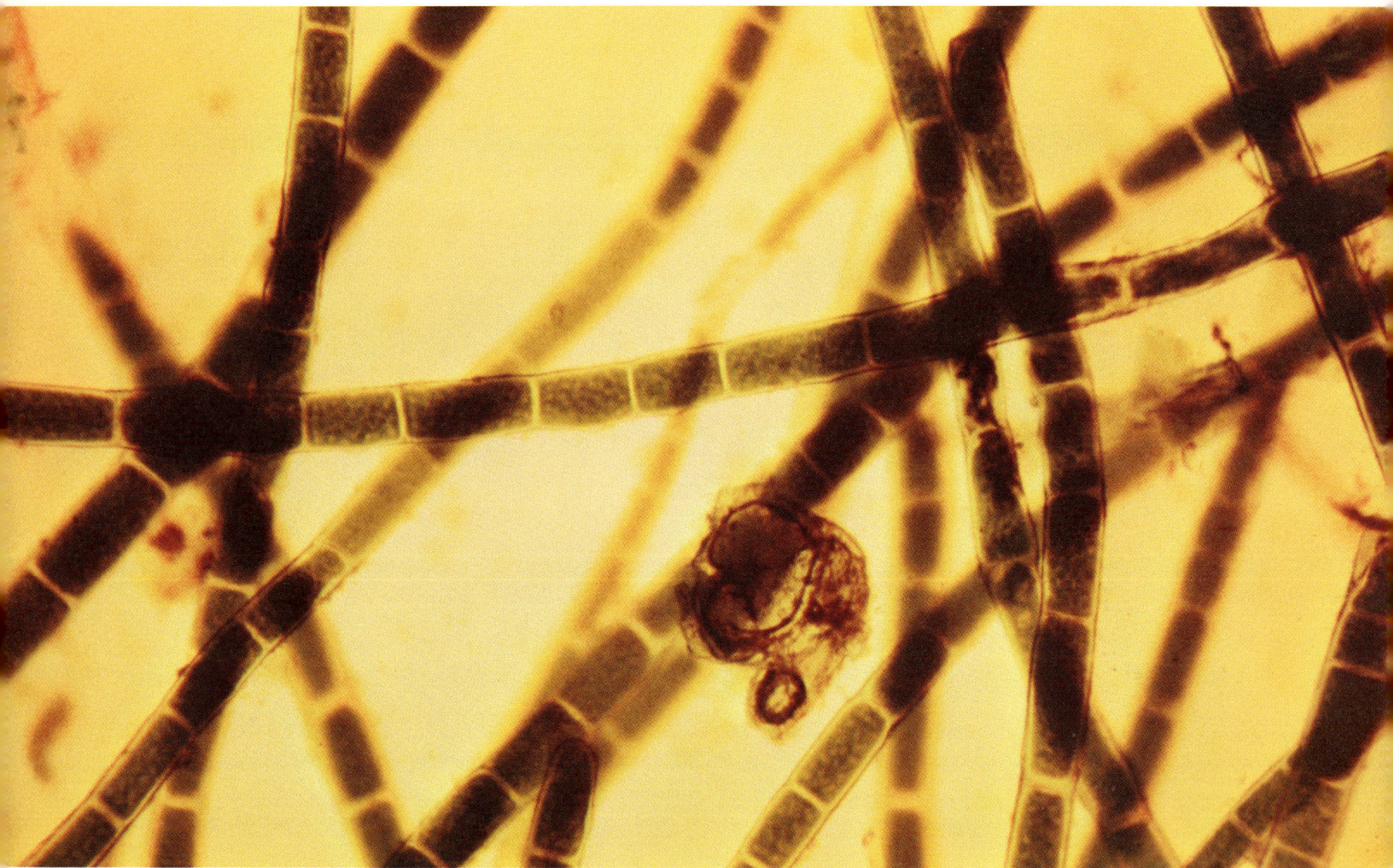

Phytoplankton and Other Algae

Phytoplankton and Other Algae

Algae, the simplest of all aquatic plants, have no true leaves, roots, or stems. They include free-floating communities of microscopic cells called phytoplankton (from the Greek, meaning "plant wanderers"), bottom-dwelling communities of both unicellular and colonial microscopic benthic algae, and the macroscopic seaweeds.

Algae, like all green plants, are primary producers. That is, they use the energy of the sun (through photosynthesis) to produce organic material and oxygen from carbon dioxide and water. Most of the synthesized organic material is used by the algae for maintaining metabolic processes, growth, and reproduction, but some is stored in the cells. Algae are consumed, in turn, by animals — primary consumers — that become food for organisms higher in the food web — secondary consumers — and so on).

Algae and other primary producers may be consumed while they are living, or they may be consumed after they die and begin decomposing. In the Potomac estuary, the food web primarily involves the consumption of living material. Decaying plant and animal matter, or detritus as it is usually called, is of secondary importance to the maintenance of the aquatic food web in this estuary. The most important primary producers are the microscopic phytoplankton, millions of which can be present in a liter of water.

PHYTOPLANKTON

Phytoplankton comprise an abundant and remarkably diverse assemblage of organisms. They float freely in the water column as single cells or small colonial plants, and their movements are primarily determined by estuarine circulation. At times, they may be so abundant that the water becomes colored by their cells, and massive floating mats may carpet vast expanses of the water surface.

Phytoplankton are constantly growing, reproducing, and dying, and their standing crop, or biomass, at any one time or in any one location is highly variable. Zooplankton — the primary consumers of phytoplankton — also have high reproductive rates, and their populations may respond rapidly to changes in phytoplankton productivity. The periods of highest zooplankton reproduction are usually triggered when phytoplankton biomass is greatest. When large populations of zooplankton graze heavily on phytoplankton, the result may be a dramatic decrease in phytoplankton biomass. Subsequently, zooplankton levels decline. Thus, seasonal zooplankton cycles are usually closely tied to corresponding phytoplankton cycles. Some large filter-feeding benthic organisms (such as oysters) also use phytoplankton for food, as do plankton-eating fish such as menhaden.

Phytoplankton are subdivided into groups by size: net plankton includes organisms larger than 60 micrometers (μm), the nannoplankton are between 5 and 60 μm, and ultraplankton are less than 5 μm. It has been estimated that nannoplankton and ultraplankton account for well over 50 percent of the total primary production in most bodies of water.[1] Hence, they play a very important role in the productivity of aquatic systems. However, because of the difficulties encountered in capturing, handling, and identifying the smaller forms, most phytoplankton surveys in the Potomac estuary have either ignored the smaller nannoplankton and ultraplankton, or have lumped them in a separate "unidentified" category. Thus, relatively little is known of their abundance, distribution, and role in the estuarine ecosystem. The following discussion is, therefore, based largely on information from surveys of net plankton and larger nannoplankton.

Factors Affecting Distribution of Phytoplankton

The temporal and spatial distributions of phytoplankton are governed by physical and chemical factors such as salinity, light, temperature, nutrients, and circulation, and by biological factors such as zooplankton grazing.[2][3] These elements interact in a complex and not fully understood fashion to produce fairly predictable gross geographical distributions and seasonal cycles of phytoplankton in any estuary.

Salinity is the major factor influencing the geographical distribution of phytoplankton species along the length of the Potomac estuary. Potomac phytoplankton communities are characteristically composed of three distinct groups of organisms, each inhabiting a specific salinity regime: 1) riverine species that can tolerate only slightly saline conditions; 2) estuarine species that primarily occur in the oligohaline and meshohaline regions; and 3) marine forms that are restricted to the higher salinities in the lower reaches of the estuary.

Within a given salinity zone, light and temperature are the major factors influencing the productivity and abundance of phytoplankton. The amount of light reaching the water surface varies with season, as does the amount of light that penetrates the water and is actually usable by phytoplankton. Because the Potomac is a turbid estuary, sufficient light for photosynthesis only penetrates a relatively short distance below the surface of the water. This portion of the water column, called the euphotic zone, extends deeper during the winter months when turbidity is lowest. Even though there is sufficient light for reproduction in winter, the phytoplankton standing crop is low because the rate of phytoplankton growth and reproduction depends strongly on water temperature. During warm months, cells may divide several times per day, while, at lower temperatures, cells divide at a much slower rate. Some may not divide for several days. Seasonal changes in phytoplankton abundance thus reflect the sequence of changing light and temperature conditions.

Concentrations of nutrients, especially of nitrogen and phosphorus, also have an important effect on the occurrence and the magnitude of phytoplankton productivity. The increase or decrease of nutrients often causes rapid increases and decreases in phytoplankton populations from one season to the next, or from one locality to another.

The rate of phytoplankton production can be measured in several ways, each based on the materials manufactured (carbohydrates and oxygen) and consumed (carbon dioxide) in the photosynthetic process.

Carbon dioxide uptake or oxygen production are usually reported as equivalent carbon produced per unit volume of water per unit time, or carbon produced per unit volume of water per unit time, or carbon produced per unit water surface area per unit time. Changes in the concentration of chlorophyll a (the major photosynthetic pigment in most plants) over time are also indicative of productivity, but more generally, chlorophyll a concentrations are used as a measure of plant biomass.

Biomass Distribution in the Potomac Estuary

In general, phytoplankton biomass is greatest in the uppermost portion of the Potomac estuary, from Washington, D.C. to Indian Head, where nutrient levels are generally highest (Figs. 3-20 through 3-23). From Indian Head downstream to Maryland Point, phytoplankton biomass gradually declines[4] because steadily increasing salinities inhibit the growth of freshwater phytoplankton populations, and salinities are too low to support any sizeable population of estuarine species. Lowest phytoplankton biomass and abundance generally occur in the transition region between the oligohaline and mesohaline salinities, from Maryland Point to slightly below Mathias Point (Figs. 3-25, 4-1, and 4-2).[4] [5] [13] Both freshwater and estuarine phytoplankton species are under stress in this area. Increasing cell counts observed at the downstream limit of this region (in the vicinity of Popes Creek, Maryland)[13] reflect the gradually increasing abundance of estuarine species.

Few surveys of phytoplankton populations or of primary productivity have been made in the mesohaline region of the Potomac estuary. Therefore, conditions there must be inferred from information gathered in other environmentally similar regions of the Chesapeake Bay. One such region is Calvert Cliffs, located on the western side of the Bay, approximately 30 miles north of the mouth of the Potomac. Over much of the year, chlorophyll a concentrations at Calvert Cliffs are comparable to those in the lower Potomac estuary.[19] It can be assumed that species found at Calvert Cliffs would be representative of those in the lower Potomac estuary.

Seasonal Abundances

As in other estuarine and marine waters,[20] the Potomac phytoplankton community follows a fairly predictable cycle of abundance from one season to the next. Typically, there are two seasonal peaks or blooms that occur throughout most of the estuary — a major one in spring and a smaller one in fall (Fig. 4-3).

During spring, relatively high concentrations of dissolved nutrients, coupled with increasing light and rising water temperatures, cause rapid phytoplankton growth and reproduction. Cells multiplying at an astronomical rate quickly deplete the dissolved nutrients, and reproduction slows down. At the same time, grazing zooplankton populations, responding to the pulse of available food organisms, begin to multiply rapidly and feed in increasing numbers on the phytoplankton

bloom. The result is a predictable decrease in phytoplankton numbers in the early summer months. Throughout the summer (when the water column is often stratified), exchange between the nutrient-rich bottom layers and the nutrient-depleted surface layers is inhibited, and phytoplankton productivity remains low. However, when wind- and tide-driven turbulence mixes the water mass in the fall, phytoplankton abundance and production increase again, although generally not nearly to the same extent as during the spring bloom. This fall bloom is terminated by decreasing temperatures and light levels, and phytoplankton abundances are lower during the winter months when both temperature and daylight are at a minimum. During the winter, a well-mixed store of dissolved inorganic nutrients builds up in the estuary for use by phytoplankton in the spring. Dramatic differences in this typical cycle in the Potomac occur from place to place or from year to year, depending on climate, nutrient enrichment, and other factors, many not entirely understood.

Seasonal fluctuations in phytoplankton abundances are usually most pronounced in the tidal fresh and oligohaline portions of the estuary (Fig. 4-3). Spring blooms in this region commonly peak at approximately 50 to 100 million cells per liter.[5] [14] [21] During late summer, total numbers of cells drop to as low as several hundred thousand per liter. Fall blooms are fairly predictable and average 5 million cells per liter.[5]

Seasonal phytoplankton cycles in the middle portion of the estuary (the low salinity portion of the mesohaline zone) generally follow those of the upper, tidal fresh portion (Fig. 4-3), although secondary fall blooms may be very slight or absent in some years, depending on freshwater input and the amount of salt intrusion. During spring blooms, total phytoplankton cell counts in most of this oligohaline-mesohaline transition region may exceed 30 million per liter, while those in the downstream portion of the region, near Morgantown, may reach over 40 million cells per liter.[13]

A somewhat different pattern of phytoplankton abundance is thought to occur in the lowermost portion of the Potomac, based on results of studies in the Calvert Cliffs area of the Chesapeake Bay.[19] Annual average abundances in this mesohaline area may be higher than in the oligohaline and oligohaline-mesohaline transition regions of the Potomac estuary, but seasonal changes in cell densities are usually smaller. Also, spring blooms tend to occur somewhat later than in the upper estuary and would most likely produce average densities of only 20 to 25 million cells per liter.[15-19]

If phytoplankton abundance is examined on the basis of biomass (i.e., chlorophyll a concentration) rather than on the basis of cell density, a somewhat different picture of distribution in the estuary emerges (see Fig. 3-25, Chapter 3). Biomass in the upper estuary is nearly twice as high as in the lower (mesohaline) regions, with minimum biomass occurring in the oligohaline-mesohaline transition zone. Such a difference could result from variations in the mean size of the phytoplankters common to these areas.

POTOMAC RIVER ESTUARY

I

ALGAE
SUMMER

PHYTOPLANKTON

Moderate to high abundances of freshwater blue-green and green algae

Proliferation of blue-green algal blooms

Gradually decreasing concentrations of freshwater green and blue-green algae

Transition from freshwater to estuarine algal forms

Increasing dominance of estuarine and marine dinoflagellates and diatoms

SEAWEEDS

General distribution of *Ulva lactuca* and *Enteromorpha prolifera*

CHESAPEAKE BAY

Scale 1:500,000

5   0   5   10 Statute Miles
5   0   5   10 Kilometers
5   0   5   10 Nautical Miles

RELATIVE ABUNDANCE

GREEN ALGAE

BLUE-GREEN ALGAE

DIATOMS

DINOFLAGELLATES AND RED FLAGELLATES AND GOLDEN-BROWN ALGAE

BLUE

Hains Point   Indian Head   Possum Point   Maryland Point   Mathias Point   Piney Point

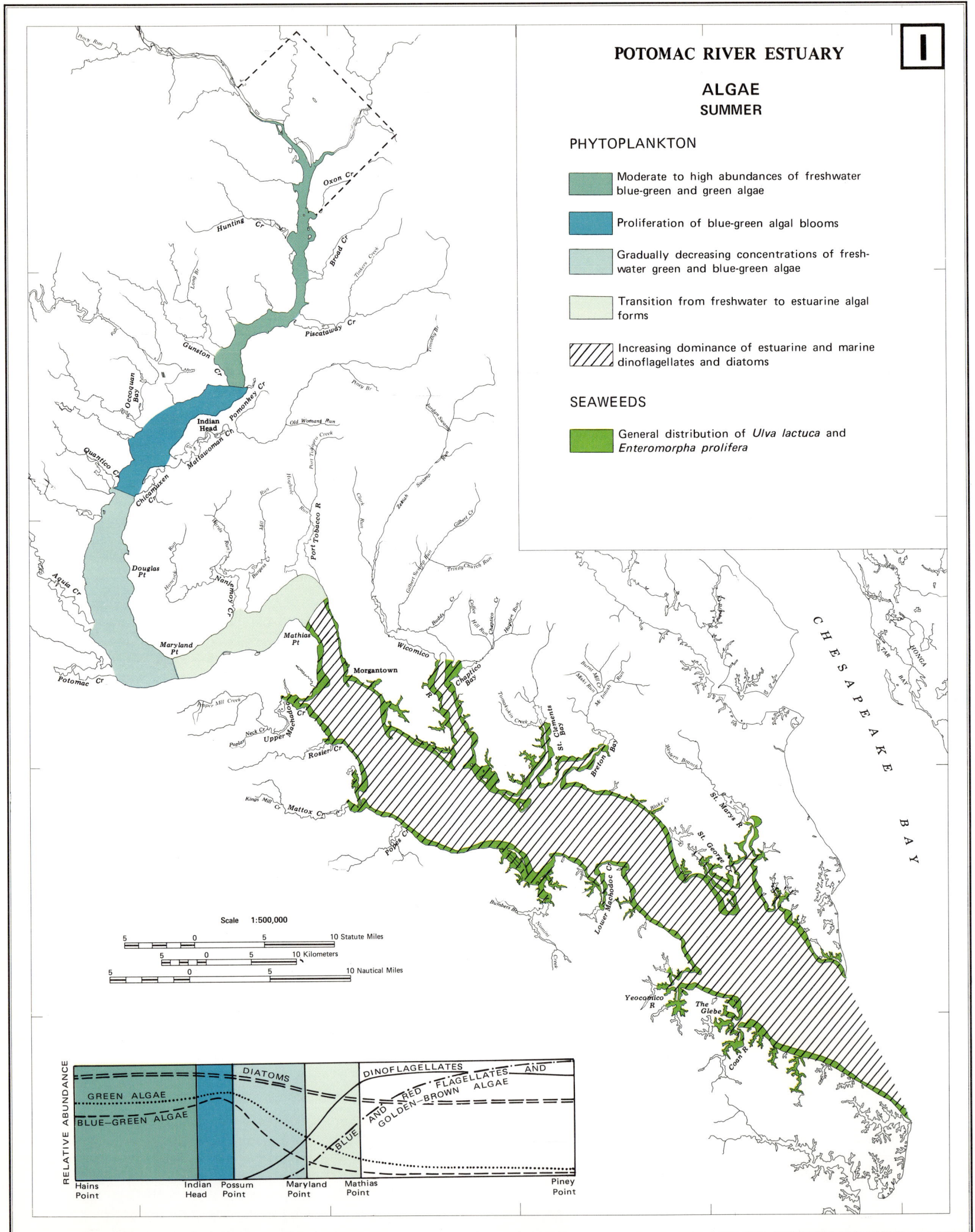

Figure 4-1. *Typical summer distributions of major phytoplankton groups and common seaweed species in the Potomac estuary. (Source: Refs. 5-12)*

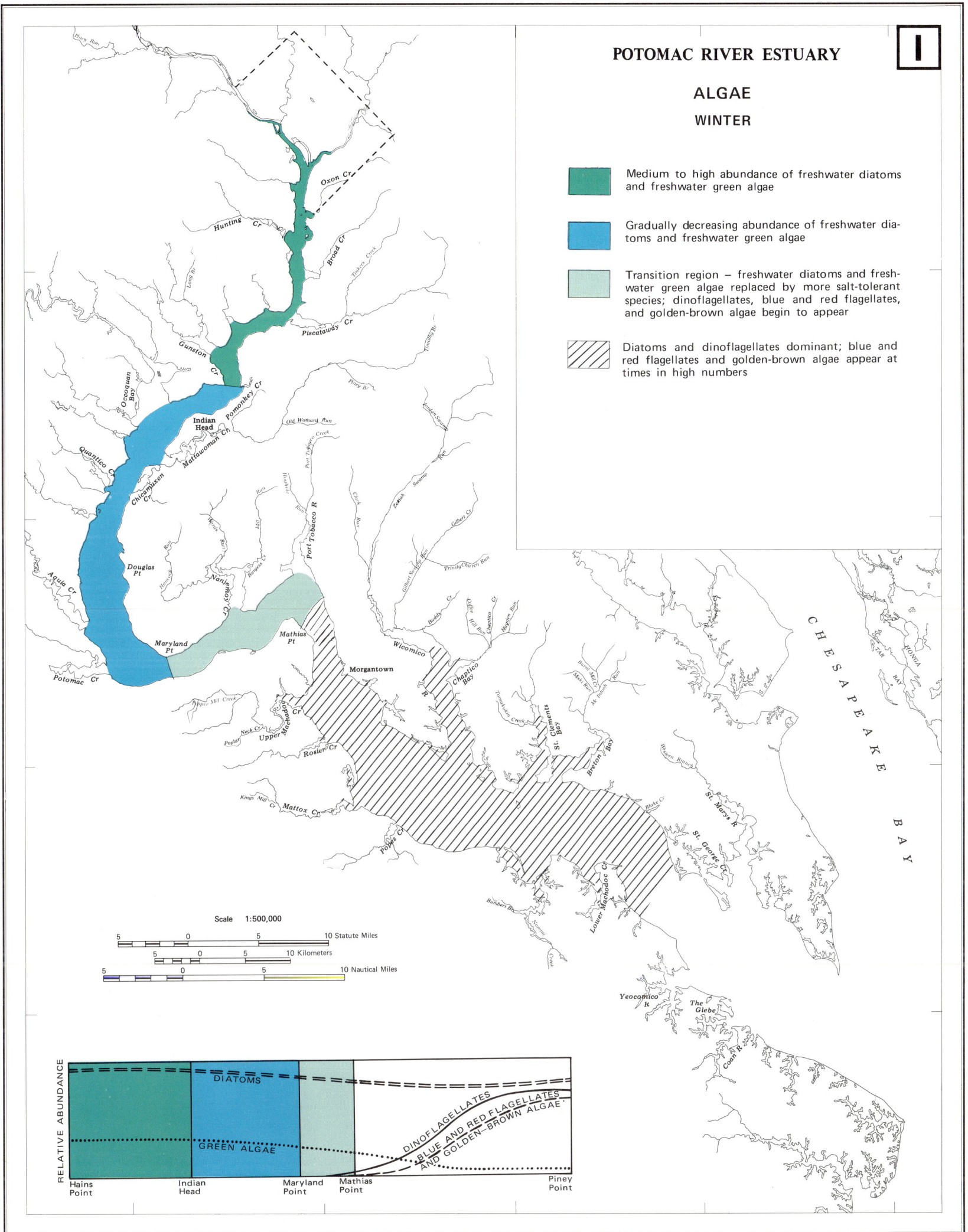

Figure 4-2. *Typical winter distributions of major phytoplankton groups in the Potomac estuary. (Sources: Refs. 5,6,13-18)*

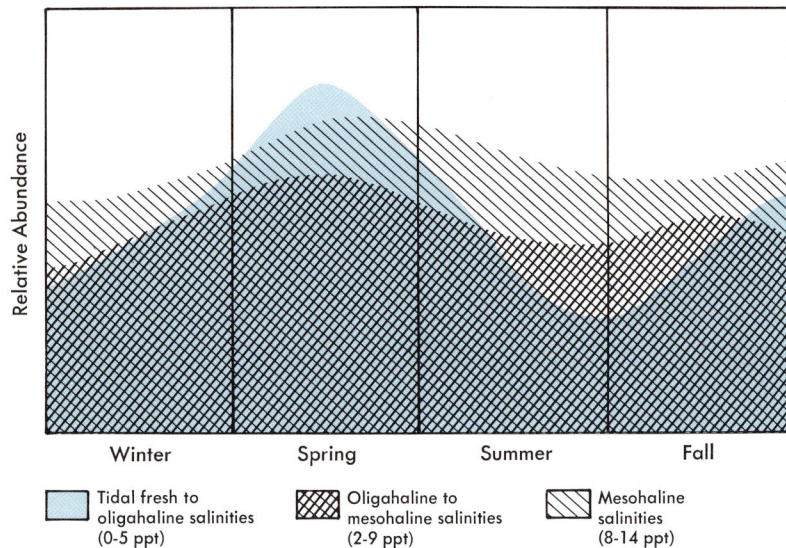

GREEN ALGAE

Carteria cordiformis

Colonial *Eudorina elegans*

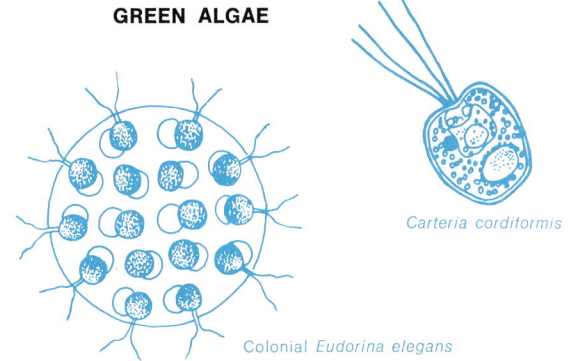

Figure 4-3. *Typical seasonal patterns of phytoplankton biomass in three regions of the Potomac estuary.*

Legend:
- Tidal fresh to oligahaline salinities (0-5 ppt)
- Oligahaline to mesohaline salinities (2-9 ppt)
- Mesohaline salinities (8-14 ppt)

Phytoplankton Classifications, Seasonal Occurrences, and Distributions

Phytoplankton communities in the various regions of the Potomac estuary are generally composed of members of eight taxonomic groups or "divisions":

Green algae	— Chlorophyta
Blue-green algae	— Cyanophyta
Diatoms	— Bacillariophyta
Golden-brown algae	— Chrysophyta
Yellow-green algae	— Xanthophyta
Dinoflagellates	— Pyrrophyta
Euglenoids	— Euglenophyta
Blue and red flagellates	— Cryptophyta

Since the salinity tolerance of a species or taxonomic group largely governs its distribution within the estuary, the species or taxonomic composition of the Potomac phytoplankton communities would be expected to change with location along the estuary. However, no single estuary-wide study of phytoplankton distribution by taxonomic group has ever been done in the Potomac. Intensive taxonomic studies have been carried out in only two locations: in the vicinity of Douglas Point, where the salinity regime changes from tidal fresh (0 to 0.5 ppt salinity) in the spring to oligohaline (0.5 to 4 ppt salinity) in summer and fall; and in the vicinity of Morgantown, where salinities typically vary from 4 or 5 ppt in the spring to 9 or 10 ppt in the fall, making it a low mesohaline area. Other intensive taxonomic work has been done at Calvert Cliffs on the Chesapeake Bay, an area environmentally similar to the lower Potomac mesohaline area and with a similar annual salinity range — 10 to 18 ppt.

Data from these three areas were collected over different seasons and years and by several organizations. Because of the great temporal and spatial variability known to occur in phytoplankton populations, findings of the studies cannot be considered completely representative of phytoplankton distributions in the Potomac estuary. However, the combination of data from all the available studies provides a general characterization of these distributions.

In the fresher portions of the estuary near Douglas Point, diatoms, green algae, and blue-green algae are the dominant forms present (Fig. 4-4). The cell densities of some of these groups in the Morgantown and Calvert Cliffs areas may be as high as or higher than those around Douglas Point (Fig. 4-5) even though proportionately, they make up smaller fractions of the total phytoplankton population. In these higher salinity areas, dinoflagellates, diatoms, blue and red flagellates, and golden-brown algae are the dominant groups, while green algae are a much less significant portion of the total community (Fig. 4-4). Altogether, hundreds of phytoplankton species have been recorded from the Potomac estuary. Appendix Table 1 lists those identified to date.

GREEN ALGAE — CHLOROPHYTA

The most obvious and striking feature of the chlorophyta or green algae is their characteristic grass-green color, created by chlorophyll pigments inside their cells. Green algae are the most prevalent phytoplankton in fresh, relatively unpolluted waters and attain their greatest abundance and diversity there. Less than 10 percent of green

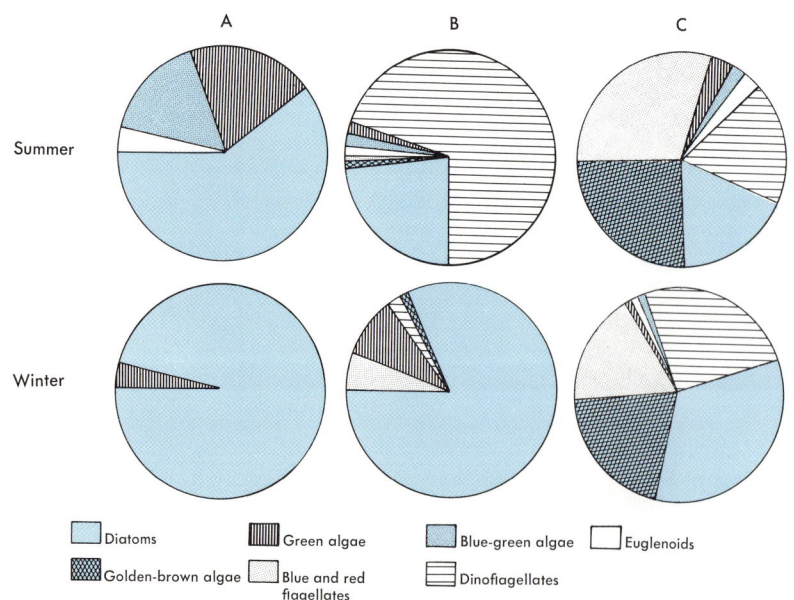

Figure 4-4. *Proportions of each algal group found in summer and winter phytoplankton populations in three salinity regimes. A—Tidal fresh to oligohaline salinites (0-5 ppt) in the vicinity of Douglas Point, 1972-74 (from data reported in Ref. 5). B—Oligohaline to mesohaline salinities (2-9 ppt) in the Mathias Point to Morgantown region, 1974 (from data reported in Ref. 13 for June through December). C—Mesohaline salinities (8-14 ppt) in the Calvert Cliffs region, which has environmental conditions similar to those of the lower Potomac estuary, 1974-77 (from data reported in Refs. 15-18).*

Legend:
- Diatoms
- Green algae
- Blue-green algae
- Euglenoids
- Golden-brown algae
- Blue and red flagellates
- Dinoflagellates

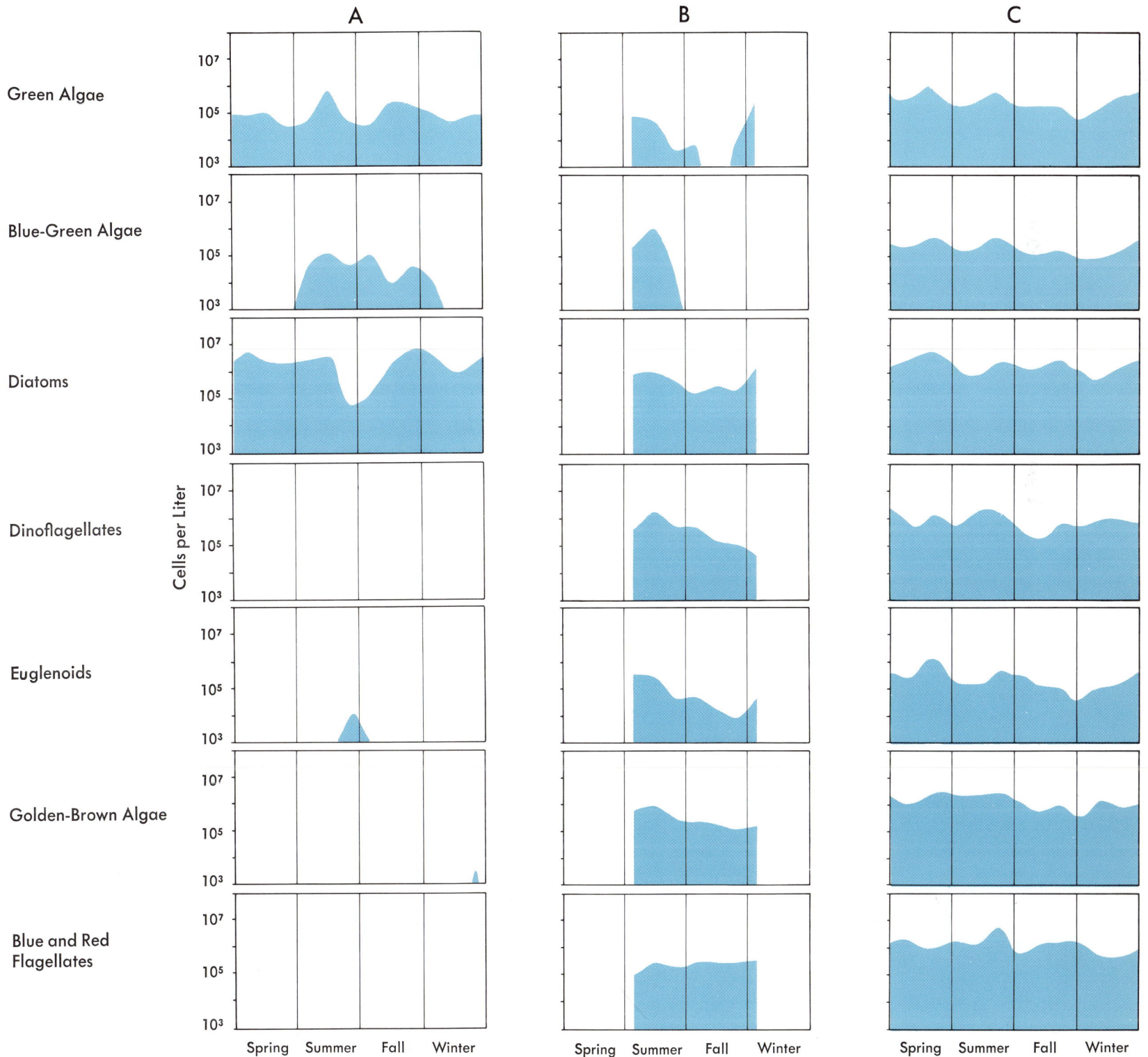

Figure 4-5. Seasonal changes in major phytoplankton groups in three salinity regimes. A—Tidal fresh to oligohaline salinities (0-5 ppt) in the vicinity of Douglas Point, 1972-74 (from data reported in Ref. 5). B—Oligohaline to mesohaline salinities (2-9 ppt) in the Mathias Point to Morgantown region, 1974 (from data reported in Ref. 13 for June through December). C—Mesohaline salinities (8-14 ppt) in the Calvert Cliffs region, which has environmental conditions similar to those of the lower Potomac estuary, 1974-77 (from data reported in Refs. 15-18).

GREEN ALGAE

Staurastrum anatinum

Sphaerocystis schroeteri

Pediastrum duplex

BLUE-GREEN ALGAE

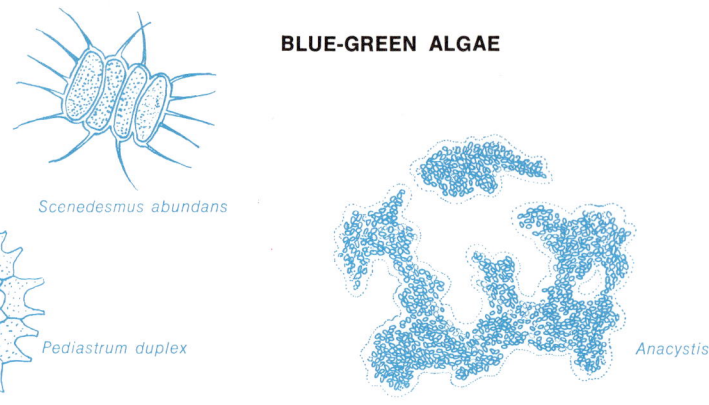

Scenedesmus abundans

Anacystis

algal species are wholly or partially marine.[22] Green algae inhabiting the Potomac estuary include small, spherical or oval unicell types, and a small number of filamentous and spherical colonial types. Although movement of phytoplankton is controlled primarily by estuarine currents and tides, many species of green algae possess one or more whip-like flagella, which enable the cells to orient themselves so they will move towards the more intense light at the surface of the water. In certain species, these flagella may form a ring, or girdle, about the cell. Nonswimming species also have adaptations that maintain them in the upper sunlit layers of the water column. For example, the cell walls of *Staurastrum* spp. are extensively branched, and large numbers of *Sphaerocystis schroeteri* cells are imbedded in buoyant gelatinous masses. Both adaptations decrease sinking rates by increasing surface-to-volume ratios, enabling the cells to remain near the surface for a longer time.

Green algae reproduce asexually by cell division (fission). Sexual reproduction, in which specialized sexual cells are liberated from the parent plant and unite in the water, also occurs. In green algae, as well as many other phytoplankton groups, sexual reproduction takes place only at the end of the normal growing season, during fall and early winter. The resultant spores remain dormant on the bottom through the winter, when environmental conditions are especially harsh,[23] and germinate when environmental conditions are more favorable for growth and reproduction.

Planktonic green algae are abundant and account for almost a third of the total phytoplankton community during the summer months in the upper tidal fresh portion of the Potomac main stem (Figs. 4-1 and 4-2), from about Washington, D.C. downstream to Chicamuxen Creek, and in the tidal fresh regions of the tributaries.[5] From 1972-74, in the vicinity of Douglas Point, they made up about 20 percent of the total phytoplankton community in summer (Fig. 4-4). In winter of these years, the green algae constituted less than 5 percent of the total community in the same area (Fig. 4-4). No distinct seasonal pattern in their abundance is evident in the tidal fresh and oligohaline region near Douglas Point (Fig. 4-5). A few green algae inhabit the Morgantown region, where they are most abundant in winter[7-9][13] (Fig. 4-5). Based on studies at Calvert Cliffs, [15-19] green algae in the lower portion of the Potomac estuary should demonstrate peak abundances in spring, with densities averaging up to a half million cells per liter. However, green algae only constitute a few percent of the total mesohaline phytoplankton community at any time (Fig. 4-4).

Over 100 species of green algae have been recorded from the Potomac estuary (Appendix Table 1), a diversity surpassed only by that of the diatoms. Freshwater communities of green algae typically contain numerous species, with no single species dominating. The three species most consistently encountered in the upper Potomac estuary are *Micractinium pusillum, Pandorina morum,* and a presently unidentified species that at times may account for over 50 percent of the total phytoplankton community.[6] Other common forms are the geometric colonies of *Pediastrum* spp., the spiny colonies of *Scenedesmus* spp., and the slender, needle-shaped cells of *Ankistrodesmus* spp.

BLUE-GREEN ALGAE — CYANOPHYTA

Blue-green algae are both benthic and planktonic, but only a few species are considered to be truly planktonic[24] (see section on microscopic benthic algae in this chapter). The blue-greens are a primitive group of organisms that lack the definite cellular organization characteristic of other algae. Although single-celled forms of blue-green algae occur, most are colonial, including forms covered either by a secreted mucilaginous sheath or by a filamentous coating.[24]

Members of the genera *Anabaena* and *Anacystis* migrate vertically from the bottom using gas-filled vacuoles for buoyancy and are invariably responsible for surface summer blooms.[25][26] As the cells from the blooms die, they and the decomposer microorganisms that feed on them aggregate to form what are known as "water flowers." [25] When congregations of billions of these tiny water flowers entangle with masses of zooplankton and other planktonic organisms near the surface, large floating mats are formed.

During the winter, many of the blue-green forms revert to a benthic existence; cells segregate and secrete protective envelopes around themselves, forming spores that pass the winter in a dormant state on the bottom. In spring, when light and temperature increase, these spores germinate and liberate new cells which once again enter the phytoplankton community.[25]

Blue-green algae are frequently encountered and are sometimes abundant constituents of summer phytoplankton communities throughout tidal fresh and oligohaline waters in the upper half of the estuary (Fig. 4-1). Although abundance of blue-green algae is generally high in the Washington, D.C. area, turbidity tends to inhibit mat formation from D.C. to an area farther downstream, near Mount Vernon.[27] Low numbers of blue-green algae are found during winter and spring (Fig. 4-5). In early summer, blue-greens begin to increase rapidly to average densities of up to 1 million cells per liter.[5] By mid- to late fall, however, numbers of blue-greens start to decline in the upper estuary (Fig. 4-5).

The largest and most massive blooms of blue-green algae usually originate near Indian Head and float downstream with freshwater flow.[27] Under normal conditions, saltier waters in the oligohaline-mesohaline transition zone inhibit the further growth of blue-green populations. For most of the year, blue-greens are a numerically unimportant group in the low mesohaline portion of the estuary near Morgantown (Fig. 4-5). Their appearance there is sporadic and usually occurs during early summer when they may constitute as much as 17 to 24 percent of the total phytoplankton cell counts.[13] In the high mesohaline section of the lower estuary, blue-green algae are not expected to constitute over 2 percent of the total phytoplankton community during most of the year (Fig. 4-4).[15-19]

DIATOMS — BACILLARIOPHYTA

Diatoms are the most ubiquitous and abundant of all phytoplankton in the Potomac estuary. Unlike all other algae, they each have a thin, often highly sculptured, siliceous external skeleton which encloses the cell like an intricately carved glass jewel box. The two halves of the frustule

DIATOMS

GOLDEN-BROWN ALGAE

Ochromonas mutabilis

NAKED DINOFLAGELLATES

YELLOW-GREEN ALGAE

Coscinodiscus

Rhizosolenia

Tribonema bombycinum

Gymnodinium

Exuviella marina

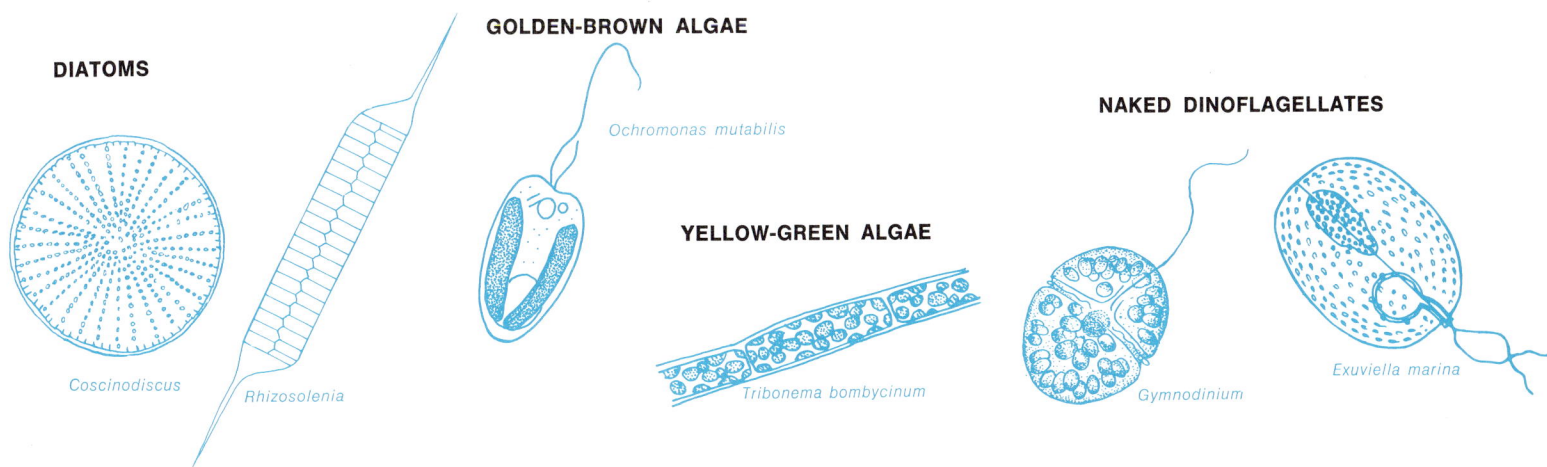

(cell wall) of a diatom fit together like a pair of valves, one inside the other. Diatom skeletons have evolved into a diverse array of shapes, which enables these nonswimming cells to remain suspended in the water column for comparatively long periods of time. Broad flattened discs, as in the genera *Cyclotella* and *Coscinodiscus,* and long needle-shaped cells, as in the genus *Rhizosolenia,* have large surface-to-volume ratios, and therefore do not sink as quickly as more compact forms. Other species have evolved ridges of bristles growing out from the perimeter of the cells, as in the genus *Stephanodiscus,* or parachute-like mucous envelopes, as in the genera *Thalassiosira* and *Fragilaria.* Both of these adaptations slow the rate at which cells sink out of the euphotic zone. Fat globules, which are commonly present in cells, also help to increase buoyancy. Vertical movements of diatoms may result from turbulence. Those that sink permanently below the euphotic zone may serve as food for filter-feeding organisms and detritus eaters.

In reproduction, the two overlapping halves of the diatom separate, and one new siliceous valve is formed in each daughter cell. With progressive divisions, average cell size becomes smaller and smaller, since the smaller valve of a parent diatom becomes the larger valve of one of the daughter cells. When the cells become very small, the diatom throws off its old shell, increases three to four times in size, secretes a membrane around itself, and develops a new and larger shell. This auxospore, as it is called, provides a mechanism for the restoration of original cell size. At that time, normal reproduction begins again.[22][23] Under adverse conditions, certain planktonic diatoms may enter a resting stage, during which the cell falls to the bottom where it remains until suitable conditions return.

During the cooler months (late fall through early summer), when temperatures are below 18 to 20° C, diatoms dominate the phytoplankton community in the upper tidal fresh and oligohaline portions of the estuary (Fig. 4-4). In the spring and late fall, diatom densities in this area generally reach their annual maximum (Fig. 4-5).[5][6] Populations are at lower levels in late summer and early fall.

In the transition region between oligohaline and mesohaline waters, diatom abundances are slightly lower than they are upstream in the tidal fresh and oligohaline region, although the seasonal patterns are similar (Figs. 4-1 and 4-2).[7-9][13] They dominate the phytoplankton community in winter and make up about 20 percent of the community in summer (Fig. 4-4). Studies in the mesohaline salinities at Calvert Cliffs have shown that diatoms are the dominant phytoplankton group in all seasons except summer.[15-19] One could expect a similar trend in the lower Potomac estuary.

GOLDEN-BROWN ALGAE — CHRYSOPHYTA

The tiny microflagellates comprising the golden-brown algae of the division Chrysophyta have only recently been the subject of any extensive research. This component of the nannoplankton makes a significant contribution to the phytoplankton community in the meso-haline portions of the estuary (Fig. 4-4), where densities may exceed 1 million cells per liter during much of the year (Fig. 4-5).

YELLOW-GREEN ALGAE — XANTHOPHYTA

The yellow-green algae typically include only benthic forms and are therefore only infrequently encountered in the phytoplankton. *Tribonema* sp. is a ubiquitous filamentous yellow-green algae that is found in floating masses and growing on submerged surfaces and on rooted aquatics in the Potomac estuary during the cooler months of the year. *Vaucheria* sp. inhabits the bottoms of well-aerated portions of the estuary and the exposed portions of mud flats.

DINOFLAGELLATES — PYRROPHYTA

Dinoflagellates are small, unicellular organisms often found in great abundance in high mesohaline waters, such as those of the lower half of the Potomac estuary. Typically, they are active swimmers, propelling themselves in a spiral path up and down in the water column by means of two flagella protruding from grooves located along the body wall. Many species of dinoflagellates have a light-sensitive organ called an "eyespot," which directs them towards the brighter surface waters. It is there that the largest numbers of dinoflagellates are found.[23] Most species are photosynthetic. However, colorless, nonphotosynthetic forms occur that ingest organic food particles.[22][23]

Most dinoflagellates in the Potomac estuary fall into two orders: the Peridiniales, or armored forms, and the Gymnodiniales, or naked forms. The armored forms found in the Potomac include the more elaborate and striking dinoflagellates (members of the genera *Ceratium, Peridinium,* and *Gonyaulax*) whose cell walls bear mosaics of glass-like hardened plates that may project out into spikes or wing-like points. In fresher waters, these projections may be extensive, increasing surface-to-volume ratios and assisting in flotation of the cells. Naked forms lack these plates and typically include the less specialized forms such as those in the genera *Gymnodinium* and *Katodinium.*[22][23] *Exuviella* and *Prorocentrum* are two genera of dinoflagellates found in the Potomac estuary that are structurally dissimilar from members of the orders Peridiniales and Gymnodiniales.

Dinoflagellates reproduce by splitting in two along their longitudinal axis. Each half forms a new daughter cell which quickly grows to the size of the parent cell. In contrast to diatoms, successive generations do not become smaller because both daughter cells are initially the same size.[22][23]

Although dinoflagellates occur in fresh and brackish waters, they seldom account for more than 1 percent of the total phytoplankton community in the tidal fresh and oligohaline portions of the Potomac estuary, such as the Douglas Point area (Fig. 4-4). In the low mesohaline region near Morgantown, however, dinoflagellate populations become more prominent (Figs. 4-1 and 4-2). Sporadic increases in certain dinoflagellate species from spring through fall cause extensive mahogany-colored surface blooms where cell densities may be as high as 8 to 11 million per liter. In spring and fall, when water temperatures are low, these blooms are usually dominated by the naked flagellate, *Katodinium rotundatum.* In warmer months, *Gymnodinium splendens,* another naked species, is generally the dominant form.[7-9][13] Dino-flagellates make up nearly 70 percent of the phytoplankton community

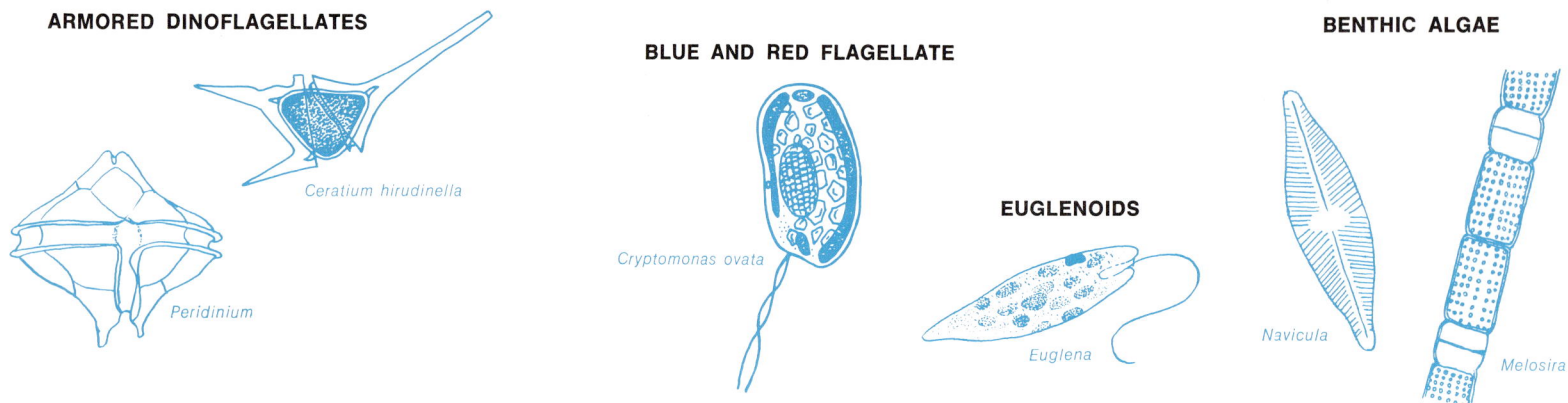

ARMORED DINOFLAGELLATES

Ceratium hirudinella

Peridinium

BLUE AND RED FLAGELLATE

Cryptomonas ovata

EUGLENOIDS

Euglena

BENTHIC ALGAE

Navicula

Melosira

82 in this region in summer, but only around 2 percent in winter (Fig. 4-4).

In mesohaline waters near Calvert Cliffs, where salinities are typical of those in the lower Potomac estuary, dinoflagellates generally exhibit two seasonal blooms — a winter high when densities can exceed 1 million cells per liter and a summer peak when cell counts may surpass 3 million per liter. [15-19] During both of these peaks, species of the genera *Katodinium, Gymnodinium,* and *Prorocentrum* are the dominant dinoflagellates and may account for up to 25 percent of the total numbers of phytoplankton cells (Fig. 4-4). [15-19]

EUGLENOIDS — EUGLENOPHYTA

The euglenoids are one-celled flagellates that can become so abundant locally that the water takes on the color of the dominant species — green in the case of *Euglena viridis* and brown for species of the genus *Trachelomonas.* All species recorded in the Potomac estuary are members of the family Euglenaceae. Most euglenoids are naked, freely swimming forms, which usually move toward light by using one or two flagella. [2,28] Some species (e.g., *Trachelomonas* spp.) are not naked, and the flagellum protrudes through a rigid, open-ended encasement enveloping the cell. [22,23]

Euglenoids reproduce by dividing along their longitudinal axis, either while the cells are swimming, as in species of *Eutreptia,* or after dropping to the bottom and secreting a mucilaginous envelope around themselves. At times, cells may temporarily encyst and undergo a resting period on the bottom. [23]

Euglenoids are found during late summer and early fall in moderate numbers in the tidal fresh and oligohaline portions of the estuary (Fig. 4-5). During late summer to early fall in 1972-74 near Douglas Point, over 20,000 cells per liter were measured, but they did not make up any appreciable portion of the phytoplankton population in that region at any time of the year. [5]

Although generally considered a freshwater group, euglenoids sporadically appear in substantial numbers in the mesohaline reaches of the Potomac estuary (e.g., at Morgantown) [7-9,13] and the Chesapeake Bay (e.g., at Calvert Cliffs). [15-19] Locally, their densities may range from about 2 to 50 percent of the total phytoplankton cell counts. However, even at peak cell densities (usually in the warmer months), this group rarely accounts for more than 1 percent of the total Potomac phytoplankton community (Fig. 4-4).

BLUE AND RED FLAGELLATES — CRYPTOPHYTA

The members of the small and not well-defined plant division, Cryptophyta, are sometimes considered to be related to the dinoflagellates, which they closely resemble. While a few species are colorless, most tend to be pigmented either blue-green, yellow, or reddish-olive. All members of this group found in the Potomac estuary belong to the order Cryptomonadales and are free-swimming, oval or oblong cells (such as *Cryptomonas ovata*). Species may be naked or sheathed in a cellulose membrane. They reproduce by simple cell fission after the cells enclose themselves in mucilage during a resting stage. [22,23]

Cryptophytes are found only occasionally in the upper tidal fresh to oligohaline portions of the Potomac estuary. They are likely to be most abundant in the lower third of the estuary (Figs. 4-1, 4-2, and 4-5). In the middle reaches of the estuary, they are at times common during the summer, and they make up 3 to 4 percent of the total phytoplankton crop in winter (Fig. 4-4). [13] In high mesohaline salinities, cryptophytes are the dominant organisms of the phytoplankton community in the summer, and generally account for around 20 percent of the cells in cooler seasons (Fig. 4-4). [15-19]

MICROSCOPIC BENTHIC ALGAE

The bottom of the estuary provides a variety of substrates for the growth of an algal flora as numerous and diverse as that in the water column. Benthic algae range in size from microscopic unicells to larger colonial forms and provide food for many invertebrates and bottom-oriented fish. [29,30]

Microscopic benthic algae consist mainly of unicellular and colonial blue-green algae and diatoms, although a few species of other groups, such as some golden-brown algae, are also found. Typically, these microscopic forms are found on surfaces of partially or wholly submerged rocks, driftwood, metal objects, and other debris. Blue-green algae form the dark greenish line frequently seen on mud banks of the intertidal zone, while diatoms frequently form brown films on the surface of sediments in shallow water. Some genera, such as the blue-green *Oscillatoria,* form long, matted strings of cells attached directly to the substrate. Others, such as the blue-green *Nostoc* spp., enclose themselves in thick gelatinous masses that lie on the substrate surface. Certain diatoms, such as species of *Gomphonema, Navicula,* and *Licmophora,* may grow at the end of long stalks of excreted mucus that extend some distance above the substrate.

Algae that grow attached to other submerged plants are referred to as epiphytes. Many microscopic benthic algae, such as the diatoms *Cocconeis* spp., are epiphytic, growing on seaweeds and rooted aquatic plants in marshes and other habitats. Algae that characteristically live on rocks and stones are called lithophytes. However, most of the algal species commonly referred to as lithophytes are actually epiphytes living on already established lithophytic plants.

A large number of the microscopic algae found on the bottom of the Potomac estuary are only temporarily benthic. Some are planktonic forms that have been deposited onto the bottom by estuarine currents. Other normally planktonic types, such as the motile green algae *Chlamydomonas* spp. and the golden-brown alga *Chromulina paescheri,* lose their flagella and sink to the bottom to reproduce. [23,31] After dividing vegetatively and forming gelatinous masses of cells, they sprout new flagella and again assume a planktonic existence. Certain phytoplankton become part of the microscopic benthic flora during their resting stages, primarily during the winter when conditions in the surface waters are most harsh. Many of these encyst in a self-secreted mucilage until warm spring waters dissolve the mucus and liberate the cells from the substrate. [23,31]

SEAWEEDS

Ectocarpus tomentosus

Polysiphonia

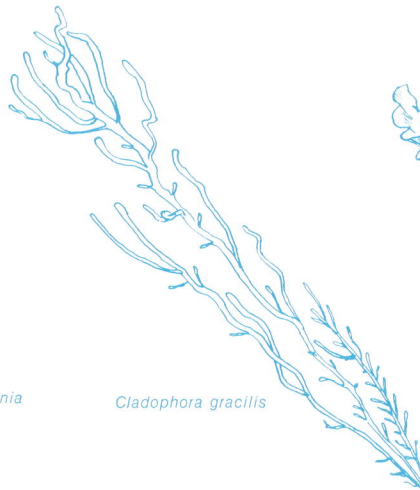

Cladophora gracilis

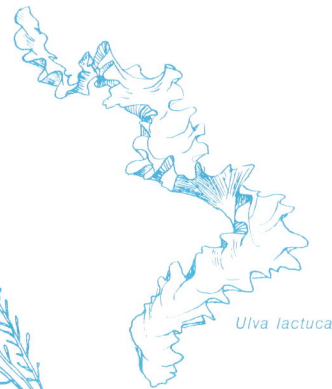

Ulva lactuca

The composition of microscopic benthic algal communities, like that of the phytoplankton, changes with salinity from the head of the estuary to the mouth. Summer communities in the tidal fresh portion of the estuary are dominated primarily by chains of the colonial diatom *Melosira* sp., gelatinous blobs of the blue-green algae *Anacystis* spp. and *Anabaena* spp., and filaments of the blue-greens *Oscillatoria* spp. and *Schizothrix* sp.[32] Often, large plaque-like growths of the golden-brown alga *Chromulina pascheri* crowd out other species. Phytoplanktonic greens (Chlorophyta), such as *Cosmarium* spp. and *Closterium* spp., can frequently be found attached to the substrate surface. In early spring and fall, species of the diatoms *Navicula, Fragilaria,* and *Asterionella* are abundant among the larger attached algae. At various times, euglenoids (*Euglena* sp.) and other planktonic flagellates — such as the green algae *Gonium* sp. and the dinoflagellate *Gyrodinium* sp. — also appear in moderate to low numbers in the microscopic benthic flora.[32]

The higher salinities of the middle portion of the estuary are associated with a significant reduction of blue-green algae and with greater dominance of diatoms in the microscopic benthic flora.[33] In winter, as the abundance of other algal groups declines, diatoms become increasingly dominant. Dinoflagellates and euglenoids may also be found as temporary inhabitants in this middle region of the estuary.[33] Very little is known about microscopic benthic algae in the lower estuary.

MACROSCOPIC ALGAE OR SEAWEEDS

The term "seaweeds," as it is used here, applies not only to marine-oriented species but also to all macroscopic algal forms found in the Potomac estuary. The vast majority of seaweeds comprises three main groups of algae: Chlorophyta (greens), Phaeophyta (browns), and Rhodophyta (reds).

Most seaweeds grow attached to a solid substrate. They generally attach themselves at the beginning of their growth by adhesive, root-like holdfasts or by basal discs made of specialized attachment cells.[25] Sandy and soft mud substrates are generally unsuitable for seaweeds. Most species are lithophytic or epiphytic. Other substrates to which seaweeds attach in sandy and muddy areas include the dead shells of molluscs; the hard, calcarious coverings of barnacles; the carapaces of crabs; the skeletons of hydroids and bryozoans; and many submerged man-made structures. Since light penetration in the Potomac estuary is limited, seaweeds are found in shallow, nearshore zones, particularly in waters less than 12 feet deep.[34]

The specific role of seaweeds in the estuarine ecosystem is not well understood, but it is known that these forms provide habitats for bacteria, protozoans, and small invertebrate organisms that are eventually consumed by larger animals. Seaweeds also serve as a substrate for many epiphytic microscopic benthic algae, which then become available to grazers such as snails and other molluscs. Along with the bacteria that they attract, the tissues of dead seaweeds contribute to the detrital supply of the estuary.

Knowledge of the ecology of seaweeds in the Potomac estuary is meager. Extensive and routine systematic collections of Potomac seaweeds have not been made, and only information about the summer seaweed communities is available.[30] However, it is known that macroalgal communities exhibit seasonal changes throughout the Chesapeake Bay area. Seaweeds may be present all year long or may appear and disappear in the benthic flora from season to season, depending on their patterns of reproduction.[35] Among the seaweeds, generations alternate between sexual and asexual forms. These forms may be similar in appearance or may be quite different in size, shape, and associated structures. The degree of their variability is related to the complexity of the life history of the particular species.[25] [31] Like other forms of algae, seaweeds develop their sexual reproductive organs during the last part of their growing season. Within the Chesapeake Bay system, the seaweed community tends to exhibit two peaks of reproductive activity — one in summer to early fall, and one during winter.[35]

Certain seaweeds have become an increasing problem for humans. For instance, decaying *Ulva lactuca* and *Enteromorpha* spp. contribute to the degradation of water quality in certain parts of the estuary by creating high nutrient loadings.[34] Of all nuisance "weeds" in the Potomac, *Ulva lactuca,* or sea lettuce as this species is aptly described, ranks as one of the worst offenders. Because their holdfast attachments are easily broken, vast quantities are frequently uprooted by currents and cast ashore during warm months. Substantial windrows of rotting sea lettuce accumulating on shorelines release noxious hydrogen sulfide gas and often become the breeding areas for mosquitoes.[34]

Problems with sea lettuce have been encountered along the middle to lower portion of the Potomac, especially in Popes Creek, Virginia, and in Lower Machodoc Creek and at the mouth of the Wicomico River. The areas along the shoreline between Cobb Island and Piney Point, Maryland, and between Popes Creek and Lower Machodoc Creek are considered to be regions of potential *Ulva* growth under proper conditions.[34]

Chapter 4 — References

 1. Van Valkenburg, 1972
 2. Fogg, 1965
 3. Raymont, 1963
 4. U.S. Environmental Protection Agency, 1965-1975
 5. Ecological Analysts, Inc., 1974
 6. Dahlberg, 1973
 7. Academy of Natural Sciences of Philadelphia, 1968
 8. Academy of Natural Sciences of Philadelphia, 1971b
 9. Academy of Natural Sciences of Philadelphia, 1972
10. Mulford, 1972
11. Mulford and Van Valkenburg, 1973
12. Flemer, 1973
13. Mountford, K., 1977
14. Ecological Analysts, Inc., 1978
15. Kachur, 1977
16. Morrill, 1975
17. Morrill and Kachur, 1976
18. Kachur, 1978
19. Mountford, K., et al., 1976
20. Russell-Hunter, 1970
21. Simmons and Armitage, 1972
22. Chapman, 1964
23. Fritsch, 1956
24. Dawson, 1966
25. Fritsch, 1952
26. Pheiffer, 1976
27. Lear and Smith, 1976
28. Bold, 1973
29. Ott, 1972
30. Krauss and Orris, 1972
31. Prescott, 1968
32. Spoon, 1975
33. Reimer, 1977
34. U.S. Department of the Army, 1973
35. Zaneveld and Barnes, 1965

Wetlands and Submerged Vegetation

Wetlands and Submerged Vegetation

Wetlands include all marshes, swamps, and other land-water interface areas that receive enough moisture and sunlight to support growths of specially adapted vegetation. Wetland plants are dependent upon permanent submersion in standing water or upon periodic inundation by tides or rainfall. They range in their dependence on water from the completely submerged rooted aquatic plants that grow along the shorelines to the emergent grasses, reeds, and broad-leaved plants of lowland marshes, and to the shrubs and trees of inland swamps (Fig. 5-1). Although submerged rooted aquatics frequently grow in association with emergent marsh vegetation, they also grow in large isolated beds in open water. Therefore, they are treated separately and in more detail in the last section of this chapter.

WETLANDS

Wetlands border the shorelines along most of the main stem and all tributaries of the Potomac estuary. They range in size from tracts of several hundred acres to narrow borders only a few yards wide. There are about 15,000 acres[1] of wetlands (of which approximately 8,700 acres are coastal wetlands[2]) along the Maryland portion of the Potomac estuarine watershed, but they constitute only a small fraction (just over 2 percent) of Maryland's Potomac estuarine drainage area. In contrast, wetlands constitute almost 33 percent of the entire land area of Dorchester County on the lower Eastern Shore of Maryland.[3] These Eastern Shore wetlands are the major contributors to the nutrient supply of the estuarine regions they border. Potomac estuary wetlands, on the other hand, contribute relatively little to the primary production of the Potomac estuarine system, being vastly overshadowed in this role by the estuarine phytoplankton (see Chapter 4). Nevertheless, the many scattered wetlands along the Potomac play a substantial role in the energy and nutrient balance of localized communities.

In the open waters of the Potomac estuary, the bulk of the detritus (decaying organic material) flushed out of the wetlands is not utilized directly as a food source, but must first be converted into available forms by decomposer microorganisms (Fig. 5-2). As wetland plants die, they are mechanically broken down into smaller and smaller pieces by wave action and animal activity. Decomposer microorganisms, such as bacteria and fungi, colonize the material derived from the wetlands and, in the process, release inorganic nutrients that can be used by phytoplankton and other wetland plants. Many of these decaying plant particles and the microorganisms that proliferate on them are ingested by filter-feeding and detritus-eating organisms such as zooplankton, oysters, and other benthic invertebrates. Parts of the detrital load may remain in the system for long periods and, by continually decomposing and releasing nutrients, may support a detrital food chain year-round.

The wetland habitat supports a diverse group of animals. Some spend all of their lives in wetland areas, feeding on the abundant detritus, phytoplankton, zooplankton, and small invertebrates. The tangles of rooted aquatic plants provide cover and food for small forage fishes. These plants as well as molluscs, worms, other small invertebrates, and fish inhabiting wetlands are consumed by a wide variety of birds, mammals, reptiles, and amphibians. In spring and through summer, wetland waters serve as nursery grounds for some important sport and commercial fish and shellfish, such as alewives, menhaden, and blue crabs.

Wetlands benefit the estuary in other ways. Aquatic plants, which are alternately submerged and exposed by tides and rainfall, act as buffers against the force of storm tides and waves, thus impeding shoreline erosion. Continual tidal flushing transports nutrients to open water areas beyond the wetland. Wetlands also act as filtering basins that collect large amounts of sediments and pollutants originating upstream. Water from land runoff and tides is percolated through and temporarily stored in the wetland substrate. As the stored water is gradually washed out, the diluted waterborne contaminants are flushed downstream and out of the estuary in less harmful concentrations.

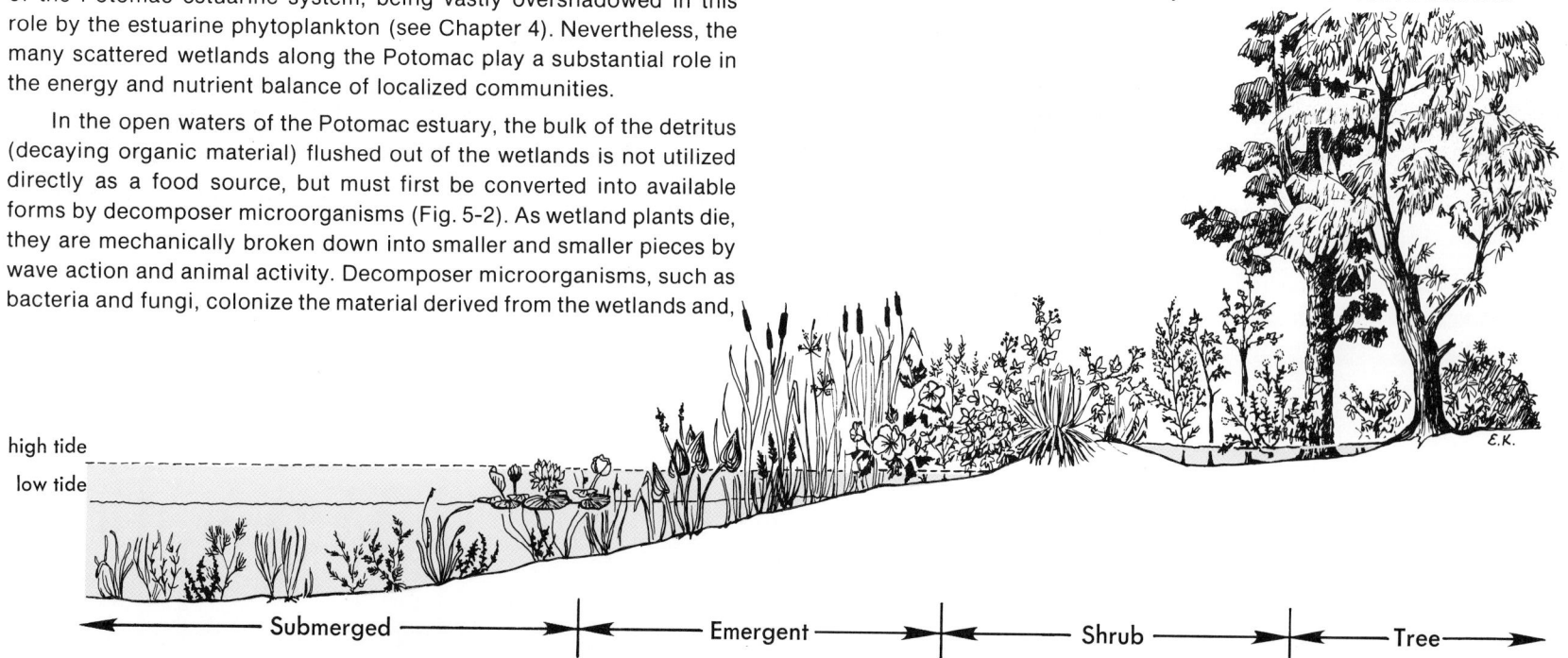

high tide

low tide

Submerged ——— Emergent ——— Shrub ——— Tree

Figure 5-1. *Schematic diagram of the nearshore environment, showing the relationship of various rooted plant types to water.*

Wetlands are never static. They are continually being formed and eroded by two major natural processes: coastal submergence and the sedimentation-erosion process. Coastal submergence, resulting from sinking land masses, gradually rising ocean levels, or both, has been causing inundation of coastal lowlands along the Atlantic Coast of the United States at a rate of 6 to 12 inches per century during recent geological times.[3] Deposition of tributary sediments in areas of slower moving currents gradually builds up layers of silt and organic material to form shoals where aquatic plants can take root, grow, and multiply, creating new wetlands from the open waters along estuarine coasts. In contrast to these gradual building processes, severe storms and floods can quickly erode wetlands that may have taken hundreds of years to form.

Wetlands have generally been thought of as wastelands in comparison with farmlands, forests, and cities. Until the passage of the 1971 Wetland Act in Maryland, many marshes, swamps, and beds of submerged rooted aquatics along the Potomac and its tributaries were altered and destroyed, acre by acre, for parking lots, housing, marinas, agriculture, industry, and disposal of dredged spoil and solid wastes. For the 25-year period from 1942 to 1967, wetland losses in the State of Maryland due to these activities exceeded 23,000 acres.[3] Since 1971, very few alterations of wetlands have taken place.[4]

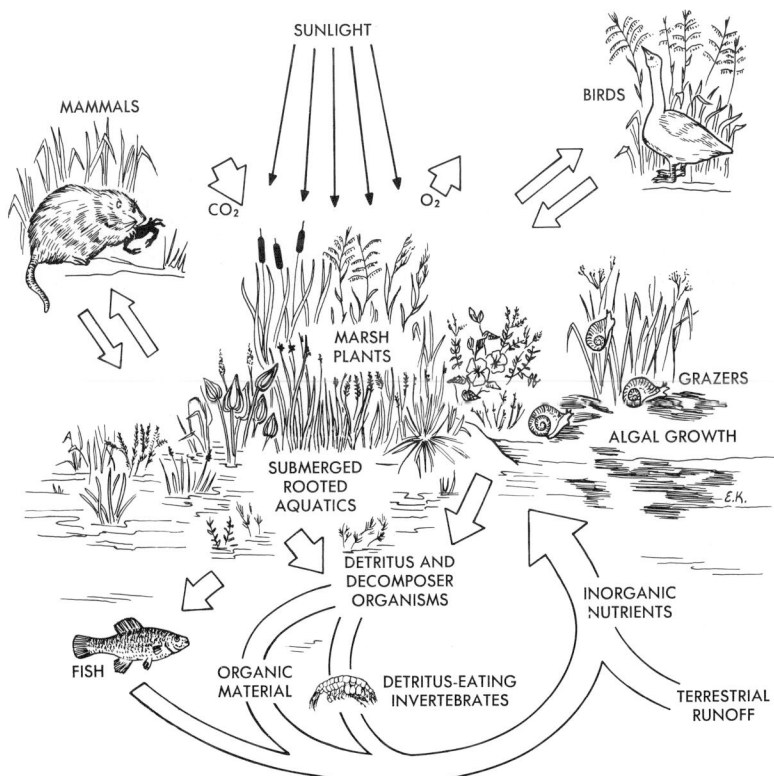

Figure 5-2. *Schematic diagram of the major cycling of organic matter and nutrients in wetlands.*

Factors Affecting the Distribution of Wetland Plants

Physical and biological factors work together to determine the plant associations in wetlands. The two most important factors are salinity and the degree of submersion caused by tides, stream flow, or rainfall. Different wetland plant species normally exist under different degrees of submergence and various concentrations of salt. In the progression from freshwater marshes and swamps to salty lowland marshes, the estuarine plant associations change noticeably. Because only those plants that are physiologically adapted to higher salinities are able to survive in tidally influenced brackish and saltwater marshes, there is less species diversity and less competition in these areas than in fresher water areas. For instance, in saltwater marshes along the lowermost portion of the Potomac estuary, saltmarsh cordgrass, which is a euryhaline species, has only a few salinity-tolerant competitors and typically grows in nearly pure stands over many acres along the main stem and the tidal tributaries. As salinity decreases in upstream areas, the saltmarsh cordgrass gradually gives way to freshwater grasses, rushes, rose mallow, and big cordgrass.

Potomac wetlands include a variety of types, from quiet, nontidal freshwater swamps of the upper tributaries, to the saltwater tidal marshes of the lower main stem (Folio Map 4). The more than 80 plant species common to wetlands in the Potomac region are listed in Appendix Table 2 with the salinity tolerances of each. Appendix Table 3 lists the occurrence of major plant species found in the wetlands bordering the Maryland shore of the Potomac estuarine drainage.

Classification of Potomac Wetlands

Wetlands are generally typed according to the associations of the plants that occur within their borders. Plant associations are difficult to fit into a discrete system of classification because their vegetative character continuously varies. A single plant species that is characteristic of one kind of marsh may also occur in association with different plants in another type of marsh. Therefore, there is a great degree of overlap in plant associations among wetland types.

The wetlands along the Maryland portion of the Potomac estuary have been classified into types by the Maryland Department of Natural Resources (MDNR). Their system of classification was formulated for an inventory* of state wetlands during 1967 and 1968 by selecting features from previous state and federal classification systems.[3] This system is used on Folio Map 4 and Appendix Table 3. Numbered wetlands on the map and table are those that were surveyed by the MDNR during its Wetland Habitat Inventory.[1] They are numbered consecutively from Washington, D.C., to the mouth of the estuary. Only wetlands larger than 5 acres were surveyed and typed by the State of

---

* A more comprehensive inventory of Maryland wetlands is scheduled for completion by MDNR during 1979. This will include a county-by-county appraisal of large, detailed aerial photographs of all Maryland wetlands, which will provide a more accurate source for determining the distribution of wetland types along the Potomac and other Maryland watersheds. A draft version (Ref. 2) has been completed.

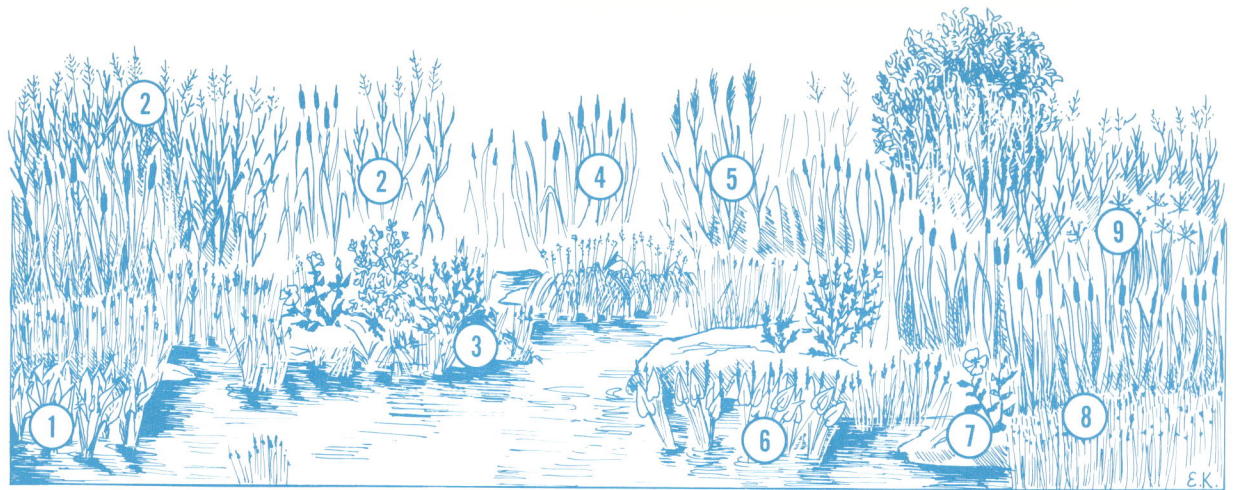

1. arrow arum
2. big cordgrass
3. hightide bush
4. cattails
5. saltmarsh cordgrass
6. pickerel weed
7. rose mallow
8. threesquares
9. bulrushes

Typical Plants in Coastal Shallow Fresh Marshes

Maryland. This does not imply, however, that wetlands smaller than 5 acres are unimportant.

Some Maryland wetlands that were not typed or numbered during the 1967-68 MDNR survey are shown on Folio Map 4, arbitrarily designated by the same classification as closely adjoining wetlands. Virginia wetlands were not given type designations on the map.* Usually, more than one (and sometimes as many as three or four) wetland types are found in any single wetland area along the Potomac estuary, but only the dominant type was mapped. In most cases, the total acreage in each mapped wetland area is dominated by a single marsh type. Where two types occur in nearly equal percentages, both types are shown. For example, both Types 6 and 12 are included in Wetland No. 4.

Of the 20 wetland types classified by the MDNR during the 1967-68 Wetland Habitat Inventory, seven (Types 5, 6, 7, 12, 13, 14, and 16) were found along the Potomac estuary.[1] Four additional wetland types (Types 1, 2, 3, and 4) have been listed in a report on anadromous spawning areas in tributary streams along the Maryland side of the Potomac.[11] However, since these were not dominant in the Potomac region, they were not shown as major wetland types on Folio Map 4.

There are two broad categories of wetland types in the Potomac estuary: coastal and inland. Coastal wetland types are found at low elevations along the tidal waters (fresh or saline) of the main stem and its tributaries. Inland wetland types occur along the nontidal fresh-water portions of tributaries.

COASTAL WETLANDS

Four types of coastal wetlands constitute the major portion of Potomac coastal wetland acreages:

Type 12 — Coastal shallow fresh marsh
Type 13 — Coastal deep fresh marsh
Type 14 — Coastal open fresh marsh
Type 16 — Coastal salt meadow

Shrub swamps (Type 6) and wooded swamps (Type 7), which are ordinarily regarded as inland types, are also considered to be coastal wetlands if they are adjacent to the upland sides of freshwater coastal marshes.

Coastal shallow fresh marshes generally lie along the tidal fresh and oligohaline portions of tributary creeks, rivers, and baylets. In the Potomac estuary, they are found along the main stem, from Broad Creek just below Washington, D.C., to Popes Creek, Maryland, just above the Rt. 301 bridge. Downstream along the main stem, salt concentrations become too high, and many plants in this type of association cannot survive. Few marshes of this type are found below the Rt. 301 bridge, except in the tidal fresh portions of some tributaries, such as the Wicomico and St. Marys rivers, and St. Clements and Breton bays (Folio Map 4).

At high tide, an average of 6 inches of water may cover the substrate in coastal shallow fresh marshes. During the spring and summer, the soil remains waterlogged, even during the lowest tides. Plant species in Potomac marshes of this type include alder, arrowheads, arrow arum, sedges, cattails, threesquares, bulrushes, smartweeds, and big cordgrass. Saltmarsh cordgrass and saltmeadow cordgrass were also found in these marshes during the 1967-68 MDNR inventory.[1]

Coastal shallow fresh marshes are a vital feeding and resting habitat for resident and migratory waterfowl. Shorebirds and waders, as well as mammals such as muskrats and nutria, also depend on these marshes for their principal food supplies. The spawning adults, larvae, and juveniles of many fish species such as banded killifish, carp, and chain pickerel inhabit the shallow, nutrient-rich waters.

---

* Tidal marsh inventories for all Virginia counties bordering the Potomac estuary are now available from the Virginia Institute of Marine Science, Gloucester Point, Virginia (Refs. 5-10).

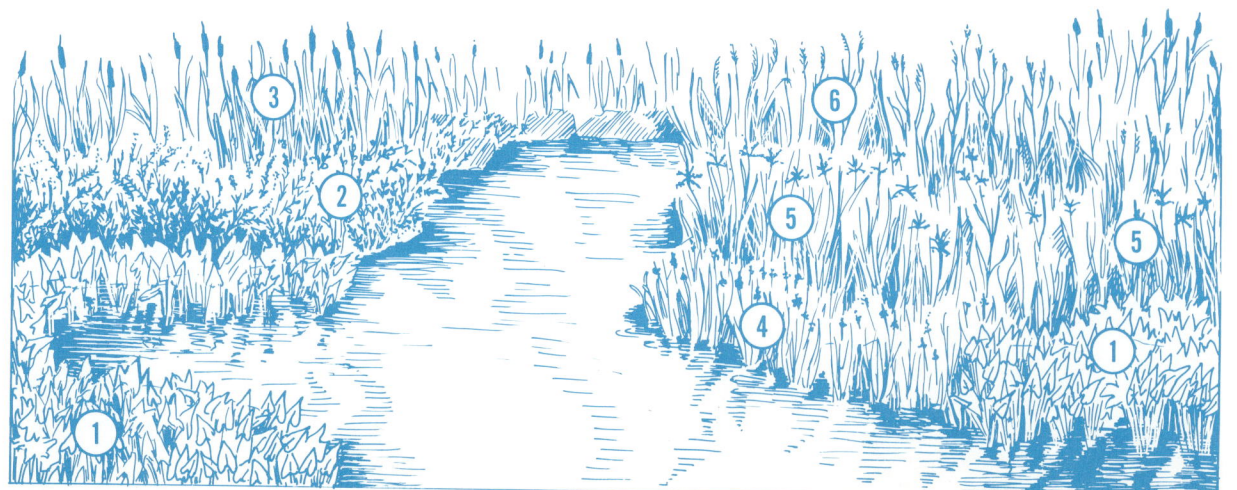

1. arrow arum
2. smartweeds
3. cattails
4. threesquares
5. bulrushes
6. saltmarsh cordgrass

Typical Plants in Coastal Deep Fresh Marshes

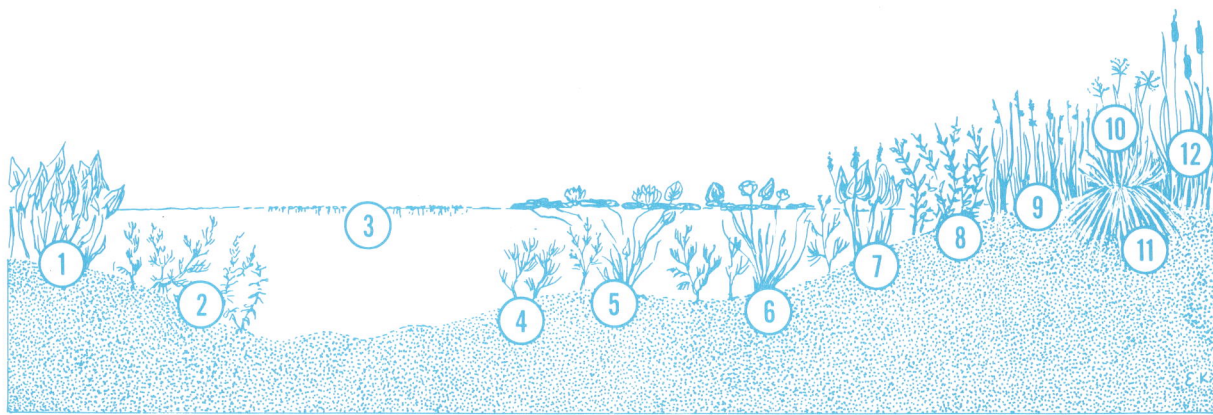

Typical Plants in Inland Open Freshwater Wetlands

1. arrow arum
2. Eurasian water milfoil
3. duckweed
4. pondweeds
5. water lily
6. spatterdock
7. pickerel weed
8. smartweeds
9. threesquares
10. bulrushes
11. sedge
12. cattails

Coastal deep fresh marshes can be found along the tidal meanders of tributaries leading to estuarine bays and sounds. In the Potomac estuary, only one wetland was found where this type was dominant — Wetland No. 6 between Broad and Piscataway creeks on the Maryland side (Folio Map 4). These areas are generally covered at high tide with 1 to 3 feet of water in all seasons. Arrow arum, cattails, smartweeds, threesquares, bulrushes, and saltmarsh cordgrass are typical of the vegetation that dominates these freshwater habitats in the Potomac estuary. During their fall and winter migrations, waterfowl feed on the plentiful submerged rooted aquatics that are present.

Coastal open fresh marshes include shallow portions of open water along tidal rivers. On the Potomac estuary, the single example of this type (Wetland No. 157) is at the mouth of the St. Marys River. The depth of these areas may vary, and the vegetation may be scarce due to turbidity. The Potomac wetland of this type has threesquares, bulrushes, cordgrasses, and hightide bush. Type 14 wetlands are frequented by feeding and resting waterfowl and by some fish species such as small killifishes.

Coastal salt meadows dominate along the borders of open waters on the Maryland side of the Potomac estuary main stem and lower tributaries, from Popes Creek, Maryland, to the mouth. Although these slightly elevated meadows are rarely covered by tides, the soil is consistently waterlogged. Periodically, during excessive spring or storm tides, a few inches of water may cover the floors of the meadows. Vegetation in the Potomac wetlands of this type includes saltmeadow cordgrass, threesquares, bulrushes, swamp rose, cattails, saltmarsh cordgrass, big cordgrass, and hightide bush. Potomac coastal salt meadows are used as feeding areas by migratory waterfowl, but they are not used as intensively as the fresher water types.

Coastal Salt Meadow

INLAND FRESH WETLANDS

Potomac inland wetlands are characterized by their submergence in nontidal fresh water. Three types dominate in the Potomac estuary:

Type 5 — Inland open freshwater wetland
Type 6 — Shrub swamp
Type 7 — Wooded swamp

Inland open freshwater wetlands usually occur as submerged and emergent vegetation in shallow waters and along fringes of open bodies of nontidal fresh water. Along the Maryland side, they generally grow at the uppermost sluggish extremities of tributary creeks and small runs from Washington, D.C., to Gilbert Creek (a tributary to Wicomico River). Vegetation may be relatively sparse or almost absent, especially in turbid or leaf-stained waters. Typical plant associations in these areas in the Potomac contain various combinations of pondweeds, water lilies, Eurasian water milfoil, cattails, and a variety of grass and algal species.

Inland open freshwater wetlands offer quiet resting and feeding spots for migrating waterfowl populations in late fall and winter. Borders of such areas occasionally serve as nesting and brooding sites for those wood ducks, black ducks, and mallards that remain in the Potomac area after others of the same species have migrated to their northern breeding grounds in spring.

Shrub swamps are growths of small trees, shrubs, and grasses along the peripheries of sluggish streams, especially along the landward ridges of many coastal wetlands. The only wetland in the Potomac estuary where this type is dominant is located along the upper elevations of Broad Creek (Wetland No. 4, Folio Map 4). The floors of these brushy swamps may be covered with up to 6 inches of water during the spring and summer.

Shrub Swamp

Plant species found in shrub swamps include small trees such as red maple; shrubs such as dogwoods, button bush, rose mallow, tag alder, and black willow; and many small rooted aquatic marsh plants, including arrow arum, threesquares, bulrushes, saltmarsh cordgrass, and narrow leaf cattail. These areas provide resting and feeding sites for mallards, black ducks, and other puddle ducks during spring and fall migrations. River otters and muskrats are permanent residents. Upland animals — raccoons, foxes, squirrels, opossums, and deer — visit these swamps in search of food and refuge. Many reptiles, amphibians, and resident and migratory songbirds also inhabit shrub swamps.

Wooded swamps consist of dense growths of deciduous trees along sluggish streams, on flood plains of tributary streams or rivers, and in poorly drained upland spots, usually in association with shrub swamps. Depending on the proximity of the swamp to the body of water it borders, these wetlands may be covered with a few inches to over a foot of water.

The largest areas of wooded swampland in the Potomac are found in the mid to upper estuary in large tracts surrounding Piscataway and Mattawoman creeks, Gilbert Swamp Run, and Zekiah Swamp (Folio Map 4). Of the wooded swamps in Maryland, these are among the most

extensive and most used by wildlife.[1] Wooded swamps also dominate the highly elevated wetlands in the vicinity of Maryland Point. Along the lower third of the estuary, they are found only at the heads of tributary creeks. A few of these small swamps lie at the most inland tips of the Western Branch of St. Marys River and at the most upstream portion of Canoe Neck Creek above St. Clements Bay.

Trees found in Potomac wooded swamps include red maple, ash, sweet gum, pin oak, black gum, willows, and alders. Smaller undergrowth includes arrowwood and smartweeds. Wooded swamps provide habitat for numerous animals. Migratory puddle ducks use them for resting and feeding. These swamps especially suit the needs of native populations of black and wood ducks for nesting, brooding, and nightly roosting. Many other resident and migratory waterbirds and songbirds frequent wooded swamps. Tall trees provide important nesting sites for bald eagles and great blue herons. Deer, opossum, and foxes from upland areas frequently visit. In wooded wetland areas, where water levels remain consistently high throughout the year, game fish species, including largemouth bass, pickerel, yellow perch, catfish, and white perch, may abound.

BEDS OF SUBMERGED ROOTED AQUATICS

Submerged rooted aquatic plants form an integral part of many wetland types, often growing adjacent to emergent vegetation. However, these underwater plants also form independent stands in shallow, tributary waters. Submerged rooted plant beds in the Potomac system (see Folio Map 4) provide habitat, cover, and food for a variety of aquatic fauna. Table 5-1 lists the most common submerged rooted aquatics found in the Potomac estuarine region and their ecological roles.

At one time, there were extensive beds of luxuriant rooted aquatic vegetation throughout the entire Potomac estuary.[14] In the early 1900s, however, imported nuisance species (discussed below) proliferated and eventually replaced many of the more desirable rooted species. Recently, increased siltation and nutrient loading have greatly degraded the water quality of the upper Potomac estuary, resulting in the destruction of most aquatic beds throughout the Potomac main stem.[13][14]

Today, little or no submerged rooted vegetation exists in the upper portion of the Potomac, from Washington, D.C., along the Maryland side to Mattawoman Creek or along the Virginia side almost to Mathias Point where healthy beds of wild celery, southern naiad, and sago pondweed were once abundant.[14] Downstream, from Mattawoman Creek to Maryland Point, fairly extensive beds of wild celery, southern naiad, waterweed, and other native plants still occur. But, because these waters remain turbid, beds of rooted aquatics are limited to a narrow zone along the shoreline and to shallow flats too small to be shown on Folio Map 4. From Maryland Point to the Wicomico River, pondweeds, wild celery, southern naiad, and waterweed may grow along shallow flats, especially in portions of the Nanjemoy Creek and

Wooded Swamp

Table 5-1 — **Ecological Roles of Submerged Rooted Aquatic Vegetation Found in the Potomac Estuary**

Species	Importance to Waterfowl	Relationship to Other Organisms
Eelgrass[a] (*Zostera marina*)	Favorite food of coastal waterfowl.	Excellent cover for shedding blue crabs.
Curly pondweed (*Potamogeton crispus*)	Excellent food for ducks in freshwater areas; tender rootstalks, seeds, and tuber-like winter buds eaten.	More resistant to pollution than many other rooted aquatics; may become weedy pests. Good food and shelter for fish, especially early spawning species (e.g., carp and goldfish). Also eaten by beaver and muskrats.
Redhead grass (*Potamogeton perfoliatus*)	Excellent duck food; slightly enlarged winter stems, rootstalks, and seeds eaten, especially by redhead and baldpate.	May create weedy pest problems in small, localized areas. Occasional food for bluegills, other fish, and mammals.
Sago pondweed (*Potamogeton pectinatus*)	Nutlets and starchy tubers provide excellent food for ducks.	Tubers provide food and shelter for young fish.
Clasping-leaf pondweed (*Potamogeton richardsonii*)	Sometimes important food for waterfowl (e.g., for black duck, canvasback, redhead, other ducks, and geese).	Good food and cover for fish. Harbors numerous insects eaten by fish.
Widgeon grass (*Ruppia maritima*)	Best all-round food for ducks in brackish-water areas.	Excellent shelter and food for fish.
Horned pondweed (*Zannichellia palustris*)	Seeds and slender threads of foliage often eaten.	Fair food for fish.
Southern naiad (*Najas guadalupensis*)	Stems, foliage, and nutlets among most important food for waterfowl, especially mallards; seeds strained out by baldpate.	Good shelter and harbors small invertebrates eaten by fish.
Common waterweed (*Elodea canadensis*)	Overall low-grade food for ducks; foliage eaten in small amounts, poor seed producer.	May create weedy problems similar to water chestnut and Eurasian water milfoil in closed or semi-enclosed bodies of water.
Wild celery (*Vallisneria americana*)	Excellent food for waterfowl, especially canvasbacks; all parts eaten, especially roots and tuber-like winter buds.	Attracts marsh birds, shorebirds, and muskrats. Harbors numerous insects and other invertebrate organisms. Provides good shade, shelter, and food source for fish.
Lesser duckweed (*Lemna minor*)	Important food item for wood duck, blue-wing teal, and black duck during early growth stages in summer before they switch to seeds of other plants in fall.	Harbors insects eaten by fish. Excessively shady and blocks light from other plants. Poor shelter.
Waterstargrass (*Heteranthera dubia*)	Attractive to some waterfowl in localized areas.	Provides food and shelter for fish.
Coontail (*Ceratophyllum demersum*)	Considerable value as food for ducks; seeds and, occasionally, foliage eaten.	Important food for muskrats. Shelters grass shrimp and other small invertebrates. Good shelter for small fish. Harbors insects that are valuable as fish food.
Water chestnut (*Trapa natans*)	No food value, but harbors small invertebrates such as snails and crustaceans that are eaten by waterfowl.	Poor food source and poor shelter for fish.
Eurasian water milfoil (*Myriophyllum spicatum*)	Generally unimportant as food for waterfowl, but harbors snails and other small organisms eaten by waterfowl; fruits occasionally eaten.	Occasional food for muskrats. Good shelter and source of insects eaten by fish.

(a) Although formerly a dominant species, eelgrass disappeared from the Potomac estuary, as it had from other Chesapeake Bay regions. Since eelgrass has returned to other Bay areas, it may become reestablished in the Potomac (Ref. 12).

Sources: Refs. 13-17.

the Port Tobacco and Wicomico rivers, in Cuckold Creek, and along the Potomac main stem near Neale Sound. On the Virginia side of the estuary, between the mouths of Choptank Creek and Popes Creek, rooted aquatics still occupy the narrow shoals of most tributaries. Smaller beds of rooted aquatics also line many of the narrow shoal areas along St. Clements and Breton bays. Along the lower portion of the Potomac, less salinity-tolerant vegetation is gradually replaced by a few salt-tolerant species such as redhead grass and widgeon grass.[11]

At times, the proliferation of undesirable aquatic vegetation has interfered with the utilization of estuarine waters by man and wildlife. In the past, two introduced species, water chestnut and Eurasian water milfoil, developed over such large areas of the Potomac that they became widespread nuisances, hampering navigation, spoiling recreational areas, and destroying habitats otherwise suitable for fishes, birds, and other estuarine wildlife.

Water chestnut is believed to have been first introduced into the Potomac River from Europe and Asia as an ornamental aquarium plant.[16] In 1900, it was common in the goldfish ponds on the Washington Mall, and by 1923, had migrated as far downstream as Oxon Creek, opposite Alexandria, Virginia. By the mid 1930s, water chestnut had infested over 10,000 acres of the upper estuary, from Little Falls to just below Quantico, Virginia.[18] The downstream spread of this freshwater plant was halted there, apparently by increasing salinities. When mosquitoes, which found the huge floating mats to be excellent breeding grounds, became an additional annoyance for residents of the Washington metropolitan area, Congress directed the U.S. Army Corps of Engineers to keep the Potomac and its tributaries from Washington, D.C., to Maryland Point clear of this nuisance species. Using underwater mowers that can cut the weeds from the bottom and remove the fragments, the Corps effectively controlled these pests in the Potomac. Although water chestnut still grows in many areas of the Potomac, it is not a large-scale nuisance problem. Only annual checks of potential water chestnut problem areas (see Folio Map 4) and occasional hand removal of localized growths are necessary today.

The lacy **Eurasian water milfoil,** like the water chestnut, is an immigrant from Europe and Asia, believed to have been introduced into the country with aquarium fish around the turn of the century. It was first identified in the Potomac in 1933.[18] Unlike most of its innocuous American relatives, this exotic variety of milfoil tolerates a wide range of physical and chemical conditions. Eurasian water milfoil grows best in fresh waters, but can establish itself in salinities over 12 parts per thousand.[13] It will grow in clear waters that are a few inches to over 9 feet deep, and in various bottom types, from the firmest sands to the finest silts.[18]

It is a perennial plant that spreads rapidly from region to region, carried by drifting currents. Small fragments broken off from stalks near the surface can root and rapidly establish an entirely new water milfoil bed. Boats and boat trailers often carry fragments from one launching site to the next. It is also spread by floating seed heads and by the ingestion and subsequent excretion of seeds by waterbirds.

After its initial introduction into the estuary and a period of somewhat slow growth, water milfoil populations began to explode.[14] By the early 1960s, densely tangled mats of Eurasian water milfoil carpeted nearly all of the shorelines of the Potomac and its tributaries from Indian Head to Piney Point.[13] At times, mats actually became so thick that ". . . birds stroll[ed] about on top of [them] and scarcely wet a foot." [18]

During the early 1960s, attempts were made to control Eurasian water milfoil infestations with herbicides and underwater mowing. Although these measures produced good results locally, they were, for the most part, only temporary. However, as quickly as it had proliferated, water milfoil began to decline dramatically in 1963 in many areas. Its virtual disappearance by 1965 has been attributed to a virus,[19] which turned the plants brown and caused their decay.

Today, water milfoil grows in localized areas, especially in tributaries and in the main stem along the middle portion of the Potomac estuary from Nanjemoy Creek to Nomini Bay (Folio Map 4). Since it is expected that a new surge could occur at any time when conditions become suitable, close scrutiny of potential problem areas continues, and control measures are taken whenever necessary.

Chapter 5 — References

1. Maryland Department of Natural Resources, 1967-1968
2. Jack McCormick & Associates, Inc., 1977
3. Metzgar, 1973
4. Sipple, 1978
5. Doumlele, 1976
6. Moore, K.A., 1975a
7. Moore, K.A., 1975b
8. Moore, K.A., 1975c
9. Silberhorn, 1975
10. Silberhorn, in press
11. O'Dell, King, and Gabor, 1973
12. Stevenson, 1977
13. U.S. Department of the Army, 1973
14. Stewart, 1962
15. Fassett, 1957
16. Anderson, 1972
17. Uhler, 1977
18. Rawls, 1964
19. Bayley, Rabin, and Southwick, 1968

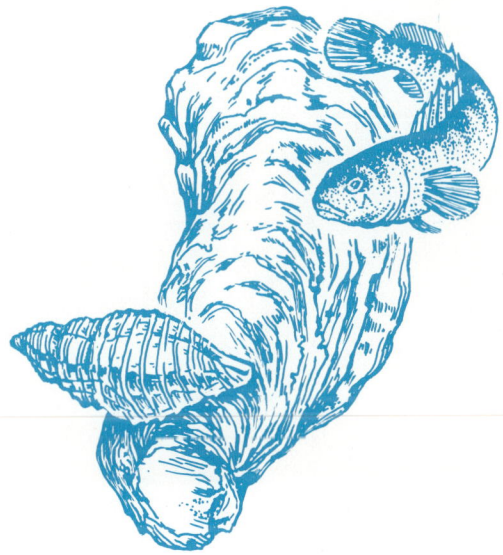

Zooplankton

Zooplankton

The animals found in and along the Potomac estuary can be broadly categorized as either vertebrates, which have backbones (for example, fish, mammals, and birds), or invertebrates, which have no backbones (for example, crustaceans, worms, and clams). Invertebrates far outnumber vertebrates, both in terms of the species present as well as in the abundance of individuals. (Appendix Table 4 lists pelagic invertebrate species found in the Potomac.) Pelagic invertebrates are those living in the water column. They range from planktonic forms such as zooplankton, whose distributions are largely governed by currents, to nektonic or free-swimming forms, such as shrimps. Figure 6-1 depicts the major types of pelagic invertebrates found in Potomac estuarine waters. These forms will be discussed in this chapter. The planktonic eggs and larvae of fish (ichthyoplankton) are vertebrate members of the zooplankton and will be discussed in Chapter 8.

Estuarine environments such as the Potomac generally support denser zooplankton populations than any oceanic habitat. As previously described, freshwater flow and runoff provide large nutrient inputs that support luxuriant phytoplankton crops capable of maintaining dense estuarine zooplankton populations. However, relatively fewer species of zooplankton have been able to adapt to the rigorous and variable estuarine conditions than to the more stable oceanic habitat.

The zooplankton are grouped by size into three major categories: microzooplankton, mesozooplankton, and macrozooplankton.[1] Microzooplankton are defined as the zooplankton that pass through nets or screens with a mesh size of 202 micrometers (μm). Mesozooplankton are organisms retained by a mesh size of 202 μm. Larger, more mobile pelagic invertebrates, which are usually retained by a 505-μm mesh, are often called macrozooplankton or macroplankton (see Fig. 6-1).

Microzooplankters that are smaller than 60 μm are mentioned only briefly in this chapter. These organisms are primarily protozoans — unicellular animals of such diverse phylogenetic groups as Foraminifera, Radiolaria, and Tintinnidae. Many protozoans are voracious consumers of bacteria, and some may also graze on phytoplankton; others utilize dissolved organic nutrients directly. Protozoa provide food for larger microzooplankton and play a significant role in energy transfer in the food web. They may be planktonic, but more often they are concentrated at the sediment-water interface where bacteria are most numerous.[2] Both protozoa and bacteria have been the subject of specialized microbiological studies, but little is known of their specific distribution in the Potomac. (For further information on these groups in the Chesapeake Bay, see Refs. 3 and 4.) Other microzooplankters include larval stages of some polychaete worms, molluscs, and arthropods.

The most numerous and important mesozooplanktonic groups in the estuary include the copepods, cladocerans, and rotifers. Other groups such as the ostracods that are only minor components of the mesozooplankton in the Potomac estuary are not discussed here.

Macrozooplankton include the jellyfish group (hydromedusae, comb jellies, and true jellyfishes); crustaceans such as amphipods, isopods, true shrimp, mysid shrimp; insect larvae; and polychaete worms.

Zooplankton can also be classified by the amount of time they spend in the planktonic state. The term meroplankton refers to animals that are planktonic for only a part of their life cycle; holoplankton, to those that are planktonic throughout their entire life.

The meroplankton group is made up, for the most part, of the larvae of benthic invertebrates (Fig. 6-2). This group includes the larvae of polychaetes and molluscs, microscopic forms that are not easily collected intact or identified. However, arthropod larvae are easily captured and frequently occur in considerable abundance. Those most often found in Potomac estuarine samples are the early larval stages of barnacles and decapods (including shrimp and crabs).

Another meroplanktonic group (sometimes referred to as tychoplankton) includes small animals that are primarily benthic, but that move into the water column temporarily, either swept by currents or voluntarily. Many isopods, amphipods, mud crabs, mysid shrimp, and polychaete worms belong in this category. (Chapter 7 discusses the benthic stages of these organisms.) Many migrate vertically into the water column at night. Polychaete worms usually do not move far from the bottom, although they too are occasionally caught in plankton collections. During late spring and early summer, *Nereis succinea*, which are ubiquitous polychaetes in the Potomac estuary, often swarm just below the water surface where they release their gametes. Some isopods are parasitic on fish, but are free in the water prior to attaching themselves to a host, and are thus considered meroplanktonic.

Holoplankton include rotifers and the dominant planktonic crustaceans — the cladocerans and copepods. The combined numbers of these three invertebrate groups far exceed those of all other mesozooplanktonic groups. These zooplankton play the major role in the energy transfer from primary producers to higher forms of aquatic life in the Potomac, since they are the primary consumers of the rich phytoplanktonic flora of the estuary and are the basic food that nourishes larval fish on the nursery grounds.[14] [15] The fecal pellets of cladocerans and copepods drop to the bottom and are ingested by benthic organisms such as clams, oysters, and barnacles. Holoplankton play a substantial role in the recycling of nutrients.

Many fish feed exclusively on zooplankton in their early developmental stages. Laboratory studies on striped bass diets have shown that the youngest larvae, which begin feeding 5 or 6 days after hatching,[16] [17] ingested rotifers and copepod nauplii (Fig. 6-3). When they reached a length of 9 to 10 millimeters (mm), their diet shifted to cladocerans and copepodite stages of copepods. At 20 to 50 mm, juvenile striped bass prefer adult copepods, insect larvae, and mysid shrimp. As they increase in size, their food requirements continue to change, and they develop preferences for larger prey such as amphipods, worms, and small fish. Other fish, such as the filter-feeding menhaden, consume zooplankton throughout their lives.

PLANKTONIC

MICROZOOPLANKTON   TROCHOPHORE LARVAE
OF POLYCHAETES

VELIGER LARVAE
OF MOLLUSCS

ARTHROPOD LARVAE

Barnacle Nauplius   Crab Zoea

MESOZOOPLANKTON

OSTRACODS

COPEPODS

Calanoid   Cyclopoid   Harpacticoid

ROTIFERS

CLADOCERANS

increasing locomotive ability

increasing size

MACROZOOPLANKTON   HYDROMEDUSAE

COMB JELLIES

Anthomedusa   Leptomedusa

Sea Walnut

TRUE JELLYFISHES

INSECT LARVAE

SMALL CRUSTACEANS

Amphipod   Isopod   Mysid Shrimp

POLYCHAETE WORMS

TRUE SHRIMP

NEKTONIC

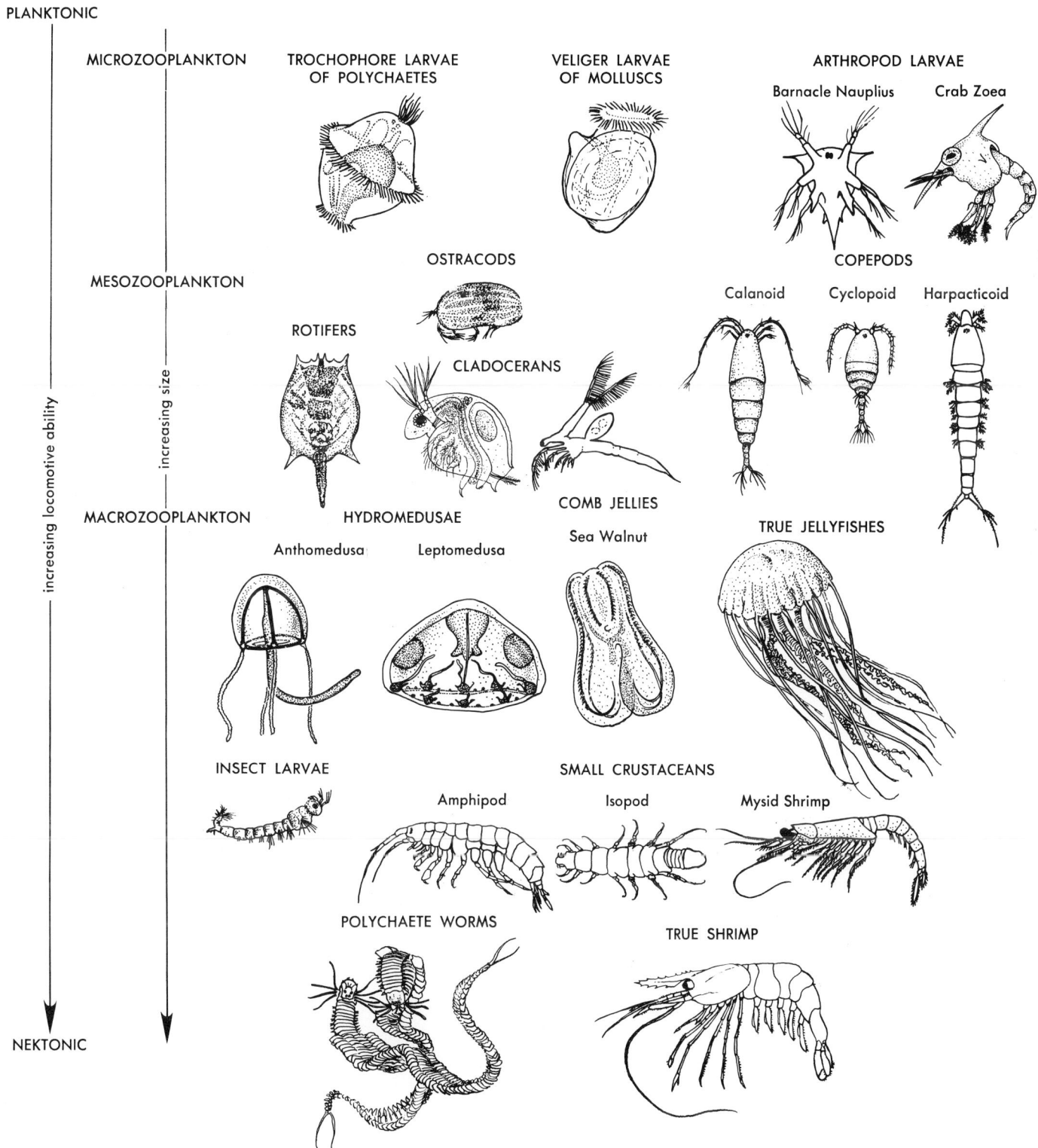

Figure 6-1. *Major types of zooplankton of the Potomac estuary.*

JELLYFISHES	POLYCHAETE WORMS	MOLLUSCS		CRUSTACEANS		
		Gastropods	Pelecypods	Barnacles	Shrimp	Crabs
		(snail)	(oyster)			
planula larva	trochophore larva	nonpelagic stages	trochophore larva	nonpelagic stages	nonpelagic stages	nonpelagic stages
polyp to ephyra stage	metatrochophore larva	veliger larva	straight-hinge veliger larva	nauplius	zoea	zoea
ephyra	post trochophore larva		umbo veliger larva	cypris		megalops

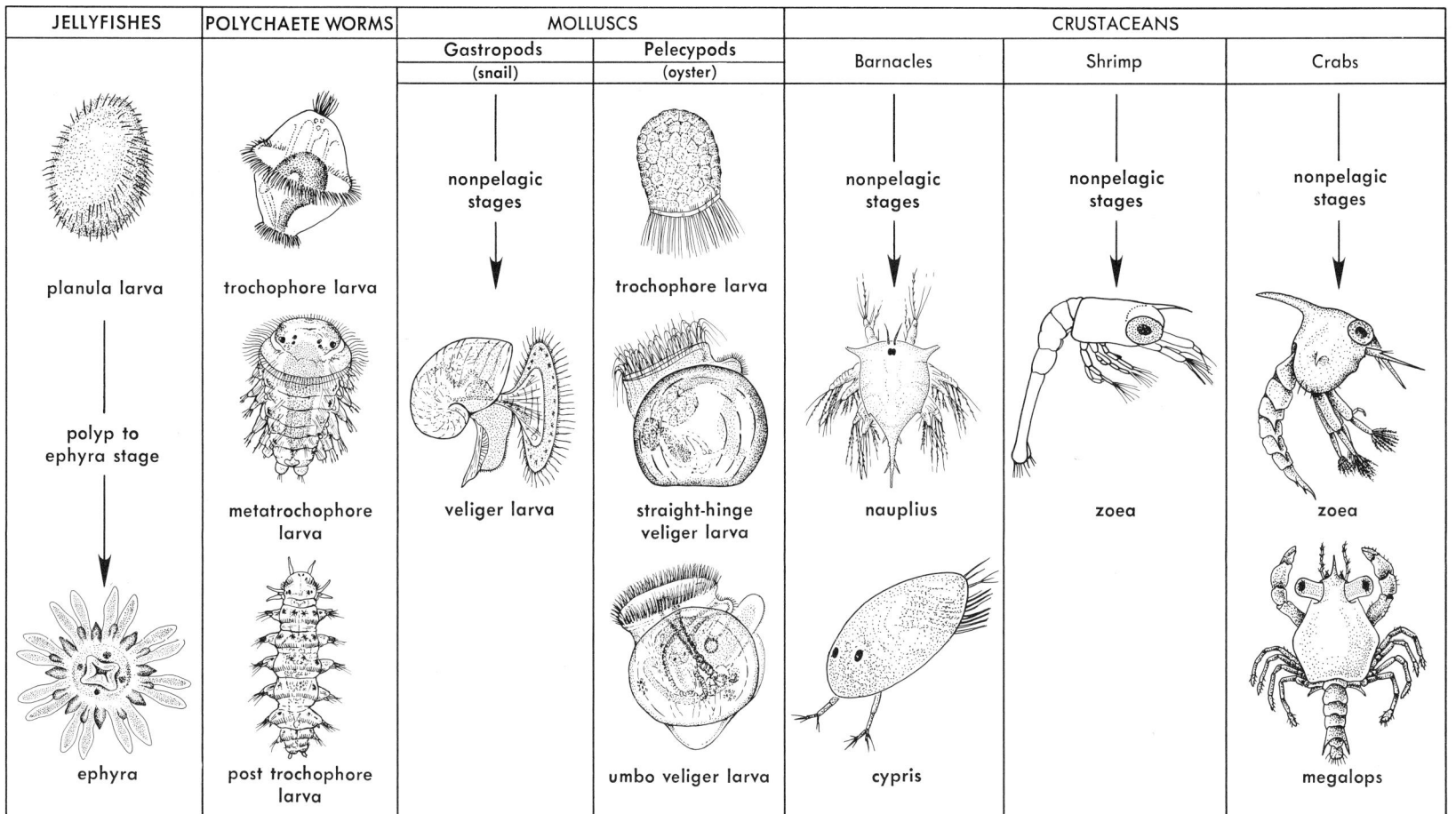

Figure 6-2. *Types of meroplanktonic larvae of some invertebrates found in the Potomac estuary. Young stages may change in appearance through their larval development and may often look radically different from the adult. Progressive stages of larval development are shown from top to bottom (drawn after illustrations in Refs. 5-13).*

FACTORS AFFECTING ZOOPLANKTON DISTRIBUTION AND ABUNDANCE

Salinity is an important factor influencing geographic distributions of zooplankton species along the length of the estuary. Rotifers and cladocerans are primarily freshwater animals, but within both groups, several species have evolved that are able to survive in the estuarine environment. Salinity tolerances vary not only among species, but also among the life stages of individual species. Salinity may also be critical to the reproductive potential of a species. Populations of species such as the fresh-to-brackish-water copepod, *Eurytemora affinis,* can survive in salinites up to 35 parts per thousand (ppt), but cannot reproduce in salinities greater than 22.5 ppt.[18] Because different species and their various life stages have different salinity requirements, specific zooplankton distributions are affected by seasonal shifts in salinity patterns.

The seasonal successions of species and the changing dominance of one group over another follow seasonal changes in temperature. For example, in the middle to upper Potomac estuary, summer populations of the copepod, *Acartia tonsa,* are replaced in the winter by *E. affinis.*

Zooplankton abundance is influenced by many other factors, including availability of food for the mature females. Egg production and early development of broods are often regulated by the seasonal changes in algal productivity. The age structure and abundances of zooplankton populations also depend on predation and competition. Competition among species for available food can cause shifts in the dominance of one form over another at any given time or place.

Many species undergo diurnal vertical migrations, moving towards the surface at dusk and back to the depths at dawn.[5][19][20] It has been suggested that this movement is an instinctive negative response to light, which may have evolved in response to food availability, predator avoidance, and/or temperature variations.[19][20] Vertical migration may provide a mechanism for retaining zooplankton populations within specific zones of the estuary. As zooplankters move up at night into surface waters, they are carried seaward by net downstream currents.

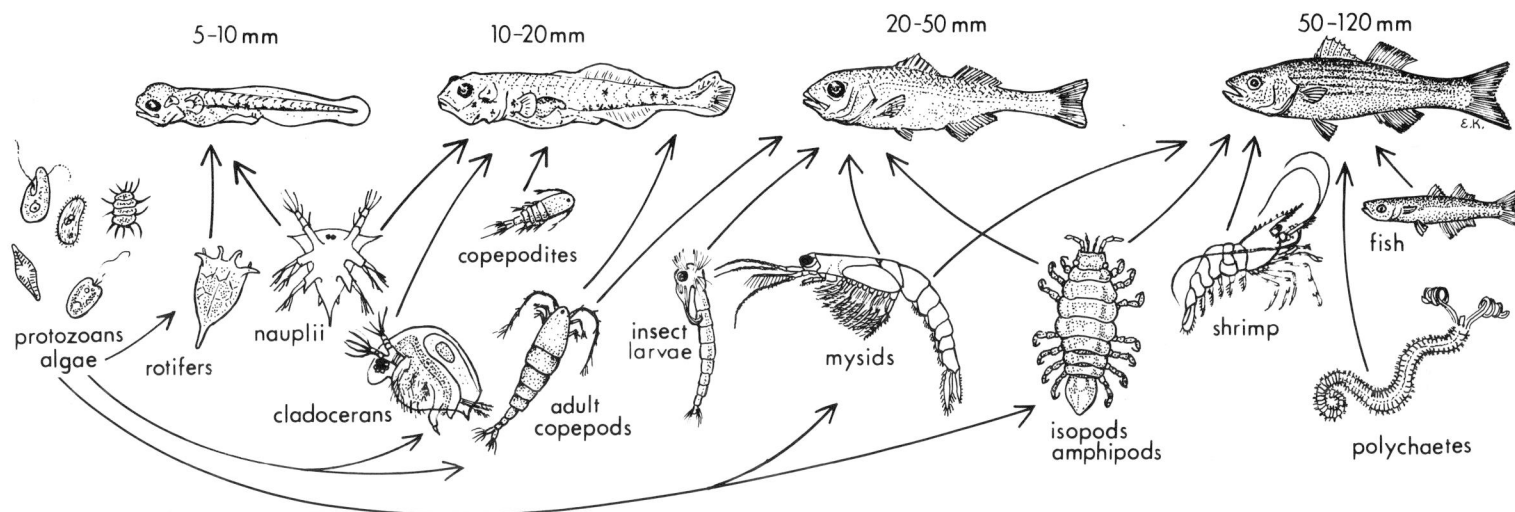

Figure 6-3. *Diet of different life stages of striped bass. (Sources: Refs. 14-17)*

As they sink lower into deeper waters during the day, net upstream flows move them back, more or less, to their original position. Without any such stabilizing mechanism, zooplankton populations could eventually be flushed out of the estuary.

In any segment of the estuary, zooplankton are not homogeneously distributed, but occur in irregular patches of varying sizes. This creates problems in sampling and in interpreting data. Thus, the patterns of abundance discussed in this chapter represent only the most generalized picture of how zooplankton may be distributed throughout the Potomac estuary.

MESOZOOPLANKTON AND MICROZOOPLANKTON

Spatial and Temporal Population Trends

Zooplankton populations in the Potomac estuary (excluding microzooplankters smaller than 60 μm) are composed primarily of the mesozooplankters — cladocerans, adult copepods, and larger rotifers — and the microzooplankters — young copepod stages, small rotifers, and the larvae of barnacles, polychaete worms, and molluscs. Standing crops are generally low in fall and winter and high in spring and summer. However, as will be seen, individual groups often depart from this general trend. The seasonal distributions of the major groups from representative salinity regimes are shown in Fig. 6-4. Since no estuarine-wide surveys have been made in the Potomac, generalized patterns of species distributions are described based on data collected from the three regions discussed below.

INDIAN HEAD TO MARYLAND POINT REGION

Characteristically, two peaks of zooplankton production occur in the region of the estuary between Indian Head and Maryland Point, where salinities change seasonally from tidal fresh to oligohaline. The first

peak occurs in spring and the second during late summer and early fall, as shown in Fig. 6-4A. In this upper part of the estuary, total zooplankton counts are often highest near the mouths of tributaries, possibly due to the very high densities of rotifers in adjacent freshwater streams washing out into the main stem.

In spring, zooplankton numbers increase in response to the abundant spring phytoplankton crop. Densities [expressed as numbers per cubic meter (m^3)] of rotifers average around 160,000/m^3, often peaking at nearly 1,000,000/m^3, while cladoceran densities can average 30,000 to 50,000/m^3. Copepod counts may reach 50,000/m^3.[19][33] Meroplanktonic larvae of benthic invertebrates are present in low abundance at these low salinities. There is a characteristic dip in copepod abundance at the end of May, due in part to the decreased availability of phytoplankton for food (see Chapter 4) and in part due to the seasonal change in the species composition of the copepod community, as discussed later. Their numbers are also reduced by predation by larval fish that are most abundant in this region in late May.

As summer progresses, copepods become more abundant than cladocerans, but rotifers remain the dominant group in the zooplankton community. Species diversity increases during the summer. Dominant species include the rotifers *Brachionus calyciflorus, Keratella cochlearis,* and *Polyarthra vulgaris;* the cladocerans *Bosmina longirostris* and *Daphnia* spp.; and the copepods *Eurytemora affinis, Mesocyclops edax,* and *Acartia tonsa.*[15][19][21]

The fall peak follows an increase in phytoplankton productivity. As winter approaches, there is an overall reduction in total numbers of zooplankton as primary productivity is reduced. Food becomes scarce, and zooplankton pass into a phase of slower population growth, reduced metabolic activity, or both. In spring, an increase in zooplankton is again observed.[15][19][21]

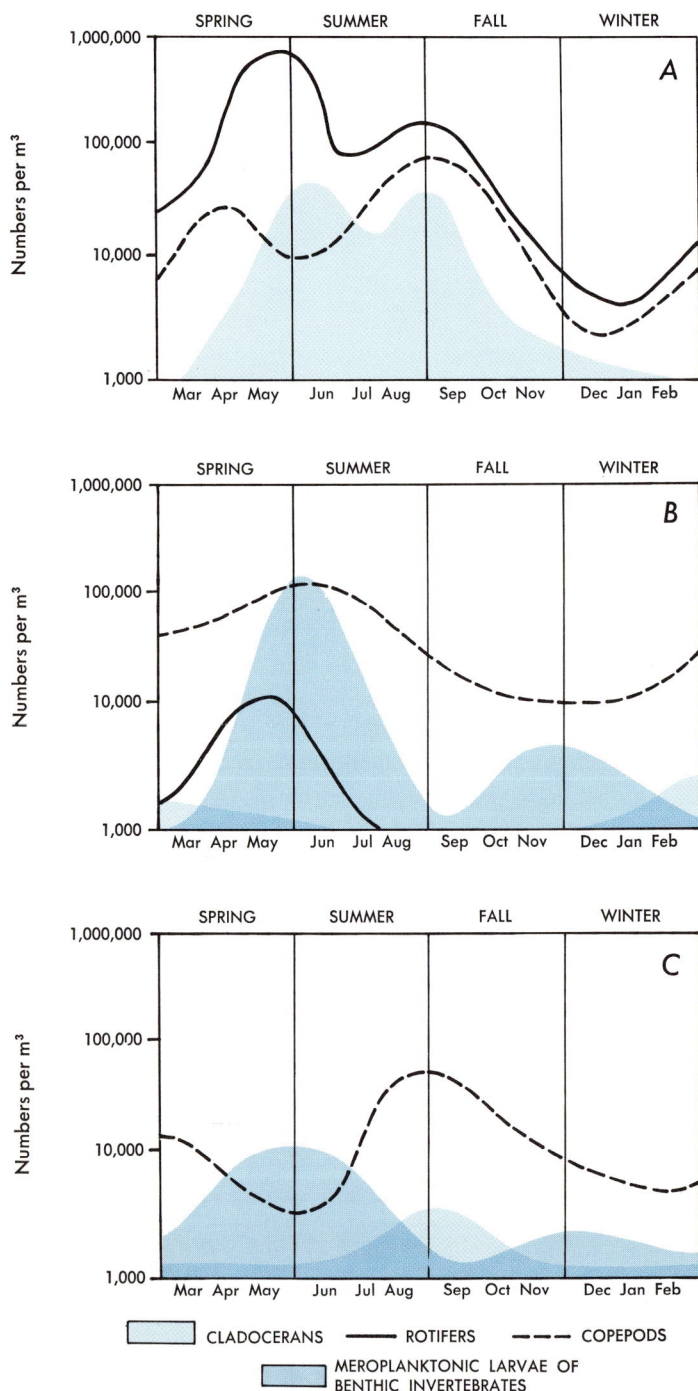

Figure 6-4. *Typical composition of dominant zooplankton groups from three salinity regimes. A—Tidal fresh to oligohaline salinities (0-5 ppt) in the vicinity of Douglas Point, 1972-74 (from data reported in Refs. 15, 19, 21). B—Oligohaline to mesohaline salinities (2-9 ppt) in the Mathias Point to Morgantown region, 1973-76, (from data reported in Refs. 15, 22-26). C—Mesohaline salinities (8-14 ppt) in the Calvert Cliffs region of the Chesapeake Bay, which has environmental conditions similar to those of the lower Potomac estuary, 1965, 1971-77 (from data reported in Refs. 27-32).*

MATHIAS POINT TO MORGANTOWN REGION

As waters become more saline in the Mathias Point to Morgantown region, the typical freshwater zooplankton of the upper estuary decrease in abundance, and estuarine-adapted groups increase in numbers (Fig. 6-4B). Salinities in this region change seasonally from oligohaline to low mesohaline (2 to 10 ppt), and the total zooplankton crop is smaller than in fresher waters. Rotifers are a relatively unimportant group in this region, occurring primarily in spring and early summer. Freshwater cladocerans nearly disappear, and estuarine cladocerans have marginal populations this far up the estuary. Copepods are usually the dominant form throughout the year, with highest numbers in spring and summer.[15 26 33] Meroplanktonic larvae of benthic invertebrates are at times the most significant group.

As is the case in the upper estuary, a few species account for over 90 percent of the total zooplankton numbers in this middle region. The estuarine copepod, *Acartia tonsa,* dominates all other species in late spring through summer and early fall, and the fresh-to-brackish-water copepod, *Eurytemora affinis,* dominates the community in winter through early spring.[22-26] Nauplii of a barnacle species (which is presumed to be *Balanus improvisus*[26]) make up the next greatest percentage of total zooplankton. Barnacle and clam larvae exhibit two population peaks: a primary one in May through June and a secondary one in October through November.[26] Rotifers are represented primarily by *Brachionus calyciflorus* and cladocerans by *Bosmina longirostris* and *Podon polyphemoides.*[15]

THE MESOHALINE REGION

Few zooplankton collections have been made in the lower Potomac estuary, so knowledge of what occurs there is based on extrapolations from information gathered in other mesohaline waters of the Chesapeake Bay (Fig. 6-4C). Seasonal abundances and distributions at Calvert Cliffs on the western side of the Bay (approximately 30 miles north of the mouth of the Potomac) are assumed to be similar to those in the lower Potomac, since habitat factors (e.g., salinity and temperature) are similar in these two areas. Lowest salinities in this region generally range from 5 to 10 ppt, and highest salinities from 12 to 18 ppt.

As in the Mathias Point to Morgantown region of the Potomac, copepods are the dominant zooplankton group at Calvert Cliffs, followed by meroplanktonic larvae of benthic invertebrates.[30-32] Rotifers are virtually absent, and cladocerans are represented by only one or two estuarine species. At Calvert Cliffs, copepod densities reach moderate levels in early spring but drop off to a low in early summer when the major zooplankton predators, comb jellies (ctenophores), are numerous. Copepod abundances peak in late summer or early fall. Populations of the meroplanktonic larvae of benthic invertebrates have two peaks: a larger one in early summer and a smaller one in late fall. The early summer peak is made up primarily of barnacle nauplii and polychaete larvae. The fall peak is composed mostly of barnacle nauplii.

Assuming that conditions in the lower Potomac approximate those at Calvert Cliffs, total zooplankton abundance in the lower Potomac estuary should be less than in the upper estuary and should peak later. Species that are common in the salinities of the lower Potomac are the copepods, *Acartia tonsa, Acartia clausi,* and *Eurytemora affinis,* and the cladoceran, *Podon polyphemoides.*[29-32]

Distributions of Major Groups
ROTIFERS

Rotifers are small organisms of the phylum Rotifera, ranging in length from 0.02 to 1.5 mm. Most species, however, are 0.2 to 0.6 mm long and are barely visible to the unaided eye.[34] Rotifers are characterized by a crown of rapidly beating cilia surrounding the mouth, which create localized currents that force small phytoplankters, protozoans, and detritus into the mouth. Because of their small size, rotifers only feed on planktonic particles that are smaller than 20 μm in diameter.[35] In turn, only the smaller invertebrates, recently hatched fish larvae, and nonselective plankton-eaters such as menhaden and comb jellies feed on rotifers.

Rotifers are most abundant and diverse in fresh waters. Of 1,700 species described worldwide, less than 5 percent are recorded from brackish to marine waters.[6] Few species tolerate salinities greater than about 5 or 6 ppt. Rotifers, like many other zooplankton groups, display some diurnal vertical migration, but overall, tend to favor surface waters.[21] Overwintering populations reproduce sexually, but when the waters start to warm in spring, females reproduce parthenogenetically (by asexual reproduction).[6] Although broods are small (usually only 15 or so), the time required for an egg to develop into an adult egg-producing stage is short — no more than a day or two under spring and summer temperatures (Table 6-1). This rapid reproductive process compensates for the low numbers of young and maintains abundant population levels.

About 80 species of planktonic rotifers have been recorded from the Potomac estuary (see Appendix Table 4).* The most common are *Brachionus calyciflorus* and *Keratella cochlearis,* although other species such as *Polyarthra vulgaris* or *Synchaeta* sp. may be more numerous locally. *Brachionus calyciflorus* is usually the most abundant rotifer and is distributed from Washington, D.C., downstream at least to Morgantown (Fig. 6-5).[15 26 33] *Keratella cochlearis* is apparently more limited in its distribution. It appears from Piscataway Creek down to Morgantown, but as a typical plankter of the estuarine interface zone (between tidal fresh and oligohaline waters), its center of distribution falls between Indian Head and Maryland Point.[19 33] *K. cochlearis* appears to be more adapted to tributary habitats than to those of the main stem. Densities in some of the major tributaries of the upper estuary reach up to 250,000/m³, whereas mainstem densities typically range from 5,000 to 20,000/m³.[33] *K. cochlearis* is found throughout the year, and, like *B. calyciflorus,* its numbers peak in spring, level off in summer, and stay at a low level in winter.[19 21 33] In early spring, *K. cochlearis* approaches *B. calyciflorus* in abundance in regions where their distributions overlap, and eventually becomes more abundant as the seasons progress into summer and fall.

Table 6-1 — **Time Periods for Generations of the Three Major Zooplankton Taxa from Egg-to-Egg Stages**[a]

Water Temperature (°C)	Approximate Date In Potomac [b]	No. of Days for Egg-to-Egg Generation		
		Rotifers	Cladocerans	Copepods
10	April, November	5-7	20-24	28-32
20	May-June, September	2-3	7-8	13-15
25	July-August	1.25-1.75	5.5-6.5	7-8

(a) Adapted from Table 1 in Ref. 35.
(b) Months when water temperatures in Column 1 occur in the Potomac.

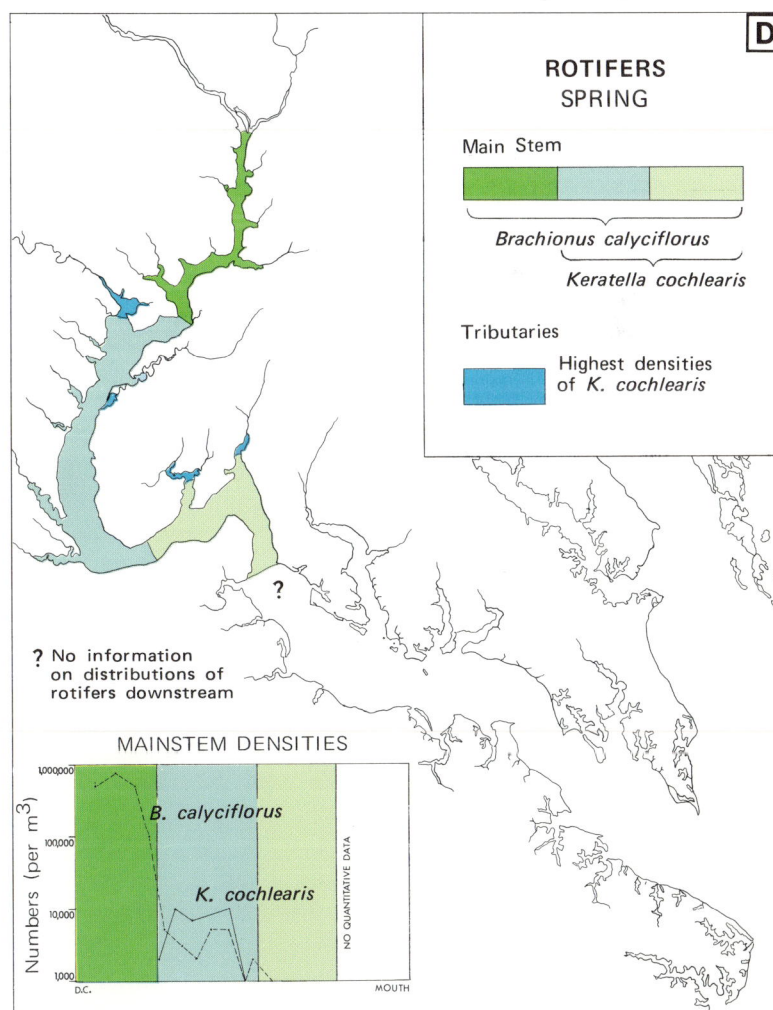

Figure 6-5. *Typical spring distributions of two common rotifers in the Potomac estuary. Graph shows average mean densities found along the estuary during various zooplankton studies during 1975-76 (from data in Refs. 15, 33).*

*Other rotifers that are sessile components of the benthic epifauna are not included in Appendix Table 4.

POTOMAC RIVER ESTUARY

I

CLADOCERANS

SPRING

Bosmina longirostris	*Daphnia* spp.	*Podon polyphemoides*
50,000		
35,000	4,500	
20,000	10,000	
10,000	1,000	
5,000	20	
3,000		
100		25
		100
		250
		400
		800

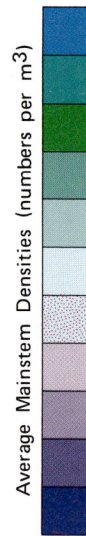

Average Mainstem Densities (numbers per m³)

No data available for tributaries

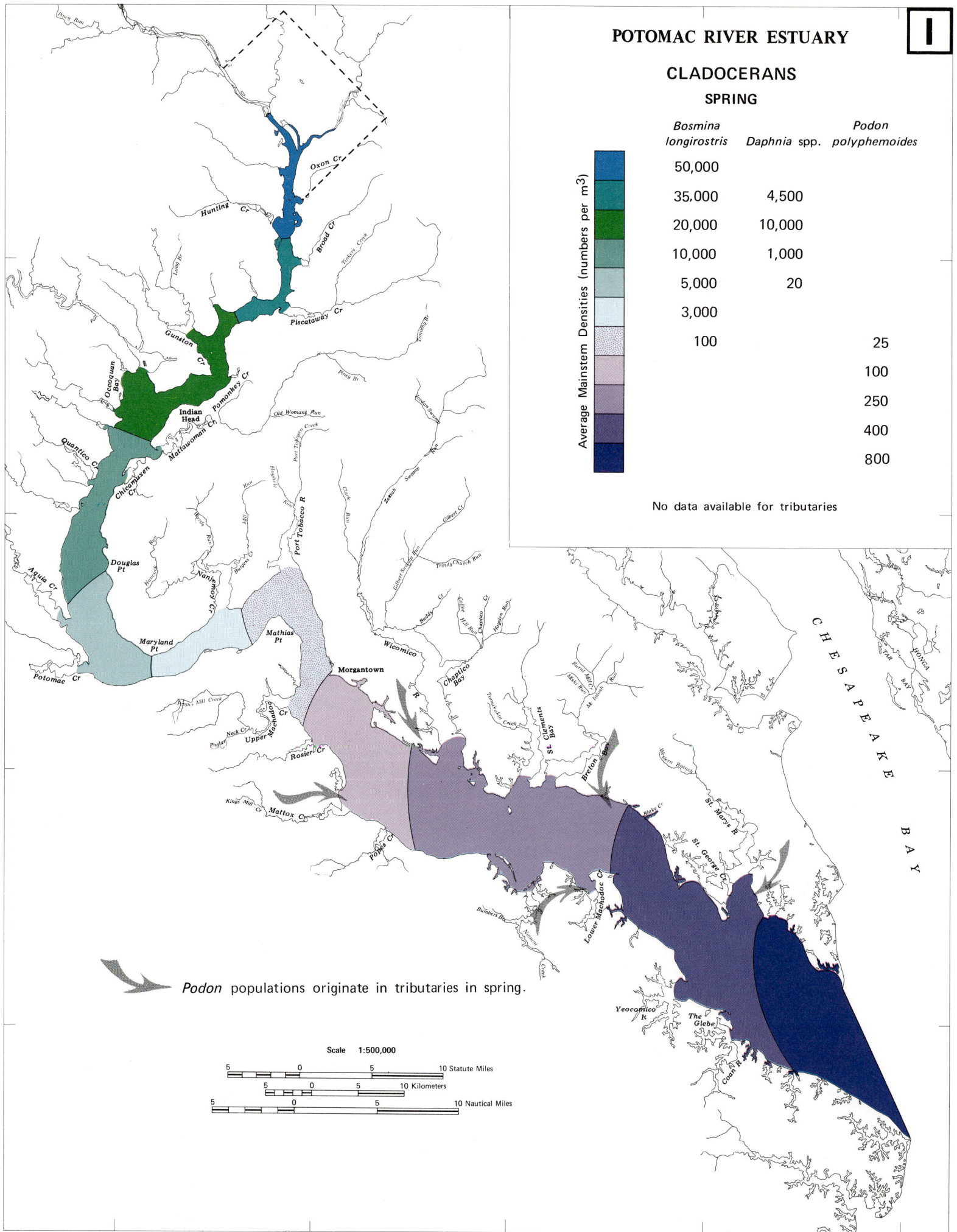

Podon populations originate in tributaries in spring.

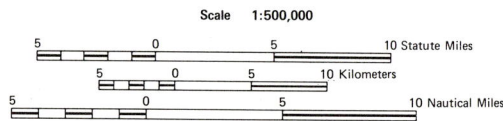

Scale   1:500,000

5   0   5   10 Statute Miles

5   0   5   10 Kilometers

5   0   5   10 Nautical Miles

CHESAPEAKE BAY

Figure 6-6. *Typical spring distributions and average mean densities of three common cladocerans in the Potomac estuary. Average densities in upper two thirds of estuary are from data taken in 1972-73 (Refs. 19, 21) and 1976 (Ref. 33). Densities of Podon polyphemoides in the lower estuary are extrapolated from data taken in 1964-66 in the Chesapeake Bay near the mouth of the Potomac and in other Bay tributaries (Refs. 28, 38).*

CLADOCERANS

Cladocerans (or water fleas) are the most primitive of the planktonic crustaceans that inhabit the Potomac estuary. Most species are generally the same size as their relatives, the copepods, and have somewhat rounded bodies and extremely large eyes for their size. An exception, *Leptodora kindtii,* has an extended body and large wing-iike appendages (see Fig. 6-1). This species has been recorded in the Potomac estuary and reaches 11 mm in length.[36] In the Potomac estuary, the most numerous and diverse cladoceran populations are found in freshwater areas. However, cladoceran species can be found throughout the entire salinity range of the estuary.

Most cladocerans are filter feeders. Movement of their appendages produces currents of water which carry various food items (phytoplankton, protozoans, detritus, and bacteria) to their mouths for filtering and ingestion. Their reproduction is parthenogenetic throughout most of the year. The population turnover is rapid, with new broods appearing every 5 to 6 days in summer (Table 6-1). Unlike many other crustaceans, which liberate their larvae into the water column, the hatched young of cladocerans are brooded in egg pouches where they develop into an adult-like form before they emerge.[6] Seasonal abundances of cladocerans are dependent on water temperatures. Winter populations are low, but when water temperatures reach 6 to 12° C, active reproduction begins.

By far, the most abundant cladoceran in tidal fresh and low brackish waters in the Potomac is *Bosmina longirostris.*[15 33 37] Year-round, the densities of this rotifer are greatest in tidal fresh waters around Washington, D.C., and gradually decline downstream.[15 19 21 33 37] The greatest downstream extension of *B. longirostris* occurs in spring and summer. In spring, densities average 50,000/m³ at D.C., gradually decreasing to less than 100/m³ at Morgantown (Fig. 6-6).[26 33] By June, *B. longirostris* reaches its peak density in the Potomac, with densities averaging 100,000/m³ from D.C. downstream to Broad Creek, and 70,000/m³ at Indian Head.[33] Below Maryland Point, summer densities of *B. longirostris* do not increase over those reached during the spring.

Cladocerans of the genus *Daphnia* are also common in the tidal fresh portion of the estuary. (Most investigators in the Potomac estuary have grouped all species of *Daphnia,* so species distinctions are not made here.) *Daphnia* species are more restricted than *B. longirostris* in distribution, both seasonally and spatially (Fig. 6-6). They have only been found in abundance in the Potomac estuary during the spring and early summer, and then in numbers substantially less than those of *B. longirostris. Daphnia* spp. are most abundant in the Gunston Cove and Occoquan Bay region, and they appear in only small numbers, if at all, at the head of the estuary. They may be less tolerant of poor water quality conditions than are *B. longirostris,* but little information exists to verify this, and other factors may be limiting. *Daphnia* spp. are somewhat less tolerant of salt than *B. longirostris,* and their populations are sparse or absent below Maryland Point.[19 26 33]

Other, less numerous but frequently identified Potomac freshwater cladocerans are *Leydigia quadrangularis, Leptodora kindtii, Chydorus* sp., and *Ilyocryptus* sp. *Leptodora kindtii* is found in the Potomac estuary from Washington, D.C., down to about Mathias Point, appearing in spring when temperatures reach 10° C and remaining in that region throughout the summer.[36] This species has been found in salinities up to 6 ppt, but only reproduces in fresh water.

Podon polyphemoides is the most abundant cladoceran found in higher salinity portions of the Potomac estuary. Very little, however, is known about seasonal changes in the abundance of this species in the Potomac. Studies conducted in similar tributaries of the Chesapeake Bay[28 39] suggest the following seasonal trend. In April, *P. polyphemoides* populations begin to proliferate in salinities around 8 ppt. By May, the population centers in these tributaries move into the Chesapeake Bay where they merge and form a single large population, with greatest densities about mid-Bay.[39] Small winter populations have been recorded at Morgantown, close to their upstream salinity limit.[26]

COPEPODS

Copepods, like cladocerans, are small crustaceans. Most adults are about 1 mm long, just visible to the naked eye. Copepods are considered to be the most abundant group of multicellular animals found throughout the oceans and estuaries.[6] They dominate the zooplankton in the mesohaline portions of the Chesapeake Bay.

There are three major suborders of copepods: Calanoida, Harpacticoida, and Cyclopoida, easily differentiated by the body shapes of the adults (Fig. 6-1). Calanoid copepods have long antennae, each with 23 to 25 segments. The major body articulation is between the fifth and sixth body segments.[40] The calanoid suborder contains the most abundant of all Potomac copepod species, *Acartia tonsa* and *Eurytemora affinis.* Harpacticoid copepods have cylindrical bodies with the major body articulation between the fourth and fifth segment, and relatively short antennae with only eight or so segments.[40] *Scottolana canadensis* and *Halectinosoma curticorne* are the most common harpacticoid copepods in the Potomac estuary. Cyclopoid copepods are usually smaller than those of the other two suborders. They have a carapace that is appreciably wider than the abdomen and moderately long antennae with about seventeen segments.[40] Of the cyclopoid copepods, *Mesocyclops edax* and *Cyclops vernalis* are the species most often encountered in the upper tidal fresh waters of the Potomac estuary.[37] *Oithona colcarva* (formerly *Oithona brevicornis*) is the most common in the mesohaline regions of the lower estuary.[41]

Although copepod numbers are high in most estuaries, diversity is low, and usually only a relatively small number of species are common. Of nearly 50 species identified from the Potomac estuary (Appendix Table 4), only 9 or so are regularly collected. In oligohaline and mesohaline waters, a single species, *A. tonsa,* often accounts for 95 percent or more of total copepod numbers. The overall seasonal distributions of six common copepods in the Potomac estuary are shown in Fig. 6-7.

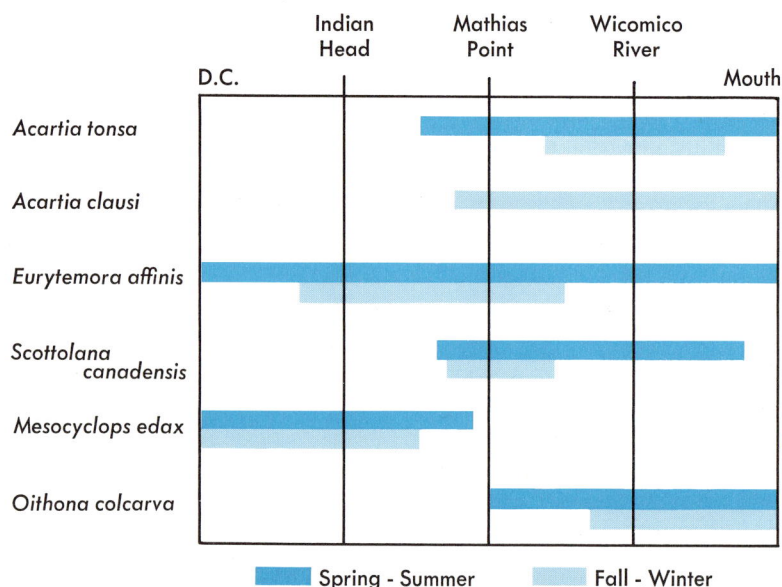

Figure 6-7. *General range and major seasonal distribution of six common copepods of the Potomac estuary (from compilation of data in Refs. 14, 15, 19, 26, 30-33, 37).*

Copepods can either filter-feed on small particles (to about 5.0 μm) or can selectively feed on larger particles.[35] As swimming copepods move through the water, they create a series of vortices which bring small particles toward a filter arrangement formed by a series of bristle-bearing feeding appendages (setae). Many copepods prefer feeding on larger particles, choose from a variety of prey, and can sense the presence of food.[35] These characteristics create a competitive advantage for many copepods over rotifers and cladocerans when food is sparse. Most copepods obtain their energy from plant material, feeding either on phytoplankton or on detritus derived from decomposing marsh grasses. (It has been estimated that in the Patuxent River, Maryland, at least half of the summer planktonic primary production is consumed by *A. tonsa*.[42]) Some copepods are omnivores and also feed on protozoans and bacteria. Copepods, in turn, are eaten by many other animals and are a primary food source for many larval and juvenile fish. Copepod fecal pellets may also provide nutrition to certain benthic species.

Copepods are capable of reproducing and maturing to another egg-producing generation within a week to a month; the variability in time depends on the temperature and the species (Table 6-1). *A. tonsa* requires 7 days at 25.5°C and 13 days at 15.5°C.[42] Eggs hatch into nauplii (the earliest larval stages), which progress through six or so moltings to a copepodite stage. Five more molts are needed before the adult stage is produced. At any one time, a copepod population typically consists of a combination of different life stages.

Quantitative comparisons of copepod abundance data in the Potomac are difficult to make because data are derived from a number of sources, and, in some investigations, individuals from different life stages are enumerated separately. In others, only adults are counted, or perhaps all stages are lumped together. In addition, densities of life stages do not indicate their relative biomass contribution.

In tidal fresh and oligohaline salinities from Washington, D.C., to Maryland Point, *Eurytemora affinis* is the dominant winter-spring copepod. The reported salinity preference of 5 to 12 ppt[41] for *E. affinis* apparently does not hold true in the Potomac estuary. Spring salinities in regions of the estuary where this species is most abundant are oligohaline (0.5 to 5 ppt) (Fig. 6-8). A comprehensive study[33] found that the highest numbers of *E. affinis* characteristically occurred in the vicinity of Maryland Point in spring, with average densities at times reaching 30,000/m³. In summer, *E. affinis* densities declined and averaged about 1,000/m³. Another study[19] found that average densities of *E. affinis* could be as high as 10,000/m³ in the vicinity of Maryland Point in August.

Acartia tonsa may appear in tidal fresh water, although rarely in salinities below 2 or 3 ppt. In spring, *A. tonsa* is found upstream as far as Aquia Creek, but only in low densities.[19] In summer and fall, as higher salinity waters move upstream, it is abundant as far upstream as Sandy Point (at about nautical river mile 65).[19][33]

A widely distributed freshwater cyclopoid copepod is *Mesocyclops edax*, a species considered to be relatively rare in the Chesapeake Bay, but found in high numbers during sampling of the upper Potomac estuary in 1972 and 1973.[19] It was the fourth most abundant species collected in the Douglas Point region and was found during every month, with a peak average density of 11,000/m³ in July and August.[19] Maximum densities were found at the farthest upstream station, off the mouth of Mattawoman Creek (32,100/m³).[19]

In the region of the estuary from approximately Maryland Point downstream to Morgantown, in which salinities range seasonally from oligohaline to mesohaline, *Eurytemora affinis* and *Acartia tonsa* are usually the most abundant copepods. There is a distinct seasonal alternation of dominance between these two species. *E. affinis* is more abundant in late winter and spring, and *A. tonsa* is dominant in summer and fall (Fig. 6-8). A complete annual cycle of *E. affinis* and *A. tonsa* abundances during 1975 in the Morgantown region is shown in Fig. 6-9A. In March, *E. affinis* densities (including both adults and copepodites) averaged around 38,000/m³ over the entire region from Mathias Point to lower Cedar Point, while *A. tonsa* was extremely sparse.[26] By May, with warming waters, *A. tonsa* began to replace the declining *E. affinis*. *A. tonsa* numbers were high throughout summer and early fall. Other studies at Morgantown have shown that, in summer and fall, *A. tonsa* numbers ranged from less than 1,000/m³ to about 70,000/m³, while spring numbers of *E. affinis* ranged from 1,000 to 120,000/m³.[22-25]

Scottolana canadensis, supposedly an epibenthic harpacticoid copepod, has often been identified in late spring plankton collections at Morgantown where it is sometimes more abundant than either *E. affinis* or *A. tonsa*. It occurs primarily in salinites of about 5 to 10 ppt (and sometimes up to 15 ppt). It overwinters in the lower salinities[41] (Fig. 6-7)

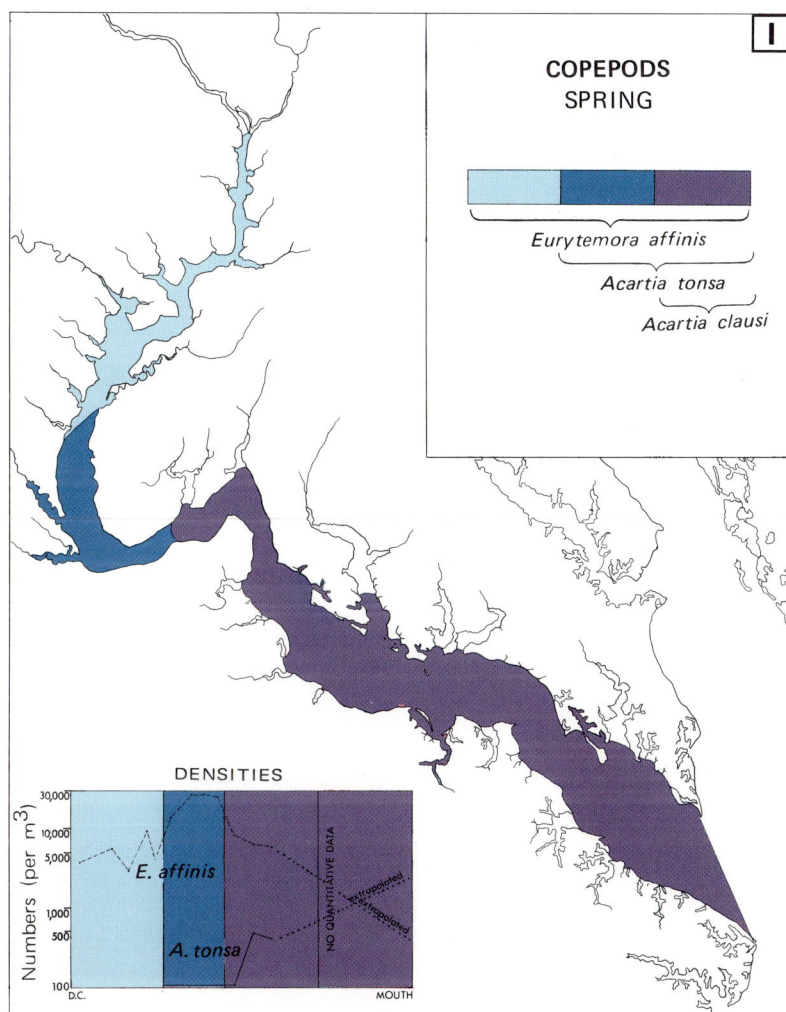

Figure 6-8. *Typical spring and summer distributions of three common copepods in the Potomac estuary. Graphs show average mean densities (excluding nauplii) found along the estuary during several zooplankton studies in 1975-76 (from data in Refs. 15, 33). Densities in lower segment of the estuary are extrapolated from data taken at Calvert Cliffs (Refs. 26, 32, 43).*

and has been found in nearshore waters near Lower Cedar Point and some midestuary tributaries.[14] Spring densities at Morgantown may reach 200,000/m³ (averaging about 6,000/m³),[23][26] but, by August, they have decreased to less than 500/m³.[44] *Oithona colcarva,* which is usually found in higher salinities, appears sporadically in significant numbers in the Morgantown region.[37]

In the high mesohaline region of the lower Potomac estuary, *Acartia tonsa* is the dominant copepod species.[41] In similar mesohaline waters at Calvert Cliffs, it is present throughout the year, with the largest populations occurring towards midsummer and fall.[27] An often observed reduction in numbers in spring or early summer has been attributed to predator grazing (primarily by ctenophores and menhaden) and to other factors such as reduced phytoplankton densities and species composition.[42][45] Usually, a late summer to early fall peak is followed by a reduction in numbers in winter. The center of the

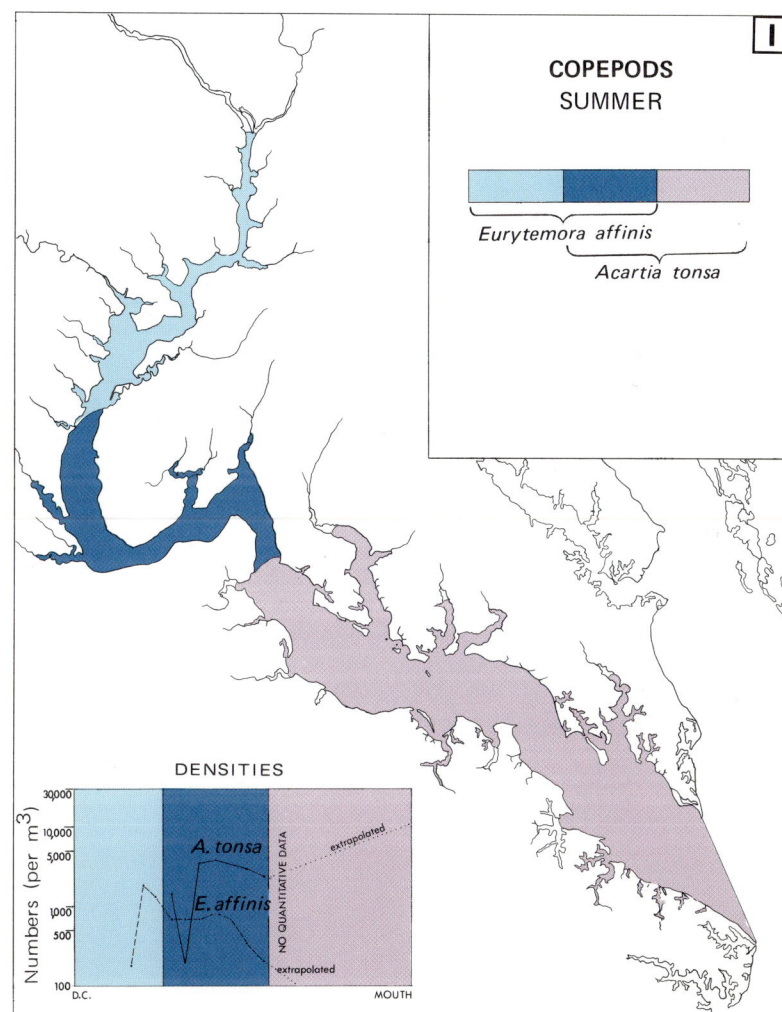

overwintering population has been reported in salinities between 5 and 12 ppt.[41][42] In the Potomac, the same salinities in winter would be found near Colonial Beach.

In mesohaline salinities like those at Calvert Cliffs, there are dramatic and predictable seasonal population shifts between the closely related species *A. tonsa* and *Acartia clausi* (Fig. 6-9B).[27][30-32] These seasonal shifts closely parallel those observed in the lower Chesapeake Bay and other East Coast estuaries where *A. clausi* replaces *A. tonsa* as the dominant copepod during winter months.[29][46][47] The sudden disappearance of *A. clausi* in these areas in late spring or early summer is attributed to its inability to reproduce and ultimately survive in warmer temperatures rather than to competition with emerging *A. tonsa* populations.

Since *A. clausi* is less tolerant of lower salinities than *A. tonsa,* its highest winter abundances would be expected to occur towards the

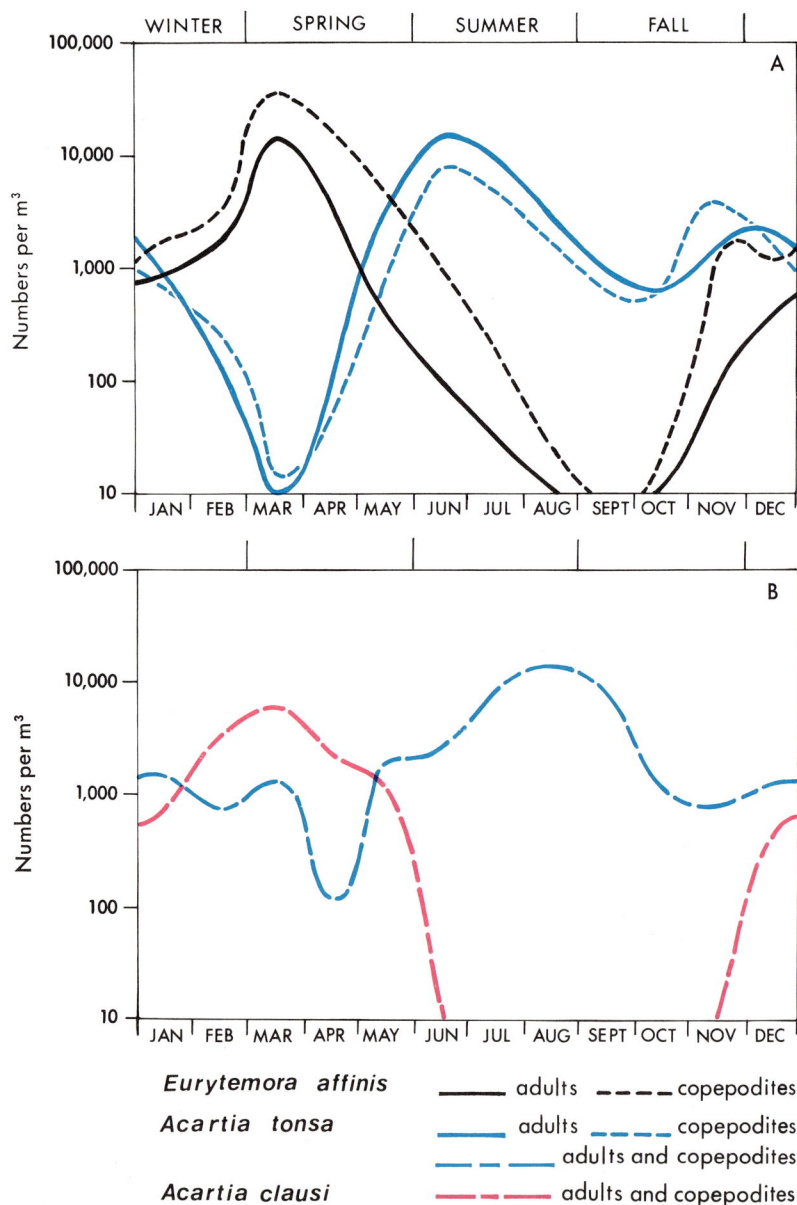

Figures 6-9. *Typical seasonal shifts in abundance of different species of copepods in different salinity regimes. A—*Eurytemora affinis *and* Acartia tonsa *in oligohaline-mesohaline salinities, mean monthly abundances at 5 stations from Mathias Point downstream to lower Cedar Point, January-December 1975 (from data reported in Ref. 26). B—*Acartia tonsa *and* Acartia clausi *in mesohaline salinities (8-14 ppt) at Calvert Cliffs, mean monthly abundances, May 1974-December 1976 (from data reported in Ref. 27).*

mouth of the Potomac. *Scottolana canadensis* is also found in mesohaline waters at least down to the Wicomico River.[41] *Oithona colcarva* is a more important component of the total zooplankton community in the lower Potomac estuary than in the oligohaline and low mesohaline regions. It is considered a fall-winter plankter, being replaced by *Oithona similis* in spring and summer.

MEROPLANKTONIC LARVAE

The most prevalent meroplankton in the Potomac estuary include larvae of barnacles, crabs, shrimps, polychaete worms, molluscs, jellyfishes, and hydroids. These forms are primarily found in the plankton in the mesohaline regions of the Potomac, downstream of Mathias Point (Fig. 6-4). Since direct identification of several of these larvae to the species level is extremely difficult, the taxonomic classification of many larval species is inferred from known distributions of the adult forms.

Barnacle Larvae

Barnacles release planktonic nauplii from eggs brooded in their mantle cavity. Nauplii have characteristically shaped triangular carapaces (Figs. 6-1 and 6-2). They develop through a meta-nauplioid stage and eventually change into cypris larvae, which resemble adult ostracods. Cypris larvae settle to the bottom, attaching themselves to any appropriate hard substrate by means of a stalk, or peduncle. These animals then metamorphose into adults and spend the remainder of their lives as sessile organisms.[6][20]

Barnacles occur rarely in fresh and low salinity areas.[48] The region between Maryland Point and Morgantown appears to be the upper estuarine limit for substantial barnacle settlement or reproduction.[26][37] The presence of barnacle nauplii in this region seems to be influenced by the amount of freshwater flow. From Morgantown downstream, barnacle nauplii are abundant and are present throughout the year. Two periods of peak abundance have been observed at Morgantown, the first and largest one in May and June with mean densities approaching 20,000/m³, and a second in October and November with densities of 3,000 to 8,000/m³.[26] In July and August, barnacle larvae were found to be the fourth most abundant group of zooplankton, exceeded only by the nauplioid, copepodite, and adult stages of copepods. During periods of reduced copepod abundances, barnacle larvae may be the most abundant zooplankton group.[26]

A similar two-season peaking was also evident in the mesohaline waters of the Chesapeake Bay at Calvert Cliffs.[31] In studies conducted in other regions of the Chesapeake Bay (Broad Creek on the Bay's eastern shore; and Solomons, Maryland, and the Rhode River on the western shore), barnacle setting peaked in June and July in salinities of 2 to 18 ppt.[49][50] This presumably followed periods of high abundances of planktonic nauplii in May and June. Moderate setting continued in these regions throughout the summer and fall, with little evidence of a distinct second fall spawning peak.[49][50] It is conceivable that a fall increase in nauplii (as has been observed at Calvert Cliffs[31]) is not always followed by successful setting.

Barnacles, which are algal feeders, have been known to release their nauplii in response to phytoplankton blooms. In May and June of 1975, a peak in barnacle larvae closely followed a diatom bloom in the Morgantown area.[26] Nauplii themselves also serve as food for young stages of commercial fish species such as striped bass and white perch.

Other smaller estuarine fishes such as naked goby and bay anchovy also feed on barnacle nauplii.

There are four species of barnacles commonly found in the Chesapeake Bay: *Balanus improvisus, Balanus eburneus, Balanus balanoides,* and *Chthamalus fragilis.*[51] *B. improvisus* is the most common barnacle in the Potomac and the most common intertidal barnacle of the upper Chesapeake Bay.[49] It is a truly estuarine species, found in salinities between 2 to 20 ppt (Fig. 6-10) and proliferating in salinities between 5 and 11 ppt.[48] *B. eburneus* apparently replaces *B. improvisus* as the dominant form in salinities greater than about 12 ppt.[48] Presumably, barnacle nauplii collected in the lower estuary below the Wicomico River would contain a high percentage of this species.

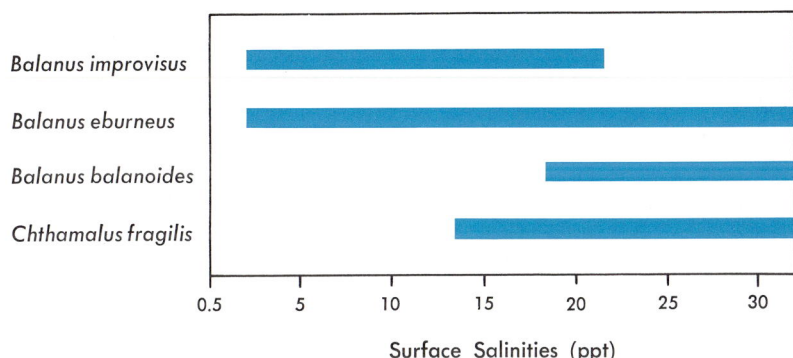

Figure 6-10. *Salinity range of Chesapeake Bay barnacle species (adapted from Fig. 2 in Ref. 48).*

B. balanoides has been found in salinities as low as 18 ppt and so would likely be found at the mouth of the Potomac. Another species, *Balanus amphitrite* has been identified from benthic samples near Morgantown.[52] However, this species was out of its usual range there, and its reported presence at Morgantown may have resulted from its transport on cultch or seed oysters from other regions. There is no record of larval stages of *C. fragilis* in the Potomac estuary, but its existence there is presumed because it has been collected close to the mouth of the Potomac.[48]

Decapod Larvae

Organisms belonging to the order Decapoda are characterized by five pairs of legs and include crabs, shrimp, crayfish, and lobsters. Decapods successfully inhabit many aquatic habitats (from fresh to oceanic) as well as the terrestrial environment. Most species produce eggs that develop into pelagic larvae, and most larvae pass their nauplioid stage within the egg and hatch as zoeae. Zoeae metamorphose through various stages until an adult-like, post-larval form emerges.

Larval decapods taken in zooplankton collections from the Potomac estuary have seldom been identified to the species level. Furthermore, most surveys have been conducted in the lower salinity

portions of the estuary where there is little decapod spawning. Consequently, knowledge of decapod larvae inhabiting the Potomac must be inferred from distributions in other nearby areas. Figure 6-11 plots distributions of decapod larvae that are presumed to occur in the Potomac, based on their seasonal occurrences in similar salinities in the York and Pamunkey Rivers of the Virginia portion of the Chesapeake Bay,[54] [56] the Patuxent River in Maryland,[45] and some North Carolina estuaries.[53] [55] As can be seen, optimum salinities for larval development of most species are greater than 15 to 20 ppt. Consequently, major concentrations of larvae would be found outside of the Potomac estuary in the more saline waters of the lower Chesapeake Bay.

The only species of shrimp zoeae that are likely to be found in the Potomac in any abundance are those of the genus *Palaemonetes*, probably towards the mouth of the estuary below the Wicomico River. Mud crab zoeae are found frequently in the Potomac: during summer months, *Neopanope texana sayi* zoeae have been collected in abundance in the Morgantown region;[26] *Rhithropanopeus harrisii* has greatest larval abundances in waters of 0 to 10 ppt.[54] Larvae of three species of fiddler crabs are typically found in the oligohaline to mesohaline waters. From known distributions, *Uca minax* would also be found upstream in tidal fresh waters of the Potomac; *Uca pugnax* and *Uca pugilator* would be found in salinities from about 5 to 35 ppt.[3] [54]

Adult decapod distributions do not always coincide with the distribution of their larvae. Also, the presence of larvae does not necessarily imply that spawning occurred in the same area. Planktonic zoeae are at the mercy of the currents, and most are oriented towards the bottom where net upstream currents may carry them away from downstream spawning sites.

Other Invertebrate Larvae

Some polychaete worms release planktonic larvae, called trochophores. These are top-shaped or spherical microscopic larvae that have a characteristic band of cilia around the middle and a smaller circle of cilia about the anus (Figs. 6-1 and 6-2). In the Potomac, trochophores of polychaete worms are abundant at times, contributing significantly to the total zooplankton numbers. They are, however, seldom identified to species level. They are largely restricted to brackish and higher salinities, and are rarely found upstream of Maryland Point.

Molluscs have planktonic larvae (Fig. 6-2) which are widely distributed by currents. Little is known of the relative densities of mollusc larvae in the Potomac estuary because those collected are difficult to preserve and identify.

Planula and ephyra larvae of jellyfishes (see Fig. 6-2) are rarely identified or counted in most zooplankton studies. Planula larvae are microscopic round-to-oval forms that are covered with cilia. They are pelagic for only a day or two before they settle onto a firm substrate and metamorphose into a sessile polyp stage. Ephyrae are small aquatic forms that develop into the medusoid stage.

		DISTRIBUTIONS BY SALINITIES	SPAWNING DATES	
		0 5 10 15 ppt	Range	Peak
Shrimps	*Penaeus sp.(a)	>22	Jun-Oct	Jul
	Palaemonetes sp.		May-Sep	Jul
	Palaemonetes pugio (b)		May-Sep	Jul-Aug
	Palaemonetes vulgaris		May-Sep	
	*Ogyrides limnicola	15-25	May-Nov	Jul-Sep
	Crangon septemspinosa (b)	20-25	Jan-Dec	May
Marsh Crabs	Sesarma cinereum	20-26	Jun-Sep	
	*Sesarma reticulatum	15-20	Jun-Oct	Jul-Aug
Spider Crab	Libinia sp.		Jun-Oct	Jul-Sep
Fiddler Crab	Uca sp.		Jun-Oct	Jul
Oyster Crabs	Pinnotheres maculatus	20-25	Jun-Oct	Aug-Sep
	Pinnotheres ostreum	15-25	Jun-Oct	Jul
Blue Crab	Callinectes sapidus	20-32+	Jun-Nov	Jul-Aug
Lady Crab	Ovalipes ocellatus	>25	Jun-Oct	Sep
Mud Crabs	Eurypanopeus depressus (b)	>26	May-Oct	Jun-Jul
	*Neopanope texana sayi	20-25	Jun-Oct	Jun-Sep
	*Panopeus herbstii	20-30	Jun-Sep	Jul-Aug
	Rhithropanopeus harrisii (b)		May-Oct	Jul-Sep

* No Potomac records, but presumed to occur in Potomac estuary.
(a) Distribution in North Carolina estuaries (Ref. 53).
(b) Also distribution in Patuxent River, Maryland (Ref. 45).

■ Greatest abundance   ▢ Reduced abundance

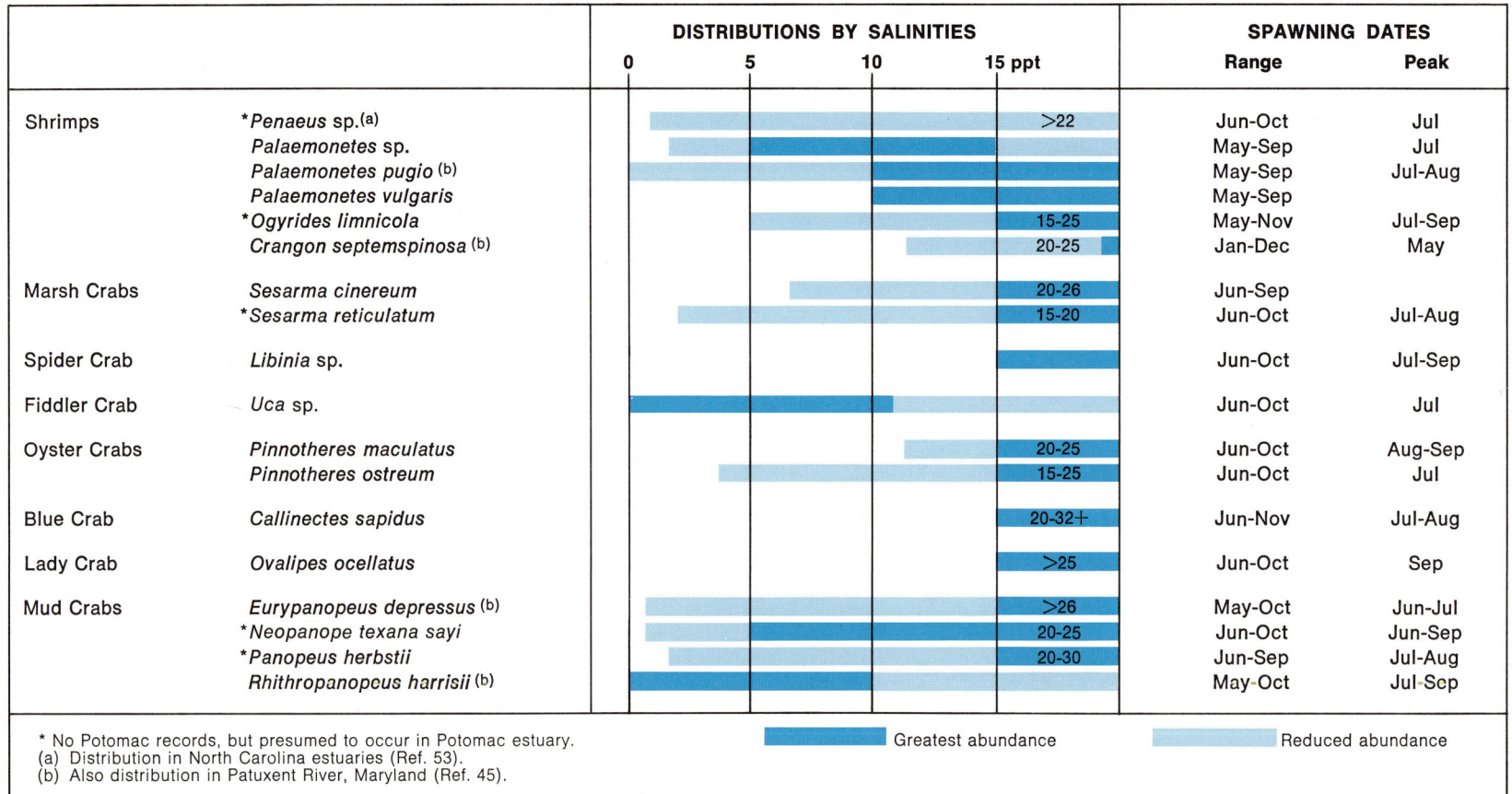

Figure 6-11. *Distribution of pelagic decapod larvae in the York and Pamunkey rivers of the lower Chesapeake Bay and other similar estuaries. Optimum salinities for larval development are in parentheses. (Sources: Refs. 45, 53-56).*

MACROZOOPLANKTON

A variety of larger invertebrates that are easily seen without the aid of a microscope exist within the tidewaters of the Potomac estuary. Such invertebrates are part of what is termed the macrozooplankton (or macroplankton) community, a group which also includes fish eggs and larvae. The principal macroplankton invertebrates of the Potomac are macrocrustaceans, insect larvae, and jellyfishes.

Macrocrustaceans

Mysid shrimp, true shrimp, and certain amphipods and isopods are all macrocrustaceans that spend at least part of their lives in the pelagic habitat. Most are important in the estuarine food web, serving as a component in the diets of larger predators such as fish and birds.

MYSID SHRIMP

Mysid shrimp (in the order Mysidacea) resemble true shrimp in overall appearance. In the Potomac, the only mysid is *Neomysis americana*. It is commonly called the opossum shrimp because it broods its eggs in a specially adapted marsupium or pouch, located just below the thorax. At hatching, juveniles have all their appendages and are adult-like in appearance. At maturity, opossum shrimp may reach 10 mm or more in length.

Neomysis americana ranges from the Gulf of St. Lawrence to Virginia, and is the most abundant mysid in inshore areas of the western North Atlantic.[57] In the mesohaline regions of temperate estuaries, this species is often the most numerous macroplankton organism. In some Chesapeake Bay tributaries, such as the Patuxent River, it is most abundant in salinities of 5 to 7 ppt.[45] In other estuaries, such as Delaware Bay, greatest numbers have been observed in salinities of 10 to 25 ppt.[58,59] Limited data are available on the distribution of the opossum shrimp in the Potomac. It is abundant at Morgantown,[60] but its extension upstream is restricted by the position of the freshwater-saltwater interface. This species has been found only occasionally above Maryland Point.[19]

Opossum shrimp are found in mid-Atlantic estuaries during all seasons and exhibit population peaks two or three times per year. These increases are closely related to brooding periods and to the release of juveniles into the population. Seasonal peaks represent distinct generations — one or two short-lived summer generations and a longer-lived overwintering generation. Populations consist of distinct size groups, reflecting the time of the year of hatching. Summer forms are mature at 6 to 8 mm, while longer-lived overwintering mysids reach maturity at 10 to 11 mm.[57]

Marked variations in population densities from year to year[58] contribute to the wide differences in the reported distributions and abundances of opossum shrimp. Furthermore, daytime collections do not always reveal their actual abundance, since mysids live primarily near the sediment-water interface in the day and rise into the water column at night to feed.[59] Also, low winter densities found in collections do not necessarily reflect true patterns of abundance. Mysids apparently seek deeper waters during colder periods, which makes them difficult to collect.

Although they are considered a fairly shallow water species, opossum shrimp have been collected at Morgantown in high densities in waters of 60 feet (18 m).[60] It is reported that they prefer sandy areas with low silt content,[58] but they were collected at the 60-foot depths near Morgantown where bottom sediments are primarily soft muds.[60]

AMPHIPODS AND ISOPODS

Although amphipods and isopods are primarily epibenthic, the densities of many species in the water column can be high during diurnal migrations. Those that are most often encountered in the water column are listed in Appendix Table 4.

Amphipods are most frequently collected in nighttime plankton samples. *Gammarus fasciatus,* a freshwater form, has been found as far downstream as Morgantown; *Gammarus mucronatus,* an estuarine species, is encountered from the mouth upstream at least to Morgantown.[61] (For more detailed information on the benthic distributions of these species in the Potomac, see Chapter 7.) Considering their large size (with weights approximately 200 times the weight of the copepod *Acartia tonsa*[62]), the contribution of the amphipods to the planktonic biomass can be significant during their population peaks. Since they graze on epifaunal communities, and are in turn eaten by fish such as striped bass, white perch, spot, and croaker, amphipods serve as a link between the lower trophic levels and higher nektonic life forms.

Parasitic isopods such as *Lironeca ovalis* have a pelagic stage before they attach to a host fish.

TRUE SHRIMP

In the Potomac estuary, only two groups of shrimp are found in considerable numbers — the grass shrimp and the sand shrimp.

Grass Shrimp

Grass shrimp of the genus *Palaemonetes* are the most abundant shrimp species in the Chesapeake Bay. Of the three *Palaemonetes* species likely to occur in the Potomac, only two, *Palaemonetes intermedius* and *Palaemonetes pugio,* have actually been recorded. Since it has been recorded in the lower Patuxent River,[63] *Palaemonetes vulgaris* would also presumably be found in the lower Potomac. *Palaemonetes pugio* is recognized as the most common *Palaemonetes* species in the Chesapeake Bay. It is euryhaline, ranging from oligohaline to poly-

haline regions.[3] In the Potomac, *P. pugio* can be expected upstream at least to Maryland Point or a little beyond.[64] *P. intermedius* is a mesohaline organism[3] and has been found as far upstream as Morgantown.[65]

In summer, grass shrimp are found in the shallows around submerged vegetation. In winter, they disappear from the shore zone, presumably to seek warmer waters. It is not known whether they move downstream to higher salinities in winter as sand shrimp do.

Sand Shrimp

From studies of similar areas in other estuaries,[43] [54] [66] [67] it can be inferred that the sand shrimp, *Crangon septemspinosa,* would be one of the most common macrocrustaceans in the mesohaline portion of the Potomac estuary. It has a tendency to burrow into sandy bottoms during the day and to become planktonic at night.[43] [66] Sand shrimp are distributed along the Atlantic from Newfoundland to Florida, but the Delaware Bay and Chesapeake Bay region is probably the southern limit for any substantial populations of this species. From known salinity distributions,[66] it can be presumed that sand shrimp would occur as far upstream as Maryland Point, but highest densities would occur only towards the mouth of the estuary.

In studies of the Chesapeake Bay and the Delaware Bay,[56] [67] it appears that sand shrimp move into the shore zone during midspring when temperatures exceed 10° C, and move back into channel areas in the fall when water temperatures again fall below 10° C. Apparently, they can only survive in low salinity waters under higher temperatures. When the water reaches 5° C or lower, they tend to move farther downstream into higher salinity waters.[56] [67]

Sand shrimp have a single extended breeding season. Since optimum spawning occurs only in salinities of 18 ppt or greater (Fig. 6-11), egg-bearing females are likely to be sparse in most of the Potomac. Females with eggs may be present from November through June, with overall peak numbers occurring in March and April. In the York and Pamunkey river systems, larvae from breeding females first appear in December and are found through June, with highest numbers occurring during May.[54]

Insect Larvae

Many insects have aquatic larvae that are quite unlike the adult forms. Insect larvae are segmented, usually ranging in length from about 5 to 70 mm, with most from 10 to 30 mm.[34] Commonly called "nymphs," they may live in the water for only a few days or for many months before eventually metamorphosing into adults and emerging from the water.[34] The aquatic larval stages of most insects live on the bottom or are closely associated with bottom detritus, aquatic plants, debris, or stones. However, some larvae swim and are part of the macroplanktonic fauna. (See Appendix Table 4 for benthic and pelagic species.)

Members of the order Diptera (which includes flies, mosquitoes, and midges) are the only insect larvae caught in significant numbers in fresh-to-brackish waters of the Potomac estuary. Their larvae are most abundant in fresher waters downstream to Maryland Point; however, they have been collected in low numbers in salinities as high as 10 ppt.[19]

Chaoborus sp. (mosquitoes) are the most common dipterans collected in zooplankton samples. They have two seasonal peaks in abundance, the first in March and the second in August, as indicated by samples taken between Mattawoman Creek and Maryland Point[19] over a period of a year. Dual peaking indicates there are two generations of chaoborids each year; one hatches during winter and early spring, matures in early summer, and produces the eggs for the second generation, which, in turn, breed in winter to produce the spring generation of larvae.[19]

The only other group of insect larvae collected regularly in the Potomac are those belonging to the midge family Tendipedidae (Chironomidae). Numerous tendipedid species have been identified from benthic samples in the Potomac estuary (see Appendix Table 4); some of these species are pelagic. The pelagic forms are found throughout the year and have the highest densities in summer when several generations are produced.[19]

Jellyfishes

The term jellyfish, used in the broadest sense, encompasses three types of gelatinous or jelly-like macroplanktonic animals — hydromedusae, true jellyfishes, and comb jellies (ctenophores). Hydromedusae and true jellyfishes belong to the same phylum (Cnidaria) and are meroplanktonic — their medusoid life stage floats in the water, alternating with a sessile benthic polyp generation (Fig. 6-12). Comb jellies (in the phylum Ctenophora) are holoplanktonic.

HYDROMEDUSAE

Although they are generally only a few millimeters in diameter and are transparent, hydromedusae can be seen without the aid of a microscope because light reflects off their bells and tentacles. They are released to float free from parent hydroid polyps (of the class Hydrozoa), which remain attached to various bottom substrates. (For a description of hydroid to medusoid generations, see Chapter 7.) Only the Potomac species of hydroids that produce medusae are listed in Appendix Table 4. There are two basic types (Fig. 6-1): bell-shaped anthomedusae with a few tentacles and only a narrow opening under the bell (characterized in the Potomac estuary by *Bougainvillia rugosa* medusae), and flattened umbrella-shaped leptomedusae with multiple tentacles and broad subumbrella surfaces (exemplified by *Clytia longicyatha* medusae). Hydromedusae are fragile, and thus are easily destroyed in sampling and sorting procedures. As a result, identification is difficult, and their actual numbers are most likely much greater than might be suggested by most zooplankton collections. No specific studies on hydromedusae have been conducted in the Potomac estuary, but they are found in other estuarine regions, principally during summer months. In the Potomac, species that bear planktonic medusae do not occur in fresh water[34] and are rarely found in oligohaline waters.

Hydromedusae feed indiscriminately on small crustaceans and other plankton. Whether they are prey for other animals or have any other ecological importance is not known.*

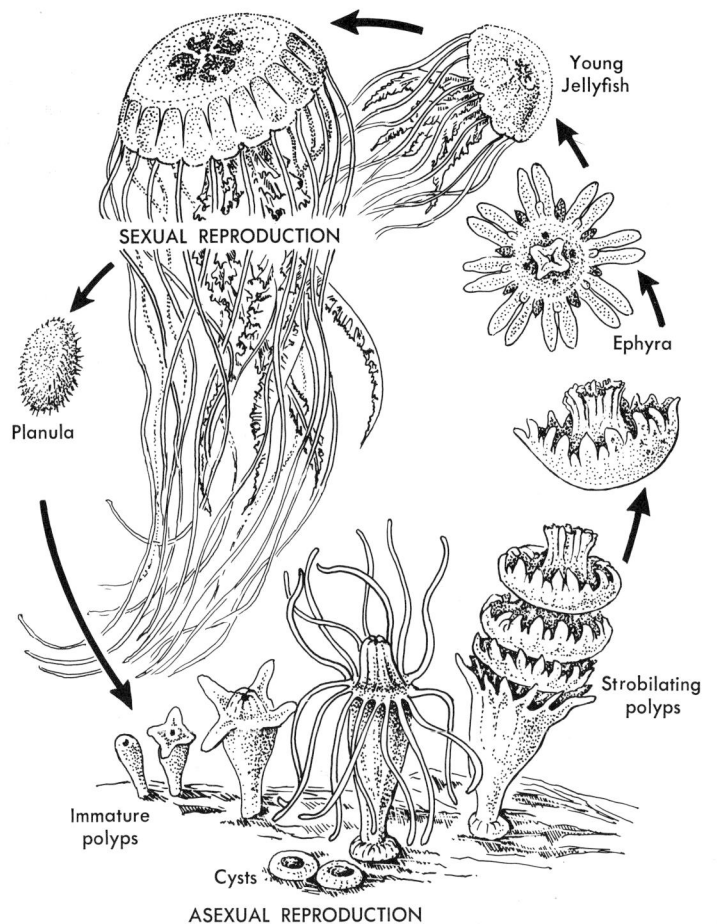

Figure 6-12. *Life cycle of the sea nettle from polyp to adult. (Source: Ref. 7)*

TRUE JELLYFISHES

True jellyfishes belong to the class Scyphozoa. The adult jellyfish represents only one of the life stages of scyphozoans. The large, jelly-like bells are the sexually mature stage of the life cycle. At spawning, sperm are set free into the water and enter the female as it pumps water through the interior cavity of the medusa. The fertilized eggs develop into planula larvae (Fig. 6-12), which eventually escape from the bell into the water to become part of the microscopic meroplankton. After a day or two, the larvae settle down on a suitable, hard substrate (preferably the undersides of clean oyster shells)[69] to develop into small (no more than a few millimeters high) polyps. Tentacles form around the margins of a central mouth, and the polyp begins a more stable and longer-lived existence than its medusoid counterpart.[7] Polyps are able to reproduce many times over by asexual budding and can encyst to survive through periods of environmental stress. Polyps periodically undergo strobilation (asexual body division) to produce the medusoid sexual stage. When strobilation is completed, a series of flattened

---

*For further information on the taxonomy and seasonal occurrences of hydromedusae in the southern Chesapeake Bay, see Ref. 68.

juvenile medusae (called ephyrae) are released into the water column from the top of the polyp. In a period of two months, ephyrae develop into adult medusae and thus complete the cycle. Through these intricate adaptations, jellyfishes have assured their reproductive success in the estuary.

Medusae are capable of some self-propulsion, which is accomplished by the rhythmic contraction and expansion of the bell, and they frequently make vertical migrations. They are nevertheless at the mercy of tidal currents and wind action, which may transport them for greater distances, creating dense concentrations in some areas and none in others.

Jellyfishes are carnivorous and opportunistic feeders, entangling and capturing prey stunned by the hundreds of stinging cells (nematocysts) on their tentacles.[7] Prey are gradually passed up through the mouth into the interior of the bell where they are digested. Half-digested food items may often be seen inside a jellyfish. Sea nettles (*Chrysaora quinquecirrha*) are known to prey heavily on comb jellies.[70] All jellyfishes are also major predators on zooplankters, which are pumped into the bell along with the water.

Jellyfish medusae provide food for a few organisms such as crabs and certain fishes.[58 70] Interestingly, a symbiotic relationship is often established between juvenile fishes and medusae. In the Chesapeake Bay, small harvestfish, *Peprilus alepidotus*, are often seen swimming among the tentacles of sea nettles. The medusae provide both a haven from predators and a readily accessible food source to the young fish, which may have some immunity to the toxin in the nematocysts.[71] Ephyrae are eaten by sea anemones, harvestfish, and barnacles.[70 72] Polyps are grazed on by nudibranchs (primarily *Cratena pilata* in the Chesapeake Bay).[7]

Distribution of Jellyfishes in the Potomac

There are three species of jellyfish recorded from the Potomac estuary, the infamous summer sea nettle, *Chrysaora quinquecirrha;* the innocuous but also highly abundant winter jellyfish, *Cyanea capillata;* and the moon jellyfish, *Aurelia aurita,* which is an occasional visitor.

The **sea nettle** is the most familiar and, in the summer, the most abundant of all Potomac jellyfishes. Their medusae are milky white, sometimes with dark red radiating stripes on the bell. All sizes may be found simultaneously as the summer progresses, the maximum with bells approximately 250 mm in diameter.[58] Sea nettles do not tolerate salinities less than about 5 ppt[69 73 74] and are generally found in salinities of 7 to 20 ppt. Sea nettle abundances increase downstream with increasing salinities.

On the basis of bottom salinities, overwintering cysts (which can tolerate salinities down to 5 ppt[69]) can be expected to be found throughout the estuary up to the vicinity of the Port Tobacco River (Fig. 6-13). In two Chesapeake Bay tributaries, the Patuxent River[69 73] and the York River,[72] polyps begin to emerge from cysts when temperatures reach 18 to 22° C in late May and early June. Small medusae first appear in smaller tributaries where the salinities are greater than 7 ppt. Over a period of 15 to 30 days, they gradually appear in the main stem.[69] Ephyrae are released in greatest numbers in the spring, although they

may continue to be liberated throughout the summer.[74] There is a general downstream movement of most sea nettles as the summer progresses, although a few may be moved upstream by currents. Greatest numbers are usually observed in July. Spawning apparently has begun by that time and continues through the summer and into early fall. Since medusae do not generally tolerate reduced fall temperatures, sea nettles are scarce by September, and by November, all medusae have died, leaving their progeny as newly-set polyps.

During most summers, when sea nettle populations reach their zenith, their presence makes swimming in Potomac waters virtually impossible. Many a swimmer has been startled by the sting from the long tentacles of these drifting animals.

Winter jellyfish are the most abundant jellyfish during the winter months, but, because they do not appear in the estuary until December and are gone by the end of May, they are not usually considered a nuisance species. Winter jellyfish are orangish-brown and are approximately the same size or slightly larger than sea nettles. They tolerate lower salinities than sea nettles and may be found in the main stem as far upstream as Indian Head and in tributaries to the extent of tidewaters (Fig. 6-14). This species, however, does not reproduce in the Potomac. Polyp and cyst stages are formed only in salinities greater than 20 ppt (and have only been observed in various areas throughout the lower Chesapeake Bay[69]). The seasonal sequence of life stages of the winter jellyfish is opposite to that of the sea nettle. Planula larvae are released from medusae, and polyps are set from February to May. In October, when water temperatures fall below 15° C, polyps no longer remain encysted, and they undergo strobilation.[69]

Moon jellyfish are the largest of the Potomac jellyfishes. The moon jellyfish is readily distinguished by its large size (bell diameter usually about 300 mm), by the absence of long tentacles, and by its four prominent, deeply colored, horseshoe-shaped gonads, which are easily seen through the surface of the bell. Moon jellyfish are distributed in the same mesohaline to polyhaline regions as sea nettles, first appearing somewhat later in the summer and remaining until fall.[7 77] They occur in far greater numbers in the lower Bay than in the Potomac. Their polyps and ephyrae are found only in higher salinities; strobilation occurs in late May to early June.

COMB JELLIES

Comb jellies are classified in a separate phylum (Ctenophora) from cnidarians, although many of their characteristics are similar. However, they have no tentacles (or a reduced number of tentacles), no stinging nematocysts, and no polyp phase. Comb jellies have cilia which are usually arranged in long comb-like bands up and down their bodies, giving them their common name. They can move through the water by moving their cilia as well as by pulsation. Comb jellies are represented in the Potomac by two species: the mermaids purse, *Beroë ovata,* and the sea walnut, *Mnemiopsis leidyi.*

Beroë ovata is generally distributed in the Chesapeake Bay only within mesohaline waters where salinities are greater than 15 ppt,[70 77] but it has been found near Morgantown in the Potomac estuary in late

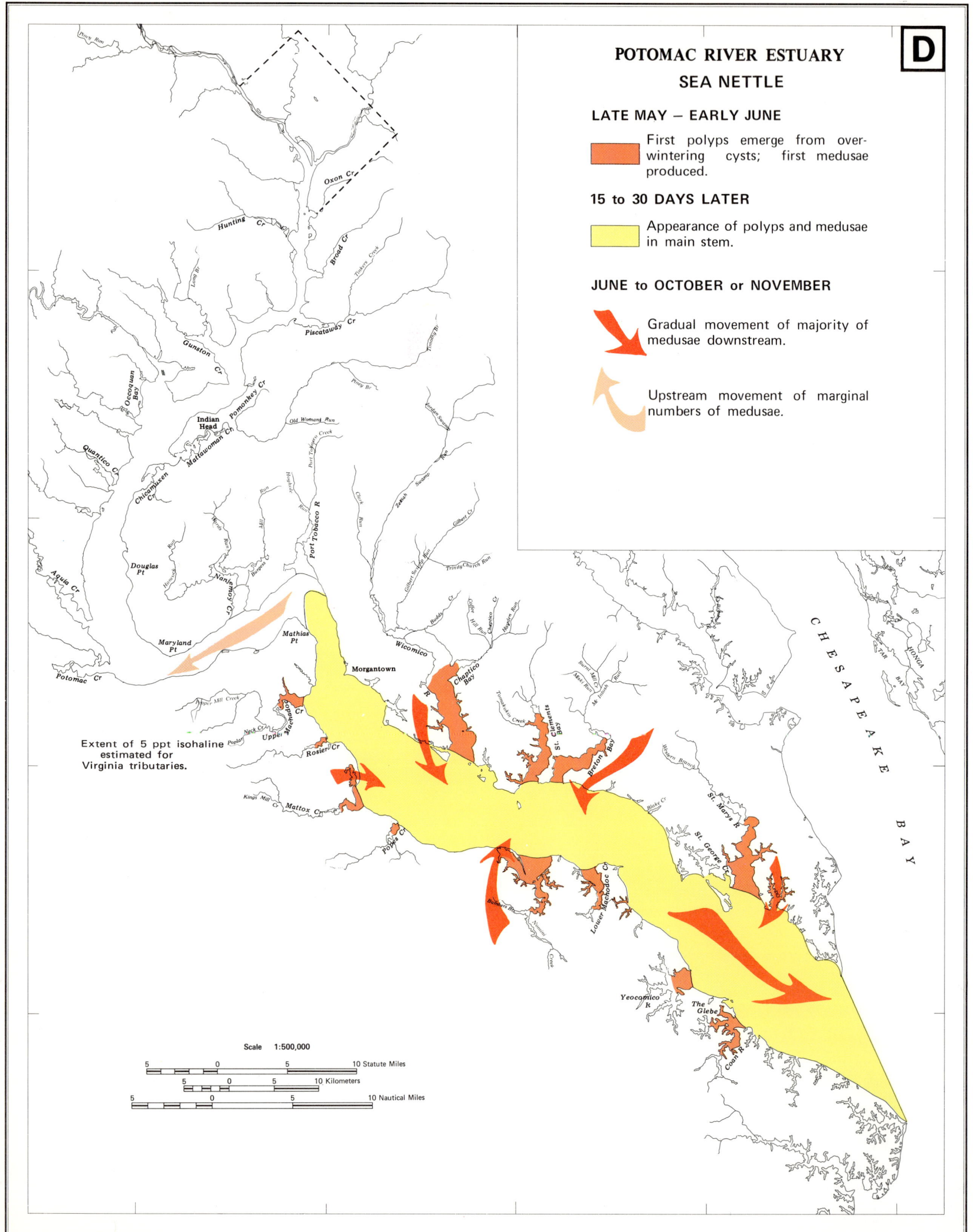

POTOMAC RIVER ESTUARY

SEA NETTLE

D

LATE MAY — EARLY JUNE

First polyps emerge from over-wintering cysts; first medusae produced.

15 to 30 DAYS LATER

Appearance of polyps and medusae in main stem.

JUNE to OCTOBER or NOVEMBER

Gradual movement of majority of medusae downstream.

Upstream movement of marginal numbers of medusae.

Extent of 5 ppt isohaline estimated for Virginia tributaries.

CHESAPEAKE BAY

Scale   1:500,000

5   0   5   10 Statute Miles

5   0   5   10 Kilometers

5   0   5   10 Nautical Miles

Figure 6-13.   *General seasonal distribution of life stages of the sea nettle in the Potomac estuary, based on distributional records from the Potomac estuary and other regions of the Chesapeake Bay (Refs. 69, 72-76). The predicted occurrence of sea nettle polyps above Morgantown is based on known salinity tolerances (Ref. 73).*

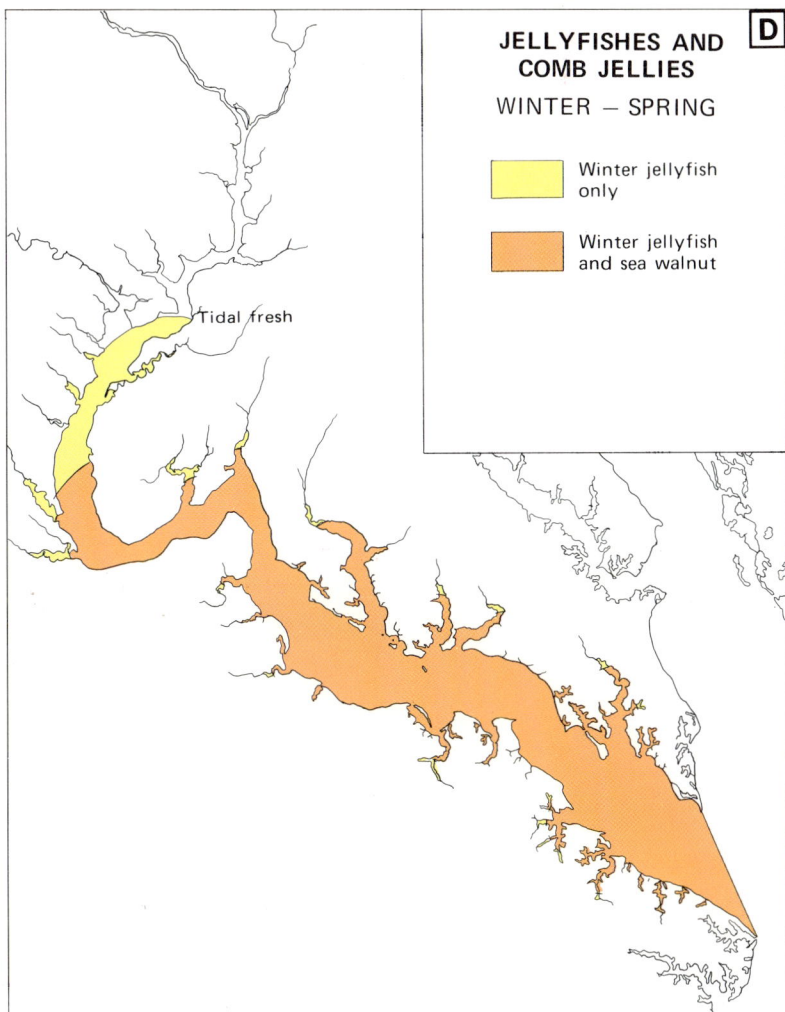

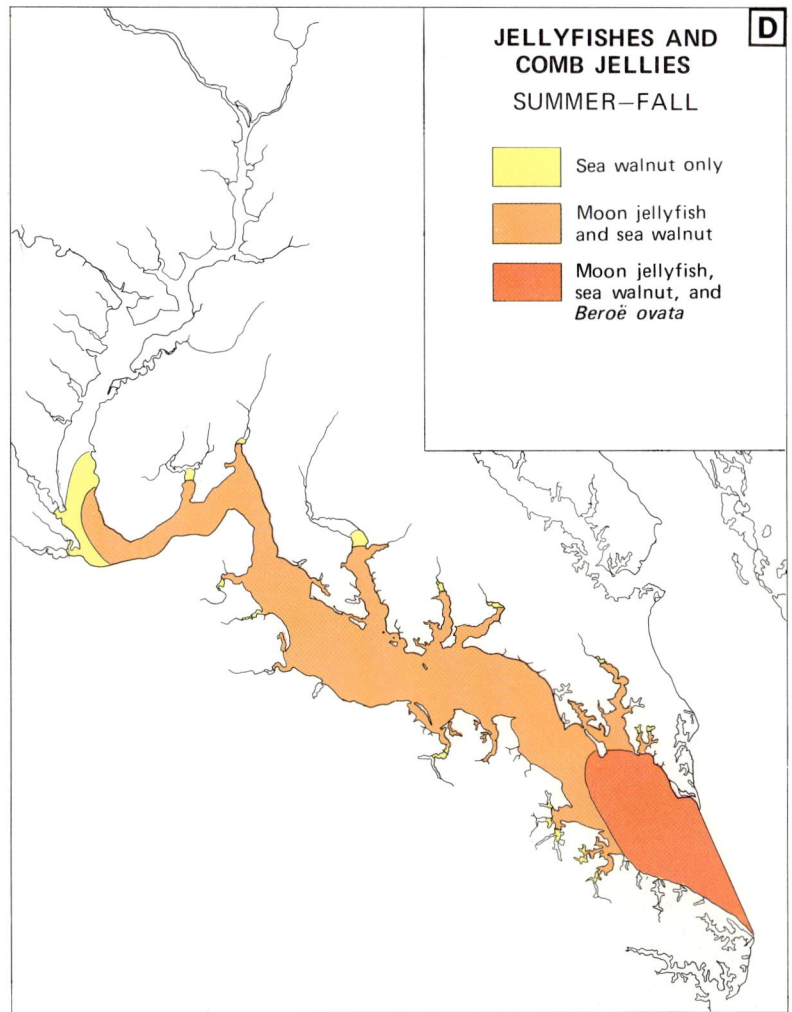

Figure 6-14. *General seasonal distributions of winter jellyfish, moon jellyfish, and comb jellies in the Potomac estuary from distributional records in the Potomac estuary and other regions of the Chesapeake Bay. (Sources: Refs. 3, 69)*

winter in salinities lower than 15 ppt.[37] Little else is known of the distribution of this species in the Potomac.

Sea walnuts are very common in the Potomac estuary. These transparent and fragile creatures are virtually invisible in the water during daylight. At night, however, with any slight disturbance, they luminesce with a soft green light. At any one time, groups of individuals of various sizes congregate together. Most range in size from 50 to 80 mm, but some sea walnuts may reach 250 mm.[58] Since sea walnuts clog plankton nets when they are most abundant, they are a great nuisance to scientists attempting to collect other forms of plankton.

Sea walnuts occur in salinities as low as 4 or 5 ppt, but are generally more abundant at slightly higher salinities. Their distribution in the Potomac is shown in Fig. 6-14. In late summer or fall when freshwater flow is low, they may occasionally be found up to Douglas Point.[19] Major concentrations, however, occur only below Maryland Point. They are found in equal abundance in both the tributaries and the main stem. Sea walnuts are present throughout the year, but are most abundant in late spring to early summer, persisting in considerable numbers into the fall.[43,60] They are found at all depths in patchy concentrations, but at night they seem to prefer surface waters. During rough weather, or in low salinities, they tend to inhabit deeper waters.[66,70]

Because they filter-feed indiscriminately on phytoplankton and zooplankton and are at times extremely abundant, sea walnuts have a tremendous impact on the ecosystem. It has been demonstrated that they are capable of consuming an average of 470 copepods per hour.[78] At this rate, estimated ctenophore densities in the Patuxent River[42] were sufficient to consume 31 percent of the total *Acartia tonsa* biomass daily, accounting for 52 percent of this copepod's daily summer mortality.[78] On the other hand, it has been documented that sea walnuts serve as a food source for *B. ovata* and for the sea nettle, *Chrysaora quinquecirrha.*[79] Thus, one jellyfish species may at times influence the population levels of another. Ctenophores have few other predators, although harvestfish and butterfish have been known to eat them.[70,71]

Chapter 6 — References

1. Biological Methods Panel Committee on Oceanography, 1969
2. Lackey, 1967
3. Wass, 1972
4. Colwell, 1972
5. Gross, 1972
6. Meglitsch, 1972
7. Schultz and Cargo, 1971
8. Rasmussen, 1973
9. Barnes, 1968
10. Galtsoff, 1964
11. Boyce Thompson Institute for Plant Research, 1977
12. Sandifer, 1972
13. Lippson, A.J., 1973
14. Loos, 1975
15. Boynton et al., 1977
16. Tatum et al., 1966
17. Sandoz and Johnston, 1966
18. Van Vaupel-Klein and Weber, 1975
19. Ecological Analysts, Inc., 1974
20. Hardy, A., 1965
21. Dahlberg, 1973
22. Heinle et al., 1973a
23. Heinle et al., 1973b
24. Heinle, H.S. Millsaps, and Lawson, 1974
25. Heinle and K.V. Millsaps, 1973
26. Sage and Olson, 1977a
27. Olson and Sage, 1978
28. Bosch and Taylor, 1970
29. Jacobs, 1978
30. Sage, 1976
31. Sage and Olson, 1977b
32. Sage and Bacheler, 1978
33. Maryland Department of Natural Resources, 1974-1976.
34. Pennak, 1953
35. Allan, 1976
36. Chambers, Burbidge, and Van Engel, 1970
37. Sage, Summerfield, and Olson, 1976
38. Bosch and Taylor, 1967
39. Bosch and Taylor, 1973
40. Miner, 1950
41. Heinle, 1973
42. Heinle, 1966
43. Wakefield, 1977
44. Heinle, H.S. Millsaps, and Lawson, 1973
45. Herman, Mihursky, and McErlean, 1968
46. Jeffries, 1962
47. Conover, 1956
48. Gordon, 1969
49. Branscomb, 1976
50. Shaw, 1967
51. Van Engel, 1972a
52. Pfitzenmeyer, 1974
53. Williams and Deubler, 1968
54. Sandifer, 1973
55. Knowlten, 1970
56. Haefner, 1976
57. Wigley and Burns, 1971
58. Delaware Coastal Management Program, 1976
59. Hopkins, T.L., 1965
60. Mihursky, 1973
61. Academy of Natural Sciences of Philadelphia, 1971b.
62. Cronin, Daiber, and Hulbert, 1962
63. Cory, 1967
64. Pfitzenmeyer, 1976
65. Krueger and Fuller, 1977
66. Browne et al., 1976
67. Price, 1962
68. Calder, 1971
69. Cargo and Schultz, 1967
70. Calder, 1972c
71. Mansueti, 1963
72. Cones and Haven, 1969
73. Cargo and Schultz, 1966
74. Calder, 1972b
75. Littleford, 1939
76. Rice and Powell, 1970
77. Calder, 1972a
78. Bishop, 1967
79. Burrell, 1968

Benthic
Invertebrates

Benthic Invertebrates

Sediments and the surfaces of submerged objects such as rocks and pilings provide habitats for many species of organisms, collectively known as benthic invertebrates. Crabs, clams, and oysters are some of the larger and more familiar examples of the benthic fauna. Also important to the estuary are the many smaller benthic organisms such as worms, sponges, snails, and shrimp-like crustaceans. These organisms have significant roles in energy and material flows throughout the food web because they are the primary foods of many finfish, waterfowl, and blue crabs.[1][2][3]

Most knowledge about the benthic invertebrates of the Potomac is limited to organisms that are larger than 0.5 millimeters (mm). The animals in this group are collectively called the macrofauna, or macroinvertebrates. Smaller organisms are called meiofauna [between 0.5 mm and 63 micrometers (μm)] and microbiota (smaller than 63 μm).[4][5]

The meiofauna consist of a broad variety of taxonomic groups such as nematodes, harpacticoid copepods, kinorhynchs, and tardigrades, as well as juveniles of some macroinvertebrate species. Millions of meiofaunal organisms may occur in only a few square centimeters of Potomac bottom.[4][6] Because of their small size and difficulties associated with identifying them, this class of invertebrates has not been intensively studied in the Potomac.[7][8][9]

The microbiota are extremely abundant unicellular organisms. Billions occur in each milliliter of estuarine sediment, and include bacteria, fungi, protozoans, and blue-green algae.[10] These organisms control the chemistry of the sedimentary environment.[11][12] They are also the initial colonizers and primary decomposers of most of the detrital material settling to the floor of the estuary.[13] In the sediments, the microbiota (primarily bacteria) convert organic materials into inorganic nutrients by decomposing detrital matter. Recovered nutrients are then recycled into the water through various resuspension and diffusion processes and become available for use by primary producers.[5][11][12][14] Many macroinvertebrates do not directly utilize detritus but, instead, obtain their nutrition by eating the meiofauna and microbiota that occur on detrital particles.[13] These smaller benthic organisms form an important link between benthic and planktonic food webs.[5] But, as with the intermediate sized meiofauna, detailed information on the Potomac microbiota is scarce.

The discussions in the rest of this chapter are devoted to the benthic macroinvertebrates inhabiting the Potomac estuary. Included in this group are a wide variety of organisms that have developed diverse adaptations for survival in the estuarine environment. (Some characteristic assemblages of macroinvertebrates inhabiting the Potomac estuary are illustrated in Figs. 7-1 and 7-2.) The first part of the discussion will describe the adaptive strategies that are favorable to the survival and propagation of different benthic macroinvertebrates under various environmental conditions. The second part will describe the major taxa that are found in the Potomac estuary.

HABITAT PREFERENCES, FEEDING MODES, AND REPRODUCTIVE STRATEGIES

Benthic macroinvertebrates can be divided into two broad groups, based on whether they live in or on the bottom: the infauna and the epifauna, respectively.[15][16]

Infauna

Infaunal organisms live in and burrow through the bottom sediments; most have relatively limited mobility. Like most organisms, they need oxygen to live, but only the upper layers of estuarine sediments contain oxygen. Depressed oxygen levels occur in the deeper sediment layers where naturally produced toxic compounds such as hydrogen sulfide also occur.[5][11][12] These deeper sediments often contain a bountiful food supply, and the deeper an infaunal organism lives in the sediments, the less likely it is to be eaten by predators. Consequently, many infaunal organisms have a variety of behavioral, morphological, and physiological adaptations that allow them to inhabit the deeper sediments. Some infaunal species construct burrows through which they circulate water containing oxygen. These burrows may either be temporary spaces created as the organism moves through the sediment (such as the burrows of species 5 and 6 in Fig. 7-3), or they may be relatively permanent, leathery, tube-like structures constructed from body secretions that are sometimes combined with sand, mud, or other debris (species 1 through 4 in Fig. 7-3). Some species have long hose-like body parts, called siphons, which extend from within the sediments to the overlying water mass. Water containing oxygen is pumped through the siphons and across respiratory surfaces and other body tissues (species 5 and 6 in Fig. 7-3). Many infaunal organisms have alternate metabolic pathways that allow them to survive for short periods (several hours to several days) at depressed oxygen levels.[18][19]

Many infaunal organisms ingest the materials deposited on the bottom, including the sediments.[5] Most of these deposit feeders selectively feed on particles of a particular size range that contain relatively high concentrations of preferred organic materials. However, some ingest sediments indiscriminately.[5][20] Since the digestion process of deposit feeders is not efficient, only a small portion of the ingested materials — the microbiota and meiofauna occurring on the sediments or on detrital particles — is used as food.[13] Thus, deposit-feeding infauna must consume large quantities of sediments to obtain sufficient nutrition for growth and reproduction. This requirement results in considerable mechanical reworking of the sediments — a process referred to as bioturbation.[5]

Deposit-feeding infauna do not usually reingest material that has recently passed through their guts, because they package the undigested and undigestable materials in mucus-bound pellets that are not in the size range preferred for ingestion. Undigested microbiota grow rapidly within the pellets, breaking down the binding. The pellets become smaller and smaller, and by the time they are reduced to a size that is likely to be reingested, microbial growth has increased their nutritional value for the infauna to a relatively high level.[5][20][21]

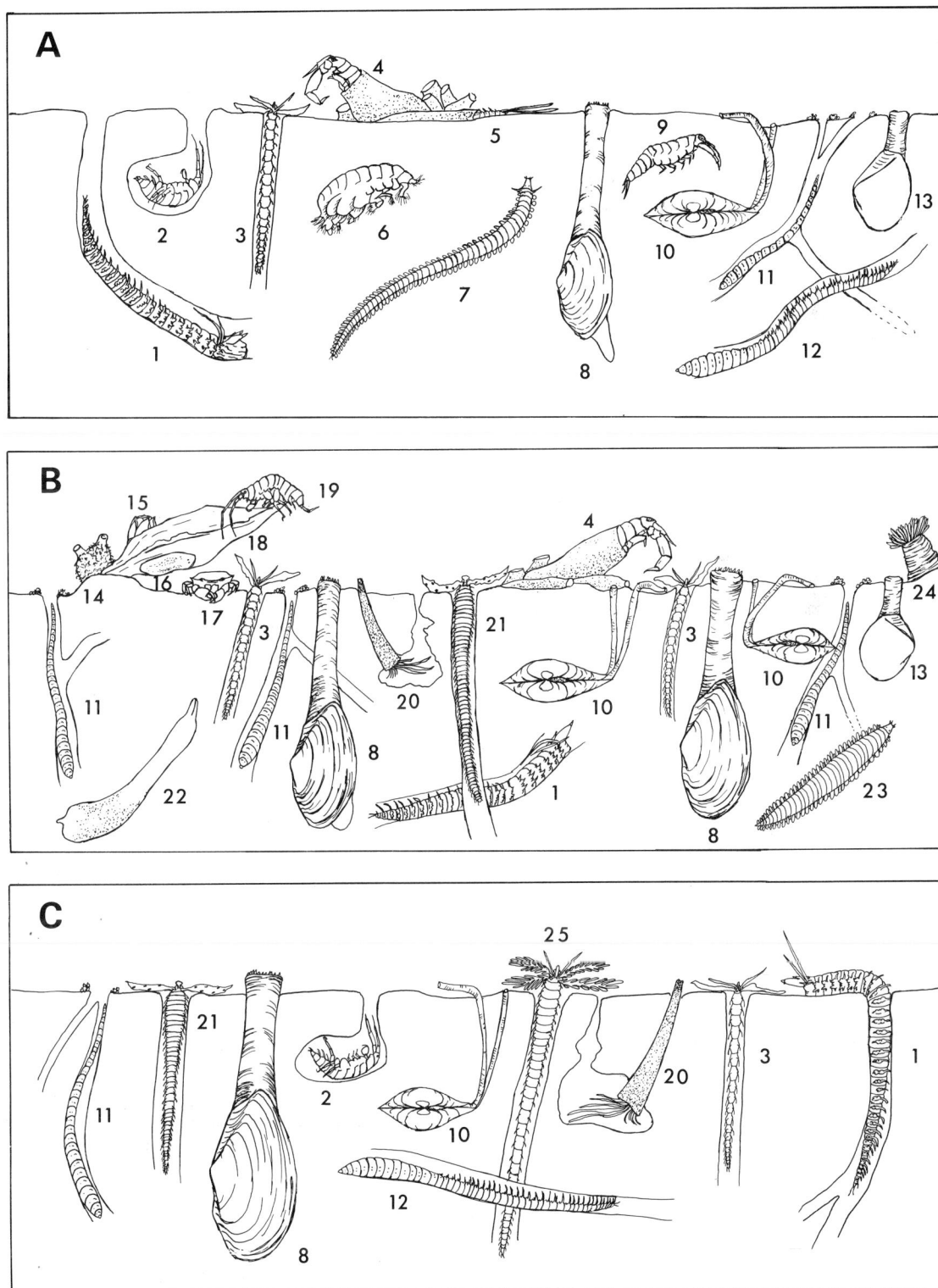

1. *Nereis succinea* (burrowing polychaete)
2. *Leptocheirus plumulosus* (tube-building amphipod)
3. *Streblospio benedicti* (tube-building polychaete)
4. *Corophium lacustre* (tube-building amphipod)
5. *Polydora ligni* (tube-building polychaete)
6. *Lepidactylus dytiscus* (burrowing amphipod)
7. *Eteone lactea* (burrowing polychaete)
8. *Mya arenaria* (soft-shell clam, burrowing clam)
9. *Monoculodes edwardsi* (burrowing amphipod)
10. *Macoma balthica* (burrowing clam)
11. *Heteromastus filiformis* (tube-dwelling polychaete)
12. *Scoloplos fragilis* (burrowing polychaete)
13. *Mulinia lateralis* (coot clam, burrowing clam)
14. *Molgula manhattensis* (sea squirt, fouling organism)
15. *Balanus improvisus* (barnacle, fouling organism)
16. *Stylochus ellipticus* (flatworm)
17. *Rhithropanopeus harrisii* (mud crab)
18. *Crassostrea virginica* (American oyster)
19. *Melita nitida* (amphipod)
20. *Pectinaria gouldii* (tube-building polychaete)
21. *Scolecolepides viridis* (tube-building polychaete)
22. *Micrura leidyi* (burrowing proboscis worm)
23. *Eteone heteropoda* (burrowing polychaete)
24. *Diadumene leucolena* (sea anemone, fouling organism)
25. *Paraprionospio pinnata* (tube-building polychaete)

Figure 7-1.   *Schematic representation of frequently encountered benthic macroinvertebrates of the higher salinity regions of the Potomac in relation to various soft sediment habitat types: A — sand, B — muddy-sand, C — mud.*

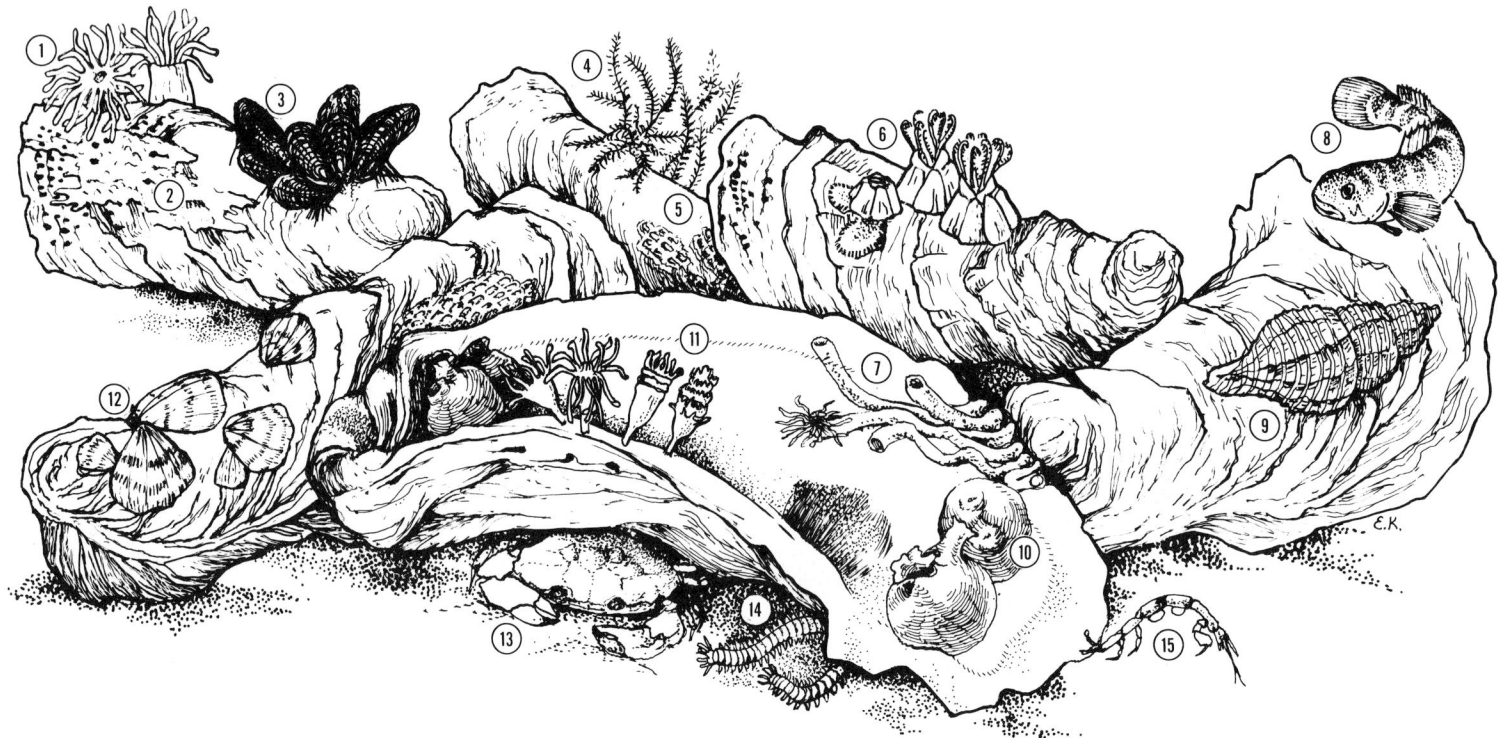

1. *Diadumene leucolena* (sea anemones)
2. *Cliona truitti* (boring sponge)
3. *Brachidontes recurvus* (mussels)
4. *Obelia dichotoma* (hydroids)
5. *Membranipora tenuis* (bryozoans)
6. *Balanus improvisus* (barnacles)
7. *Sabellaria vulgaris* (tube-building polychaetes)
8. *Gobiosoma bosci* (young naked goby)
9. *Urosalpinx cinereus* (oyster drill)
10. *Molgula manhattensis* (sea squirts, tunicates)
11. *Chrysaora quinquecirrha* (enlarged jellyfish polyps)
12. *Crassostrea virginica* (juvenile American oysters, spat)
13. *Rhithropanopeus harrisii* (mud crab)
14. *Nereis succinea* (burrowing polychaete worm)
15. *Caprella penantis* (amphipod)

Figure 7-2. *A representative oyster community of the Potomac estuary.*

Some infaunal organisms, especially those with siphons, obtain their food by filtering suspended organic matter from the water. To obtain sufficient nutrition for growth and reproduction, these organisms must filter relatively large quantities of water [22] — some as much as 20 to 30 liters of water per day — from which significant quantities of suspended materials are removed.[23] Not all the filtered material is selected for ingestion. Particles of certain sizes, with low nutritional value, are bound in mucus and discharged back to the environment without being eaten. These rejected particles are called pseudofeces. Undigested materials that pass through a filter feeder's gut are also bound in mucus and compacted before they are discharged to the environment. This binding and compacting process increases the density and size of filtered material, which assists in depositing it on the bottom. This activity is part of the process of biodeposition.[3]

Epifauna

Epifaunal organisms move about on the estuarine floor (e.g., crabs, shrimp, and snails) or are sessile and live firmly attached to a hard substrate slightly off the bottom (e.g., oysters and mussels, Fig. 7-2). Mobile epifauna are generally predators, scavengers, or grazers, and their mouth parts and appendages are generally adapted for capturing, holding, and ingesting prey; for tearing apart decomposing food items; or for scraping submerged surfaces to remove food items.[16] Sessile epifauna generally filter-feed in a manner similar to the filter-feeding infauna, and sometimes have a hard outer covering that protects them from predators or adverse conditions. Only a few epifaunal organisms are deposit feeders.

Reproduction

Most benthic invertebrates only live for one or two years. They replenish their populations through asexual or sexual reproduction.[3] Asexual reproduction takes place without the production of eggs and sperm (gametes), while sexual reproduction involves the union of sperm and egg (fertilization). An individual may be male or female; or both sexes may occur in one individual (hermaphroditism). Fertilization may involve an exchange of gametes between different individuals (cross-fertilization), or the sperm and egg may come from the same individual (self-fertilization).

Two general sexual reproductive modes occur among benthic macroinvertebrates: 1) gametes or partially developed embryonic stages (usually referred to as larvae) are released into the plankton where further development occurs, and 2) embryonic stages are retained by the adult and protected as they grow — a process generally referred to as brooding.[16] [24-27] In addition to the reproductive mode, other factors determine reproductive success, with the more important ones being tolerances of developing embryos and juveniles to physical and chemical conditions; behavioral adaptations which ensure that developing embryos and juveniles will encounter habitats where survival will be high; and biological interactions such as competition, predation, and disease. The production of planktonic larvae ensures a broad distribution of species that are sessile as adults, and provides a means through which relatively immobile species can compete for new habitats or repopulate areas where they previously experienced high mortalities during adverse conditions.[24][26][28]

Planktonic larvae rarely look like the adult organism. (The planktonic stages of a few representative benthic macroinvertebrates inhabiting the Potomac are illustrated in Fig. 6-2 in Chapter 6.) The number of developmental stages and the duration of planktonic development varies from species to species.[24][28]

Because planktonic larval stages are frequently more sensitive to fluctuations in environmental conditions than are the adult stages, they are generally considered to be the critical stage in the life cycle.[24][29] To ensure survival of the species, benthic macroinvertebrates have evolved numerous behavioral and physiological adaptations so that early life stages will be produced and released at times when environmental conditions (e.g., temperature, salinity, light, and food) are favorable for their growth and development.[30-33] For example, some barnacles release larvae only when particular planktonic food items are available to the reproductive adults, indicating that the spring plankton bloom is about to occur.[33] Thus, the rapidly growing and developing barnacle larvae are not released until they are likely to encounter the kinds and quantities of food required for completion of their embryonic development.

If embryonic development is totally planktonic, usually less than one percent of the fertilized eggs develop to settle to the bottom and take up a benthic existence.[3][24][35] Predation on planktonic stages by other zooplankton or fish, and exposure to unfavorable physical and chemical conditions are generally the major sources of larval mortalities.[35-38] A suitable habitat must be encountered before the larvae can successfully take up a benthic existence,[34] but unfavorable displacements by currents can prevent the larvae from encountering such habitats. As a result, larvae of most benthic macroinvertebrates have evolved behavioral adaptations that help them locate habitats where survival will be high.[5][34] Two examples of these adaptations are the migration of larvae away from light, which brings them closer to the bottom, and the delay of metamorphosis into adults until a favorable habitat is encountered.[34] Because of the low survival in the plankton, large numbers of eggs (1×10^4 to 1×10^9 per female per breeding season) must be produced to ensure population stability.[24] Exceptionally successful reproduction occurs when conditions are favorable for spawning, development, and settling, and large numbers of juveniles of the same age are produced.

Many benthic macroinvertebrates, especially the mobile epifauna, do not have planktonic developmental stages and brood their young throughout embryonic development. In most species, the developing embryonic stages are retained by the adult in a specialized body structure or pouch (for example, as in the amphipod in Fig. 7-4).

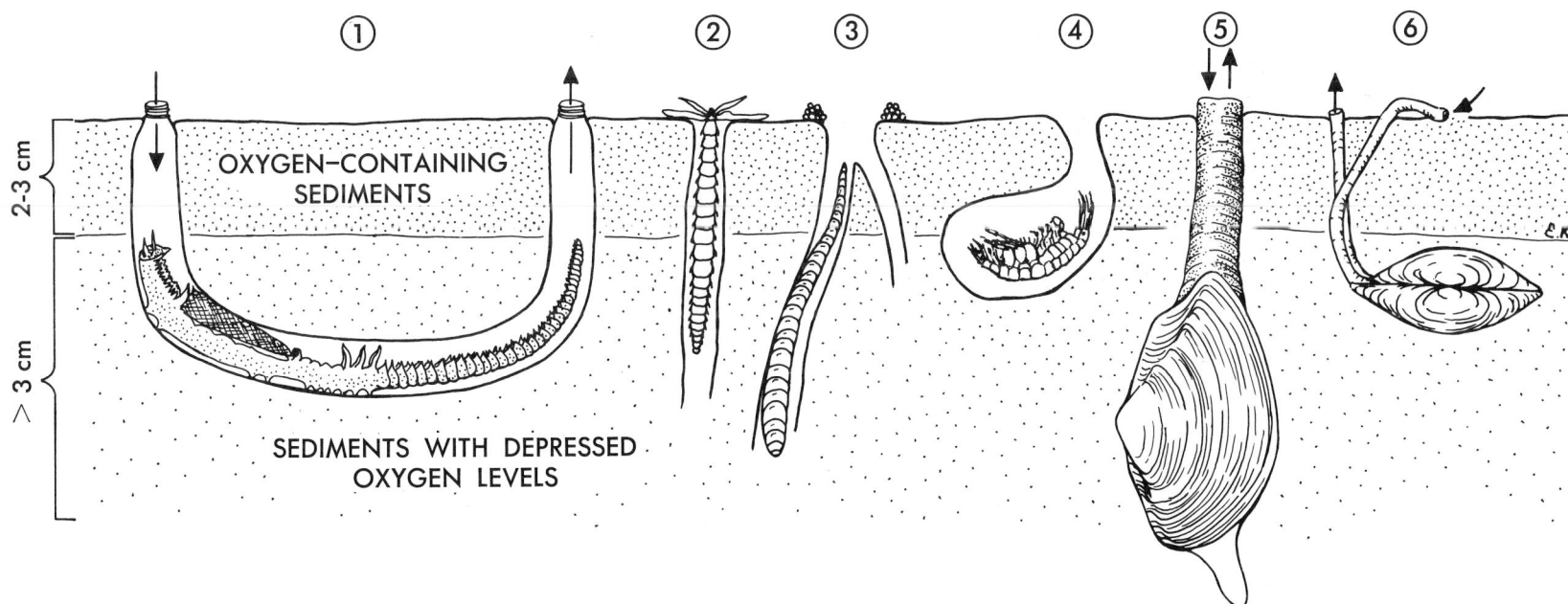

Figure 7-3. *Burrows of a few representative macroinvertebrates inhabiting the Potomac: 1:* Chaetopterus variopedatus, *2.* Streblospio benedicti, *3.* Heteromastus filiformis, *4.* Leptocheirus plumulosus, *5.* Mya arenaria, *6.* Macoma balthica. *(Source: Ref. 17)*

120

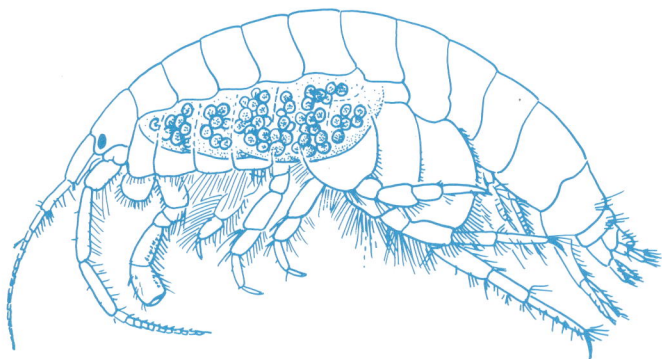

Figure 7-4. *A representative Potomac amphipod (*Leptocheirus plumulosus*) brooding eggs (drawn from Ref. 39 and a preserved specimen).*

Depending on the species, the duration of brooding ranges from several days to several weeks. Brooding adults are generally better able to evade predators, find food, and avoid unfavorable environmental conditions than are planktonic larvae. Thus, brooding provides some protection to embryonic stages from unfavorable physical and chemical conditions as well as from interactions with other species. Compared to species with planktonic larvae, fewer eggs (1×10^2 to 1×10^3 per female per breeding season) are required for the maintenance of population levels.

When environmental conditions are favorable, species that brood their young can rapidly increase in abundance.[25] However, because brooded young generally quickly settle to the bottom following their release, brooding species have limited dispersal capabilities and colonize new habitats more slowly than species with planktonic larval stages.

Many species use combinations of the two general reproductive modes discussed. The most frequently observed combination is to brood young during the earliest developmental stages and then release them into the plankton after most embryonic development has occurred. This provides for protection of the young during their most vulnerable stages when highest mortalities would occur if they were in the water column, and provides for limited dispersal through a planktonic phase. Species that use this combination of reproductive modes produce about 1×10^3 to 1×10^5 eggs per female per breeding season, a number that is intermediate between the number of eggs produced by species with completely planktonic development and species with completely brooded development.[24]

Only limited information is available on the seasonal patterns of reproduction and on the external factors controlling these patterns for benthic macroinvertebrates inhabiting the Potomac estuary. However, a summary of reproduction times for some of the benthic macroinvertebrates found in the Potomac (see table in upper right-hand corner of Folio Map 5) was generated using information from other regions of the Chesapeake Bay or other temperate zone estuaries. This table shows that some species are reproductively active during all seasons and that major peaks in reproductive activity occur during the warmer months (spring through fall).[40–46]

The duration of the reproductive period varies from species to species (see table in upper right, Folio Map 5). Most species inhabiting the Potomac are reproductively active over several months, and they frequently have more than one reproductive peak annually. For example, the commercially harvested soft-shell clam, *Mya arenaria,* has a summer and a fall peak.[40]

Differences in seasonal reproductive activity among benthic macroinvertebrate species contribute to the large variations observed in adult standing stocks over the annual cycle.[44] There is only limited information on seasonal variation in standing stocks of benthic macroinvertebrates in the Potomac. However, in environmentally similar regions of the Chesapeake Bay, infaunal standing stocks increase primarily from late fall through spring.[44 47 48] The fall reproductive period generally results in a more lasting increase to adult standing stocks than does the spring reproductive period,[48] primarily because larvae that settle to the bottom in the fall have several months to grow and mature before their major predators (crabs and bottom-feeding finfish such as spot) become abundant in late spring. In contrast, many more infaunal larvae settling to the bottom during spring (Folio Map 5) are eaten before they develop the means to evade the large numbers of bottom-feeding fish and crabs inhabiting the Chesapeake Bay during summer. As a result, standing stocks of these species rarely increase greatly in summer.[48] For infaunal populations inhabiting depths greater than 9 to 10 meters, the low dissolved oxygen concentrations that develop at these depths during summer become an additional factor causing summer declines in the standing stocks.[45]

Standing stocks of Chesapeake Bay epifaunal communities are highest during the summer and fall and lowest during winter.[49] Apparently, standing stocks of epifaunal organisms are not as affected by predation from seasonally abundant finfish and crabs as are standing stocks of infaunal organisms.[49] Temperature — through its influence on growth rate, asexual reproduction, and development — appears to be the major factor influencing abundances of epifaunal organisms.[49]

It is difficult to group benthic macroinvertebrates into distinct categories based on reproductive and feeding modes or habitat preferences. For example, many species of crustaceans live under stones, shells, and other debris, but do not burrow into the substrate. These organisms are members of both the infauna and the epifauna. Some species, such as clams of the genus *Macoma,* are deposit feeders in muddy habitats and filter feeders in sandy habitats. It should thus be recognized that the adaptationally defined categories discussed above are general and only provide a framework that assists in understanding distributional patterns and ecological relationships between the benthos and the rest of the estuarine food web.

Factors Affecting Distributions

Benthic macroinvertebrates are not uniformly distributed over the bottom; rather, distributional ranges are generally determined by physiological tolerances of species to physical and chemical conditions.[50][51] Along the length of the Potomac and in other estuaries, the various physical and chemical factors affecting benthic macroinvertebrate communities are salinity range, type of substrate, currents, dissolved oxygen concentrations, and pollutant concentrations.[3] Biological interactions, such as competition for space and food among the epifauna, predation by bottom-feeding predators on the infauna, and diseases (such as MSX, an oyster disease caused by the microscopic protozoan, *Minchinia nelsoni*) further shape the nature of benthic assemblages.[48][52] Since many benthic animals are sessile or have limited mobility for most of their lives, environmental conditions such as salinity only have to deviate for a short time from what a species can physiologically tolerate to prevent that species from successfully inhabiting an area or to eliminate those present.

Salinity range and type of substrate are the major chemical and physical factors controlling the occurrence and distribution of benthic macroinvertebrates in estuaries.[3][5] Both must be suitable for an organism to inhabit an area. In the Potomac, most benthic macroinvertebrates (like other groups of aquatic animals) fall into three general categories, based on their salinity tolerances — 1) the euryhaline marine species, which tolerate a broad range of salinities; 2) the true estuarine species; and 3) the freshwater organisms.[3] Most marine benthic macroinvertebrates cannot survive in salinities below 25 ppt (parts per thousand) and do not occur as far up the Chesapeake Bay as the Potomac estuary.[3] Euryhaline marine benthic macroinvertebrates, however, can tolerate salinities as low as 15 to 18 ppt, and in the Potomac these species inhabit the deeper channels near the mouth, which are polyhaline in the fall and during periods of low flow.[3][53] A few extremely tolerant euryhaline benthic macroinvertebrates even range as far upstream as the oligohaline zone near Douglas Point.[54–58] Estuarine species generally have peak abundances in the middle or upper reaches of estuaries, between salinities of about 2 to 10 ppt.[59] In the Potomac, the estuarine species are abundant from the mouth up to Mathias Point.[3][52][54][55] Freshwater species are most abundant at the head of the Potomac estuary near Washington, D.C. However, a few range downstream to Douglas Point.[54][57][60] The fewest species of benthic macroinvertebrates occur in the transition region between the tidal fresh and oligohaline waters from Indian Head to Maryland Point.[52][54][57] Figure 3-10 in Chapter 3 gives the seasonal distribution of bottom salinities along the estuary. In addition, a small map of bottom salinities is provided on Folio Map 5. This information can be used to locate regions of the estuary where major changes in the makeup and abundances of benthic assemblages occur.

Within a particular salinity range, type of substrate is the major physical factor controlling the distribution of most benthic macroinvertebrates.[3][5][27][53][55][56] Folio Map 3 summarizes the distribution of sediments in the Potomac. Mud is the predominant sediment type throughout most of the estuary, particularly in the upper portion. Sandy sediments occur mostly near the mouth, below Breton Bay. In the middle regions of the estuary, muds are frequently mixed with sand, especially in the shoal areas along the Virginia side. Information on sediment types is useful in predicting the general kinds of benthic organisms one would expect to find in any particular region of the Potomac estuary.

Muddy sediments generally support relatively large bacterial populations that provide a bountiful food supply for deposit-feeding infauna.[5] Filter feeders generally do not inhabit muddier environments because mechanical reworking by deposit feeders alters the physical characteristics of the sediment surface.[5][27] These alterations interfere with settlement and survival of larvae following reproduction and interfere with feeding activities.[5][27] In the sandy sediments of most temperate zone estuaries, filter-feeding infauna are usually numerically dominant and deposit feeders occur only in low numbers, possibly because sandy sediments generally do not have sufficient quantities of organic material to support large populations of deposit-feeding infauna.[3] The Potomac estuary and most of the Chesapeake Bay are unusual in that tube-building deposit feeders, which rely on organic material that settles out on the surface of the sediment, are successful in their sandy habitats along with filter feeders.[44][46][53][56][61]

Currents, dissolved oxygen concentrations, and pollutant concentrations also affect macroinvertebrate distributions. Currents influence the distribution of planktonic larval populations over a section of the bottom, and also determine the physical characteristics of the sediments and the food supply available to immobile benthic organisms.[3][5][62] Planktonic larvae frequently suffer high mortalities when they settle in regions with low dissolved oxygen concentrations,[3][5][47] such as those that occur during the summer in the tidal freshwater regions at the head of the estuary or at depths greater than 10 meters in the lower estuary. Organic wastes or toxic substances often occur near metropolitan and industrial discharge sites, and can affect the makeup of benthic assemblages.[63]

Although the distribution of benthic macroinvertebrates is primarily controlled by substrate type and salinity, biological processes such as competition and predation also play a role.[1][3][5][64–67] For example, summer and fall standing stocks of most shallow burrowing clam species in the Chesapeake Bay are largely a function of the intensity of predation by bottom-feeding finfish and crabs.[48]

General features of benthic distributional patterns can be determined with the aid of Folio Map 5. This map presents four informational units, which in part summarize the preceding discussions: 1) a large map of the distribution of benthic habitats throughout the estuary, based on salinity ranges and types of substrates (in the center); 2) a tabular summary of important ecological information on some Potomac species (upper right); 3) a small map (middle right) summarizing the distribution of bottom salinities (showing extremes for tidal fresh and polyhaline), which can be referred to for information on the location of the upstream or downstream extents of salinity ranges; and 4) a

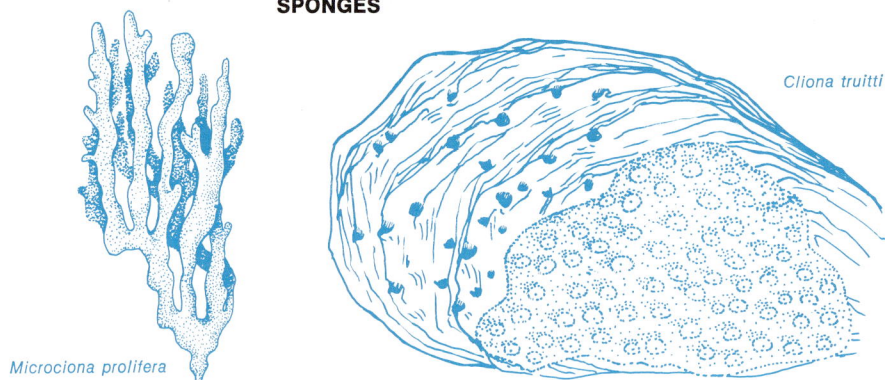

SPONGES

Cliona truitti

Microciona prolifera

schematic presentation of the major benthic habitats occurring in the Potomac estuary (lower left).

Benthic habitats were mapped rather than assemblages because there are relatively large portions of the Potomac estuary for which survey data are not available. Also, the abundances of benthic organisms fluctuate erratically from year to year, due to seasonal and annual variations in environmental factors, predominantly salinity and temperature. A distributional map based on abundance data would thus have limited predictive value, and its period of usefulness would probably be relatively short. However, a table of the essential features of the benthic macroinvertebrate assemblages likely to be encountered in the Potomac is provided on Folio Map 5. This table, along with the mapping of habitats, can be used to predict the types of benthic organisms that one might find in any particular region of the estuary. This map will retain its predictive usefulness as long as the salinity ranges and distributions of substrates in the Potomac do not change substantially.

MAJOR GROUPS OF BENTHIC MACROINVERTEBRATES

The remainder of this chapter briefly describes some of the general characteristics and distributional patterns of abundant groups of benthic macroinvertebrates found in the Potomac estuary. The order chosen for discussion generally follows the taxonomic ordering of groups given in Appendix Table 5. This table lists all the species of benthic macroinvertebrates that have been recorded from the Potomac estuary or that are presumed to occur there because of their known distribution in environmentally similar regions of the Chesapeake Bay.

Phylum Porifera — Sponges

Sponges are found from the freshwater portion of the Potomac to its confluence with the Chesapeake Bay.[60][68][69] These organisms are predominantly epifaunal and vary greatly in size and shape. They are composed of masses of loosely associated cells around a maze of internal canals that open to the exterior. The entire mass of most sponges is supported by spine-like structures called spicules, which assist in keeping the internal canals open. Water is circulated through the internal canals, bringing in food such as plankton, detritus, and dissolved organic compounds. Food particles and dissolved organic substances are engulfed or absorbed by the cells lining the interior canals and are passed on to interior cells where digestion occurs. Waste material is discharged into the water circulated through the canals and is transported away.

Sponges reproduce sexually and asexually. In sexual reproduction, sperm and eggs are released into the water. Both cross-fertilization and self-fertilization occur. Asexual reproduction usually involves outgrowths of the body (called buds), which eventually separate from the parent and become new sponges. Also, fragments that are accidentally broken from adults will frequently grow into new sponges. Under adverse environmental conditions, such as cold weather, some sponges produce small packets of cells, called gemmules, that have a hard outer covering. Gemmules grow into new sponges when environmental conditions become favorable.[70][71]

Sponges are grazed upon by some snails, nudibranchs, and other epifaunal organisms. Their branching colonies form crevices where many kinds of organisms hide, and they frequently harbor a complex association of worms, shrimp-like crustaceans, and other invertebrates.[17][71][72]

Three of the most abundant sponges found in the Potomac estuary are encrusting marine types. The redbeard sponge, *Microciona prolifera,* is a brightly colored species found attached to rocks, pilings, or other submerged objects near the mouth of the Potomac. It has only rarely been collected from the mesohaline region farther upstream.[68][73] The bright orange-red structures of redbeard sponges grow to a height of almost six inches in marine environments.[71] However, specimens found in the Potomac are not this large because they are near the lowest salinity levels they can tolerate.[68][69] The dull yellow-orange species, *Haliclona permollis,* is sometimes found growing with the redbeard sponges and occurs both as an encrusting and an erect branching species.[70] This species occurs as far upstream as Posey's Bluff Oyster Bar near Breton Bay in mesohaline salinities.[68][69] *Cliona truitti,* a boring sponge, lives in holes it bores in oyster shells. During initial colonization, it resembles warts on the oyster shell.[74] As the colony grows, it may completely cover the shell.[68] Shells weakened by *C. truitti* are probably more susceptible to crab predation than normal oyster shells. *Cliona truitti* have been reported as far upstream as Sheepshead Oyster Bar near St. Clements Bay in the mesohaline region.[68]

Phylum Cnidaria — Sea Anemones, Hydroids, Jellyfishes

The phylum Cnidaria consists of three classes of organisms, all of which occur in the benthic fauna of the Potomac during some phase of their life cycle: 1) the sea anemones (Anthozoa); 2) the colonial and solitary hydroids (Hydrozoa); and 3) the true jellyfishes (Scyphozoa). The planktonic phases of hydroids (hydromedusae) and the true jellyfishes were discussed in Chapter 6. Anemones do not have a medusoid phase during their life cycle.

Although a few freshwater cnidarians occur in the Potomac, the majority of those present are marine and estuarine species.[52] Cnidarians are symmetrical organisms. They have a mouth that is surrounded by tentacles used in feeding and a central digestive cavity that does not have an anus. The tentacles are frequently armed with specialized cells, called nematocysts, that assist in capturing food by paralyzing prey, and may also be used in warding off predators. Cnidarians feed on a broad range of prey, including small fish, amphipods, and plankton.[17][71][72]

Many hydroids and most true jellyfishes have two alternating morphological forms: the polyp and the medusa. Anemones have a planula larva and a polyp form. Polyps of all three groups are generally

SEA ANEMONES

Diadumene leucolena

HYDROID

Bougainvillia rugosa

FLATWORM

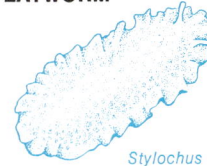

Stylochus ellipticus

PROBOSCIS WORM

Cerebratulus lacteus

attached to hard substrates. Polyps may be solitary or may form plant-like colonies, which are often mistaken for filamentous algae, bryozoans, or sponges. As discussed in Chapter 6, the medusoid form of cnidarians is generally planktonic and may be flat and saucer-shaped, or deep and bell-like. The stinging sea nettle, *Chrysaora quinquecirrha,* is an example of the medusoid form. However, the sea nettle also has a benthic polyp form that is an epifaunal component of benthic associations on rocks, shells, and other submerged hard objects.

Sea anemones have broad cylinder-like bodies. A sea anemone can anchor itself to almost any hard substrate, but can also move about slowly. Tentacles surrounding the mouth adhere to and transport plankton and detritus into the mouth. In addition, larger prey, such as small fish, are captured and incapacitated by nematocysts on the tentacles. The immobilized prey are then pushed into the gut by the tentacles and digested. The most frequently encountered sea anemone in the Potomac is *Diadumene leucolena*. It is primarily found during the summer months as far upstream as the oligohaline-mesohaline transition zone near the Rt. 301 bridge.[53 56 58 68]

Hydroids are found throughout the Potomac estuary, from the tidal freshwater regions to its confluence with the Chesapeake Bay.[52 54 57 60 68] These organisms grow alone or as branched, plant-like colonies that vary in size and shape from minute, vine-like stalks to large, bush-like structures. Many colonies are prolific and cover submerged objects of all kinds. Hydroids are grazed upon by large macroinvertebrates and small fish. Small epifaunal organisms often use the branches of dense colonial species as a refuge.[17 58 68 69 71 72]

Two frequently encountered species of hydroids found in the Potomac are *Cordylophora lacustris* and *Bougainvillia rugosa*.[52 57 60 69] *C. lacustris* is a profusely branching colonial hydroid that occurs in the oligohaline and tidal fresh regions of the Potomac as far downstream as Douglas Point.[57] During periods of warm weather, conditions for hydroid growth are favorable, and dense colonies of *C. lacustris* (20 to 100 mm high) are frequently observed. During fall, when conditions for growth become less favorable, these colonies become brown, weed-like mats. This hydroid produces no planktonic medusae; instead, the parent colony broods its embryonic developmental stages and releases larvae that, after a short period, metamorphose into polyps.[75]

Bougainvillia rugosa is also a branched colonial hydroid, and bushy growths of this species are most abundant in the mesohaline and polyhaline regions of the Potomac estuary during the summer months.[75] *B. rugosa* produces bell-shaped, planktonic medusae in which embryonic developmental stages are brooded for a short time before larvae are released into the water. After encountering a suitable habitat, the larvae metamorphose into polyps.[75]

Jellyfish characteristics, distributional patterns, and life cycles have already been discussed in Chapter 6. In the Potomac, polyps of the common sea nettle, *Chrysaora quinquecirrha,* probably occur as far upstream as the Port Tobacco River.[76]

Phylum Platyhelminthes — Flatworms

Flatworms occur from the freshwater portion of the Potomac to its confluence with the Chesapeake Bay. However, they are particularly abundant in the freshwater regions from Piscataway Creek to the Little Falls Dam, where they are one of the major predators of protozoans and other small organisms.[60] Most flatworms are small (less than 20 mm long) and have soft, flattened, and unsegmented bodies. They have a simple digestive system, consisting of a mouth that also serves as an anus. Almost all flatworms are hermaphrodites. However, self-fertilization is rare. No free-living larval stages are produced.[17 60 71 72]

Stylochus ellipticus is a frequently encountered flatworm that occurs from the oligohaline-mesohaline transition region near the Rt. 301 bridge to the mouth of the Potomac. This omnivorous species is generally associated with oyster shells and creeps over the bottom by means of tiny cilia on its underside to search for small prey or dead animals.[56 58 68]

Phylum Rhynchocoela — Proboscis Worms

Proboscis worms are unsegmented, slender, often colorful worms, which sometimes resemble the flatworms. However, they have a true digestive tract with a mouth and an anus. Most proboscis worms are carnivorous and are equipped with a tubular proboscis, a tongue-like structure which can be quickly extended to twice their body length to grab prey. These worms feed on a variety of small invertebrates by impaling them on the tip of the proboscis or by wrapping the proboscis around and holding the prey with the aid of a sticky mucus.[17 71 72] Reproduction in this group is predominantly sexual. Fertilized eggs are either extruded in gelatinous masses into the water where embryonic development takes place, or the young are brooded in the adult's body. Proboscis worms have extraordinary powers of regeneration; fragments of adults frequently grow into complete worms.[77] In the Potomac estuary, proboscis worms are most abundant in the higher salinities near the mouth.[53 56 57 61 68 73]

Micrura leidyi is the most abundant proboscis worm found in the Potomac. This species occurs from the oligohaline-mesohaline transition zone near the Rt. 301 bridge to the mouth of the Potomac.[53 56 68] It has a reddish colored body and a pale head and is slender, generally about 25 to 30 times longer than it is wide. *Cerebratulus lacteus* is a whitish burrowing species, often tinged with yellow or pink, and has only been observed in the near polyhaline salinities of the lower Potomac.[73] During spawning, which occurs in late spring and summer, *C. lacteus* turns a deep red and swarms to the surface of the water where fertilization occurs.[71]

Phylum Bryozoa — Bryozoans

Scientists frequently consider the phylum Bryozoa as being two separate phyla, the Endoprocta and the Ectoprocta. Although there are structural differences that justify grouping the Endoprocta and Ectoprocta separately, the two groups are discussed as one in this Atlas

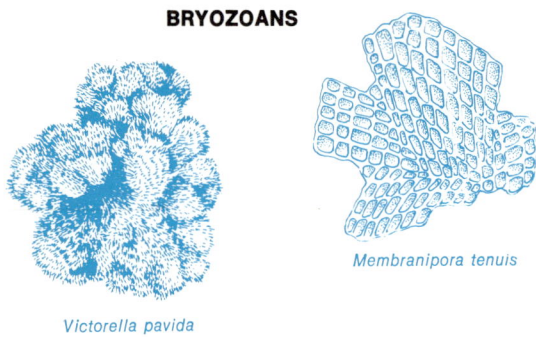

BRYOZOANS

POLYCHAETES

Polydora ligni

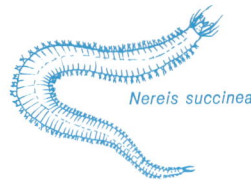

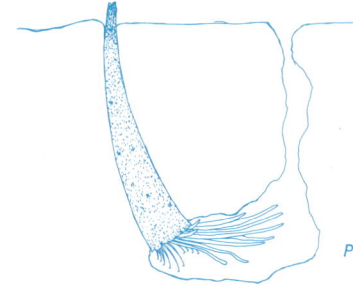

Nereis succinea

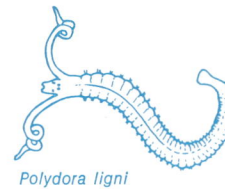

Membranipora tenuis

Heteromastus filiformis

Victorella pavida

Pectinaria gouldii

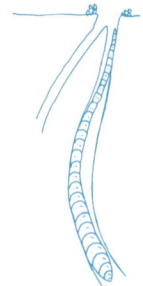

124 because they are similar in their outward appearance and functional roles in the Potomac estuary. Bryozoans are generally important components of epifaunal communities and inhabit hard substrates in all segments of the Potomac.[52] [53] [56–58] [60] [68] They are especially successful in regions with strong currents, and dense populations generally occur on intake and discharge structures of industrial users of Potomac water, particularly power plants. These growths are expensive to retard or remove.

The basic structure of the bryozoan colony is a horseshoe-shaped body crowned by ciliated tentacles. Although solitary individuals occur, most bryozoans are colonial, and because of their appearance, they are frequently mistaken for algae or hydroids.[17] [71] [72] The cilia on the tentacles create currents of water from which small planktonic organisms and detritus are collected for food. Bryozoans reproduce sexually and asexually. Following fertilization, small ciliated larvae develop and swim away from the adult. These larvae eventually attach to a suitable substrate and metamorphose into an adult. Budding is the primary form of asexual reproduction and is the major means by which bryozoan colonies spread and produce dense mats after initial colonization by a few larvae.

Victorella pavida is an abundant bryozoan species that occurs in dense mats, 3 to 6 mm high, from the oligohaline region to the mouth of the estuary.[52] [56] [58] Two abundant encrusting bryozoan species that occur in the same regions are *Membranipora tenuis* and *Electra crustulenta*. These two species inhabit a variety of hard substrates and are especially abundant on oyster shells in the mesohaline region.[58] [69] *Bowerbankia gracilis* is a short, gray to pinkish, vine-like species which primarily occurs in the mesohaline region of the estuary below lower Cedar Point.[68] *Pectinatella magnifica* is a bryozoan species that is frequently encountered in tidal fresh waters. It generally forms thick, brownish colonies on twigs and other debris in shaded waters between Douglas Point and Little Falls Dam.[57] [60]

Phylum Annelida — Segmented Worms

Annelids are segmented cylindrical worms that are abundant throughout the Potomac estuary. Three classes of annelid worms are represented in the Potomac fauna: Polychaeta (bristle worms), Oligochaeta (aquatic earthworms), and Hirudinea (leeches).

Polychaetes are one of the most abundant and diverse groups of benthic animals in the Potomac estuary. They have an important function in almost every habitat and are members of the infauna and the epifauna.[17] [71] [72] All major feeding types are represented in the Potomac polychaete fauna. Polychaete worms have foot-like appendages called parapodia on each body segment. Mobile epifaunal species have well developed parapodia and use them for propulsion, as gills, and in food collection (species 1, Fig. 7-1).

Most polychaetes occurring in the Potomac estuary have pelagic reproductive stages. Their larvae are abundant in the plankton throughout most of the year, usually peaking during the summer months.[78] Some polychaetes have embryonic stages that are brooded along the side or in the body of the female, or that live inside burrow tubes or in gelatinous pear-shaped egg masses attached to the bottom, and have only a short planktonic existence.[25] Some polychaetes have a reproductive stage called the epitoke, during which normally bottom-dwelling adults metamorphose into free-swimming, active forms that swarm in great numbers to the surface of the water where the release and union of gametes occurs. *Nereis succinea*, a ubiquitous Potomac polychaete, has frequently been observed swarming during late spring and early summer.

Certain polychaete species in parts of the United States are commercially harvested and sold as "bloodworms" to sportfishermen for bait. In the Potomac, polychaetes rarely attain a size large enough to be used for this purpose. However, they are an essential component of the Potomac food web and are the preferred food of many fishes and crabs. In addition, their burrowing, feeding, and tube-building activities greatly influence the physical properties of the sediments.[5]

Heteromastus filiformis is a small infaunal polychaete that occurs from the oligohaline-mesohaline transition zone at about the Rt. 301 bridge to the confluence of the Potomac with the Chesapeake Bay. This deposit-feeding species does not have elaborate parapodia and is generally most abundant in sandy or muddy-sand habitats in the near polyhaline or high mesohaline regions.[53] [56–58] *Polydora ligni* is a tube-building polychaete that also does not have elaborate parapodia. It has tentacle-like appendages on its head that are used in feeding, and is abundant on most firm substrates from the oligohaline region to the mouth of the Potomac.[53] [56] *Streblospio benedicti, Paraprionospio pinnata,* and *Scolecolepides viridis* (species 3, 25, and 21 in Fig. 7-1) are small infaunal tube builders with anterior feeding appendages similar to those of *P. ligni*. These species are most abundant in the mesohaline region, but occur from the oligohaline zone to the mouth of the Potomac.[53] [56–58] [61] *Pectinaria gouldii* is a relatively large, tube-building deposit feeder that carries its tube (constructed of sand grains that have been cemented together) with it as it moves through the sediments. This polychaete occurs only as far upstream as Piney Point, in high mesohaline salinities.[53] In the Potomac, the common bloodworm, *Glycera dibranchiata*, is a small organism that is limited to the high mesohaline and near polyhaline regions.[52] [73] It has jaws that are capable of inflicting a bite that feels roughly like a bee sting. However, it is not a predator but deposit-feeds on detrital material.

Oligochaetes are mostly infaunal organisms and have not been particularly well studied in the Potomac. However, in oligohaline and tidal fresh waters of other Chesapeake Bay estuaries, they have important ecological roles in the food webs of organically enriched habitats where they are frequently very abundant. They function as consumers of organic material in the bottom sediments and as major recyclers of nutrients from the sediments to the water.[79] [80] Little information is available on the importance of oligochaetes as food items for organisms in higher trophic levels.

GASTROPOD EGG CASES

OLIGOCHAETE

Aeolosoma sp.

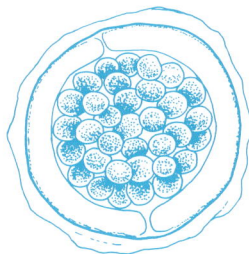

Gelatinous attached

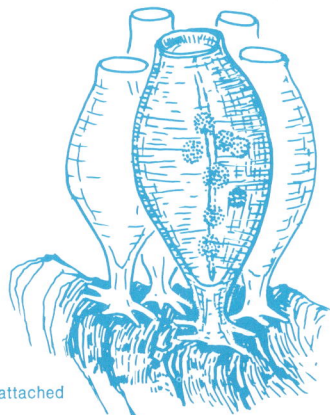

Leathery attached

GASTROPODS

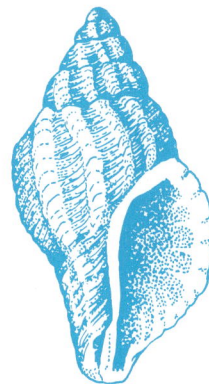

Urosalpinx cinereus

Crepidula fornicata

The segments of oligochaetes look alike from head to tail. They do not have parapodia, but do have small inconspicuous spines along the sides of their small, slender bodies. Oligochaetes feed on decaying plant material and on bacteria in organic-rich muds by ingesting the sediment.[17 72 81]

Oligochaetes reproduce sexually and asexually. Asexual reproduction is the principal reproductive mode during warm months, whereas, at colder temperatures, the principal mode is sexual. Oligochaetes are hermaphroditic, but self-fertilization is rare.[17 71 72 81] No planktonic larval stages occur; rather, fertilized eggs are initially brooded in a protective cocoon along the body of the adult. Shortly after fertilization, the cocoon slips off the adult and remains on the bottom for the duration of embryonic development. Young oligochaetes hatch from the cocoon as immature adults.[81]

Approximately 25 species of oligochaetes have been reported from the Potomac.[53 54 56 57 60] For the most part, they are freshwater organisms, and only a few occur as far downstream as Maryland Point.[54 56 57] Only one species, *Peloscolex gabriellae*, has been frequently collected from the mesohaline and polyhaline regions of the estuary.[53 82]

Leeches are highly specialized, darkly colored annelids that have distinct suckers at each end of their body and one under the head region. Superficially, leeches resemble flatworms, but these two groups are easily distinguished because leeches have segmented bodies, and flatworms do not. Leeches always have 33 segments, although exterior rings may make some of the segments difficult to see. Most leeches are ectoparasites and use their suckers to extract blood from fish or other invertebrates. After a blood meal, they leave their hosts and live for several weeks on the bottom before attaching to another host. Leeches only reproduce sexually and do not have pelagic larval stages.[17 71 72] They are hermaphroditic, but only cross-fertilize. All larval development takes place in cocoons that are attached by the adult to rocks, plants, sediments, or to a living host.

Three families of leeches are represented in Potomac benthic fauna. However, because leeches are not particularly abundant, they have not been well studied, and the role of this group in the Potomac ecosystem is unclear.[57 68 69]

Phylum Mollusca — Snails, Slugs, Clams, and Oysters

Molluscs are one of the most abundant and diversified groups of macroinvertebrates inhabiting the Potomac estuary and occur in all salinity ranges and all habitats. In the Potomac, molluscs are of two basic types: the snail-like forms, or gastropods, and the clam-like forms, or bivalves.

Gastropods have a single, frequently coiled shell. These organisms are generally slow-moving and have a soft, unsegmented body that is separated into a head, foot, and mantle (shell-making) region. The head generally has at least one pair of well-developed tentacles. Most

gastropods use a tongue-like apparatus, the radula, to scrape algae, bacteria, and other food material from rocks, debris, or the surface of sediments.[17 71 72] A few gastropods use the radula to bore small holes through the shells of sessile prey, through which they eat the soft tissues. The oyster drill, *Urosalpinx cinereus,* which occurs in the lower Potomac in salinities above about 15 ppt, preys on oysters and young clams in this manner.[69] Its boring of shells is assisted by an accessory organ which secretes chemicals that dissolve and soften the shell.[84]

Most gastropod species are hermaphroditic; however, self-fertilization is usually prevented by morphological and behavioral adaptations. Gastropods have two larval stages during embryonic development. The first stage, which is called a trocophore, occurs inside a protective capsule or egg mass, and is rarely planktonic. The sizes, shapes, and textures of the protective capsules vary from species to species. Some are simply gelatinous coatings around the eggs, whereas others are elaborate, leathery structures attached to submerged objects. The second larval stage (or veliger) of freshwater snails also occurs within the protective capsule, but veligers of snails inhabiting saline waters are usually planktonic.[17 71 72]

Prosobranch snails (gill-breathing gastropods which have their mantle cavity in the front) are common inhabitants of the more saline regions of the Potomac.[52 53 68 73] At some time during their development, individuals in this subclass of snails grow a horny, plate-like operculum (cover) attached to the back of their broad foot. The operculum is used to close the shell opening during adverse conditions or when these snails are being attacked by predators. Most prosobranchs are gregarious, shallow-water forms.[17 71 72 85]

The small, conical-shaped marsh periwinkle, *Littorina irrorata,* is a common prosobranch that inhabits high mesohaline and near polyhaline marshes of the Potomac estuary.[68] This species leaves trails of mucus behind as it scrapes microbiotic food organisms from the surfaces of marsh grasses.[85] The slipper-shells, *Crepidula fornicata,* are small prosobranch snails found in the lower estuary and occur in stacks, with the oldest member, a large female, on the bottom of the stack.[69 73] *C. fornicata* can undergo a unique sex reversal in which young males develop into sexually viable females. However, the oldest female in the stack discharges hormones into the water that prevent such sex reversals from occurring until after her death.[17 71] Mud snails, *Nassarius obsoleta,* are sometimes locally abundant in the high mesohaline and near polyhaline regions of the Potomac estuary.[68 73] This gregarious snail species is among the most active of aquatic scavengers and has well-developed chemoreceptors used for locating decomposing material. Its mouth is located at the end of a protruding snout, and by using a retractile proboscis, *N. obsoleta* feeds on small prey or decaying organisms.[85]

The opistobranch snails (snails with poorly developed shells and slug-like forms) are not abundant in the Potomac. A typical species of this subclass of snails that is sometimes encountered in the near polyhaline and high mesohaline regions is *Haminoea solitaria.*[53] This

NUDIBRANCH

Cratena pilata

PULMONATE SNAIL

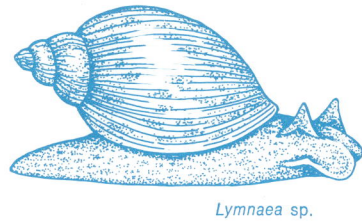

Lymnaea sp.

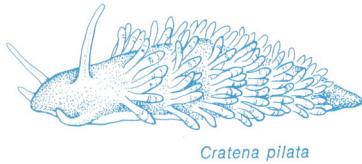

OYSTER LARVAE

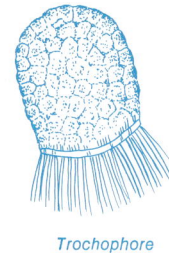

Trochophore

Veliger

species has not been reported upstream of Piney Point.[53] The wing-like extensions of *H. solitaria* envelop its shell. This species swallows small prey alive and crushes them with gizzard-like plates lining the gut; it also eats algae. *Odostomia impressa* is an opistobranch snail that morphologically resembles the prosobranch snails, but has no radula. It feeds by sucking body fluids from other molluscs with a long tubular proboscis. *O. impressa* is frequently abundant on oyster bars in mesohaline and near polyhaline regions of the Potomac and has been reported as far upstream as the Cobb Island Oyster Bar.[53 68]

Three species of nudibranchs, or sea slugs, including *Cratena pilata*, occur in the mesohaline and near polyhaline regions of the Potomac estuary. These organisms cast off their shells during development and do not have gills. Respiration occurs directly through their body walls. Nudibranchs are relatively mobile and graze on a wide variety of other invertebrates, including sponges, sea squirts, bryozoans, and hydroids.[17 71 72 85]

Pulmonate snails (gastropods that breathe atmospheric gases through a modified lung-like sac) predominantly inhabit the freshwater regions of the Potomac.[52 57 60] Two genera, *Lymnaea* and *Ferrissia*, are fairly abundant.[60] These species are important secondary consumers in tidal freshwater benthic communities, and are important food items for fish, small mammals, and waterfowl.[60]

Bivalves are molluscs that have two shells. This class of molluscs includes oysters and soft-shell clams, two of the most abundant and commercially important species inhabiting the Potomac estuary. The bodies of bivalves are enclosed in two hinged shells or valves. When the muscles that are attached to the inside of the shells contract, the two valves close, protecting body tissues from adverse conditions or predators. The two shells of most bivalves are approximately equal in size. However, in some, one valve is flatter or larger than the other (as they are in oysters). A burrowing bivalve has a well-developed fleshy foot that is used when the organism moves through the substrate (see species 8, Fig. 7-1). Some infaunal bivalves use their foot to move up and down in a relatively permanent burrow.[86] In epifaunal bivalves that are fairly sessile (such as mussels or oysters), the foot is small or absent.[16 17 71 72]

Most bivalves have two tubular siphons, one which brings water and food particles into the body cavity and another which discharges water and wastes. Siphons vary greatly in size and morphology, depending on whether the species is a filter feeder or a deposit feeder. The deep-burrowing, deposit-feeding species generally have the longest siphons, and filter-feeding species have the shortest ones (as illustrated in Fig. 7-1 by species 10 — a deposit feeder — and species 13 — a filter feeder). Siphons of epifaunal bivalves such as oysters may be poorly developed.

Sexes are generally separate in bivalves, but some species are hermaphroditic, and self-fertilization can occur. Most marine bivalves shed their eggs and sperm into the water. Fertilization and embryonic development occur in the plankton. Bivalves have two planktonic stages: a short-lived trochophore stage, and a longer-lived veliger stage

which eventually settles to the bottom and metamorphoses into an adult-like juvenile. Only a few marine bivalves brood their embryonic stages. This practice is more common in freshwater bivalves.[17 72]

Two species of bivalves — the American oyster, *Crassostrea virginica,* and the soft-shell clam, *Mya arenaria* — are commercially harvested in the Potomac. Because of their economic importance, the distributional patterns and biological characteristics of these two species are discussed first. The distributional patterns and biological characteristics of the remaining species of bivalves are discussed according to their taxonomic order.

The American oyster is the most sought-after bivalve in the Potomac estuary and the Chesapeake Bay. It is long, narrow, and has a rough, heavy, grayish shell. One shell is generally smaller and flatter than the other, although the shape of either valve may vary greatly depending on environmental conditions.[23] Oysters are found in the Potomac from the Port Tobacco River to the mouth. Within this range, oysters occur on scattered rocks or bars in waters generally less than about 8 meters deep.[68 87 88] Folio Map 6 shows the locations and names of charted oyster-producing areas in the main stem of the Potomac estuary. In addition, Fig. 7-5 shows the locations of charted oyster bars in Maryland tributaries.

American oysters have a relatively complex life cycle. Juvenile oysters (12 to 16 weeks old) are bisexual, and their gonads contain both eggs and sperm. Cold winter temperatures stop development of the eggs, and at the end of the first growing season, gonads of most juveniles contain only sperm. However, some of these immature individuals are hermaphrodites and females. Hermaphroditic oysters are capable of self-fertilization, although this is probably a rare event. By completion of the second growing season, the number of females and males in any particular yearclass or population is approximately equal. During the winter months, when oyster metabolism is lowest, their gonadal tissues are small and underdeveloped. With the advent of warm temperatures, gamete development begins, and by the end of June, oysters are full of developing sperm and eggs, and their gonadal tissues have a milky appearance.

Spawning can occur at temperatures between 15 and 30°C, and in the Potomac estuary lasts for about 2 months.[87] Peak spawning occurs when water temperatures reach about 22°C.[87] A rapid increase in temperature between 20 and 25°C is apparently critical to initiate release of sperm by males, which subsequently causes the females to spawn.[23] During a single spawning period, many millions of eggs are released into the water by a single female.[23] Fertilization occurs in the water. Planktonic development proceeds through two larval stages, the trochophore and veliger and lasts for 2 to 6 weeks. At the completion of their planktonic development, the larvae migrate to the bottom in search of a suitable substrate on which to metamorphose into juvenile oysters.

Consistently high oyster recruitment only occurs on oyster bars near the mouth of the Potomac in high mesohaline and near polyhaline

WICOMICO RIVER

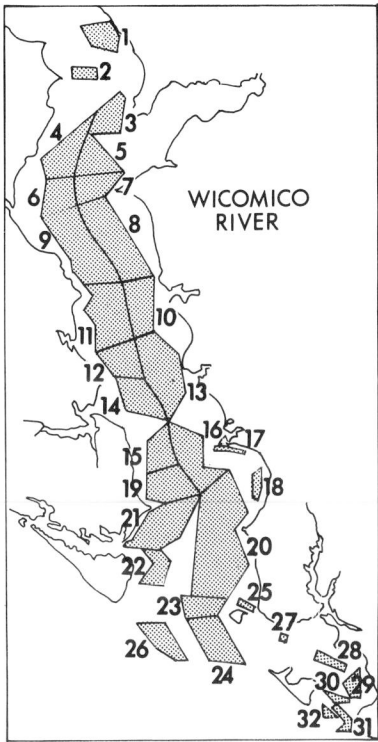

1. Key
2. Stoddard
3. Cohouck
4. Wicomico Lumps
5. Chaptico Lumps
6. Mills West
7. Mills East
8. Russell
9. Joes Lump
10. Manahowic Creek
11. Windmill
12. Fenwick
13. Bramleigh Creek
14. Wicomico Middleground
15. Charleston Creek
16. White Point
17. White Point Hollow
18. Blakistone
19. Lancaster
20. Bluff Point
21. Rock Point
22. Shipping Point
23. Mouth of River
24. Bullock
25. St. Margaret
26. Cobb Point
27. Bullock Island
28. St. Catherine
29. Cheseldine
30. Hackley Creek
31. Waterloo
32. Silver Spring

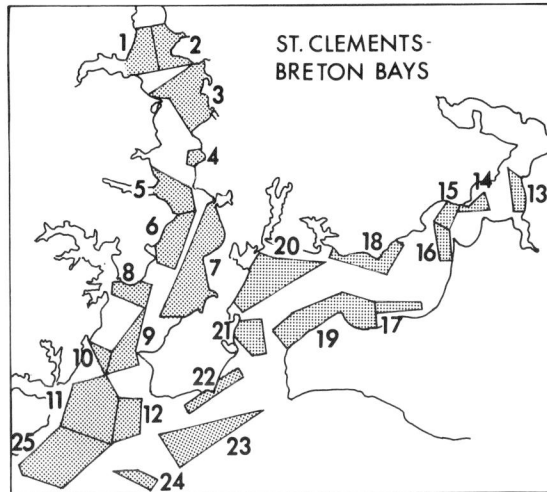

ST. GEORGE CREEK

1. Chadwick
2. Long
3. Shehan
4. Rollin
5. Cedar Point
6. Swan
7. Tarkhill
8. Straits
9. Milbourne Shore
10. Island Shore
11. Goose Point
12. Hurdle
13. St. George

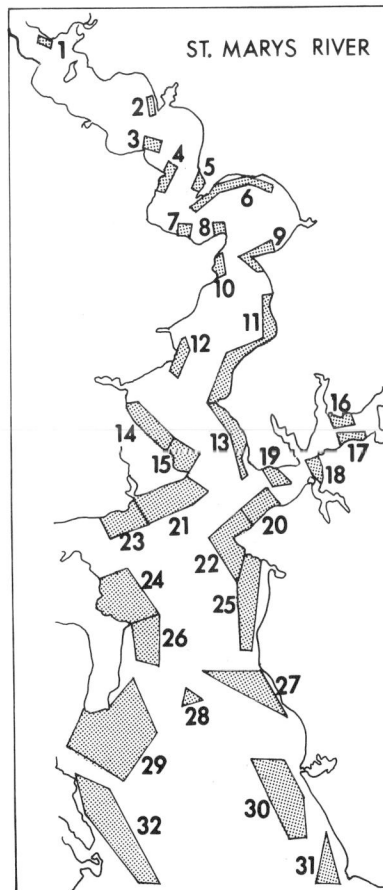

ST. CLEMENTS-BRETON BAYS

1. Harry Jacks
2. Guest Marshes
3. Reed Point
4. Abell
5. Mileys Creek
6. Bluff Woods
7. Chapel Point
8. Canoe Creek
9. Newtown Flats
10. Old Wreck
11. Horse
12. St. Clement Entrance
13. Mouldy Creek
14. Island
15. Paw Paw Hollow
16. Lovers Point
17. Gough
18. Stony
19. Black Walnut
20. Railway
21. Bretons Bay
22. Blue Sow
23. Heron Island Sound
24. Heron Island Reef
25. Dukehart Channel

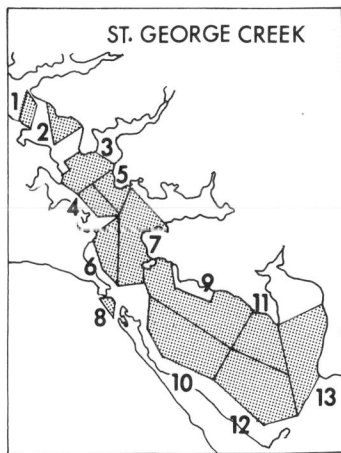

ST. MARYS RIVER

1. Tippity Witchity
2. Martin Point
3. Bryan
4. Short Point
5. Horseshoe
6. Horseshoe Bend
7. Biscoe
8. Pagan
9. Seminary
10. West St. Marys
11. Gravelly Run
12. Portobello
13. Rosecroft Hollow
14. Cooper Creek
15. Coppage
16. St. Inigoes North
17. St. Inigoes South
18. Jones
19. Kennedy
20. Raleys Shore
21. Thompson Creek
22. Priest
23. Carthagena Creek
24. Langley Hollow
25. Goad
26. Edmund
27. Fort
28. Middleground Lump
29. Cherry (Cherryfield)
30. Chicken Cock
31. Sedge Point
32. Mouth of Creek

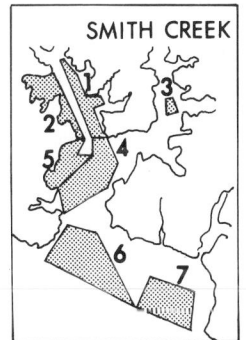

SMITH CREEK

1. Jutland
2. Graves
3. Dunbar
4. Smith Creek
5. Barnes Point
6. Old Hare
7. Calvert Bay

Figure 7-5. *Locations of charted oyster bars in Maryland tributaries of the Potomac estuary. (Source: Ref. 89)*

American oyster, *Crassostrea virginica*

salinities and is generally higher on the Maryland side than on the Virginia side due to prevailing salinity and current patterns.[68 87 88] The uppermost oyster bars (above the Rt. 301 bridge) generally experience the lowest recruitment, and middle bars, near the Colonial Beach area, have moderate recruitment.

American oysters can live under a wide range of environmental conditions and can tolerate salinities from about 5 to 35 ppt.[23 90] However, their best growth and reproduction occur in mesohaline and polyhaline salinities, and populations of oysters found in salinities below 5 to 7 ppt are highly stressed.[23] For this reason, many scientists believe that the Hawks Nest and Beacon Bar oyster populations, which are located above the Rt. 301 bridge in salinities below 5 to 7 ppt (see Folio Map 6), consist of a special strain or distinct population because of their remarkable tolerance to fresh water.[91] At salinities higher than about 15 ppt, oyster predators and diseases are more prevalent.[68] The mesohaline region of the Potomac estuary is thus an optimum habitat for oyster growth and reproduction because it is relatively free of oyster predators and diseases.[23 68 87 88] As a result, the extensive mesohaline region of the Potomac estuary has historically been one of the most productive oyster–producing areas of the Chesapeake Bay (see Chapter 10).

Temperature is a very important factor in oyster survival, growth, and reproduction. In addition to controlling the onset of spawning, temperature affects the survival of eggs and the development of embryonic stages in the plankton. Oyster larvae will not develop in temperatures below 15°C, and their development is abnormal above 40°C.[92] Above 42°C, all body functions are reduced to a minimum.[23] Adult oysters cease feeding between 3 and 7°C. Below 3°C (temperatures which frequently occur in the Potomac estuary during the winter months[23]) oyster metabolism is reduced to its lowest level.[23]

Since the late 1800s, oyster abundances in the Potomac have declined tremendously.[68 87] Productive bars today represent only a small fraction of those present during the early part of this century.[68 87] In an attempt to supplement productivity of natural populations, young newly settled oysters (called spat or seed) are dredged up from areas where recruitment is high, but growth rate is low (mostly from near the mouth), and moved to upstream areas where recruitment is low, but growth rate is high. Another method used in the Potomac estuary to increase productivity of oyster producing areas is to dredge up buried shell from depleted or dead bars and move it to areas where reproductive success and settlement of oyster larvae are high. Much of the dredged shell comes from the St. Marys River.[91] The timing of shell transplants is critical. Ideally, shell should be transplanted shortly before the peak settlement of oyster larvae so that the maximum quantity of clean surface is available on which young oysters can settle.[23]

The rains during and preceding Hurricane Agnes produced extremely low salinity levels throughout the Potomac estuary during the summer of 1972 and destroyed 70 percent of the oyster bars.[88 91 94 96] This catastrophe shows the susceptibility of oyster populations in the

optimum growth areas, such as the Potomac's mesohaline region, to periods of high freshwater runoff. Oyster bars above Swan Point suffered 100 percent mortalities.[88 91 93–96] Commercial harvesting of oysters above the mouth of the Wicomico River was immediately halted by the Potomac River Fisheries Commission.[91] Because the storm hit the Potomac estuary during the normal spawning season, essentially no oyster recruitment occurred in 1972. Recruitment in 1973 remained low, also due to heavy rains and low dissolved oxygen concentrations which occurred throughout much of the mesohaline region.[91 93 94 96] It was not until 1974 that a reasonably successful spat set occurred, and oyster populations in the Potomac began to recover from the effects of Hurricane Agnes.[91] Oyster bars that were closed due to Agnes were reopened to commercial harvesting in 1976.[91]

The soft-shell clam has historically occurred in large numbers in the Potomac estuary.[88] This species has two, long retractile siphons encased together in a tube. In the Potomac, soft-shell clams spawn twice a year, once in spring and again in fall.[40 88] Embryonic stages are planktonic, and recruitment success varies radically from year to year, with a strong recruitment only about every 10 to 15 years in the Potomac.[40 88] Soft-shell clams have a patchy distribution on firm sand or sandy-mud bottoms from the mouth of the estuary to the Rt. 301 bridge (see Folio Map 6).[73 88] Although they inhabit salinities as low as 5 ppt, commercially harvestable numbers occur mostly in the lower reaches of the estuary, above salinities of 8 ppt.[73 88] With the development and introduction of the hydraulically operated commercial harvester in about 1951, soft-shell clams were taken from the Potomac in record numbers. However, by 1967, populations of this species declined to extremely low levels.[88] Reasons for these population declines are not known, and in recent years, very few soft-shell clams have been taken from the Potomac (see Chapter 10). In 1974, no significant populations were found anywhere in the estuary.[88] Because population densities of this species fluctuate radically from year to year, only general distributional patterns have been plotted on Folio Map 6.

Two species of saltwater mussels (species of bivalves in the family Mytilidae) inhabiting the saline regions of the Potomac are the Atlantic ribbed mussel, *Modiolus demissus,* and the hooked mussel, *Brachidontes recurvus. M. demissus* has only been collected as far upstream as the high mesohaline region, and *B. recurvus* occurs from the mouth of the Potomac to the oligohaline-mesohaline transition zone near the Rt. 301 bridge.[53 56 58 68] These organisms have strong, thin, and somewhat pear-shaped shells with pearly interiors. The strongly-ribbed, yellowish-brown shells of *M. demissus* are generally embedded in mud flats and stream banks at the low-tide mark; the small, dark, curved shells of *B. recurvus* are commonly attached to pilings, shells, rocks, or other submerged objects. Although sedentary and attached by strong leathery threads, mussels are capable of some limited mobility when the threads are broken or disengaged.[85] The animal then uses its small, poorly developed foot to move to a new location. Mussels are important prey animals for raccoons and other small predatory mammals, as well as for waterfowl, fish, and crabs.

MUSSELS

Hooked mussel, *Brachidontes recurvus*

CLANS

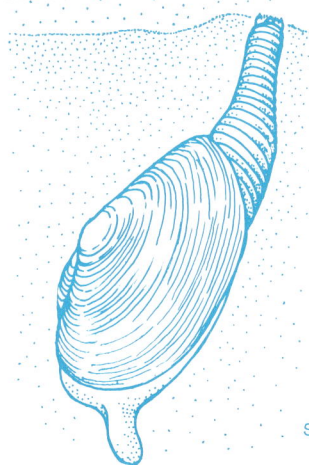

Soft-shell clam, *Mya arenaria*

Brackish-water clam, *Rangia cuneata*

A number of species of fingernail clams of the genus *Sphaerium* occur in the tidal freshwater regions of the Potomac.[57] [60] [69] They have thick, oval shells about 10 mm in diameter. *Sphaerium* spp. mostly reproduce by self-fertilization. Young are brooded in a pouch formed by the gills and hatch as miniature adults. Newly hatched juveniles are about one fourth to one third as large as the adult.

The pearly, or freshwater, mussels of the family Unionidae inhabit freshwater regions of the Potomac estuary and were, until the advent of plastics, the major source of pearl buttons used by the clothing industry.[57] [69] This family of bivalves has a unique form of brooding.[17] [72] Embryonic development proceeds in the mantle cavity through the veliger stage, after which, the embryos are released onto the water where they attach to fish. Once attached, the veligers encyst and continue their embryonic development. When development is complete, the cysts fall to the bottom, and the larvae metamorphose into immature adults.

Freshwater bivalves are eaten by many fishes, including gizzard shad, suckers, and perches. The muskrat is their primary mammalian predator. However, freshwater bivalves are also eaten by otters, raccoons, turtles, and certain amphibians.

Congeria leucopheata, or Conrad's false mussel, is an estuarine species and attaches itself by short leathery threads to hard substrates in the oligohaline and low mesohaline regions of the Potomac.[52] [53] [56] [58] [68] This species inhabits tidal fresh waters but is only occasionally observed at high mesohaline salinities.[53] *Mulinia lateralis* is a small, triangular-shaped species which is very abundant in the mesohaline and near polyhaline regions of the Potomac.[53] [68] This clam is an active burrower and inhabits both sand and mud substrates (species 13, Fig. 7-1). Both of these bivalves are eaten by bottom-feeding finfish and crabs, and experimental evidence suggests that the abundance of *M. lateralis* may be controlled by predation.[1]

The brackish-water clam, *Rangia cuneata*, is a southern species that has only been observed in the Potomac since about 1960.[73] [88] [97] [98] This species was probably introduced into the estuary with seed oysters from the James River.[73] [88] In 1964, it was abundant in the oligohaline and high mesohaline regions from about Swan Point to Colonial Beach.[73] [88] Since then, the population range has expanded as far upstream as Mattawoman Creek and as far downstream as Herring Creek.[56] [57] [88] [91] Conditions for successful reproduction of brackish-water clams are temperatures between 8 and 32°C and salinities between 2 and 20 ppt.[99] In the Potomac estuary, these conditions occur from just above Douglas Point to the mouth of the estuary. In temperate zone estuaries, the abundance of brackish-water clams fluctuates from year to year because of the high mortalities that occur during cold winters. Following the cold winters of 1976 and 1977, large declines in the numbers of brackish-water clams were noted in the Potomac.[97] [100] *R. cuneata* is not currently commercially harvested in the Potomac because of its small adult size (1 to 1.5 inches or 2.5 to 3.7 cm on the average) and undesirable taste when steamed. However, it is an

important food item in the diets of some waterfowl and small mammals (see Chapter 9).

Several species of tellin clams (of the family Tellinidae) are abundant in the Potomac estuary. This group of clams lives well below the sediment surface, and each tellin has a relatively large, well-developed foot that enables it to move about in the substrate. By burrowing vertically, tellins can escape predators, and by moving horizontally, they can search for food. Tellins filter- and deposit-feed through long, slender siphons that can be extended several times the lengths of their shells to the sediment surface. *Macoma balthica* (species 10 in Fig. 7-1) and *Macoma phenax* are two tellins that are abundant in the Potomac estuary. These are rather dull, chalky-white, asymmetrically shaped clams, that are abundant from the oligohaline region to the mouth of the Potomac, with peak abundances occurring in the high mesohaline region.[52] [53] [56] They filter-feed in sandy habitats and deposit-feed in muddy habitats. Both *Macoma* species are important food items in the diets of fish, crabs, small mammals, and waterfowl, particularly ducks (see Chapters 8 and 9).

Tagelus plebeius, the stout razor clam, is occasionally found in the mesohaline and near polyhaline regions of the Potomac.[73] It is long, somewhat cylindrical, and edible. However, it burrows so deeply and quickly into the sediment that it can only be collected with considerable effort.[86] The large, green-brown Atlantic jackknife clam, *Ensis directus*, is also an edible species that is occasionally found in the high mesohaline and near polyhaline regions of the lower Potomac.[53] It is a quick burrower and is probably the longest of the bivalves found in the Potomac estuary, sometimes reaching lengths of 12 to 18 centimeters (cm). These two species are prey items for a variety of shorebirds, fish, and crabs.[17] [71] [72] [85] [86]

Phylum Arthropoda — Barnacles, Shrimps, Crabs

Crustaceans are a class of arthropods that are represented in the Potomac by a large number of species. They are the only arthropod class that is discussed in detail. Other arthropod classes have not been sufficiently studied in the Potomac to warrant much discussion. Crustaceans are usually covered by chitinous or calcareous shells and have jointed legs and two pairs of antennae. Those with a chitinous exoskeleton increase in size through a complex physiological process called molting, during which they shed their old shell and produce a new, larger one. For a short time following molting, the new outer shell is soft, and the individual is defenseless until it hardens. Crustaceans include large organisms such as crabs and shrimps, as well as many small and inconspicuous forms. The major crustacean groups inhabiting the Potomac include: Cirripedia (barnacles); Stomatopoda (mantid shrimps); Cumacea (small shrimp-like organisms); Tanidacea (small shrimp-like organisms); Isopoda (sow bugs); Amphipoda (beach fleas); Mysidacea (medium-sized shrimp-like organisms); and Decapoda (large shrimps and crabs).

Barnacles are sessile, epifaunal creatures that frequently encrust firm substrates in salinities above 2 ppt.[101] Pilings, docks, and the water

129

BARNACLE

Balanus eburneus

CUMACEAN

Leucon americanus

MANTID SHRIMP

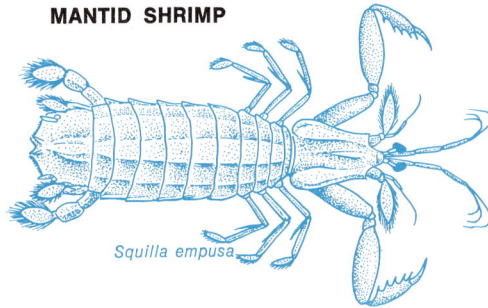

Squilla empusa

TANAID

Leptochelia rapax

ISOPODS

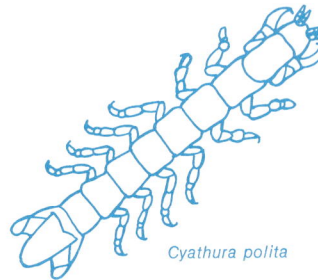

Cyathura polita

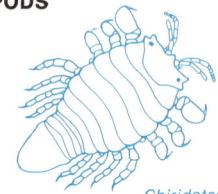

Chiridotea almyra

130 intake and discharge structures of power plants and of other industrial users of Potomac estuary water are fouled by these crustaceans.[101] It is extremely expensive to keep intake and discharge structures clean of barnacles and the species associated with them. Antifouling chemical agents such as chlorine, which are toxic to most living organisms, are often used to control their setting and growth.[102]

Barnacles are filter feeders and use feather-like appendages to remove planktonic food items from the water. Their hard outer coverings protect them from predators and adverse conditions. Reproduction in barnacles is sexual and involves both planktonic and brooded phases. Eggs are brooded in the mantle cavity. However, embryonic development through two larval stages occurs in the plankton (see Fig. 6-2 and discussion in Chapter 6). The final planktonic stage settles to the bottom, attaches to a firm substrate, and grows to adult size in only a few months.[17 72 85] The hard outer coverings of adult barnacles are composed of a series of calcareous plates that are held together by living tissue or interlocking teeth. These plates are not shed during molting, but increase in size by the continual addition of material along the edges. However, the exoskeleton covering the body on the inside is periodically shed during molting. The distribution of various barnacle species in the Potomac estuary is discussed in Chapter 6.

Mantid shrimps do not normally tolerate the estuarine salinities of the Potomac. However, one species, *Squilla empusa,* has been reported on a few occasions from the deeper channel areas near the mouth where the highest salinities occur.[69] This species, which may become more than 12 cm long, has a large pair of anterior claws for catching and holding prey such as small crustaceans or fish. These claws will wound a careless handler.

Cumaceans are small (less than 8 mm), slender, and shrimp-like. They have no eyes, and their head is fused to a short thorax. Cumaceans are infaunal filter feeders that remove minute food particles from water passed over their mouth parts by their anterior appendages. Sexes are separate, and adults are morphologically different. Males possess a long pair of antennae which the females do not have.[17 71 72] Of the three species of cumaceans reported to occur in the Potomac, *Cyclaspis varians, Leucon americanus,* and *Oxyurostylis smithi,* only the latter two are found with any regularity, and they are primarily limited to high mesohaline and polyhaline regions.[46 53 57 103]

Tanaids seldom exceed a few millimeters in length. They have pairs of appendages modified for walking, swimming, brooding young, and burrowing into the substrate.[17 71 72] Only one tanaid species, *Leptochelia rapax,* is reported to occur in the Potomac.[53 58] This species is a shallow-water form found in or on soft sediments as far upstream as the oligohaline-mesohaline transition region near the Rt. 301 bridge.

Isopods, or "sow bugs," are dorso-ventrally flattened with segmented bodies, eight pairs of walking or swimming appendages, and a telson, or tail. The sexes are separate, and the eggs and the young are generally brooded in a protective pouch under the tail or body of the female. Planktonic larval stages do not occur during embryonic development, and developing young look like adults. Isopods are generally omnivores, feeding on detritus and its associated fauna, dead organisms, and other small prey, and are themselves eaten by fish and other large predators.[17 71 72 104]

Several species of isopods are frequently encountered in the Potomac estuary. All are less than 25 mm long. *Cassidinisca lunifrons, Cyathura polita, Chiridotea almyra,* and *Edotea triloba* occur from the oligohaline to the near polyhaline region at the mouth.[53 55–57 61 68 83] *C. lunifrons* is a relatively rare, tube-building species inhabiting oyster shells and sometimes the mantle cavities of other molluscs.[55 56 83] *C. polita* is the only isopod species that is found in high numbers in the Potomac estuary.[53 55 56] It is also a tube builder, but constructs its tubes in stable sands or stable muddy sands.[104] *C. almyra* and *E. triloba* are burrowing infaunal species that primarily inhabit debris on sandy substrates.[83]

Amphipods are small, bug-like organisms (from 2 to 40 mm long in the Potomac estuary), which occur in almost every aquatic habitat, on or in all types of substrates, and in salinities from fresh water to the near polyhaline region at the mouth of the Potomac estuary. Amphipods have seven pairs of appendages used for swimming, grasping, or walking, and three pairs of paddle-like appendages used for swimming and circulating water during filter-feeding and brooding of young. Males are usually larger than females and have larger, more developed anterior claws, which they use to grasp the female during copulation. Eggs are brooded in a pouch under the abdomen of the female. No planktonic stages occur during embryonic development, and released young generally look like the adult. Amphipods are omnivores, feeding on detritus and its associated fauna, dead organisms, and small prey. They are eaten by many fishes, crabs, and other predatory invertebrates.[17 71 72 105]

Gammarus fasciatus and *Hyalella azteca* are two amphipod species that occur in the tidal freshwater regions of the Potomac estuary.[57 60] Most of the other abundant amphipod species in the Potomac occur in saline waters. *Leptocheirus plumulosus* (Fig. 7-4) is a tube-building amphipod that constructs its tube in soft sediments and is abundant in muddy-sand substrates of the oligohaline and low mesohaline regions of the estuary.[53 55–58 61] The abundance of this species decreases upstream and downstream of the middle reaches of the Potomac.[39 53 55–58 61] *Corophium lacustre* is also a tube builder and is most abundant in the middle reaches of the estuary.[53 55–58 61] However, *C. lacustre* constructs its tubes of sand and mud on shells and other hard submerged objects. *Gammarus mucronatus, Gammarus palustris,* and *Melita nitida* are three amphipod species that occur from the oligohaline zone to the mouth of the Potomac estuary. These epifaunal amphipods increase in abundance in a downstream direction.[53 105] *G. mucronatus* is most frequently encountered in floating debris on grass beds.[105] *G. palustris* is most abundant in marshy areas, especially near the base of marsh grasses.[39 106] *M. nitida* is generally most abundant among bryozoans and hydroids growing on oyster shells in areas where currents are relatively strong.[55 105]

AMPHIPODS

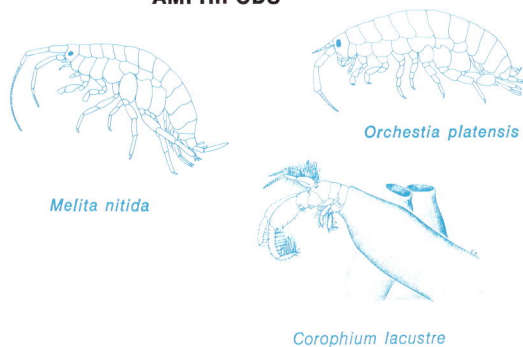

Orchestia platensis

Melita nitida

Corophium lacustre

TRUE SHRIMP

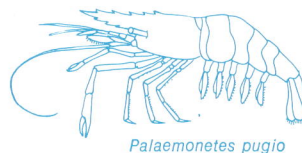

Palaemonetes pugio

MUD CRAB

Rhithropanopeus harrisii

Neohaustorius schmitzi and *Lepidactylus dytiscus* are two burrowing infaunal amphipods that occur in the Potomac.[55 57 58 105] These species remove suspended organic material from the water between sediment particles. Because they have this feeding mechanism, their distributions are limited to porous sandy sediments. *N. schmitzi* has only been observed in the high mesohaline and the near polyhaline regions of the Potomac.[105] *L. dytiscus* was recorded as far upstream as Mathias Point.[55 57 58 105] Both of these amphipods may be locally abundant in sandy habitats near the mouth of the Potomac. *Orchestia platensis* and *Talorchestia longicornis* are semiterrestrial amphipods that are found in debris from the shoreline up to 100 meters inland and that range from tidal fresh areas to the mouth of the Potomac estuary.[39 58 105]

Mysids are medium-sized (<10 mm in the Potomac), shrimp-like organisms with stalked eyes and fan-like tails. The only mysid species reported from the Potomac is *Neomysis americana,* the opossum shrimp. It is abundant in submerged vegetation and occurs throughout the estuary. The distribution of this species is discussed in Chapter 6.

Decapods are a diverse taxon, and for simplification, have been subdivided into two groups: the shrimp-like forms and the crab-like forms. Decapods have separate sexes that are easily distinguishable. Eggs are brooded in a pouch-like structure under the body of the female, frequently referred to as an apron. The planktonic larval stages of decapods are discussed in Chapter 6 (see Fig. 6-2).

Two shrimp-like forms that are frequently observed from the oligohaline region to the mouth of the Potomac are *Palaemonetes pugio* and *Crangon septemspinosa.* These two species have short-lived planktonic stages, and their distributions are discussed in Chapter 6. Most of the crab-like decapods are limited to the saline regions of the Potomac. However, some adult decapod species (e.g., the blue crab, *Callinectes sapidus*) penetrate into tidal fresh water.

Mud crabs, *Rhithropanopeus harrisii* and *Eurypanopeus depressus,* are found on all substrates from the oligohaline to the near polyhaline region at the mouth of the Potomac.[55 57 58 61 68] These species are most abundant on oyster bars where they use their large claws to crush juvenile oysters, small clams, small mussels, and any other prey they can find. Mud crabs are generally epifaunal. However, they retreat under shells or burrow in the mud to avoid being eaten by fish and other bottom-feeding predators (see species 13, Fig. 7-2). The small decapod crab, *Pinnotheres ostreum,* lives inside the shells of living oysters and clams.[68] Because it inhabits a protected environment, this decapod has lost some of the external characteristics of free-living decapods. Its shell is relatively soft, its legs are poorly developed, and it has only limited mobility as an adult. *Sesarma cinereum* is a small, inconspicuous decapod that inhabits muddy banks of the salt marshes of the lower reaches of the Potomac estuary (see Fig. 6-11 in Chapter 6).[68] This species is semiterrestrial as an adult; however, its reproduction and embryonic development is only successful at salinities greater than about 12 ppt.

The blue crab, *Callinectes sapidus,* is the largest decapod crustacean inhabiting the Potomac estuary or the Maryland portion of the Chesapeake Bay. Blue crabs have pointed shells, and mature crabs frequently reach a width of 20 cm from point-to-point.[107] However, the average width of a mature crab is only about 12 to 15 cm.[108 109] The first two legs of blue crabs are modified into large claws, which they use to capture, hold, and tear apart food items and to ward off predators. The top or back of blue crabs is brown to dark green. Their common name is derived from the azure blue on the undersides of their large claws. Blue crabs spend most of their time walking along the bottom on the tips of their pointed legs in search of food. However, their rear legs are flattened, paddle-like swimming appendages, and they are good swimmers.[107] In fact, their generic name, *Callinectes,* is derived from Greek and means "beautiful swimmer".[110] Blue crabs swim relatively long distances during migrations associated with reproduction, and they can make quick darting movements to avoid predators.

Blue crabs are omnivores, and feed on benthic macroinvertebrates, small fish, aquatic vegetation with its associated fauna, and dead organisms.[107 111 112] They, in turn, are preyed upon by large fish such as striped bass, by wading birds, and of course by man. The species name, *sapidus,* is derived from Latin, and means "tasty" or "savory". Blue crabs, like other crustaceans, are more susceptible to predation during molting when their shells are soft, and they generally hide under submerged objects or in sediments at this time. Soft crabs are harvested by man as a special delicacy. Commercial and recreational crabbers harvest millions of soft and hard crabs per year from the Potomac estuary (see Chapter 10).

Blue crab, *Callinectes sapidus*

Blue crabs have a complex life cycle that begins with the courtship and mating of the separate sexes. During courtship, male crabs strut about, waving their claws and walking on the tips of their legs, whereas, females rock back and forth and move their claws in and out. At the end of the courtship, the female turns her back to the male and may try to back under him. The male crab mounts the female and cradles her under his body until she molts. Cradling not only ensures that the female crab is protected from predators during and following molting,

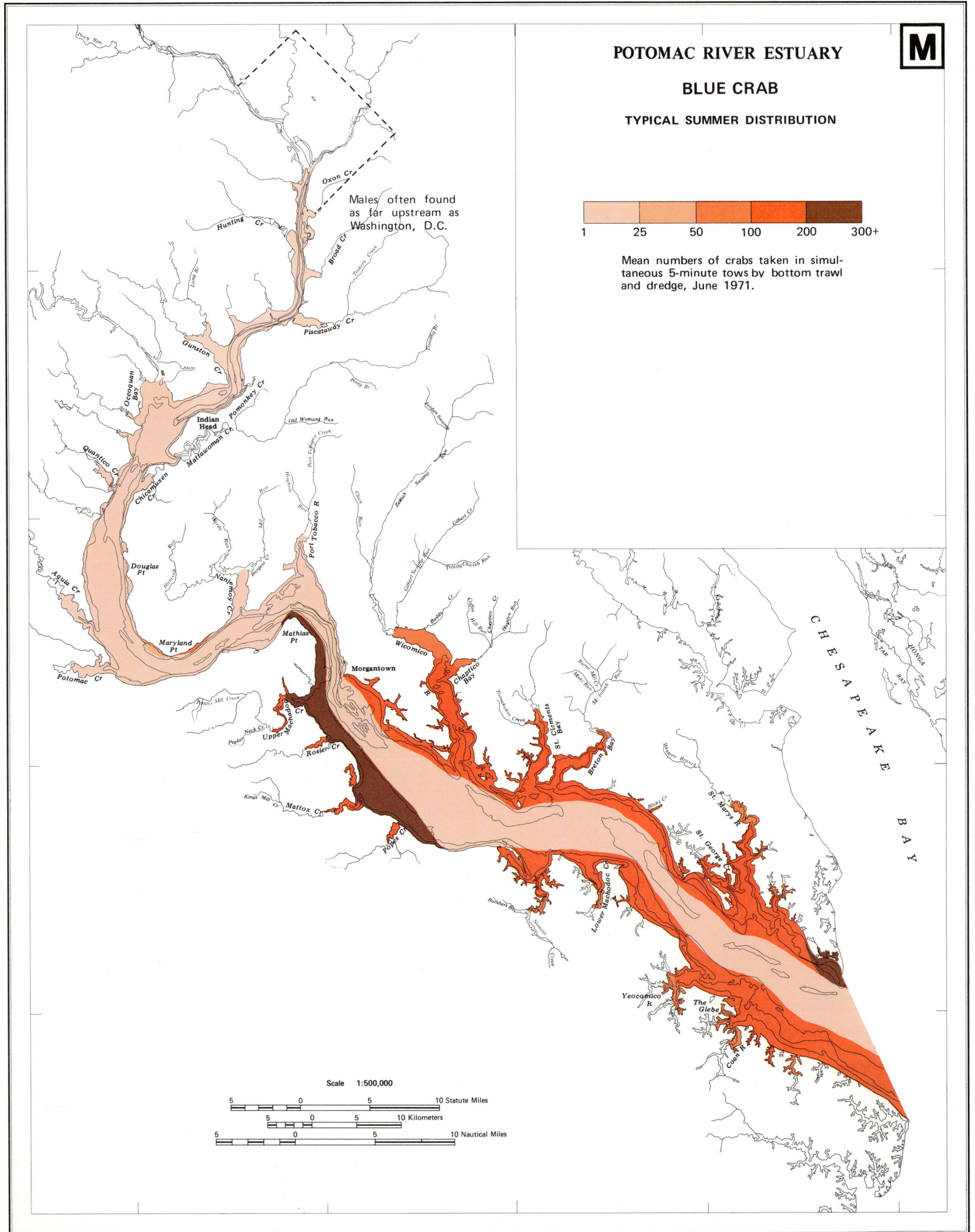

POTOMAC RIVER ESTUARY

BLUE CRAB

TYPICAL SUMMER DISTRIBUTION

Mean numbers of crabs taken in simultaneous 5-minute tows by bottom trawl
and dredge, June 1971.

Males often found
as far upstream as
Washington, D.C.

Scale   1:500,000

5   0   5   10 Statute Miles

5   0   5   10 Kilometers

5   0   5   10 Nautical Miles

Figure 7-6.   *Typical summer distribution of blue crabs in the Potomac estuary, extrapolated from numbers of crabs taken at multiple sampling stations along the entire estuary in June 1971. (Source: Ref. 69)*

POTOMAC RIVER ESTUARY

BLUE CRAB

TYPICAL WINTER DISTRIBUTION

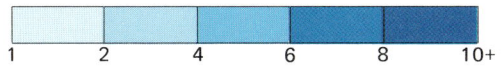

Mean numbers of crabs taken in simultaneous 5-minute tows by bottom trawl and dredge, February 1970 and March 1971.

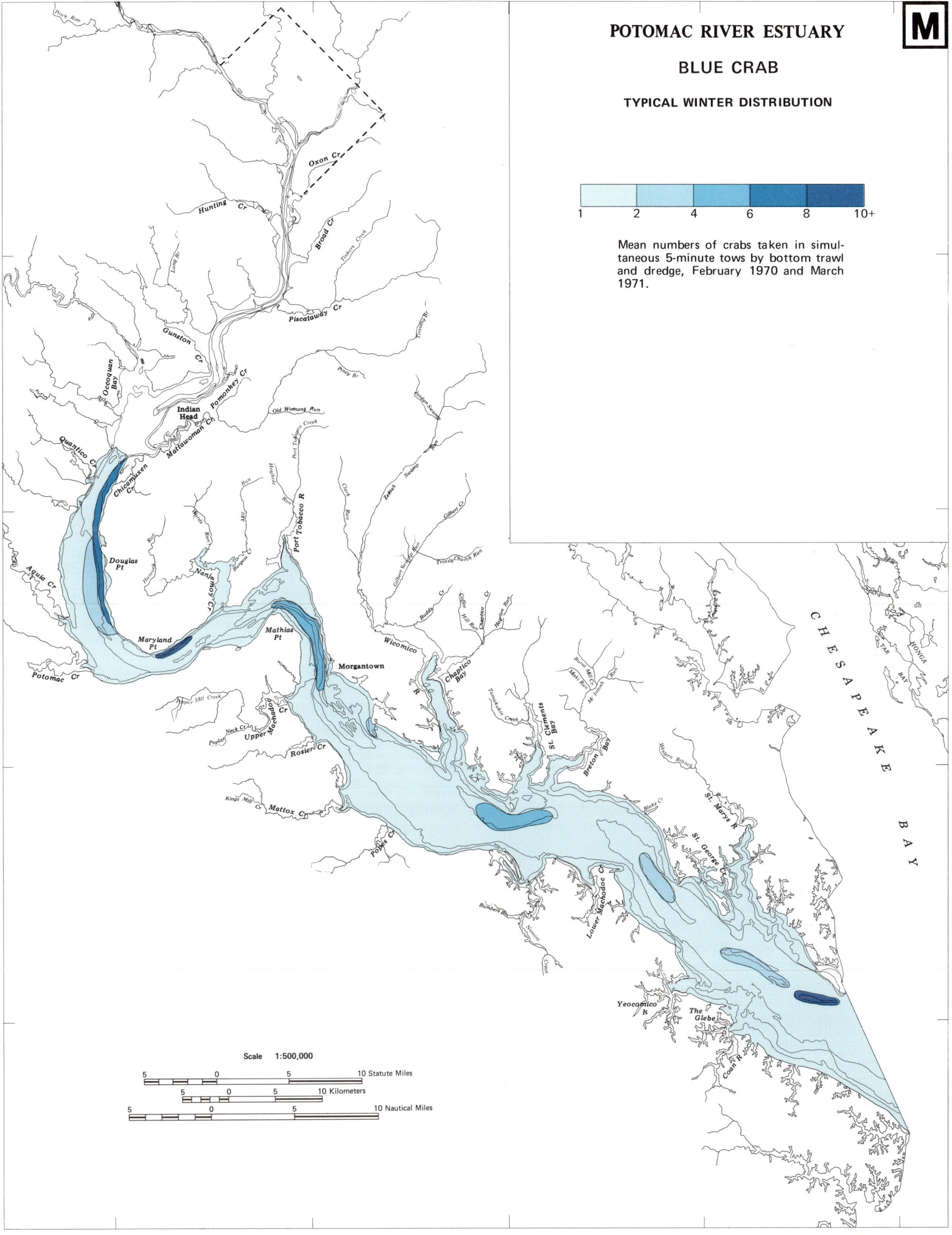

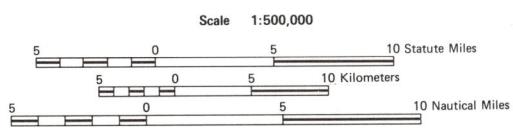

133

Figure 7-7. *Typical winter distribution of blue crabs in the Potomac estuary, extrapolated from average numbers of crabs taken at multiple sampling stations along the entire estuary in February 1970 and March 1971. (Source: Ref. 69)*

BLUE CRAB LARVAE

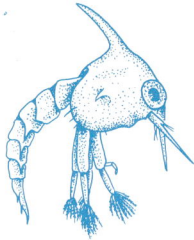

Zoea

Megalops

SEA SQUIRT

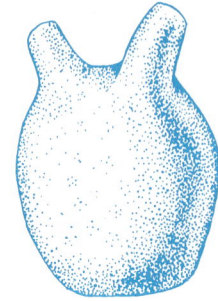

Molgula manhattensis

134 but most importantly, it ensures the presence of the male crab during the brief time when the female will accept sperm. Mating occurs from June to October, generally peaking in July and August.[107–110] [113] [114] Females mate only once, after their final molt; males may mate several times.[112] [114] After mating, Potomac females begin a long migration to the mouth of the Chesapeake Bay and higher salinity waters where spawning occurs.[108] [113] [114] Eggs can be fertilized immediately following mating. However, in the Potomac, females probably store the sperm in a specialized sac and use it later when they are closer to the spawning grounds. Stored sperm are also utilized in subsequent spawning.[112–114]

The bright orange, fertilized eggs are brooded under the body of the female. From 0.5 to 2 million eggs are brooded per female per spawning season.[108] [113] [114] Because the eggs have a spongy appearance, brooding females are frequently called sponge crabs. A planktonic stage, the zoea, emerges from the egg mass in about 15 days[109] [115] at salinities of 20 to 32 ppt; the higher salinities are optimal.[108] [109] [116] The zoea molts several times and, in about 6 weeks, settles to the bottom as a small crab-like stage called a megalops, which quickly metamorphoses into a juvenile crab.[109] [116] Juvenile crabs and megalops migrate up the Chesapeake Bay and into all its tributaries, including the Potomac. Juveniles from spring spawns reach the Potomac in the fall; those from fall spawns do not arrive in the Potomac until the following spring or summer after overwintering in the Chesapeake Bay.[109] However, during exceptionally warm years, some might reach the lower Potomac and overwinter there.[109]

Blue crabs grow quickly. Those added to the population from the spring spawn become sexually mature individuals within 12 to 16 months.[108] [109] Crabs added to the population in the fall, however, usually require 18 to 20 months to reach maturity and generally are not capable of reproduction until their second summer.[109] Blue crabs usually do not live more than three years.[109] [112]

The activity and behavior of blue crabs vary with seasonal changes in water temperature.[108] In spring, as water temperatures begin to increase, blue crabs become active. Juvenile crabs and mature males that overwintered in the deep channel areas migrate upstream and toward shore.[108] [109] Juveniles added to the population in the lower Chesapeake Bay spawning grounds during the previous fall begin arriving at the mouth of the Potomac in spring.[109] [117] Throughout the summer and warm fall months, blue crabs primarily inhabit the shallow nearshore waters (Fig. 7-6). Populations in the estuary are a mix of mature males, mature females, and juveniles. However, the proportion of males to females changes rapidly upstream of the Rt. 301 bridge, and in the upper reaches of the estuary, especially above Maryland Point, the population is mostly males.[57] [117–122]

During fall, as water temperatures begin to decrease, mature females migrate out of the Potomac estuary to the spawning grounds in the lower Chesapeake Bay. Some of the remaining juvenile crabs and mature males move into deeper channel areas where they overwinter, buried superficially in the mud.[108] [109] A few blue crabs may stay in the shallow areas until October or November, but at the first abrupt drop in temperatures, they also move into the deeper channel areas and settle into the bottom muds for the winter (Fig. 7-7).[108] [109] Blue crab populations sometimes suffer heavy winter mortalities following summers when their abundances are high and competition for food is intense.[108] [109] This may occur because many crabs are unable to obtain sufficient food to physiologically prepare themselves for overwinter dormancy.[108] [109]

Phylum Chordata — Tunicates, Sea Squirts

Although tunicates are invertebrate-like in their adult stage, they are classified as chordates because their embryonic stages have distinct chordate characteristics. Tunicates are commonly called sea squirts because of the jets of water they eject when agitated. They are sometimes abundant members of the epifaunal benthic community in high mesohaline and near polyhaline regions of the Potomac estuary.

The outer covering of tunicates is composed chiefly of cellulose and is frequently encrusted with debris. Tunicates are filter feeders. Their incurrent siphon takes water into the basket-like gill chamber where suspended particles are removed for ingestion. Waste products and filtered water are discharged through the excurrent siphon. Both sexual and asexual reproduction occurs in tunicates. They are hermaphroditic; however, self-fertilization is uncommon. The most frequently encountered species in the Potomac is *Molgula manhattensis,* which sheds its eggs and sperm into the water where fertilization and embryonic development occur.[53] [58] [68] The tadpole-like larvae attach to a suitable hard substrate and metamorphose into their adult stage.

Chapter 7 — References

1. Virnstein, 1977
2. Chao and Musick, 1977
3. Carriker, 1967
4. Mare, 1942
5. Rhoads, 1974
6. Sikora, Sikora, Erkenbrecher, and Coull, 1977
7. Higgins, 1972a
8. Higgins, 1972b
9. Higgins, 1972c
10. Colwell, 1972
11. ZoBell and Feltham, 1942
12. Jorgensen and Fenchel, 1974
13. Fenchel, 1972
14. Burchard, 1971
15. Petersen, 1913
16. Friedrich, 1969
17. Meglitsch, 1972
18. Mangum and Van Winkle, 1973
19. Hochachka, 1975
20. Fenchel, Kofoed, and Lappalainen, 1975
21. Newell, 1965
22. Jorgensen, 1955
23. Galtsoff, 1964
24. Thorson, 1950
25. Grassle and Grassle, 1974
26. Vance, 1973
27. Sanders, 1958
28. Mileikovsky, 1971
29. Calabrese and Davis, 1970
30. Kinne, 1970
31. Kinne, 1971
32. Segal, 1970
33. Barnes, 1962
34. Meadows and Campbell, 1972
35. Korringa, 1941
36. Thorson, 1946
37. Lebour, 1933
38. Loosanoff, 1966
39. Bousfield, 1973
40. Pfitzenmeyer, 1962
41. Shaw, 1965
42. Shaw, 1967
43. Orth, 1971
44. Boesch, 1973
45. Holland, N. K. Mountford, and Mihursky, 1977
46. Mountford, N. K., Holland, and Mihursky, 1977
47. Holland et al., 1977
48. Holland et al., 1978
49. Cory, 1967
50. Mills, 1969
51. Thorson, 1957
52. Pfitzenmeyer, 1976
53. Virnstein and Boesch, 1975
54. Dahlberg, 1973
55. Haire, 1978
56. Polgar, Krainak, and Pfitzenmeyer, 1975
57. Ecological Analysts, Inc., 1974
58. Krueger and Fuller, 1977
59. Boesch, 1977
60. Spoon, 1975
61. Krueger, 1977
62. Mangum, Santos, and Rhodes, 1968
63. Rosenberg, 1973
64. Maurer et al., 1978
65. Connell, 1961
66. Woodin, 1974
67. Paine, 1966
68. Frey, 1946
69. Lippson, R. L., 1969-1971
70. Fell, 1974
71. Gosner, 1971
72. Barnes, 1968
73. Pfitzenmeyer and Drobeck, 1963
74. Old, 1941
75. Calder, 1971
76. Cargo and Schultz, 1967
77. Riser, 1974
78. Olson and Sage, 1978
79. Diaz, 1976
80. Robertson, 1977
81. Galloway, 1911
82. Boesch, 1978
83. Wass, 1972
84. Carriker and Williams, 1978
85. Miner, 1950
86. Holland and Dean, 1977a
87. Beaven, 1954
88. Haven, 1976
89. Meritt, 1977
90. Castagna and Chanley, 1973
91. Norris, 1975
92. Hidu et al., 1974
93. Potomac Basin Reporter, 1974
94. MacKenzie, 1974
95. Haven and Davis, 1973
96. Dunnington, Haven, and Drobeck, 1974
97. Hopkins, S. H., Anderson, and Horvath, 1973
98. Pfitzenmeyer and Drobeck, 1964
99. Cain, 1973
100. Cory, 1978
101. Gordon, 1969
102. Branscomb, 1976
103. Van Engel, 1972b
104. Burbanck, 1967
105. Holland, 1978
106. Bousfield, 1969
107. Williams, 1974
108. Van Engel, 1958
109. Lippson, R. L., 1971
110. Warner, 1976
111. Darnell, 1959
112. Tagatz, 1968
113. Churchill, 1919
114. Truitt, 1939
115. Costlow and Bookout, 1959
116. Costlow, 1967
117. Sulkin, 1973
118. Abbe, 1977b
119. Academy of Natural Sciences of Philadelphia, 1969a
120. Academy of Natural Sciences of Philadelphia, 1970a
121. Academy of Natural Sciences of Philadelphia, 1971a
122. Academy of Natural Sciences of Philadelphia, 1972

Chapter
8

Fishes

Fishes

Among the animals of the estuary, finfish are the most important in terms of economic and recreational value. Some are harvested by man; others serve primarily as forage for predatory fishes. Most fish species hold a position near the top of the estuarine food chain, and their success and survival indicate the general state of the ecosystem. Because of their economic value, the impact of man's activities on fish populations is of great importance. Whether such impact occurs is, in part, determined by monitoring the changing distributions and abundances of fish populations, which are discussed in this chapter.

More than 100 species of fish have been found at one time or another in the Potomac estuary and its tributaries (Appendix Table 6). They range from the well-known striped bass to the small, secretive gobies rarely seen by anyone other than scientists and oystermen. As in other mid-temperate-zone estuaries along the Atlantic Coast, the Potomac has a large variety of species because it is near the northern geographical limit of southern species and the southern limit of more northern species. However, since many species are at the limits of their ranges, year-to-year fluctuations in climatic conditions can cause their numbers and distributions to vary dramatically. Additionally, the reproductive success of many other species is strongly affected by variations in environmental conditions during spawning and larval development. As a result, the composition of the finfish community fluctuates widely over periods of several years.

Besides these long-term changes in community composition, large seasonal variations occur as species move into and out of the estuary or from one area to another within it. These migrations revolve around reproduction, feeding, and overwinter survival.

Reproductive or spawning migrations are by far the most dramatic population movements. Anadromous species return from extensive oceanic travels to spawn where they themselves were spawned. Even species which spend their entire life cycle within the estuary may travel significant distances to spawn in certain preferred regions, moving upstream into tributaries or into shoal areas. Spawning migrations are generally mass movements by large portions of the sexually mature population.

Feeding migrations, in contrast, involve young, immature fish as well as sexually mature adults. Since feeding areas tend to be much more widespread and dispersed than spawning locations, feeding migrations are generally diffuse. In the Potomac, these movements occur in spring and summer, and the fish tend to move upstream and/or into shallow water.

Fish survive the severe temperatures of winter in a number of ways. Some species, such as bluefish and spot, leave the estuary entirely to overwinter in the ocean. Most resident species move out of the tributaries and shoal areas into channel depths where warmer, more constant temperatures prevail. These patterns of movement generally occur in the fall.

All migrations are linked, in some way, to changes in salinity and temperature. For example, successful spawning of most species can occur only within a certain temperature range. In many cases, the hatching success of the eggs is closely related to both salinity and temperature. (See Folio Map 7 for ranges of salinity and temperature over which common species spawn.) Survival of larvae and growth of juveniles can be a function of the availability of food, which in turn is also related to salinity and temperature. Most species are physiologically adapted to certain salinity ranges and experience stress when exposed to higher or lower levels. This close relationship between migrations and temperature-salinity patterns is the focal point of the discussion of individual species that follows.

There are five major categories of fish in the Potomac estuary — freshwater, estuarine, marine, anadromous and semianadromous, and catadromous. The first three are categorized (as are other biota) by the salinities in which they usually live (Fig. 8-1). The fourth and fifth categories consist of specialized marine or estuarine species that must migrate between fresh and estuarine or marine regions to reproduce.

Freshwater fishes are the resident inhabitants of nontidal waters of the Potomac (i.e., above Washington, D.C.) and of the freshwater streams that flow into its tributaries. Many of these species regularly descend into tidal fresh and low brackish waters (Fig. 8-2). Generally, greatest downstream penetration by adults occurs during the winter months. Spawning is usually restricted to nontidal waters, but, after hatching, larvae are often transported into tidewaters. Freshwater species are generally late-spring to early-summer spawners with extended or multiple spawning periods. Because the Potomac is a low salinity estuary, a fairly large number of freshwater species enters its waters.

Estuarine fishes are the resident species of tidal waters where salinity values range from 0 to 30 ppt. Often, these species stray into nontidal fresh water or, at the other extreme, into coastal regions of the sea. In the Potomac, they occur throughout the estuary, but the greatest numbers are found downstream from the interface zone where fresh water meets salt water (Fig. 8-3). The distribution of species fluctuates with the annual or seasonal displacement of this interface along the Potomac.

Estuarine fishes do not exhibit mass spawning migrations, although a few species move downstream to salinities greater than 5 ppt

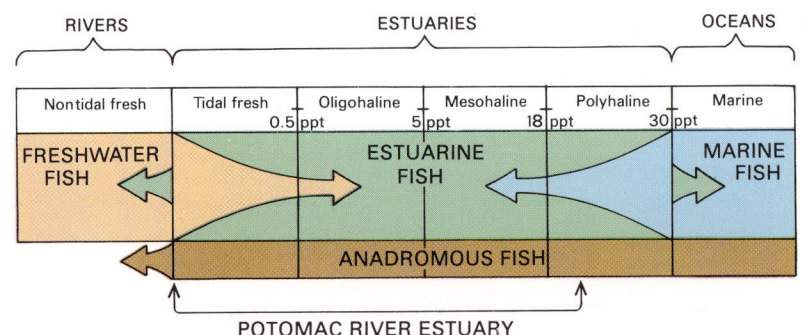

Figure 8-1. *Distributions of different types of fishes by salinity zones.*

FRESHWATER FISHES

INDIAN HEAD   MORGANTOWN   ST. GEORGE IS.
HAINS PT.   MARYLAND PT.   WICOMICO R.

Spring — MAR, APR, MAY
Summer — JUN, JUL, AUG
Fall — SEPT, OCT, NOV
Winter — DEC, JAN, FEB

TIDAL FRESH   OLIGOHALINE   MESOHALINE   POLYHALINE

0.5ppt   5ppt   10ppt   18 ppt

93   75   55   40   29   10
← RIVER MILE FROM MOUTH

Eggs   Larvae   Juveniles   Adults

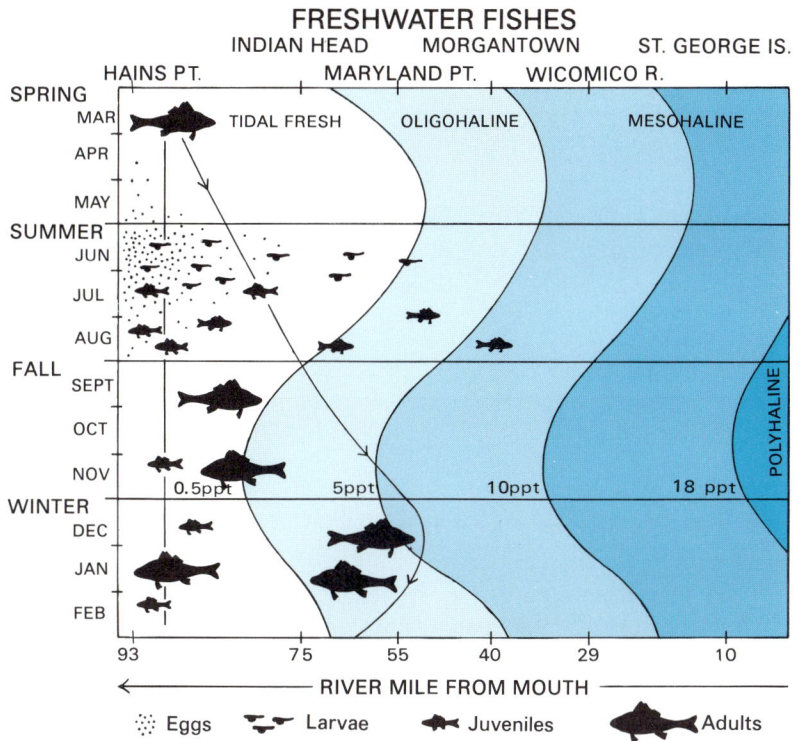

Figure 8-2. *Typical seasonal distributions of freshwater fishes in the Potomac estuary in relation to location of salinities indicated by isohalines.*

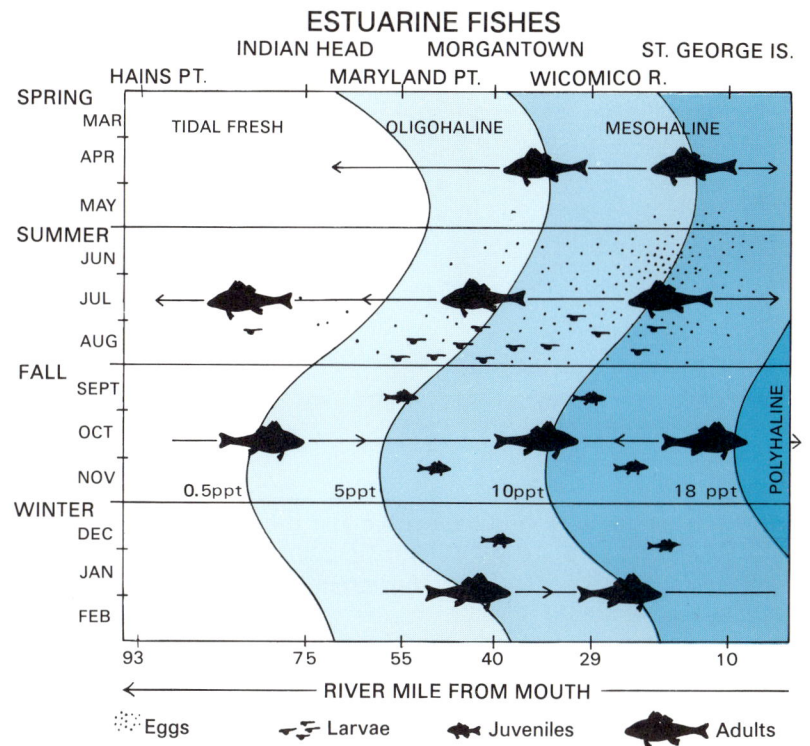

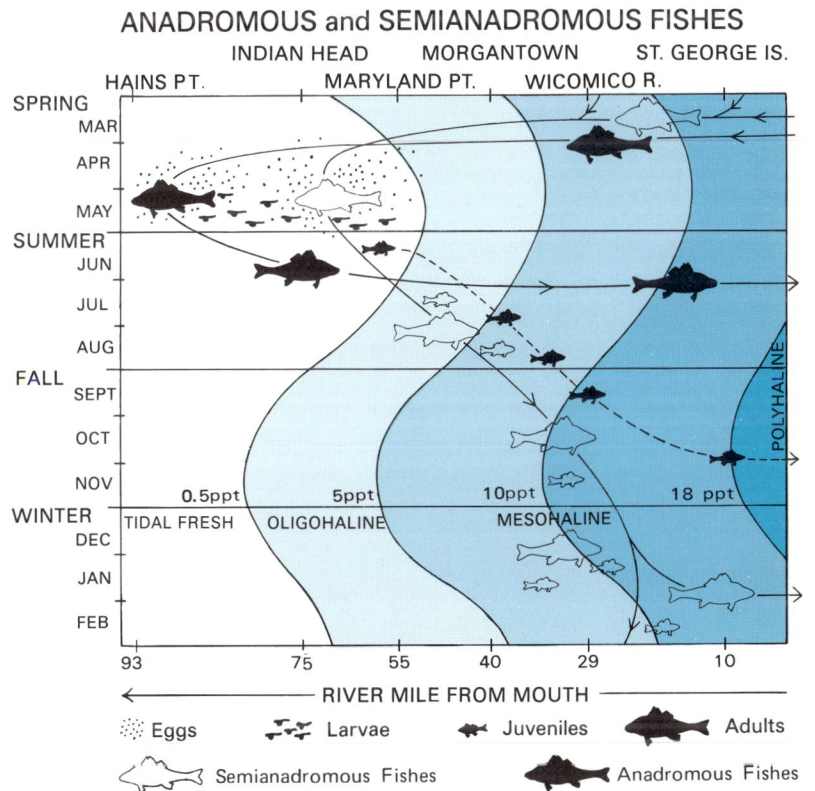

ESTUARINE FISHES

INDIAN HEAD   MORGANTOWN   ST. GEORGE IS.
HAINS PT.   MARYLAND PT.   WICOMICO R.

Spring — MAR, APR, MAY
Summer — JUN, JUL, AUG
Fall — SEPT, OCT, NOV
Winter — DEC, JAN, FEB

TIDAL FRESH   OLIGOHALINE   MESOHALINE   POLYHALINE

0.5ppt   5ppt   10ppt   18 ppt

93   75   55   40   29   10
← RIVER MILE FROM MOUTH

Eggs   Larvae   Juveniles   Adults

139

Figure 8-3. *Typical seasonal distributions of estuarine fishes in the Potomac estuary in relation to location of salinities indicated by isohalines.*

MARINE FISHES

INDIAN HEAD   MORGANTOWN   ST. GEORGE IS.
HAINS PT.   MARYLAND PT.   WICOMICO R.

Spring — MAR, APR, MAY
Summer — JUN, JUL, AUG
Fall — SEPT, OCT, NOV
Winter — DEC, JAN, FEB

late larvae or juveniles from ocean

fresh water tolerant species

few overwinter

0.5ppt   5ppt   10ppt   18 ppt

TIDAL FRESH   OLIGOHALINE   MESOHALINE   POLYHALINE

93   75   55   40   29   10
← RIVER MILE FROM MOUTH

Larvae   Juveniles   Adults

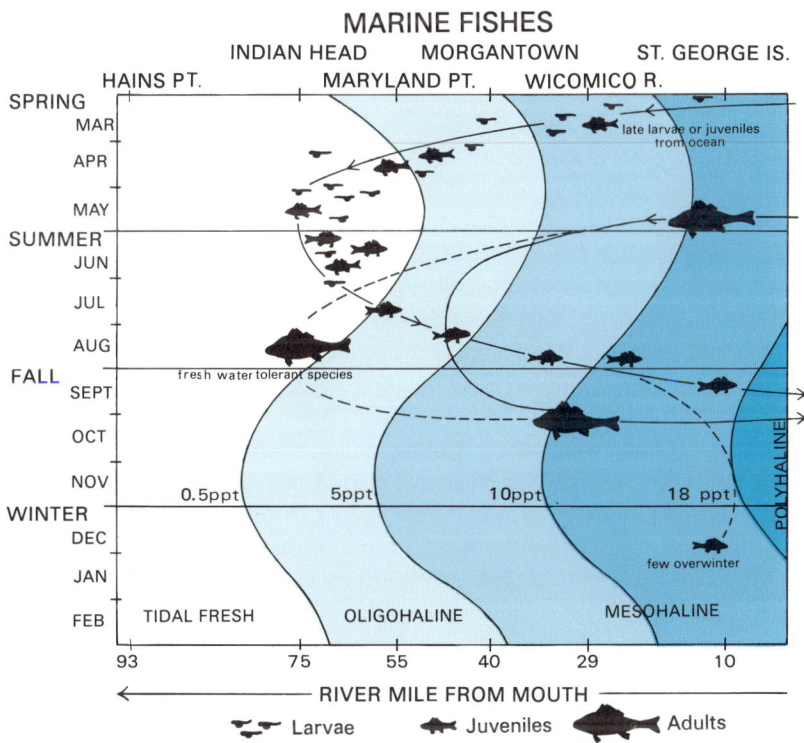

Figure 8-4. *Typical seasonal distributions of marine fishes in the Potomac estuary in relation to location of salinities indicated by isohalines.*

ANADROMOUS and SEMIANADROMOUS FISHES

INDIAN HEAD   MORGANTOWN   ST. GEORGE IS.
HAINS PT.   MARYLAND PT.   WICOMICO R.

Spring — MAR, APR, MAY
Summer — JUN, JUL, AUG
Fall — SEPT, OCT, NOV
Winter — DEC, JAN, FEB

0.5ppt   5ppt   10ppt   18 ppt

TIDAL FRESH   OLIGOHALINE   MESOHALINE   POLYHALINE

93   75   55   40   29   10
← RIVER MILE FROM MOUTH

Eggs   Larvae   Juveniles   Adults
Semianadromous Fishes   Anadromous Fishes

Figure 8-5. *Typical seasonal distributions of anadromous and semianadromous fishes in the Potomac estuary in relation to location of salinities indicated by isohalines.*

to spawn. Their larvae move upstream to lower salinity nursery areas. Most estuarine species begin spawning in late spring, often continuing throughout most of the summer, with peaks in June and July. Optimum spawning salinities for most of these fishes in the Potomac occur from below Morgantown to the mouth. Seasonal migrations are usually upstream and towards shores in spring and summer and downstream to deeper waters in fall and winter.

Marine fishes in estuaries are those that spawn and generally spend most of their lives at sea but, at certain periods, either as adults or young fish, migrate into coastal bays and rivers. In summer, adults usually move in to feed on the abundant schools of small forage fish that roam the estuaries. In the Potomac, few adult marine fish travel upstream of the bend of the river at Maryland Point (Fig. 8-4). The young of marine species that are estuarine-dependent in the early stages of their development penetrate far beyond this point. Hatched from eggs spawned in the ocean or at the mouth of the Chesapeake Bay, the larvae of these species migrate into low-salinity nursery areas. Marine young are present in spring and summer months, primarily in the stretch of the estuary between Indian Head and Morgantown and in

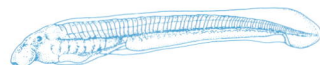
Ammocoete of brook lamprey,

Longnose gar, *Lepisosteus osseus*

Eastern mudminnow, *Umbra pygmaea*

140 upstream portions of the Potomac's major tributaries where salinities are generally 2 to 5 ppt. As water temperatures decrease, both adults and juveniles start moving out of the Potomac. Most have left by the end of October.

Anadromous fishes, such as the herrings and shad, spend most of their adult lives at sea but return to fresh water to spawn. **Semianadromous** fishes, such as white perch and striped bass, are essentially estuarine, but exhibit spawning behavior similar to that of true anadromous species. Anadromous fishes move up the Potomac estuary and its tributaries from the ocean in great numbers each spring on their way to freshwater spawning grounds. Their migration extends at least as far as the limit of tidal influence, with a few pressing on into nontidal waters (Fig. 8-5). In the Potomac main stem, major spawning areas are somewhat below Washington, D.C., in an area above the saltwater-freshwater interface. After spawning, adults migrate back downstream; most return directly to the ocean, although some remain in the Potomac estuary or the Chesapeake Bay for a while. Young-of-the-year also gradually move downstream, and most leave the estuary by late summer or fall. Semianadromous fishes migrate to their spawning grounds from the lower Potomac or the Chesapeake Bay. After spawning, they return to these same regions rather than to the ocean.

Catadromous is a term used for a special category of marine fishes — those that spend most of their adult lives in fresh water but must return to the sea to spawn. This group is represented in the Potomac by a single species, the American eel.

FRESHWATER FISHES

A wide variety of freshwater species is found in the Potomac estuary. For the most part, they are small to medium-sized, rather solitary fish that search for food along the bottom or among aquatic vegetation. Certain species are found regularly, others sporadically. The extent of their migration downstream largely depends on species-specific tolerances to increasing salinities. Freshwater fishes are rarely found in salinities greater than about 8 to 10 ppt, although some species have been found in waters with salinities of 15 to 20 ppt, the extreme limits of their tolerance. Consequently, few are ever captured in the lower half of the Potomac estuary or towards the mouths of the tributaries from the Wicomico River downstream. Although some freshwater species are found in both shallow and deep open waters of the estuary, more are found along the shoreline or in coves. Other species inhabit marshes, tidal pools, or weedy shallows where the water is still. Some are highly selective of habitat; others are more indiscriminate.

The distribution map of freshwater fishes (Fig. 8-6) is based on salinity zones and broad habitat divisions. The freshwater species found in Potomac tidewaters are listed on Fig. 8-7 (to be used with the map) and are divided into three categories:
- Fishes that are common in the main stem and the tributaries
- Fishes that are occasionally in the main stem, but primarily in the tributaries

- Fishes that are tributary oriented (strongly oriented to the tributaries) and rarely or never recorded from the main stem.

Of the 39 species in Fig. 8-7, only 13 occur frequently in the Potomac estuary. In the fresh headwaters of tributaries to the estuary, there are many freshwater species not recorded from tidal waters (but which are listed in Appendix Table 6). A complete list of freshwater fish species for the entire Potomac watershed above Washington, D.C. would, of course, be even more extensive. For convenience, the discussion of freshwater species is presented by taxonomic category; the estuarine and marine species will be discussed according to their habitat preferences. Spawning regions of the more common freshwater species are shown on Folio Map 7.

LAMPREYS

Two freshwater lampreys, the least brook lamprey and the American brook lamprey, occur in the Potomac tidewaters. They are nonparasitic, diminutive, eel-like fish only about 20 cm long. A lamprey spends most of its life (about 4 years) as a larva or "ammocoete" burrowed in sand or soft bottom deposits in clear waters. Spawning occurs in narrow streams in March and April, and the adults die after breeding. Occasionally, the ammocoetes and adults of lampreys are used as fish bait.[15]

Least brook lampreys are captured primarily in tidal fresh waters and occasionally in low brackish areas. They are apparently common in most Potomac tributaries.[13] The American brook lamprey is rare in the Potomac (recent studies have suggested that only small populations occur in the upper Anacostia River).[11] The true abundance of lampreys is difficult to evaluate because they burrow into bottom substrates, which makes them inaccessible to ordinary fish sampling methods.

GARS

A single gar species, the longnose gar, occurs in the Potomac. Longnose gars are large, primitive fish with square, armor-like scales. They predominantly occur in the shallow, weedy areas of clear streams.[16] They are also perennial residents of brackish water, but, since they are sluggish, secretive fish, they are not often observed or caught by fishermen. Longnose gars may reach a length of about 2 m,[17] but fish larger than about 75 cm are rare.[16] They are carnivores, feeding largely on small fish and amphibians.

In the Potomac, longnose gars are found both in the tributaries and the main stem, where they have been recorded down to the vicinity of Douglas Point.[5] Spawning occurs in shallow grassy areas in fresh water in May or June.[8]

HERRINGS

One freshwater herring, the threadfin shad, is sometimes found in the Potomac estuary. Originally introduced into freshwater tributaries of the Potomac estuary, threadfin shad may range down to the meso-haline zone. In the Potomac, juveniles have been collected in tributaries between Aquia and Piscataway creeks and in the main stem at

Spottail shiner, *Notropis hudsonius*

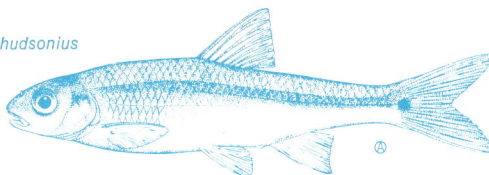

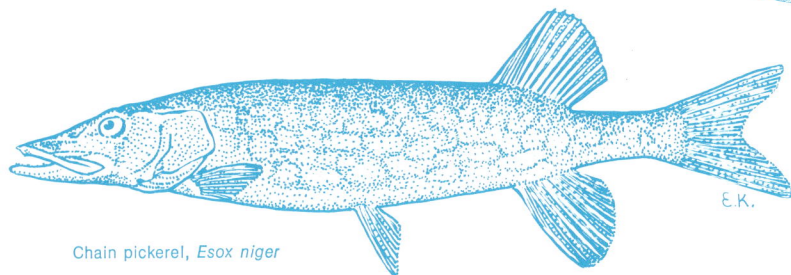

Chain pickerel, *Esox niger*

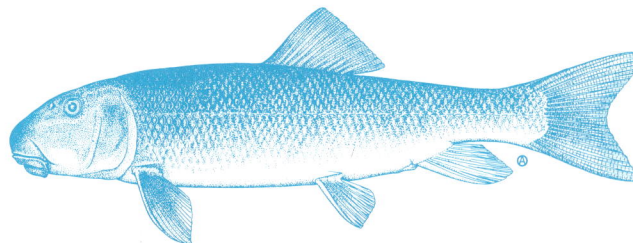

White sucker, *Catostomus commersoni*

Maryland Point.[4] Evidently, these juveniles moved into the more saline waters from their freshwater spawning areas. There are no recent records of adults in the tidal portion of the Potomac, although they have historically been listed as part of the Potomac estuarine fish fauna.[6] A recent listing of Maryland fishes[18] failed to include this species from the Potomac estuary, but juvenile records confirm its presence.[4]

MUDMINNOWS

The eastern mudminnow, the only Potomac mudminnow species,[10] is a small (5 to 7 cm), unobtrusive inhabitant of muddy-swampy regions. They often bury themselves in soft, silty bottoms in small, sluggish streams, where their brown mottled color makes them inconspicuous. They spawn in hollows formed in masses of algae, usually in March or April (see Folio Map 7).[19]

PIKES

Two pike, or pickerel, species inhabit the Potomac — the redfin pickerel and the more common chain pickerel.[3,20] Pickerel are elongated fish with extended jaws armed with many sharp teeth. The chain pickerel is the larger of the two species, normally reaching a length of about 50 cm. Redfin pickerel rarely exceed 25 or 30 cm.[7]

Pickerel live in weedy streams, hiding among the leaves or vegetation until they suddenly dart out to capture their unsuspecting prey. Chain pickerel are common in low brackish waters[21] and are more tolerant of salt water than most freshwater species. They frequently enter waters with salinities up to 12 ppt.[7] Chain pickerel are present in tributaries from the Anacostia River[9] to the St. Marys River[1] and have been captured in the main stem down to the mouth of Nanjemoy Creek.[3,22] Since they are solitary fish and are never found in great numbers, their actual abundance is unknown. Redfin pickerel are most often found high in the tributaries; however, they occasionally stray into tidal waters. In the Potomac, they have even been recorded at Morgantown[23] and in Herring Creek near the mouth of the St. Marys River.[1] Both species spawn in weedy areas in fresh and tidal fresh waters, beginning in March, and thus are among the earliest spawners in the Potomac (see Folio Map 7).[8,19]

Both species are sought by anglers, and pickerels occasionally appear in commercial catch records. Although they have delicately flavored meat, their numerous bones discourage consumption by man. Pickerel are predators on small fish (silversides, sticklebacks, and killifishes)[31] and on invertebrates. They readily assault almost any seemingly edible object in the water.

MINNOWS AND CARPS

The minnow family (which includes the carp) is the largest family of freshwater fishes in the Potomac estuary. There are 10 species of minnow recorded in tidewaters and an additional 12 in tributaries above tidewater.

Carp and its relative, the goldfish, were introduced into United States waters from Europe and the Orient. Today, carp are well established, numerous, and widespread in Maryland.[25] Goldfish have been much less successful in becoming established. The various other species of minnows are often difficult to distinguish, even for biologists. Most are silvery, with small mouths, single spineless dorsal fins, and either dark stripes along their sides or spots at the bases of their tails. They may be called minnows, shiners, chubs, or daces.

Minnows (other than carp) rarely exceed 15 cm in length and have varied habitat preferences. Some seek shallow, weedy shores and coves of quiet streams; some are found in open clear waters, and still others prefer rapids and riffles.[16,19,25] The most common and also the most widely distributed species of this family in the Potomac estuary and its drainage system are the silvery minnow, spottail shiner, golden shiner, satinfin shiner, and carp.[1,4,5,20,23] Goldfish have a similar distribution, but are only present in low numbers. The remaining species, the fallfish, creek chub, blacknose dace, and bridle shiner have only been recorded from tributaries in tidal fresh or very low brackish waters.[1,4,14,20,26]

The farthest downstream distribution of minnows occurs during the winter months.[5] In spring and summer they are mostly found in upstream spawning areas. Their demersal eggs usually stick to rocks or logs and are often found in nests constructed of pebbles or stones.[8] Spawning frequently occurs just above the line where tidal influence ends, and larvae may be carried downstream into tidal waters by currents. The silvery minnow, golden shiner, carp, spottail shiner, satinfin shiner, bridle shiner, and fallfish may also spawn in tidal fresh to low brackish areas.[8,19]

Minnows have long been used as bait by sportfishermen,[25] but there are no figures on the economic value of the bait fishery. Carp contribute sufficient poundage to commercial catches to be of economic value. Although in great demand as a food source in both Europe and the Orient, carp are not considered a particularly marketable fish in this country.

Fishes of this family are omnivores that feed on algae, insects, fish eggs and larvae, and occasionally small fish.[16,27] They in turn are preyed on by larger fishes (such as basses and pikes).

SUCKERS

The primary difference between the suckers and their close relatives, the minnows, is their sucker-like mouths, which are adapted for scavenging along the bottom. Most suckers can attain lengths of approximately 60 cm, but one species, the northern hog sucker, usually does not exceed 25 cm.[16] Suckers generally have small scales and are less colorful than minnows. They prefer open waters above clear, sandy, or muddy bottoms. Suckers move upstream in spring to headwaters to spawn and move downstream to overwinter in deeper waters.[5] (See Folio Map 7 for spawning habits and habitats.)

Suckers are rarely seen in the main stem of the Potomac. Within the tributaries, white suckers and creek chubsuckers are common residents of tidal fresh and low brackish waters.[1,9,11,13,20] The shorthead redhorse, quillback, and northern hog sucker may be found in tidal fresh regions

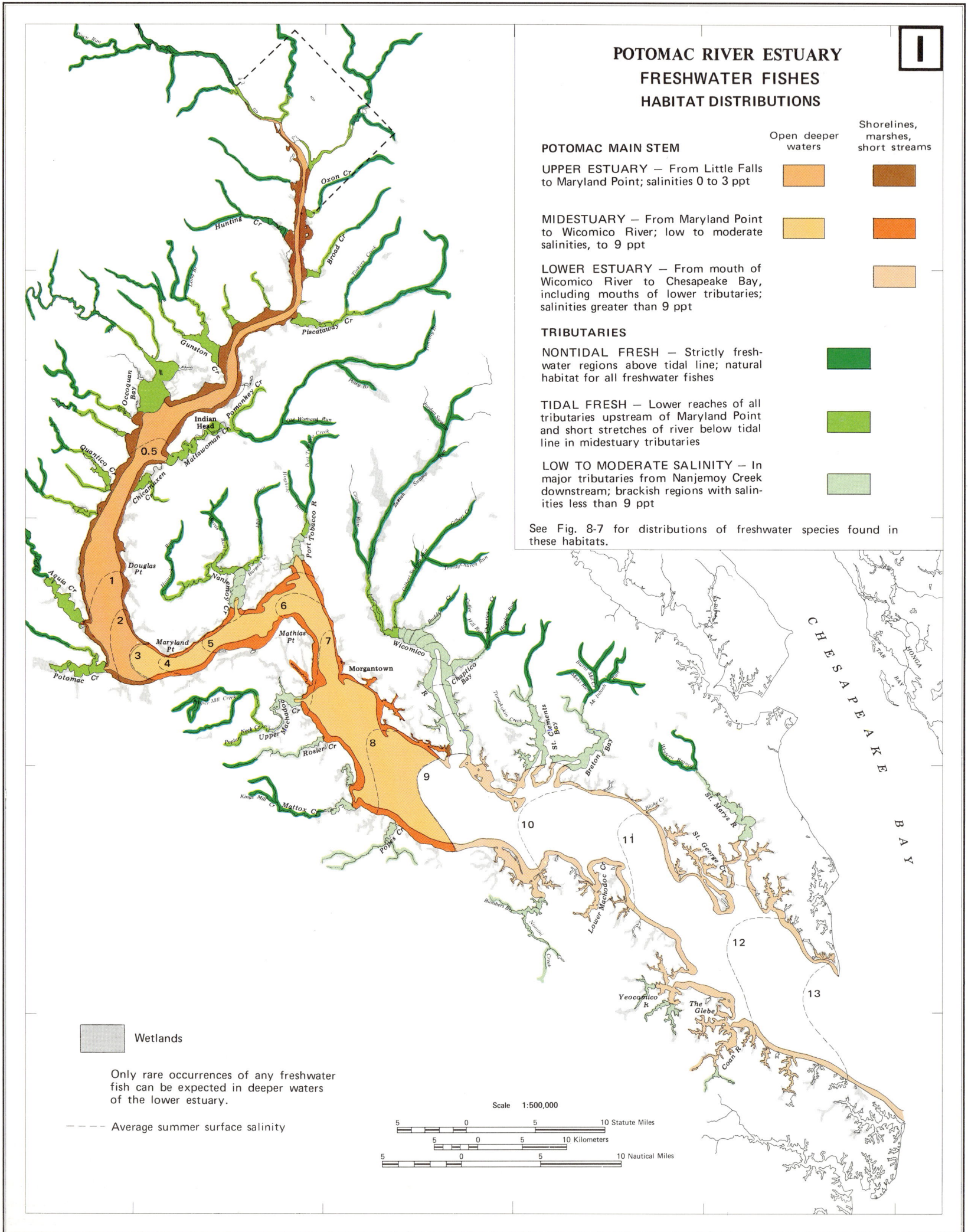

Figure 8-6. *Distributions of habitats of freshwater fishes in the Potomac estuary; see Fig. 8-7 for a list of species occurring in these habitats.*

142

Figure 8-7 presents a matrix of habitats used by freshwater fishes. Column headers are organized as follows:

	Main Stem			Tributaries			Occurrence in Maryland Tributaries, 1970-71																				
Species	Upper Estuary	Midestuary	Lower Estuary	Nontidal Fresh	Tidal Fresh	Low to Moderate Brackish	Rock Creek	Anacostia River	Oxon Creek	Broad Creek	Swan Creek	Piscataway Creek	Pomonkey Creek	Mattawoman Creek	Chicamuxen Creek	Nanjemoy Creek	Port Tobacco River	Popes Creek	Ravens Crest Creek	Wicomico River	St. Clements Bay	Breton Bay	Poplar Hill Creek	Herring Creek	St. Marys River	Jutland Creek	Whites Neck Creek
COMMON IN MAIN STEM AND TRIBUTARIES																											
*Yellow perch	■	■	■	■	■	■		■		■	■	■		■	■	■	■		■	■		■					
*Silvery minnow	■	■	■	■		■			■	■	■				■	■				■							
*Spottail shiner	■	■	■	■	■					■	■			■	■					■		■					
*Brown bullhead	■	■	■	■	■	■				■	■	■	■	■			■			■		■			■		
*White catfish	■	■	■	■	■	■					■	■		■	■					■		■					
*Channel catfish	■			■	■	■				■	■	■		■													
*Golden shiner	■	■	■	■	■	■		■		■	■	■		■	■	■	■	■		■		■			■		■
*Carp	■	■		■	■									■	■				■	■		■	■				
*Chain pickerel	■	■		■	■	■		■				■		■						■		■			■		
OCCASIONALLY IN MAIN STEM AND COMMON IN TRIBUTARIES																											
*Bluegill	■	■	■	■	■	■		■		■	■			■		■	■		■	■		■			■		
*Pumpkinseed	■	■	■	■	■	■	■		■	■	■	■	■	■	■	■	■	■	■	■	■	■	■	■	■	■	■
Threadfin shad	■			■	■																						
Goldfish	■			■				■																			
Longnose gar	■			■	■																						
*Satinfin shiner	■	■		■	■																						
Mosquitofish	■	■		■	■																						
Tessellated darter	■			■	■	■						■		■		■	■			■		■			■		
TRIBUTARY ORIENTED AND RARELY IN MAIN STEM																											
*White sucker	■			■	■		■	■		■		■		■		■	■			■							
Shorthead redhorse	■			■	■																						
Redfin pickerel	■	■		■	■																				■		
Quillback	■			■																							
Yellow bullhead	■			■			■																				
Bluespotted sunfish	■			■																						■	
Largemouth bass	■	■		■	■	■													■	■	■						
Black crappie	■			■	■									■					■	■							
Eastern mudminnow	■	■		■	■																						
*Creek chubsucker	Not Recorded in Main Stem			■	■			■						■		■	■			■	■	■	■	■	■	■	
Northern hog sucker				■																							
Redbreast sunfish				■	■		■	■									■					■	■		■		
Longear sunfish				■	■												■				■	■					
Fallfish				■	■	■		■						■			■			■							
Creek chub				■	■				■		■						■										
American and least brook lampreys				■	■																						
Blacknose dace				■	■			■				■															
Margined madtom				■	■	■														■							
Tadpole madtom				■	■																						
Bridle shiner				■																■							
Green sunfish				■			■	■																			

*Common species.

Figure 8-7. Habitats used by freshwater fishes recorded from the Potomac estuary (Refs. 1-14), and records of species occurring in some Maryland tributaries during 1970 to 1971 (Ref. 1); see Fig. 8-6 for distributions of habitats in the Potomac estuary and tributaries.

Mosquitofish, *Gambusia affinis*

White catfish, *Ictalurus catus*

Bluegill, *Lepomis macrochirus*

144   of the estuary.[9] [11] [19] The quillback is considered a rare species in the Potomac estuarine drainage.[11]

Although suckers are caught by sport anglers with some frequency, they are not a particularly appetizing food fish, and there is no commercial fishery for them. Like minnows, suckers feed voraciously on a wide variety of bottom fauna.

CATFISHES

Catfishes have distinctive whisker-like barbels, smooth scaleless bodies, and sharp barbed spines at the margins of their pectoral fins. Three species (brown bullhead, white catfish, and channel catfish) are abundant in the Potomac. They are large (averaging between 25 and 50 cm), prolific, and often schooling. Two other species, yellow bullheads and margined madtoms, are occasionally found in the Potomac. Tadpole madtoms are rare. Because the madtoms are secretive fish, often burrowing in bottom muds, their abundance is difficult to assess.

Catfishes are all similar in appearance, but bullheads are distinguishable by their squared tails, channel catfish by forked tails, and madtoms by heavy, rounded tails with a flap extending forward along the mid-dorsal line. All species move along the bottom, searching for food with "tastebuds" located on their sensitive barbels.[16] [28] The barbels play an important role in searching for food since visibility is generally poor along the bottom.

Catfishes (with the exception of madtoms) are tolerant of low salinities. They suffer stress in salinities above about 10 ppt, although they have been reported from salinities as high as 15 ppt.[29] In the Potomac main stem, they have been found only to Mathias Point.[12] Monthly sampling between Mattawoman Creek and Maryland Point in 1973 and 1974[5] showed that brown bullheads were the fourth ranking species, white catfish were ninth, and channel catfish were thirteenth. In the Potomac main stem, catfishes are found in deeper channel waters, but in tributaries, they frequent shallow, weedy zones. Brown bullheads are more likely to inhabit the upper regions of the smaller streams.

Spawning occurs in fresh water in late spring and early summer (May-July; see Folio Map 7) in nest-like depressions on coarse sand or gravel.[8] Dense schools of recently hatched young catfishes are often observed moving about in swirling masses at the surface. Occasionally, these young fishes move into tidal waters. Apparently, there is an upstream movement of catfishes in the spring from the estuary to freshwater spawning sites.

Catfishes are omnivorous bottom feeders, grabbing at any form of food, be it algae, snails, worms, clams, small fish, or even dead material.[16] [28] Most catfishes take bait readily from hook-and-line fishermen and enter the commercial fisheries in substantial numbers (see Chapter 10). Catfishes are highly prized as food by many people, although they are not as popular in Maryland as they are farther south.

MOSQUITOFISH

A peculiar fish, the mosquitofish is the only member of its family (the livebearers) in the Potomac.[10] It is tiny, no more than 4 to 5 cm long, with a small superior mouth (opening at the top of the snout), stout body, and a blunt, square tail. Mosquitofish are closely related to the killifishes. The popular aquarium fish, the molly, is also a member of this freshwater family, and, like the molly, mosquitofish give birth to living young rather than discharging eggs. This makes them unique among Potomac fishes. Reproduction takes place in fresh and brackish waters from May through August.

Mosquitofish are usually found in shallow pools and tidal inlets in marshy, weedy areas where the water is quiet and often stagnant.[30] [31] Not much is known of specific Potomac distributions, although there are records from such diverse locations as the Anacostia River at the head of the estuary[9] [11] and tidal pools on St. George Island toward the mouth of the estuary.[32]

Frequently found moving about near the surface in search of insect larvae, mosquitofish are often referred to as "top minnows." As their name implies, they are voracious predators of mosquito larvae and have been used in mosquito control in certain regions of the United States.

SUNFISHES

Eight representatives of this large family appear in Potomac tidewaters; an additional four are found above tidewater in tributaries below Washington, D.C. (Appendix Table 6). This family includes not only the species generally called sunfishes, but also the freshwater basses (or black basses) and crappies (Table 8-1).

Table 8-1 — **Members of the Sunfish Family in the Potomac Estuary**

Sunfishes	Crappies	Freshwater Basses
Bluegill	Black crappie	Largemouth bass
Pumpkinseed		
Redbreast sunfish		
Green sunfish		
Longear sunfish		
Bluespotted sunfish		

Sunfishes and crappies are small (generally less than 15 cm long), deep-bodied fish with large, spreading dorsal and anal fins which give them a rounded profile. Freshwater basses are larger (commonly to 30 cm), elongate and robust, and have small fins in relation to their body size. Sunfishes are generally brilliantly colored, often with a distinct spot on a flap-like extension at the margins of the gill coverings. Crappies are less colorful but are brightly speckled with black or brown. Freshwater basses have dark green backs which shade to yellow on the belly. Their sides are marked with a series of dark bands.

Largemouth bass, *Micropterus salmoides*

Tessellated darter, *Etheostoma olmstedi*

As a group, sunfishes are found in a variety of habitats, although they generally prefer shallow, vegetated cover in fairly still waters. In the Potomac main stem, only three species of this large freshwater family tolerate oligohaline conditions, and only two of those tolerate mesohaline salinities.[2]

Pumpkinseeds are the most widely distributed and plentiful sunfish in the Potomac, and are found throughout the main stem and in all tributaries, usually close to shore.[1 3 5 13 20] They occur farthest downstream in the main stem during winter and early spring. In the shore zone between Mattawoman Creek and Maryland Point, the pumpkinseed ranked eleventh of all species collected over a year-long study in 1973.[5]

The closely related bluegills are less numerous than pumpkinseeds and less likely to be encountered in the main stem, although they have been recorded at least down to Morgantown. It appears that their tributary distribution is also widespread.[1 4]

Other sunfish species have been found sporadically in estuarine waters at various locations (Fig. 8-7). Although both largemouth and smallmouth bass are indigenous to Maryland, only largemouths are found in the estuarine portion of the Potomac. Black crappies are now dominant over white crappies, which are nearly gone from this region. At the turn of the century, the reverse was true, with white crappies outnumbering black.[31] The reason for this reversal is unknown.

All sunfishes are late spring to late summer spawners (May through August), with peak spawning in June (see Folio Map 7). Eggs are laid in shallow depressions or nests and, except for pumpkinseed and bluegill eggs, are laid in nontidal waters.[33]

Sunfishes are among some of the most highly prized freshwater game fish. Not only are they scrappy fighters when hooked, but they are also very tasty. All species are carnivores, feeding on insects, crustaceans (such as copepods, amphipods, and isopods), worms, and other fish. The largemouth bass is a particularly rapacious feeder and in many lakes and ponds, its predation controls populations of smaller fish such as pumpkinseed or bluegill.[16]

PERCHES

The yellow perch and tessellated darter* are the only members of the perch family recorded from Potomac tidal waters. The yellow perch will be discussed in the section on semianadromous species. Seven other members of this family inhabit or have formerly inhabited Potomac estuarine drainage streams above tidewater (Appendix Table 6).

Tessellated darters are small (5 to 7 cm long), elongate members of the perch family, with blunt heads, eyes set close together on top of their heads, and small inferior mouths. They are common in the Potomac and are the only darters frequently found in estuarine regions. Darters get their name from their jerky movements as they travel along the bottom in search of food. Often they may be found over riffles in rapidly flowing waters.[16] They range widely in Potomac tributaries from the St. Marys River to the Anacostia River.[9 11] However, polluted conditions in the Anacostia have seriously affected darter abundances.[11] Although preferring smaller streams, tessellated darters have also been collected in many regions of the main stem above Maryland Point.

Spawning occurs in late April to June in both nontidal and tidal fresh waters.[35] Eggs are usually deposited under rocks in shallow streams (see Folio Map 7). In the main stem, they have been found at Indian Head and Chapman Point.[4] Larvae have been collected in salinities as great as 7 ppt.

ESTUARINE FISHES

Most estuarine fish species are small and are unexploited in the sport or commercial fishery. This group includes schooling species commonly seen along the shore or around piers and a wide variety of less often encountered bottom dwellers.

There are 22 estuarine species recorded from the Potomac and another 4 presumed to occur there because of their presence in other mesohaline areas of the Chesapeake Bay region.

Figure 8-8 shows the habitats and salinity ranges for three categories of estuarine fishes in the Potomac: shallow-water fishes, pelagic fishes, and bottom-oriented fishes. The breakdown of salinities on the figure does not follow the basic Venice System of zonation, but corresponds to species groupings, as interpreted from known distributional data. Salinity distributions are shown for summer when estuarine fishes are most dispersed. The map only shows species occurrences. In general, numbers of species are greatest towards the mouth and decline in the upstream direction.

Shallow-water fishes, which inhabit the Potomac's edges, marshes, and tidal pools, are all small. They are often seen darting among aquatic vegetation or silhouetted over sandy bottoms in clear areas. These species usually show little migratory behavior.

Pelagic species are those that swim freely throughout the water column, usually in schools, and most often near the surface. These species usually have marked migrational patterns.

Bottom-oriented species are found over mud or sand bottoms throughout the estuary or among crevices formed by shells in the oyster bar community. They feed primarily on benthic organisms and may be migratory or nonmigratory.

---

*Previously Potomac records have listed johnny darter; however, recent taxonomic studies have concluded that in the Chesapeake drainage above the Rappahannock River, the eastern johnny darter or tessellated darter is the correct designation for the species in the area.[34]

POTOMAC RIVER ESTUARY

I

ESTUARINE FISHES

DISTRIBUTION BY HABITAT IN SUMMER

Open deeper waters   Shallows, marshes

UPPER TIDAL FRESH — Uppermost reaches of main stem Potomac and tributaries to extent of tidal influence

LOWER TIDAL FRESH — Generally region between Broad Creek and Indian Head and mouths of adjacent tributaries

INTERFACE — Salinity to 3 ppt in the main stem and tributaries; no discrete transition zones in lower tributaries that have sharp salinity gradients.

LOW SALINE REGION — Salinities from 3 to 7 ppt

MIDESTUARY MESOHALINE — Salinities from 7 to 10 ppt in main stem and tributaries; also includes other salinity zones in sharply gradient tributaries.

LOWER ESTUARY MESOHALINE — Salinities > 10 ppt

OYSTER BARS

Similar habitats from 3 to 15 ppt

– – – – Average summer surface salinities

Shallows, Marshes
← decreasing salinity

Nontidal Fresh	Upper Tidal Fresh	Lower Tidal Fresh	Interface	Saline Water > 3 ppt

SHALLOW-WATER FISHES

Killifishes

  Banded killifish
  Mummichog
  Sheepshead minnow
  Striped killifish
  Rainwater killifish
  Spotfin killifish
  Marsh killifish

Sticklebacks

  Fourspine stickleback
  Threespine stickleback *

Pipefishes

  Northern pipefish
  Dusky pipefish **

PELAGIC FISHES

Silversides

  Tidewater silverside
  Atlantic silverside
  Rough silverside

  Bay anchovy

Open Water
← decreasing salinity

Nontidal Fresh	Upper Tidal Fresh	Lower Tidal Fresh	Interface	Low Saline	Midestuary Mesohaline	Lower Estuary Mesohaline	Oyster Bars

BOTTOM ORIENTED FISHES

Hogchoker

  Naked goby
  Oyster toadfish
  Skilletfish
  Striped blenny
  Green goby
  Seaboard goby *
  Feather blenny

*   Presumed occurrence.

**   Dusky pipefish only at mouth of Potomac.

Scale   1:500,000

5   0   5   10 Statute Miles
5   0   5   10 Kilometers
5   0   5   10 Nautical Miles

CHESAPEAKE BAY

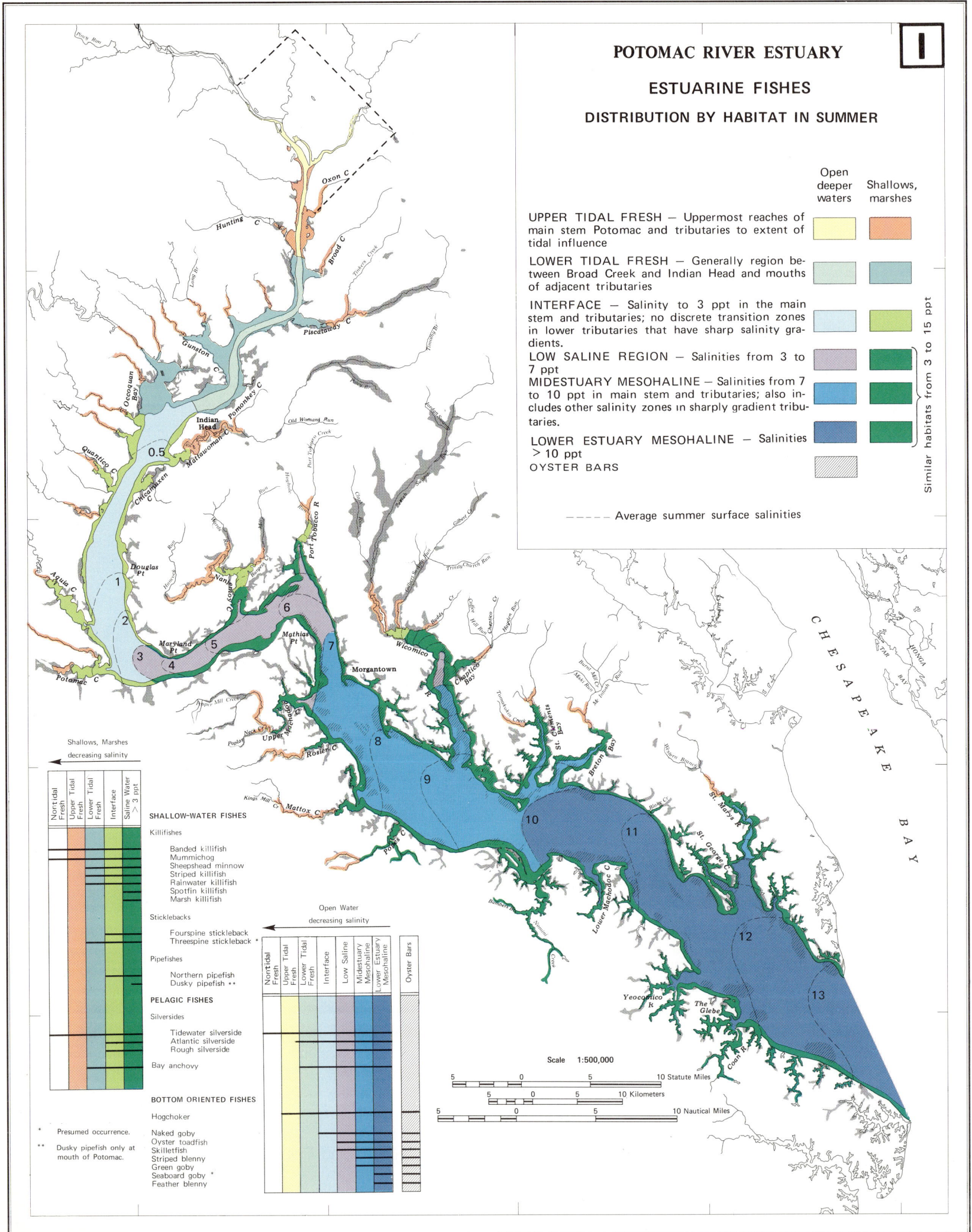

Figure 8-8. *General distribution of estuarine species according to habitat in the Potomac estuary in summer. Map is compiled from data taken in major surveys conducted by the Maryland Department of Natural Resources (Refs. 1, 3, 4, 20); the Chesapeake Biological Laboratory (Ref. 36); Ecological Analysts, Inc. (Ref. 5); The Johns Hopkins University (Ref. 5); and the Academy of Natural Sciences of Philadelphia (Ref. 23); and from Potomac distributions reported in the literature (Refs. 6, 8, 30-32, 37, 38).*

POTOMAC RIVER ESTUARY M

KILLIFISHES
RELATIVE ABUNDANCES OF COMMON SPECIES, 1971

Striped killifish

Banded killifish

Mummichog

Sheepshead minnow

Average number of individuals collected per tributary or estuary segment per 200-foot seine haul

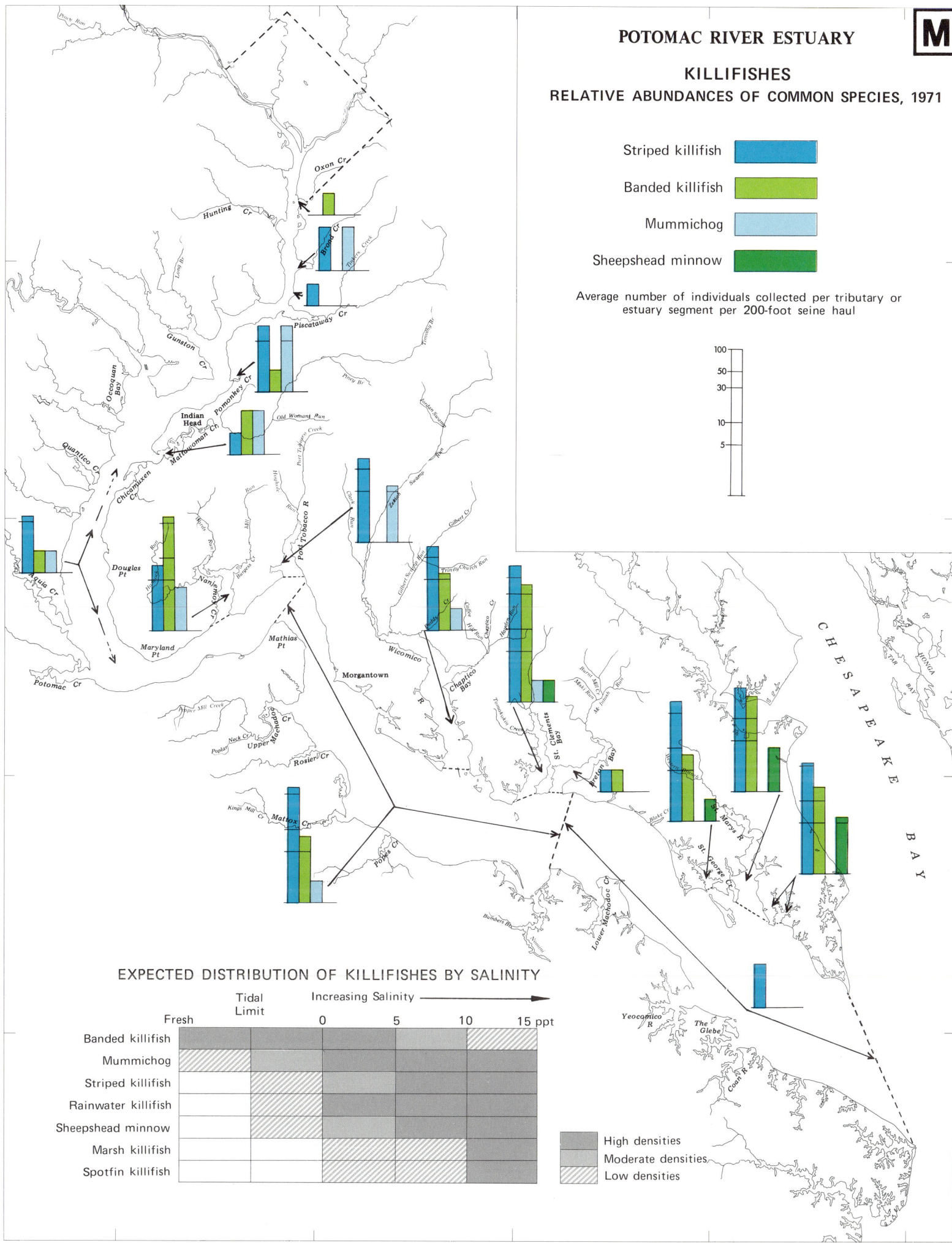

100
50
30
10
5

EXPECTED DISTRIBUTION OF KILLIFISHES BY SALINITY

	Fresh	Tidal Limit	Increasing Salinity →			
			0	5	10	15 ppt
Banded killifish						
Mummichog						
Striped killifish						
Rainwater killifish						
Sheepshead minnow						
Marsh killifish						
Spotfin killifish						

High densities
Moderate densities
Low densities

Map labels: Piney Run, Oxon Cr., Hunting Cr., Broad Cr., Long Br., Tiptoe Creek, Piscataway Cr., Gunston Cr., Pomonkey Cr., Occoquan Bay, Indian Head, Old Womans Run, Mattawoman Cr., Port Tobacco Creek, Trimble Br., Piney Br., Quantico Cr., Chicamuxen Cr., Wolf Run, Mill Run, Burgess, Nanjemoy Cr., Douglas Pt, Aquia Cr., Maryland Pt, Potomac Cr., Mathias Pt, Port Tobacco R., Clark Run, Zekiah Swamp, Gilbert Cr., Jordan Swamp, Gilbert Swamp Run, Trinity Church Run, Wicomico R., Allen's Fresh, Coffee Run, Chaptico Cr., Chaptico Bay, Hoghole Run, Upper Mill Creek, Poplar Neck Cr., Upper Machodoc Cr., Rosier Cr., Kings Mill Cr., Mattox Cr., Popes Cr., Morgantown, Trent Hall Creek, St. Clements Bay, Burnt Mill Cr., Mask Run, Mt. Island Cr., Breton Bay, Blake Cr., St. Mary's R., St. George Cr., Lower Machodoc Cr., Yeocomico R., The Glebe, Coan R., CHESAPEAKE BAY, TAR BAY, HONGA

Figure 8-9. Relative abundance of four common killifish species along the Potomac estuary main stem and tributaries in 1971. Calculated from Maryland Department of Natural Resources, Anadromous Fish Program Data Base (Ref. 1). Numbers of killifish taken within each tributary or broad mainstem segment are grouped. Table of expected killifish distributions by salinity from Refs. 2, 18, 39, 40.

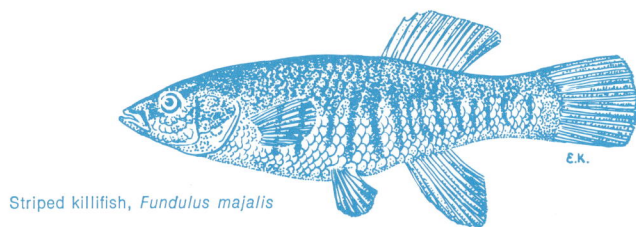

Striped killifish, *Fundulus majalis*

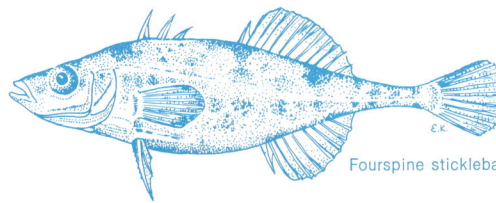

Fourspine stickleback, *Apeltes quadracus*

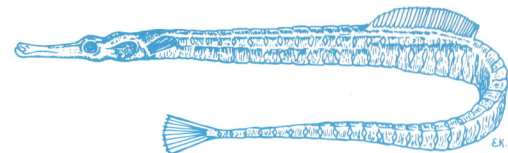

Northern pipefish, *Syngnathus fuscus*

Shallow-Water Fishes

KILLIFISHES

Killifishes, or "bull minnows," can be readily found in the shore zones of estuaries. Usually small darting schools observed along the beach consist of killifishes. They are frequently mistaken for the true minnows, but may be distinguished by their stouter bodies, more rounded, heavier tails, and protruding lower jaws. Many species have distinct striped markings along their sides.

Seven killifish species are recorded from the Potomac,[1-3 38 39] and are distributed along the Potomac estuary according to salinity, as presented in the table in Fig. 8-9. Mummichogs, striped killifish, banded killifish, and sheepshead minnows are the most common of these seven species.[3 4 22 39] Distributions found during a single survey in 1971[1 22] are shown in Fig. 8-9. Although striped and banded killifish were found to be the dominant species during this survey, other studies indicate that the mummichog is generally the most widespread and abundant species.[3 4 40] Mummichogs are usually associated with banded killifish in tidal and nontidal fresh waters and with striped killifish in more brackish regions.[4 40] In the Potomac, these three species occur from the tidal fresh waters near Broad Creek down to the mouth and in all tributaries, inlets, and tidal pools. Mummichogs and banded killifish extend farther upstream into nontidal regions of the Anacostia River and other tributaries of the Potomac estuary.[22] Sheepshead minnows, which resemble small sunfish, occur infrequently in the tidal freshwater areas of the Potomac above Indian Head.[1] There have been isolated records of sheepshead minnows and of rainwater, spotfin, and marsh killifish from nontidal fresh waters.[18]

Rainwater killifish, as the smallest members of this group (less than 4 cm long), are often mistaken for the young of other killifish species. Their distribution is similar to that of sheepshead minnows.

Marsh and spotfin killifish are considered rare species.[30] Marsh killifish have been recorded only in the vicinity of the Rt. 301 bridge.[30] There is only one record of the spotfin killifish from the Potomac estuary;[32] however, they are probably more abundant than records indicate because they can easily avoid seine nets, the major sampling gear used in surveys for shore-zone fishes.[41]

Killifishes usually respond to tidal changes, moving close to shore and into marsh areas at high tide, retreating into more open waters during ebb tides. Seasonal migratory responses have been observed for both mummichogs and rainwater killifish — an upstream movement with warming spring temperatures and a return to higher salinities in winter. Presumably, the other killifish species have similar migrations. This pattern is not always consistent, however; mummichogs observed in a small creek system made a reverse migration, moving upstream in winter.[42] In colder temperatures, activity is reduced, and killifishes tend to bury themselves in bottom muds in deeper waters.

They have an extended spawning season from May through August and sometimes into September (see Folio Map 7).[4 30 39] Peak spawning occurs from May to July in shallow waters among vegetation or over clear, sandy bottoms. Eggs are demersal. Long filaments extending from the egg surfaces usually entangle the eggs in weeds or filamentous algae. Larvae remain close to shore, and schools of small killifish can often be seen swimming among marsh grasses at high tide.

Killifishes are caught and sold as bait.[30] Ecologically, they are important as forage for larger fish and shorebirds and in controlling mosquitoes and other insects. Killifishes adapt readily to living in aquariums and make excellent experimental animals. The mummichog, in particular, has been widely used in physiological experiments on temperature and salinity tolerances.

STICKLEBACKS

Sticklebacks are small fish (5 to 7.5 cm long) easily identified by their fusiform shape and prominent dorsal spines. Fourspine sticklebacks* mingle frequently with pipefishes[31] and killifishes, particularly rainwater killifish,[43] in waters with salinities greater than 3 ppt.[24 31 44] Like killifishes, sticklebacks leave the shallows for deeper channels in winter.[2]

The threespine stickleback is presumed to occur in the Potomac estuary,[2] but it is a rare species throughout the Chesapeake Bay system. Threespines are reputedly anadromous.[2]

Both sticklebacks spawn in April and May in shallow weedy areas. Eggs are deposited in algal nests and guarded by the male.[8] Larvae of fourspine sticklebacks have been collected in salinities from 7 to 22 ppt in other locations in the Chesapeake Bay,[45] indicating that spawning in the Potomac would occur only downstream from the mouth of the Wicomico River.

PIPEFISHES

Pipefishes are odd looking pencil-shaped fish which are closely related to seahorses. Their elongated bodies are ringed with knobbed segments, and, like seahorses, they have small mouths at the ends of extended squared snouts. There are two Potomac species, northern pipefish and dusky pipefish.

Northern pipefish are often seen in close association with killifishes and sticklebacks. Like other shore fishes, they inhabit weed beds along the shores in summer, and deeper channels in winter. Inshore migration occurs in late March or early April, offshore migration in November.[31] Although known to ascend streams to fresh water, and recorded once as far upstream as Gunston Cove,[6] no pipefish were collected in the estuary above Maryland Point in three years of sampling.[5] It can be assumed that they are rare in this region and that their normal distribution in the Potomac is below the Rt. 301 bridge.[22 37]

The dusky pipefish is a polyhaline species that is found commonly in the lower Chesapeake Bay,[2] but only occasionally in the Potomac.[31]

Female pipefishes deposit their eggs in a modified fold or pouch on the abdomen of the male. After hatching, larvae remain in the pouch until the yolk is almost absorbed, at which time they are projected into the water. Because of this parental care, there are no specific spawning

---

*The fourth spine, which is the first spine on the second dorsal fin, is not obvious at first glance.

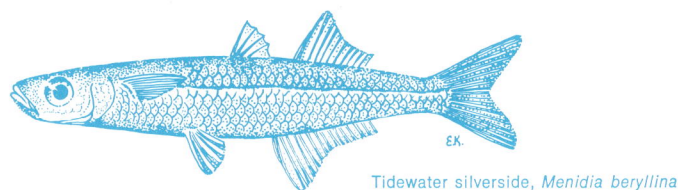

Tidewater silverside, *Menidia beryllina*

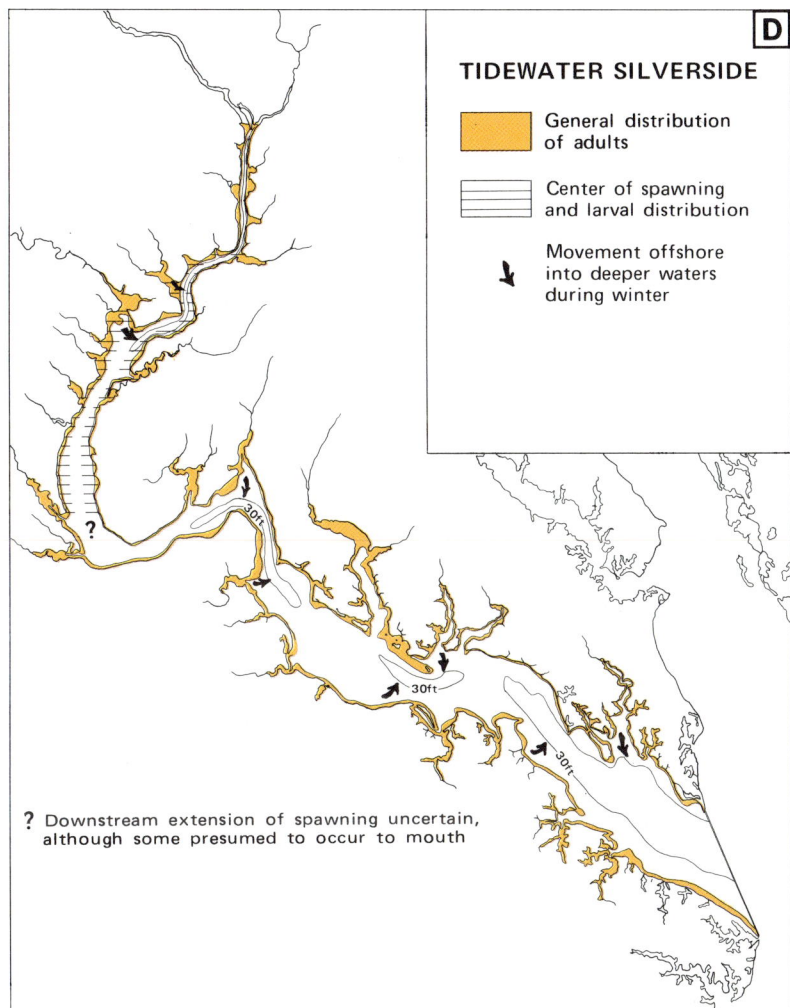

D

TIDEWATER SILVERSIDE

General distribution of adults

Center of spawning and larval distribution

Movement offshore into deeper waters during winter

? Downstream extension of spawning uncertain, although some presumed to occur to mouth

Figure 8-10. *General distribution of tidewater silversides in the Potomac estuary. (Sources: Refs. 3, 4, 20, 22)*

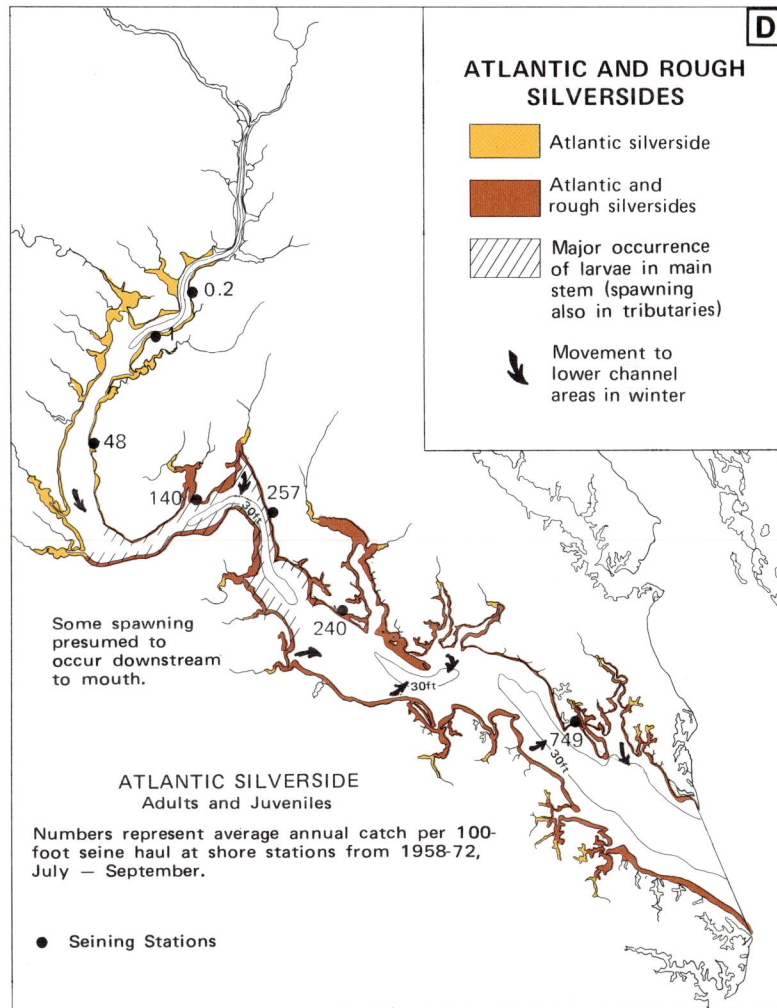

D

ATLANTIC AND ROUGH SILVERSIDES

Atlantic silverside

Atlantic and rough silversides

Major occurrence of larvae in main stem (spawning also in tributaries)

Movement to lower channel areas in winter

Some spawning presumed to occur downstream to mouth.

ATLANTIC SILVERSIDE
Adults and Juveniles

Numbers represent average annual catch per 100-foot seine haul at shore stations from 1958-72, July — September.

● Seining Stations

Figure 8-11. *General distribution of Atlantic silversides and rough silversides in the Potomac estuary (Refs. 3, 4, 20, 22). Numbers by the dots show relative abundance of Atlantic silversides along the shore zone of the Potomac main stem (derived from data in Ref. 3).*

or nursery areas, and distribution of young is similar to that of adults. Northern pipefish larvae have been collected along the mainstem shoreline near Piccowaxen Creek and in the Wicomico River.[4] In other areas of the Chesapeake Bay, northern pipefish larvae have been collected in salinities ranging from 2 to 22 ppt.[45] Spawning can extend from April to October with peaks in May and July.[8] This is the most extended spawning period of any species in the Potomac (see Folio Map 7).

Pipefishes feed primarily on copepods and other small crustaceans such as amphipods.[31]

Pelagic Fishes

SILVERSIDES

Silversides are abundant and ubiquitous in the Potomac estuary. They are small, schooling species, most often seen swimming in surface waters of inshore shoals and tidal creeks. The common name is derived from the distinct silver band on each side of the fish. There are three species of silversides in the Potomac: tidewater silverside, Atlantic silverside, and rough silverside. All serve as forage for many economically important fish species.

Tidewater silversides, the most widely distributed of the three species in the Potomac, range from Washington, D.C., downstream to the mouth (Fig. 8-10). They are most abundant in the upper tidal fresh and low-salinity portions of the estuary.[4][5] In the shore regions from Indian Head to Maryland Point, they were found to be the second most abundant species (after white perch) in net catches in a 1973 survey.[5]

Below Maryland Point, the tidewater silverside is replaced in dominance by the Atlantic silverside and rough silverside. The abundance of these two species gradually increases towards the mouth of the estuary. The Atlantic silverside ranges more often into lower

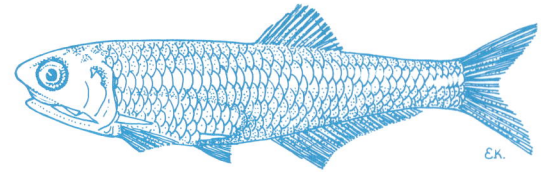

Bay anchovy, *Anchoa mitchilli*

salinities than does the rough silverside (Fig. 8-11),[31] although both species have been collected in tidal fresh waters as far upstream as Indian Head and the mouth of the Pomonkey Creek.[3 4 20 22 36]

Silversides are strongly oriented towards the shore. Seasonal movement shoreward begins as early as March, with schools remaining inshore at least into November. They probably remain near shore for most of the rest of the winter, moving to deeper waters only during the coldest periods. It is uncertain where tidewater silversides overwinter in the estuary since they are absent from the shore zone and sparse in deeper waters of upstream locations during this season.[5] It is presumed that they migrate downstream.

Silverside eggs are deposited on the bottom, attached to aquatic vegetation by long filaments.[8] As a result, few are captured in surveys of fish eggs and larvae (ichthyoplankton), and spawning distributions are determined from larval catches. Because early larval stages of silversides are difficult to identify, the delineation of nursery areas and the presumed spawning areas for each species is based on distributions of postlarval stages. All three species apparently spawn in salinities of less than 15 ppt (see Folio Map 7).[45] In the Potomac, two centers of larval concentration have been indicated from recent surveys of the shore zone (Fig. 8-12): the first in tidal fresh waters just above the freshwater-saltwater interface near Douglas Point, the second just below Morgantown.[4 36] Because of the known distribution of adults, it can be assumed that most spawning in the upstream site is by tidewater silversides, and that most in the downstream site is by Atlantic and rough silversides. Although it has generally been thought that

silversides spawn throughout the estuary, the data presented in Fig. 8-12 imply that silversides from the lower Potomac may move upstream to spawn.

BAY ANCHOVIES

Bay anchovies are abundant in the Potomac. They are small, schooling fish found in open waters along shore as well as in midstream. Bay anchovies are one of the major forage fish species in the Chesapeake Bay, particularly for economically important species such as white perch, striped bass, and bluefish. Bay anchovies are often confused with silversides, another important forage fish, because both have a distinct silver band along each side. Anchovies, however, have extremely long mouths that extend well behind their eyes. Another species, the striped anchovy, is a marine species that is occasionally found in the Potomac. It has been found as far upstream as Breton Bay[4] and into the tidal fresh zone in some Virginia tributaries.[46] However, the common anchovy species of the Potomac is the bay anchovy.

Bay anchovies occur in the main stem from the mouth at least to Broad Creek and in all adjacent tributaries up to tidal fresh water (Fig. 8-13).[4 5 20 23 36] Summer populations are spread throughout the Potomac, while winter populations are concentrated in deeper channels of the lower estuary below Morgantown. In May, when water temperatures reach 14 to 16° C, great numbers move above Maryland Point. Large schools are common up to Indian Head in the summer and fall. During these two seasons, a 10-minute tow with a trawl has often netted from 1,000 to 5,000 bay anchovies.[6] By November, when water temperatures drop below 14° C, bay anchovies again move downstream.[5]

Spawning occurs from May to October, a period when water temperatures are generally 13° C or above.[36 45] Spawning activity is most intense at water temperatures greater than 20° C, usually reaching a peak in July and tapering off in August.[8 45] Bay anchovies presumably spawn throughout the Potomac main stem and in all tributaries below Indian Head, although few data are available for the region below the Wicomico River (Fig. 8-14).[4 36] In other regions of the Chesapeake Bay, major spawning has been observed in salinities between 6 and 19 ppt, peaking between 13 and 15 ppt.[45] However, a 1974 survey in the Potomac showed that the most intensive spawning occurred in the region around Mathias Point in salinities between 5 and 9 ppt[36] (Fig. 8-15). During the following year, in July, densities of more than 500 eggs per 1,000 m³ were located at the Mathias Point to Maryland Point region in salinities of 4 to 6 ppt.[47]

After hatching, many larvae are transported upstream into a low salinity nursery area (1 to 7 ppt)[5 36 47] which, in a year with average runoff such as 1974, is located between Indian Head and the bend of the estuary at Mathias Point (Figs. 8-14 and 8-15). Tributary-spawned larvae migrate up their natal stream to low brackish regions, but nursery areas are not discrete.[4] Larvae may be concentrated in primary nursery regions, but moderate densities are found throughout the Potomac. Young bay anchovies remain in the nursery areas until cold weather and then migrate downstream with adults to overwinter in deeper waters of the lower estuary (Fig. 8-13).

Bay anchovies prey on small crustaceans such as copepods, amphipods, and mysid shrimp, and on smaller anchovies.[31]

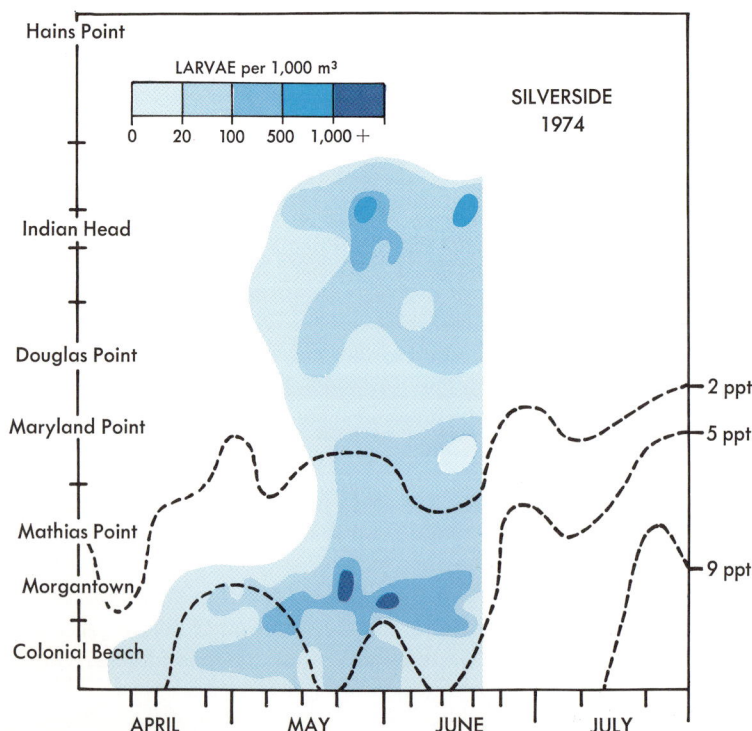

Figure 8-12. *Mean densities of silverside larvae in the Maryland shore zone of the upper half of the Potomac estuary in spring and early summer 1974 (derived from data in Ref. 4). Dashed lines show changes in 2, 5, and 9 ppt isohalines in that year.*

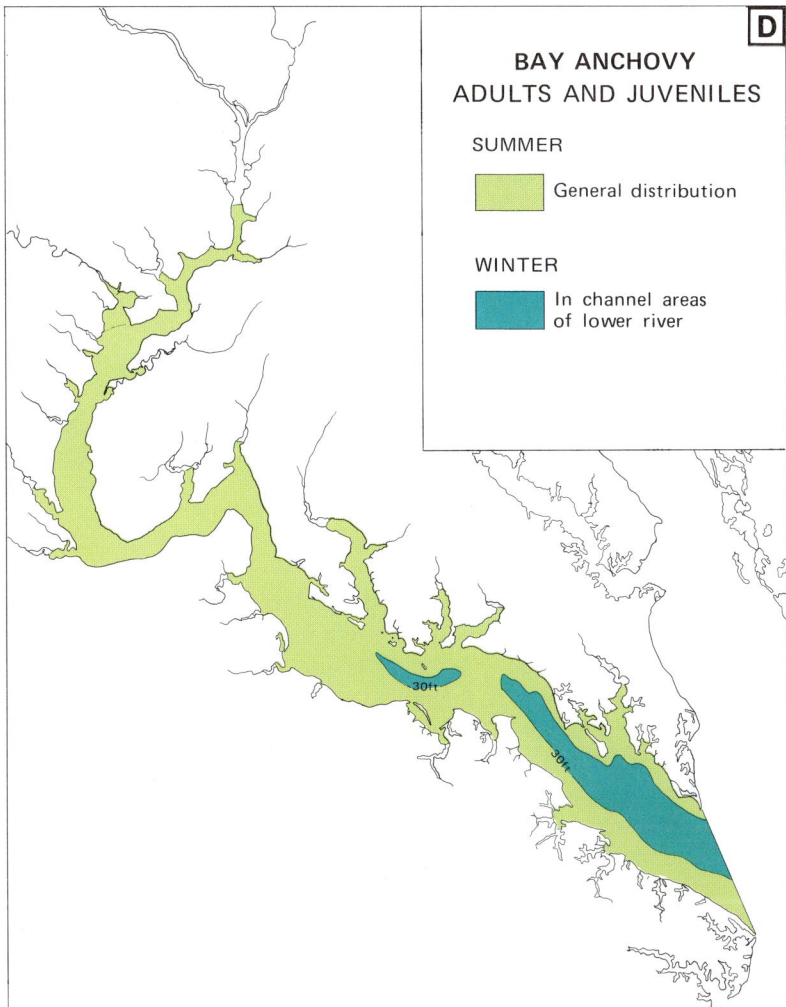

Figure 8-13. *General seasonal distribution of bay anchovy adults and juveniles in the Potomac estuary. (Sources: Refs. 1, 4, 5, 36)*

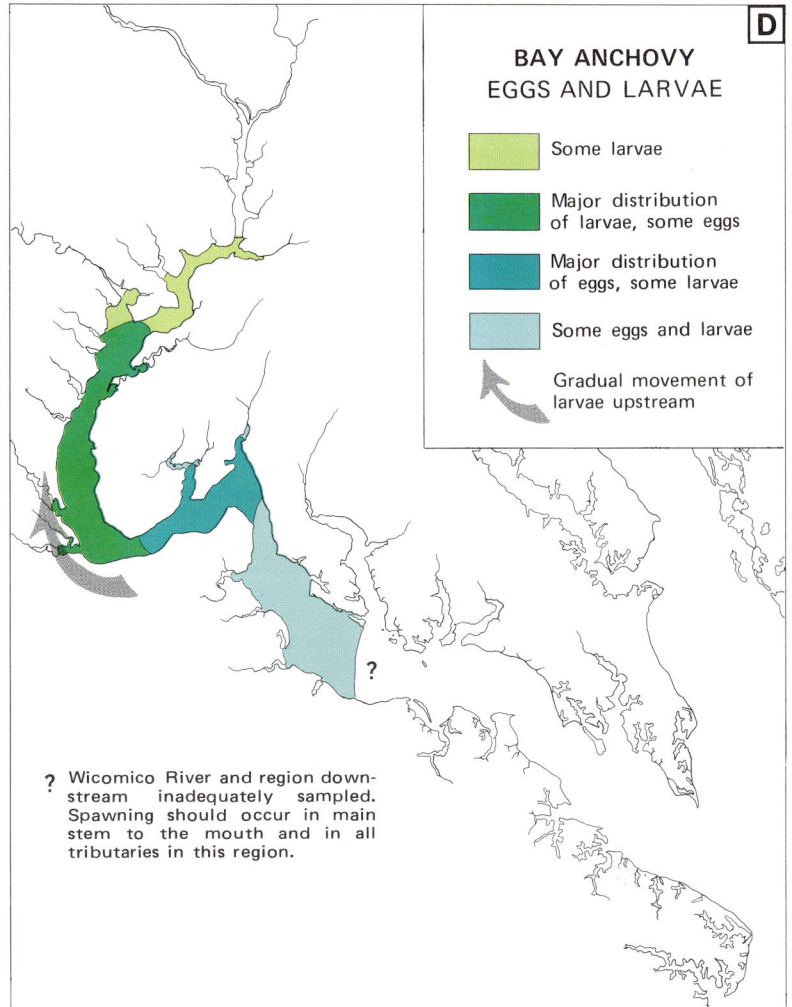

? Wicomico River and region down-stream inadequately sampled. Spawning should occur in main stem to the mouth and in all tributaries in this region.

Figure 8-14. *General distribution of egg and larval stages of bay anchovy in the Potomac estuary in a 1974 survey. (Source: Ref. 36)*

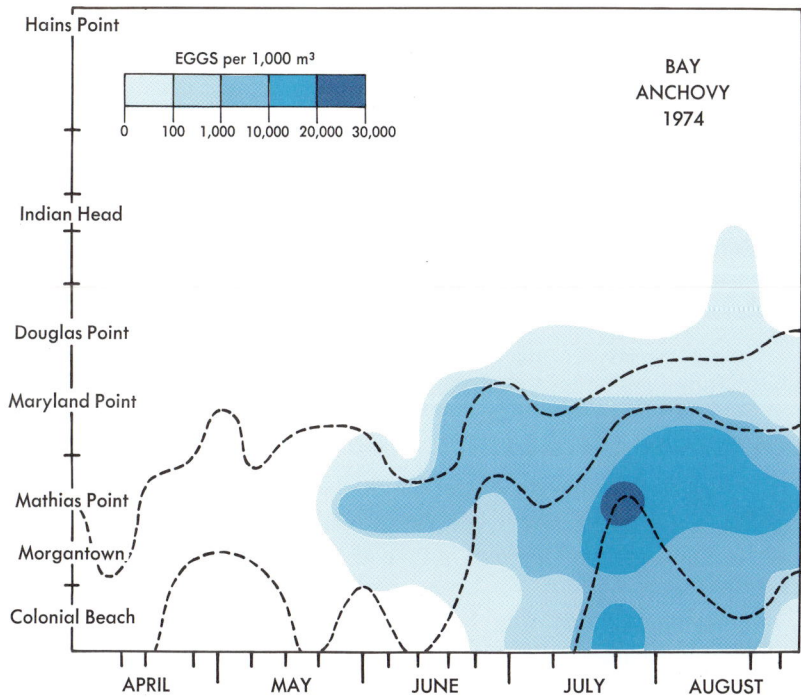

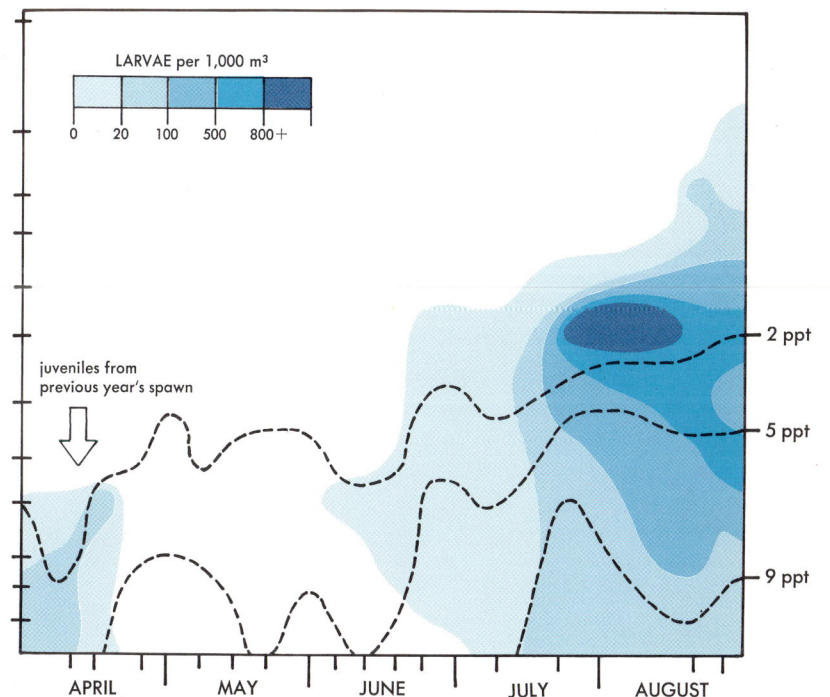

Figure 8-15. *Mean densities of bay anchovy eggs and larvae in the upper half of the Potomac estuary in spring and summer 1974 (derived from data in Ref. 36). Dashed lines show seasonal changes in 2, 5, and 9 ppt isohalines in that year.*

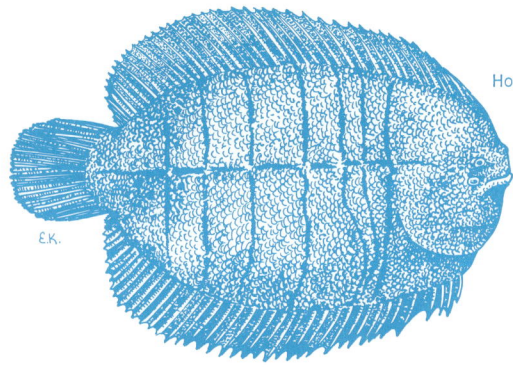

Hogchoker, *Trinectes maculatus*

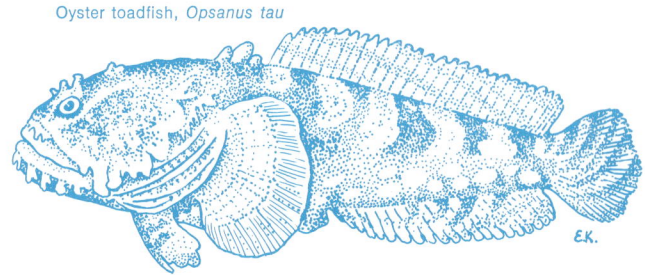

Oyster toadfish, *Opsanus tau*

Bottom-Oriented Fishes

HOGCHOKERS

Hogchokers are small flatfishes that are rarely longer than 15 to 18 cm. They are related to the flounders, which are discussed in the section on marine fishes. Flatfishes are immediately recognizable by the asymmetrical position of their eyes on the dark colored top sides of their bodies. In the larval stage, they have normal bisymmetrical eyes. With growth, the eye on the side that will eventually lie towards the bottom migrates over the top of the head, finally settling close to the other eye. In a "right-handed" flatfish, the left eye has migrated, and when the fish is viewed from the dark side with its head at the top, the mouth always faces right. Conversely, in a "left-handed" flatfish, the mouth faces left.

Hogchokers are right-handed flatfish. They are distinguishable from other right-handed flatfishes by their disc-shaped bodies and dark stripes across their backs. Although they are rarely seen because they

HOGCHOKER

Adults — major distribution

Adults and juveniles

Major spawning

Marginal spawning

Juveniles and late larvae — nursery area

A few young adults to D.C.

Figure 8-16. *General distribution of hogchoker in the Potomac estuary. Presence of various life stages is dependent on season as shown in Fig. 8-17. (Sources: Refs. 4-6, 22, 23, 36)*

are camouflaged against the bottom muds that they burrow in, hogchokers are one of the most abundant and ubiquitous fish species in the Potomac. They are distributed throughout the Potomac main stem and its tributaries from Washington, D.C., to the mouth.[1][4-6][23][32][36][37] Major distribution of adults is in the estuary below Mattawoman Creek (Fig. 8-16). In the upper estuary from Mattawoman Creek down to Maryland Point, they ranked seventh of all species caught in bottom habitats in monthly sampling in 1973.[5] In the mid-region of the estuary from Popes Creek down to Piccowaxen Creek, hogchokers ranked second in abundance (after white perch)[23] in monthly sampling during 1969. Juveniles are more abundant in low brackish and tidal fresh upstream waters; some move into nontidal regions.

There is a marked seasonal movement of adults downstream to spawning grounds in spring, mostly during May, and upstream in September and October (Fig. 8-17).

Spawning extends from the end of May to the first of September, with 80 percent of spawning activity occurring in July (Fig. 8-17 and Folio Map 7).[8][48] Although hogchoker eggs have been collected in salinities of 0 to 24 ppt, peak spawning occurs in 9 to 16 ppt.[48] In the Potomac, the major spawning area is from approximately Colonial Beach to the mouth (Fig. 8-16). Egg densities above this area are low, and no eggs have been collected above Mathias Point.[4][5][36]

After hatching, the youngest larvae remain within the area, but older larvae and juveniles migrate to low salinity nursery areas (Fig. 8-17). By the onset of winter, most juveniles are in salinities of 0 to 8 ppt.[5][48] In the Potomac, the nursery area extends from Washington, D.C., down to approximately Maryland Point and from the upper portions of many tributaries down to their mouths. The area of greatest concentrations of juveniles is uncertain but appears to be somewhere above Douglas Point in tidal freshwater regions. In spring of the following year, juveniles are displaced somewhat downstream but not as far as adults. With each year of growth, hogchokers move increasingly farther downstream in spring and summer, and less distance upstream in fall.[48]

Hogchokers are major predators on a variety of benthic organisms, and worms are apparently their favored food.[31]

OYSTER TOADFISH

The ugly and aggressive oyster toadfish are a bane of many fishermen because these fish readily take bait meant for edible fish. Their strong jaws, sharp teeth, and spiny fins make removing a hook from them a dangerous procedure.

Oyster toadfish are typically mesohaline-polyhaline species, preferring salinities greater than about 7 ppt. They have been known to enter fresh water[2] (one specimen was obtained as far upstream as Gunston Cove[6]), but they are rare above Mathias Point. Oyster toadfish are generally considered sedentary fish, although seasonal movements into and out of "home" spawning grounds have been observed.[49]

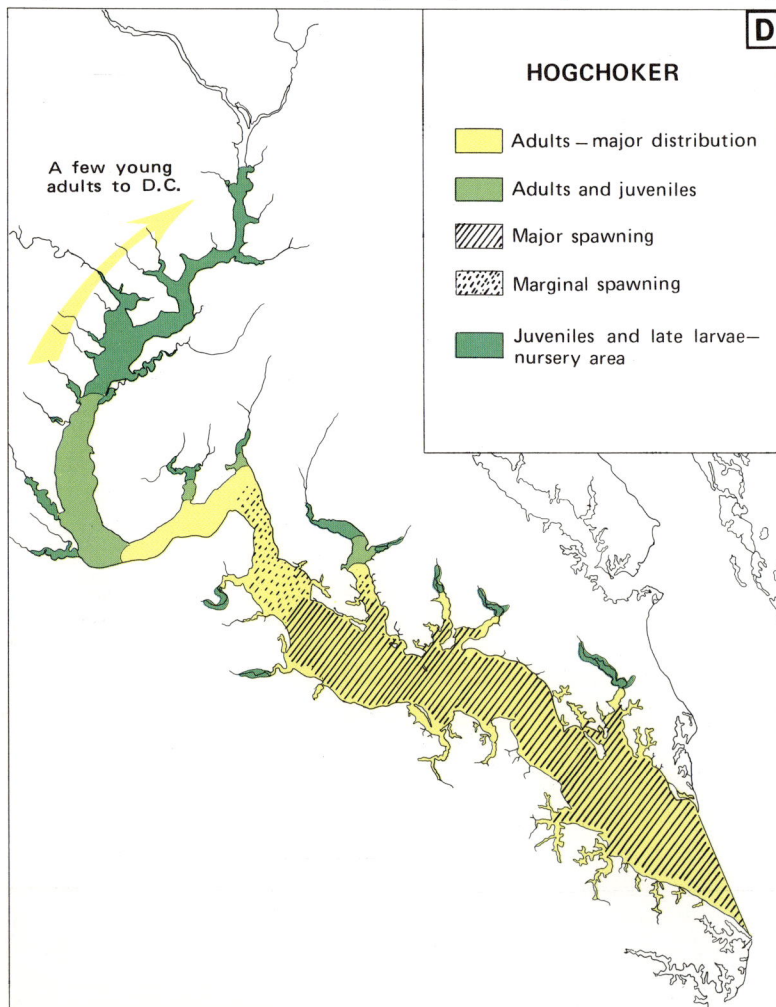

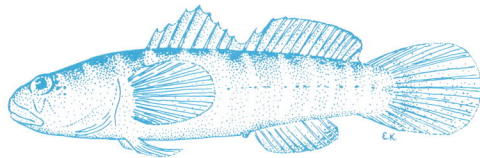

Naked goby, *Gobiosoma bosci*

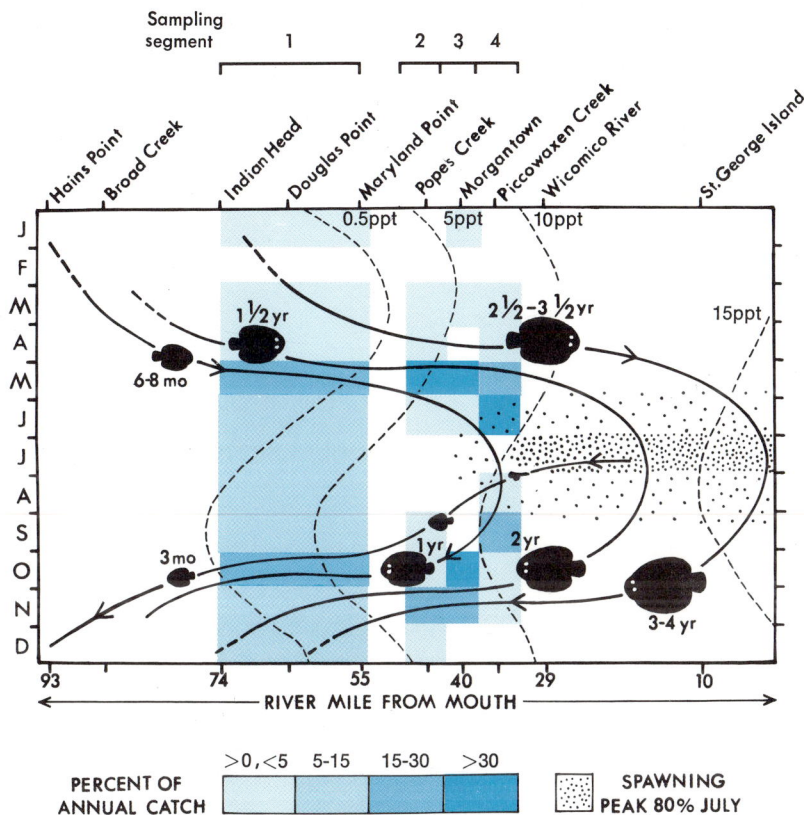

Figure 8-17. *Diagram of seasonal distributions of various age groups of hogchoker in the Potomac estuary. With increasing age, hogchokers migrate farther downstream. Percentages of annual catch of hogchoker per each month are from studies of two segments of the river — one from Indian Head downstream to Maryland Point (data from Ref. 5) and another from Popes Creek, Maryland, downstream to Piccowaxen Creek (data from Ref. 23) — and reflect the seasonal movements upstream and downstream. Dotted lines show average seasonal changes in 0.5, 5, 10, and 15 ppt isohalines.*

However, they cannot be considered strongly migratory since the maximum documented movement is only 9 km.

Spawning is prolonged, extending from April to August and sometimes into September (see Folio Map 7). Oyster toadfish eggs are the largest of any fish eggs spawned in the Potomac (about 5 mm in diameter), adhering to almost any hard substrate — stones, large shells, logs, tin cans, or pottery shards.

Oyster toadfish are predators on crustaceans such as crabs and shrimp, but take almost any other kind of food as well. They are also scavengers.

GOBIES

Two goby species, the common naked goby and the elusive green goby, have been recorded from the Potomac,[4] [50] and a third, the seaboard goby, is presumed to occur in the estuary because of its frequent association with the other two in some areas of the

Chesapeake Bay.[50] [51] Two additional species listed in Appendix Table 6, the code and clown gobies, have been recorded in mesohaline regions of the Chesapeake Bay near the mouth of the Patuxent River[50] and are presumed to occur in the Potomac as well.

Gobies, which commonly live on oyster bars, are similar to blennies and skilletfish in distribution, habitat preference, behavior, size, and, they can be distinguished from one another upon close inspection. Gobies have prominent eyes set close together on top of the head, whereas blennies have deep heads and long, wide dorsal fins, and skilletfish have broad, flat heads tapering to smaller bodies (giving them the "skillet" shape). Gobies and skilletfish have ventral fins that have evolved into sucking discs used for clinging to oyster shells, rocks, or other substrates.

From data collected on its planktonic larval and juvenile stages, the distribution of the naked goby (so-called because of its lack of scales) is better understood than that of any other oyster-community fish. Naked goby larvae accounted for 55 percent of all fish larvae collected in studies of the upper Chesapeake Bay,[45] indicating that they are one of the more common fishes of the upper Bay. Similarly, naked goby larvae are the most common in collections from the Potomac estuary.[5] [36]

In the Potomac, naked goby adults are mainly found below Maryland Point (Fig. 8-18). The extent of upstream or downstream migration by adults is uncertain, but a survey of the midestuary region around Mathias Point found no adult naked gobies during July and August, implying that they had moved away from this area[23] (presumably downstream). During the warmer spawning months, adult gobies are usually found in benthic habitats in waters 10 to 18 feet (3 to 5.5 m) deep. In colder months, they move to deeper waters, but not to depths greater than about 30 feet (9 m).[50]

Naked gobies start spawning when water temperatures reach 19° C, usually in mid-May, and continue until the first part of October; peak spawning usually occurs from mid-June to mid-July.[45] Eggs are deposited in empty oyster shells in the lower half of the Potomac, but, soon after hatching, the benthic-oriented larvae are apparently transported by upstream flows to waters of 2 to 5 ppt salinity (Fig. 8-19)[4] [36] [45] — the same low-salinity nursery areas occupied previously by anadromous fish larvae that hatched 6 or more weeks earlier.

Very little is known about distributions of other goby species. Green gobies have been recorded only from the Wicomico River[4] and near the mouth of the St. Marys River.[50] This goby is a mesohaline-polyhaline species, but prefers salinities of 11 to 13 ppt. In a study of the York River, Virginia, 86 percent of the green goby larvae were collected in salinities less than 18.5 ppt.[51] It is often found among the entwined fingers of the redbeard sponge, *Microciona prolifera*.[50]

Seaboard gobies, if present in the Potomac, would occur only in deeper waters on oyster bars or over sandy bottoms near the mouth of the estuary, where salinities are above 15 ppt.

Gobies feed primarily on small crustaceans (mainly amphipods) and annelid worms.[31]

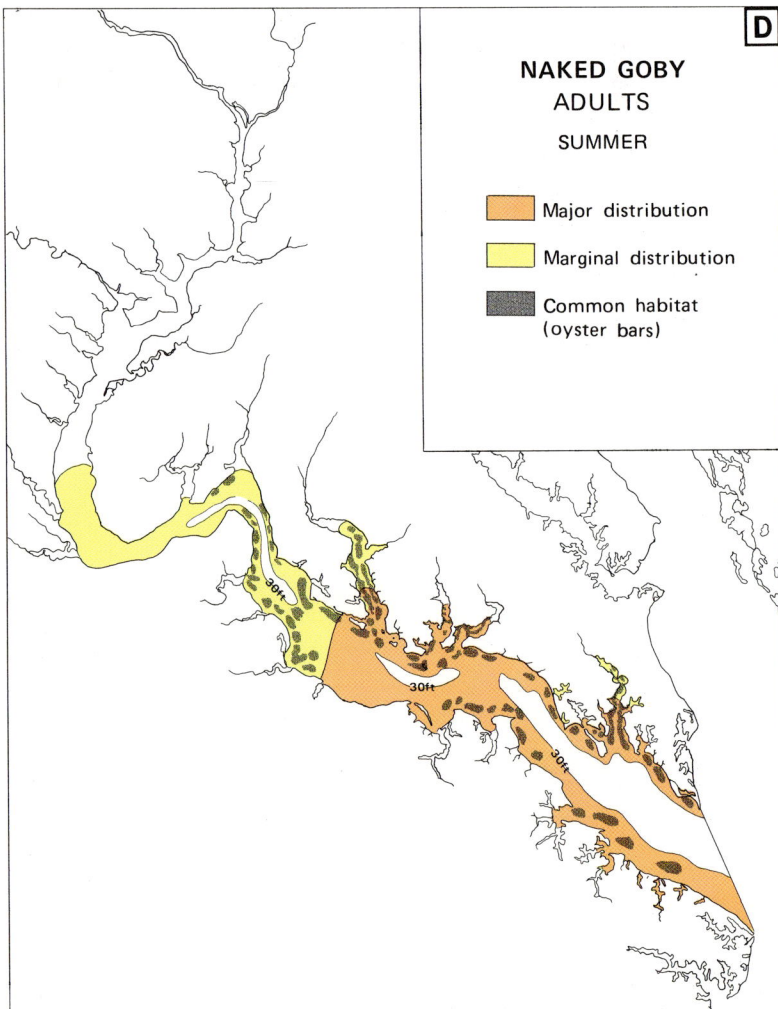

Figure 8-18. *General distribution of adult naked gobies in the Potomac estuary. (Sources: Refs. 4, 5, 23)*

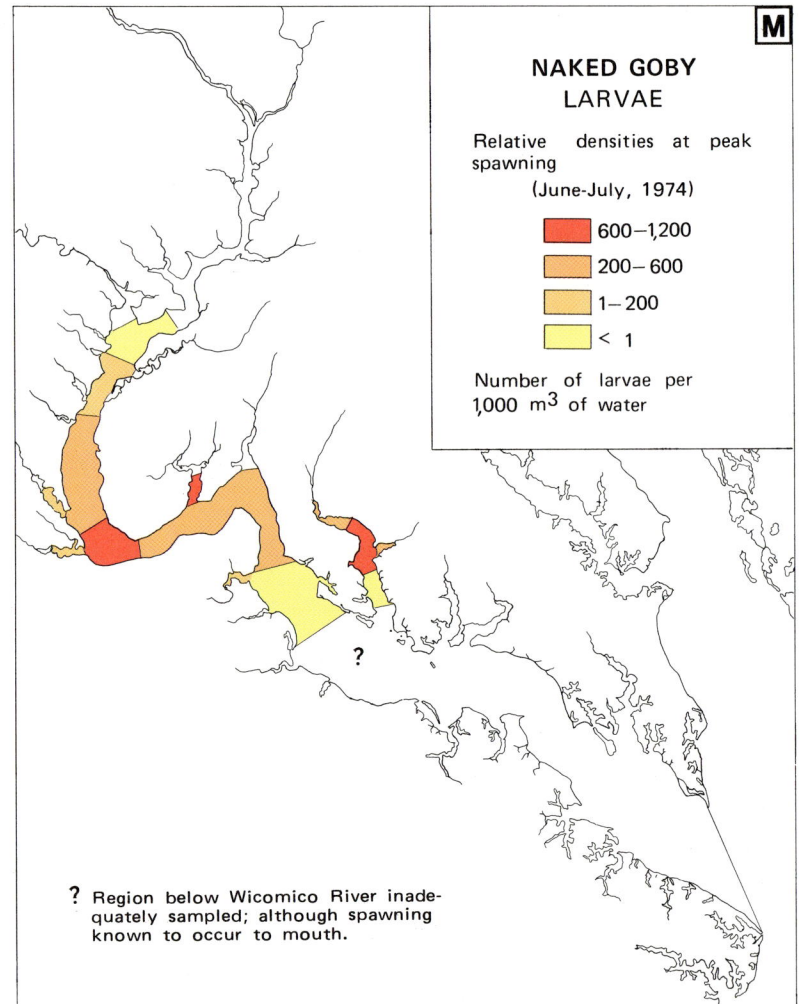

? Region below Wicomico River inadequately sampled; although spawning known to occur to mouth.

Figure 8-19. *General nursery areas of naked gobies as shown by mean densities of larvae during their peak spawning season in 1974 (derived from data in Ref. 36).*

BLENNIES

Two species of blennies, the striped blenny and the feather blenny, are residents of oyster bars in salinities of 12 to 25 ppt.[2] As with naked goby, their larvae and juveniles are found in lower salinities. Blenny larvae have been collected in the upper Chesapeake Bay in salinities of 5 to 16 ppt, with greatest numbers in 11 to 14 ppt,[45] and in waters of approximately 8 ppt in the Potomac near Morgantown.[4] So few adult blennies have been recorded in the Potomac that little can be said of their distribution or relative abundance. In other parts of the Chesapeake Bay, striped blennies are more common than feather blennies.

Blennies spawn from May to August, laying eggs in empty oyster shells and within protected crevices in other bottom debris.[8] Males

guard the area until the larvae hatch. Spawning occurs primarily in mesohaline regions, and larvae apparently move somewhat upstream.

SKILLETFISH

The skilletfish (or "clingfish") is a common and widely dispersed species in the lower half of the Potomac. Skilletfish cling tenaciously to shells, rocks, and pilings. Skilletfish often settle on oyster beds, but they are also frequently discovered in shallow, grassy, bottom habitats along with pipefishes and sticklebacks.[31] Spawning occurs from April through August, peaking in June and July (see Folio Map 7).[8] Eggs are laid in oyster shells, often along with those of blennies. Skilletfish eat mostly isopods and amphipods.[31]

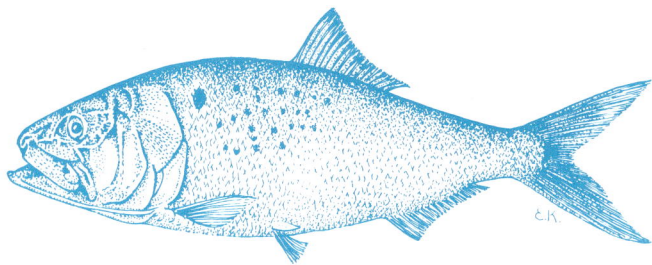

Atlantic menhaden, *Brevoortia tyrannus*

MARINE FISHES

The general distribution of the more common marine fish species in the Potomac estuary is portrayed in Fig. 8-20. Summer salinity conditions are also shown, since most marine species are present in summer. Many other marine fishes have been recorded from Potomac waters, including a large number listed as stragglers in Appendix Table 6. Conceivably, any species passing along the mouth of the Chesapeake Bay might enter the Bay and possibly move as far upstream as the Potomac. Salinity demarcations on the figure do not reflect the Venice System zonations (see Chapter 1) but are based on species groupings derived from known distributions. The number of marine species is greatest towards the mouth of the estuary and gradually declines upstream. The species listed on the map are arranged according to the general extent of their distributions up the estuary.

The marine species of the Potomac can be placed into two categories: estuarine-dependent (Atlantic menhaden, bluefish, fishes of the drum family, and flounders) and summer transient. The estuarine-dependent species are those that require the estuary for successful completion of their life cycle. They generally use it as a nursery area, although adults may also make heavy use of it for feeding. Most marine species that are important in recreational and commercial fisheries fall into this category. Summer transients are those species that are often found in estuaries in summer and that can survive equally well in oceanic or coastal waters.

Estuarine-Dependent Species
ATLANTIC MENHADEN

Atlantic menhaden look much like their close relatives, the anadromous blueback herrings and alewives, except for their proportionally larger heads. Fishermen often refer to Atlantic menhaden as "mossbunkers," "bunkers," or, incorrectly, as "alewives." They are truly pelagic fish, swimming close to the surface in very tight schools. In their frenzy to elude predator fish, they often break the surface and provide a welcome sign to the sportfishermen, who quickly set their lines for the attackers.

Atlantic menhaden are plentiful in the Potomac, ranking first among all species caught in a 1975 gill net survey of the lower two thirds of the estuary. (In this same study, striped bass ranked second, gizzard shad third.[52]) In a year-long midwater trawl survey of the nursery regions in the upper estuary, young menhaden were second in abundance after bay anchovies.[5] They generally rank first by weight in annual commercial catches.[53]

The majority of the Potomac population of Atlantic menhaden may be divided into three groups: postlarvae (2 to 4 cm long), prejuveniles up to 1 year old (4 to 13 cm long), and sexually immature fish 1 to 3 years of age (longer than 13 cm).[54] Menhaden become sexually mature after their third summer, but menhaden over 3 years old are seldom taken in commercial catches in the Chesapeake Bay (Table 8-2).[54] In the Atlantic Ocean, they may attain 10 years of age and a

Table 8-2 — Average Fork Length and Weight of Atlantic Menhaden in Samples from Commercial Chesapeake Bay Catches in 1958 (Sexes Combined) (a)

Maximum Age (years)	Fork Length (cm)	Weight (g)
1	12(b)	28
2	19	119
3	21	162
4	24	214
5	31 (b)	469
6	31 (b)	480

(a) Derived from Appendix Table 12 in Ref. 54.
(b) Only one or two specimens.

length of about 35 cm, but even there, populations of older fish are low.[54]

Atlantic menhaden are distributed from Nova Scotia to Florida,[19] with the Chesapeake Bay approximately at the center of their range. The number of immature adults that return to the estuary for their second and third summer gradually decreases upstream from the mouth of the Potomac estuary (Fig. 8-21). Populations drop off sharply above Quantico Creek, with schools only rarely observed up to Washington, D.C. Immatures move in from the Atlantic Ocean, usually arriving in the Potomac in April and May, and leave by the end of October. Some stragglers may overwinter in looser aggregations in the deeper waters of the lower estuary, but mortalities are apt to be high for these fish, particularly during harsh winters. Fall emigrants move southward in the ocean to below Cape Hatteras, North Carolina, to mingle with populations from the mid- and North Atlantic coastal waters that have also migrated south to overwinter.[56] In spring, Atlantic menhaden migrate north again, but individuals do not necessarily return to the same areas they occupied during the previous summer.[55]

Principal spawning grounds are in the Atlantic Ocean over the continental shelf. Although it is thought that the major spawning occurs off North Carolina, a significant amount occurs all along the Atlantic Coast, including the area near the mouth of the Chesapeake Bay.[19] Since Atlantic menhaden eggs have been found in the mesohaline portion of the Patuxent River (a tributary of the Chesapeake Bay) during the spring months,[45] it is possible that some marginal spawning occurs in the lower Potomac estuary. Spawning in the Potomac would have to be considered atypical, particularly since so few mature fish appear in the estuary. Spawning takes place in almost every month at some location along the distributional range of adult Atlantic menhaden, but there are generally two peaks: one in late fall to winter, another in spring. The winter period (December, January, and February) is the most productive.

Postlarvae enter coastal estuaries just before reaching their prejuvenile stage. Their ocean migrations are accomplished by unknown mechanisms, but once they are within the estuary, postlarvae are believed to be transported upstream in the higher salinity bottom

POTOMAC RIVER ESTUARY **I**

MARINE FISHES

Approximate Extent of Movement into Estuary for Species in Table Below

- To mouth of St. Marys River, to approximately 12 ppt
- To 9 ppt
- To 6 ppt
- Into interface region
- To tidal fresh water
- Into tidal fresh water

– – – Average summer surface salinities

Extent of Distributions Into Estuary	Species	Pelagic	Bottom	Shallows
	Atlantic needlefish			
	Atlantic menhaden			
	Spot			
	Atlantic croaker			
	Silver perch			
	Bluefish			
	Weakfish			
	Summer flounder			
	Winter flounder			
	Spotted seatrout			
	Northern puffer			
	Harvestfish			
	Halfbeak			
	Striped anchovy			
	Cownose ray			
	Black drum			
	Lined seahorse			
	Inshore lizardfish			
	Striped mullet			
	Northern searobin			
	Butterfish			
	Black sea bass			
	Red drum			

—— Adults and juveniles (if present)

– – – Juveniles only

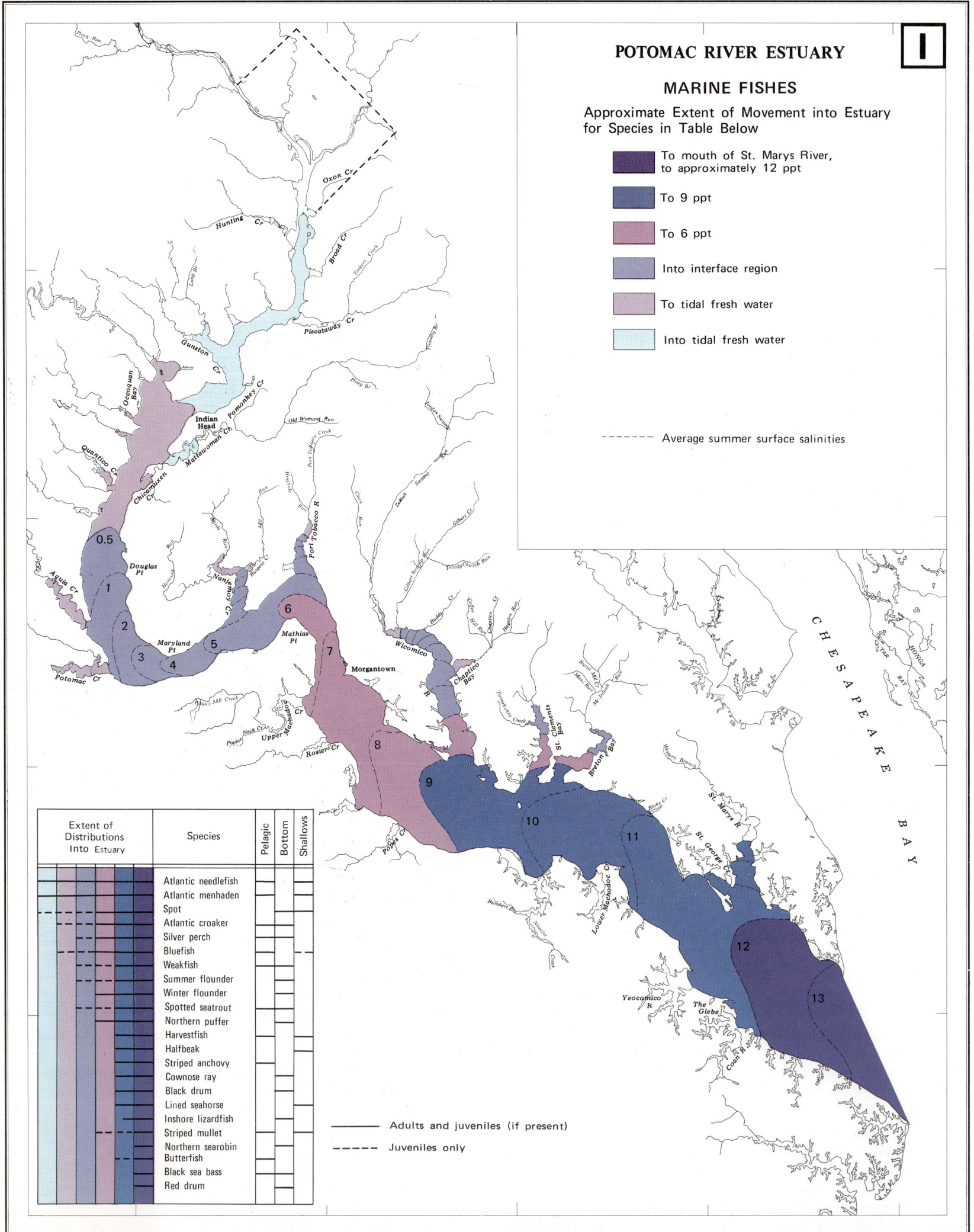

Figure 8-20. General distribution of the more common marine species according to salinity in the Potomac estuary, showing known extent of upstream movement by any life stage. All species have potential to move farther than records have shown. Map is compiled from major surveys conducted in the Potomac by the Maryland Department of Natural Resources (Refs. 1, 3, 4, 20); The Chesapeake Biological Laboratory (Refs. 12, 36, 52); Ecological Analysts, Inc. (Ref. 5); and the Academy of Natural Sciences of Philadelphia (Ref. 23); and from Potomac distributions reported in the literature (Refs. 6, 31, 32, 37).

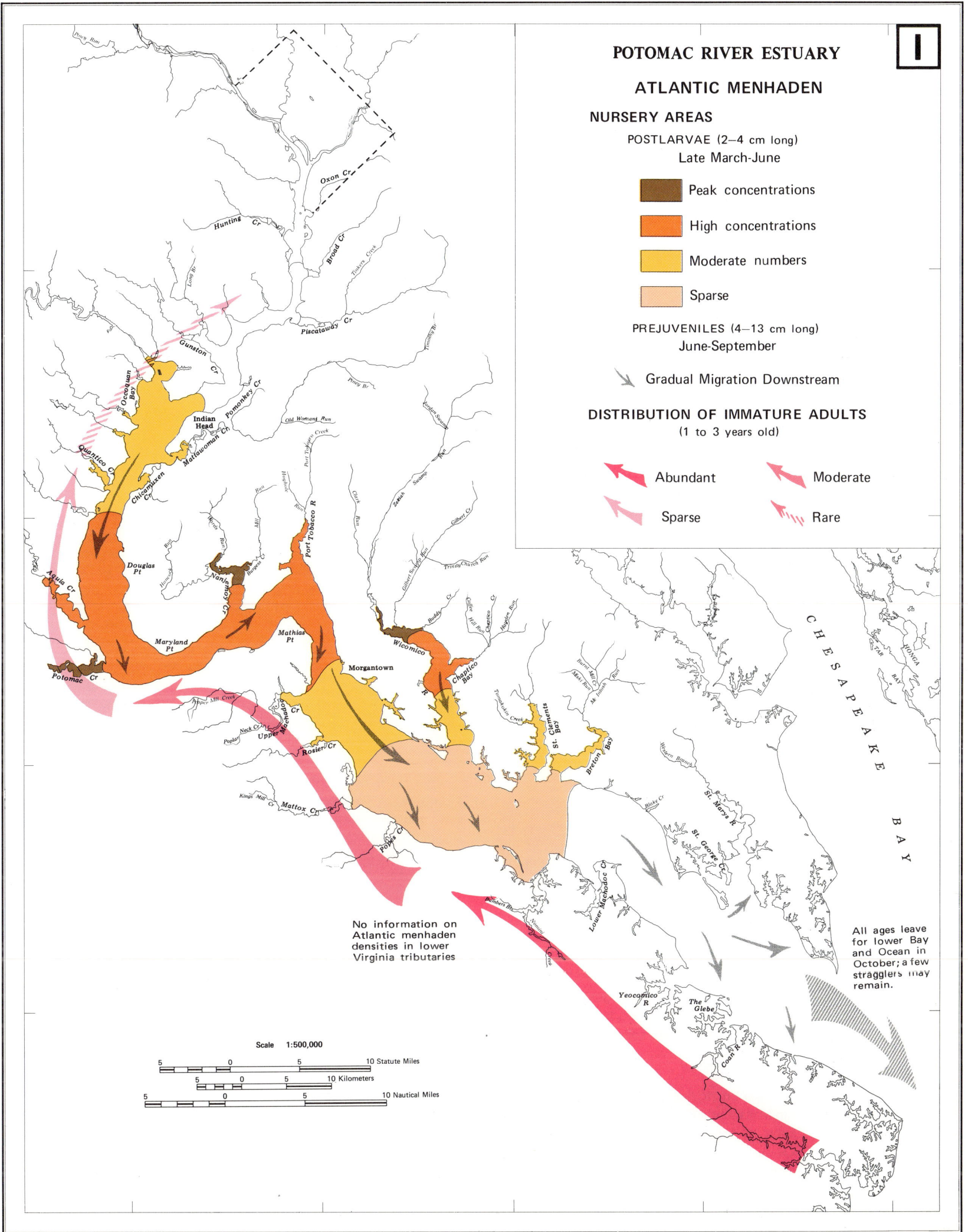

Figure 8-21. *General distribution of various age groups of Atlantic menhaden in the Potomac estuary. Postlarvae and prejuveniles are present in late winter to spring (Refs. 4, 5, 36). Juveniles and immature adults are present from spring into fall (Refs. 1, 3-5, 52, 54, 55).*

Bluefish, *Pomatomus saltatrix*

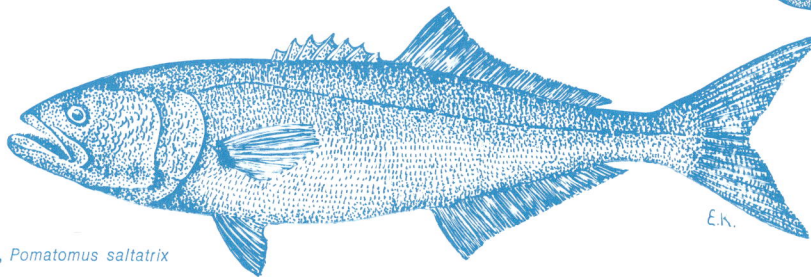

Spot, *Leiostomus xanthurus*

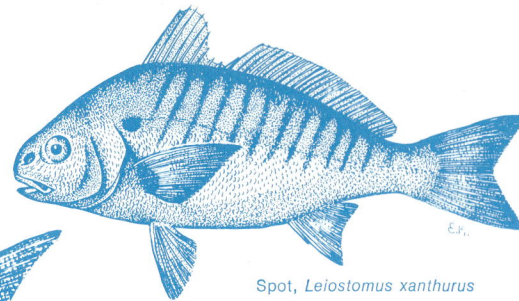

waters. The majority of young Atlantic menhaden eventually move up into the tributaries to low brackish and tidal fresh waters where salinities range from 0 to 5 ppt. They start to arrive in the nursery areas as postlarvae by late March and continue to arrive as prejuveniles or juveniles throughout the summer, leaving with older Atlantic menhaden by the end of October.

In the Potomac, the primary nursery area extends from just below Morgantown upstream to Chicamuxen Creek (Fig. 8-21). Less concentrated populations usually extend to Indian Head and occasionally as far as Broad Creek. Many tributaries are also used as nursery areas. Greatest densities have been observed in the upper estuarine segments of the Wicomico River and Nanjemoy and Potomac creeks. Aquia Creek, the Occoquan and Port Tobacco rivers, and St. Clements and Breton bays are also havens for these young stages. As summer progresses and Atlantic menhaden grow into a juvenile stage, they gradually disperse and move downstream.

Larvae and prejuveniles feed selectively on zooplankton, but, as they metamorphose into juveniles, their feeding and digestive structures undergo extensive changes. They become nonselective plankton feeders, filtering organisms out of the water that constantly sweeps through their mouths. As abundant low-level trophic feeders, Atlantic menhaden are a significant link in the food chain between the primary producers and consumers and the large predator fish such as striped bass and bluefish.[55]

BLUEFISH

Bluefish enter the Chesapeake Bay and the Potomac estuary in spring and summer to feed. They are not as dependent on estuaries as some of the drums or Atlantic menhaden, but because of the large numbers of their young that regularly invade estuaries each summer, they are categorized as estuarine-dependent species. Schools of adult blues are often seen breaking the surface as they attack forage fish. Because they are rapid swimmers and easily avoid sampling trawls, little data exist on their relative abundances. However, the large commercial and sport-fish catches indicate that they are present in considerable numbers.

Three categories of bluefish enter the Potomac: sexually mature adults approximately 45 to 60 cm long; immature adults about 23 to 29 cm (called "snappers" or "tailors"); and juveniles that are 2.5 to 3.8 cm long when they enter the estuary and about 20 cm by the time they leave.[57]

Bluefish are found from Argentina to Nova Scotia,[58] but their major distribution is from Florida to Massachusetts. Schools move seasonally, governed by the abundance of forage species (primarily Atlantic menhaden in the Atlantic Ocean). Populations between North Carolina and New York generally move offshore and south to Florida in fall and early winter, with a northerly coastal migration in spring.[57]

The larger the adult population, the earlier the runs into the Chesapeake Bay begin and the farther they range upstream. In years when numbers are low, penetration into the Bay is more limited, and few adult bluefish are caught in Maryland waters.[31] In years when populations are large, adults are caught as far up the Chesapeake Bay as Baltimore.

Adult bluefish start to enter the Chesapeake Bay in March or April[59] and arrive shortly after that time in the Potomac. Sexually mature adults leave the Bay to move into coastal waters to spawn in early summer, while sexually immature adults continue to move into the Bay. Adult bluefish are seldom found above Mathias Point (Fig. 8-22), although there are records as far back as 1915 of their occurrence near Washington, D.C.[38]

Offshore spawning occurs in two waves involving two principal populations that spawn in different areas and at different seasons: a spring spawning (April through May) in the Gulf Stream from Florida to North Carolina; and a summer spawning (June through August) over the continental shelf from Cape Hatteras to Cape Cod.[57] Juveniles from spring spawning enter the Chesapeake Bay in early summer. Juveniles from the summer spawning generally do not enter coastal estuaries, but move southward in the fall.[57] They enter the estuaries along the mid-Atlantic Coast the following spring as the immature "snappers." Juveniles may be found as far upstream as Liverpool Point,[3] just above Douglas Point (Fig. 8-22). They have also been collected in the Port Tobacco River and in Breton Bay.[4] By mid-November all stages of bluefish have left the estuary, moving offshore and southward for the winter.[59]

As high-level predators, bluefish feed voraciously on the abundant forage fish, primarily Atlantic menhaden, silversides, and bay anchovies, and on invertebrates such as crabs, mysid shrimp, and annelid worms. There is some concern that they have out-competed striped bass for forage and have contributed to the decline in numbers of striped bass.

SPOT

The spot is one of five members of the drum family often found in the Potomac estuary. The other drums include Atlantic croaker, silver perch, weakfish, and spotted seatrout. Black and red drums and southern and northern kingfish are occasional visitors, although the kingfishes are encountered less frequently than the drums (see Appendix Table 6). The family name of "drum" comes from the characteristic loud drumming or croaking sounds produced by most species.

The spot is moderately sized, with a distinct spot just behind the gill openings, a long sloping head, and a series of oblique stripes along each side. The stripes and a slightly forked tail immediately distinguish it from the Atlantic menhaden, which has a similar spot, but a strongly forked tail and no stripes. Congregating in loose schools, spot usually forage along the bottom.

Although adults rarely exceed 33 to 35 cm, spot grow rapidly, attaining a length of about 13 cm in their first year of life. Sexual maturity is reached near the beginning of the third year.[61] Spot range

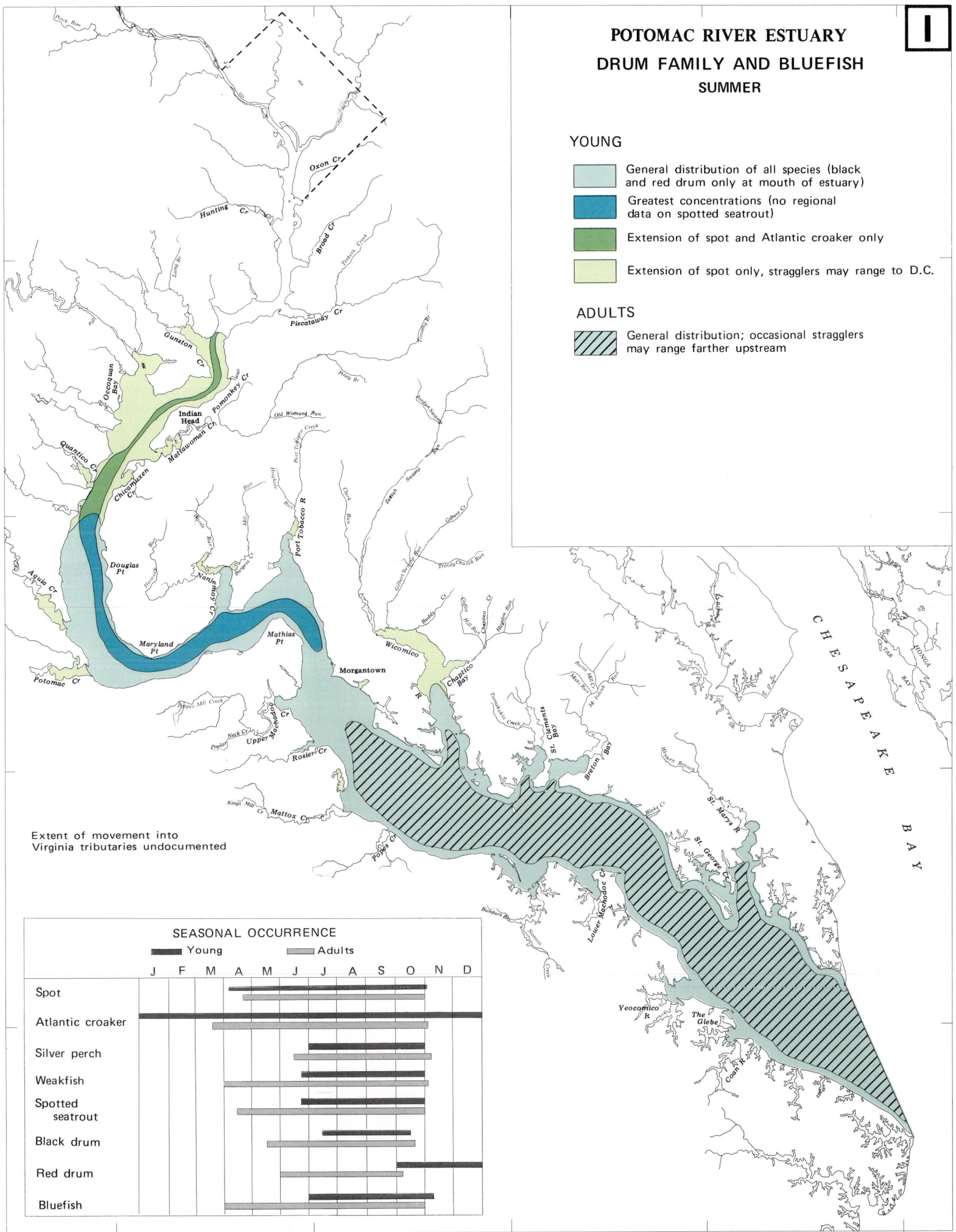

Figure 8-22. *General distribution of adults and young of estuarine-dependent species of the drum family and of bluefish in the Potomac estuary. With growth through summer, young disperse shoreward from channel nursery areas (Refs. 3-5, 20, 36, 59, 60, 65). Seasonal occurrence of adults and young compiled from Refs. 8, 31, 59-61.*

POTOMAC RIVER ESTUARY

DRUM FAMILY AND BLUEFISH

SUMMER

I

YOUNG

General distribution of all species (black and red drum only at mouth of estuary)

Greatest concentrations (no regional data on spotted seatrout)

Extension of spot and Atlantic croaker only

Extension of spot only, stragglers may range to D.C.

ADULTS

General distribution; occasional stragglers may range farther upstream

Extent of movement into Virginia tributaries undocumented

SEASONAL OCCURRENCE

Young   Adults

J F M A M J J A S O N D

Spot

Atlantic croaker

Silver perch

Weakfish

Spotted seatrout

Black drum

Red drum

Bluefish

159

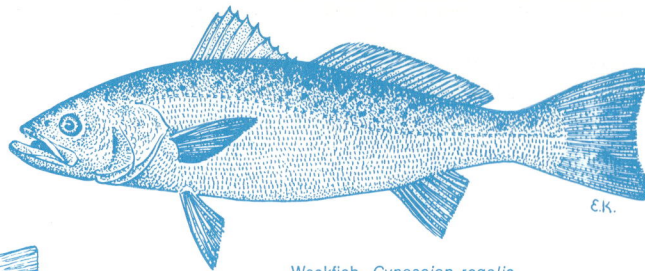

Weakfish, *Cynoscion regalis*

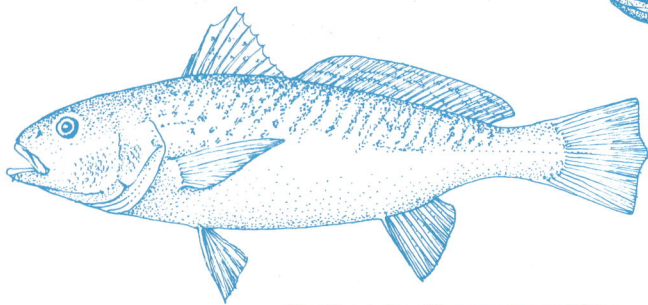

Atlantic croaker, *Micropogonias undulatus*

along the Atlantic Coast from Massachusetts to Florida, though they are less abundant in the northernmost states. Their numbers are high in the Potomac where they are the most numerous representative of the drum family. Larger and older fish prefer channel waters; spot found inshore are rarely more than 15 cm long.[62] White perch were the only species caught in higher numbers than young spot in bottom trawls made in the estuary between Mattawoman Creek and Maryland Point during monthly sampling in 1973.[5]

Spawning occurs offshore in the Atlantic from late autumn through winter (November to February).[8] Larvae 1.5 to 2 cm long generally appear in the lower Chesapeake Bay as early as April[61] and enter the Potomac by early May.[5] Some young remain in the lower reaches, but most move upstream to nursery regions between Gunston Cove and Mathias Point (Fig. 8-22).[3-5] The youngest spot in the nursery areas are in deeper channels where somewhat higher salinities prevail, but they tend to move shoreward as they grow. Nursery areas are also located in all the major tributaries adjacent to the main stem nursery region and in Allens Fresh in the Upper Wicomico River.[4] Some adults and juveniles may overwinter at the mouth of the Chesapeake Bay, but most move into the ocean and towards spawning grounds between September and November.[61]

Adults, 2 or 3 years old, which enter the estuary in May and June, are seldom found above the Colonial Beach region (Fig. 8-22). In contrast, younger fish, which arrive somewhat earlier, may be found as far upstream as Washington, D.C.[6] Their food consists primarily of benthic invertebrates, including small crustaceans, annelid worms, and small molluscs.[31] Their rapid growth in their first year, along with their abundance, suggests they may have substantial impact on benthic macroinvertebrate communities.

ATLANTIC CROAKERS

The Atlantic croaker has a heavy rounded snout and small tactile barbels on each side of its lower lip. It has a series of oblique, speckled bars across its back and sides, unlike spot which have solid bars. The sound emitted by Atlantic croakers when they are drawn from the water is the loudest made by any of the drums.[31]

Atlantic croakers are larger than spot. They attain lengths of about 50 cm and weights of about 1.5 kg, although the bulk of the market catch consists of smaller fish — about 35 cm and 0.7 kg.[31] Atlantic croakers are distributed from Massachusetts to Florida, with centers of abundance in the southern portion of their range.

Adult croakers enter the Chesapeake Bay in March or April[60] arriving in the Potomac at approximately the same time as the shad arrive. They remain in the lower half of the estuary throughout the summer and return to the ocean in September and October. Since spawning occurs offshore along the Atlantic Coast from August through December, young croakers appear on the nursery grounds earlier than the other drums.[60] Arrivals from the August spawning period may appear in the upper estuary as early as September. Large

numbers of young are found in channels throughout the winter and into April, whereas juveniles of other drum species do not generally arrive at the nursery areas until May, June, or July.

In the Potomac, young Atlantic croakers tend to be found along with spot in the same upstream channel area. Greatest concentrations of young Atlantic croakers were found from the vicinity of Mathias Point downstream to Morgantown in one survey of the estuary from Washington, D.C., to Colonial Beach (Fig. 8-22).[4] [36] Densities below this location are unknown. Since juveniles have been found regularly in salinities from 0 to 21 ppt in the upper Chesapeake Bay[45] and in other tributaries of the Bay, it is reasonable to assume that substantial numbers of these young fish are also in the lower Potomac.

Before 1950, croakers were a plentiful and popular food fish. Since then, their numbers have declined radically, and larger adults have virtually disappeared from the Chesapeake Bay. The cause of this abrupt decline is uncertain, although cold winters are strongly implicated. The Atlantic croaker is a southern species, and the Chesapeake Bay region is near the northern limit of its distribution. Colder waters in the winter nursery grounds have often caused mass mortalities of juveniles.[63] It has been postulated that predation on juveniles by overwintering striped bass in the deeper channels has also contributed to the decline of Atlantic croaker populations.[64]

Recently, there have been indications of a comeback. Large numbers of small croakers about 23 cm long, "pinheads" as they are sometimes called, are again being caught by sportfishermen, and commercial catches have gone from a few hundred pounds in 1970-72 to over 100,000 pounds (45,360 kg) in 1974.[53] Atlantic croakers, like other drums, are bottom feeders that eat benthic macroinvertebrates, including crustaceans, worms, and molluscs.[31]

OTHER DRUMS

Weakfish and spotted seatrout are closely related and similar in appearance except for the pattern of spots along their backs and sides. Weakfish have small irregular blotches that sometimes form wavy oblique lines, while spotted seatrout have more rounded spots that are larger and randomly scattered. These fish grow larger than either spot or Atlantic croaker, commonly reaching 65 to 75 cm and 2.5 to 4.5 kg. Both species may attain even greater sizes: at least 13.5 kg for weakfish and 8 kg for spotted seatrout.[31]

Weakfish, more often called "squeteague" or "gray trout," are usually more common than spotted seatrout in Potomac waters. They arrive in late March or April and usually leave by November. Although large numbers of spotted seatrout often appear sporadically in either the spring or fall, weakfish occurrence is more continuous. Both species spawn in the lower Chesapeake Bay from April to August (and sometimes into September).[60] Late larvae and juveniles are found in the Potomac estuary in the same general nursery areas as spot and croakers, but they do not migrate as far upstream.[4] [36] Unlike most other members of the drum family, both species are pelagic feeders. The

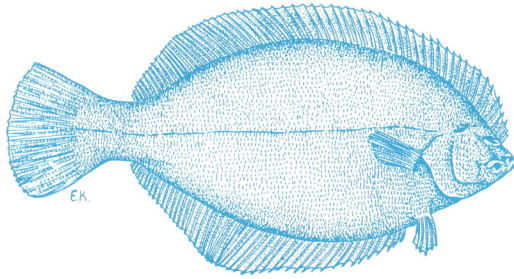

Winter flounder, *Pseudopleuronectes americanus*

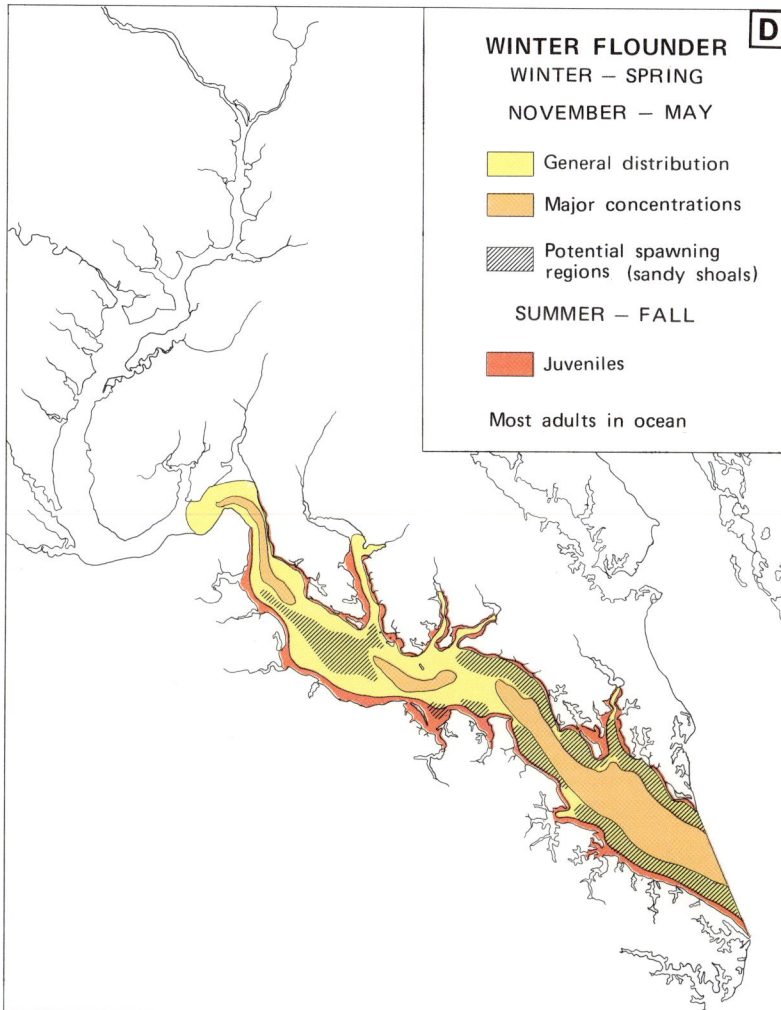

WINTER FLOUNDER [D]
WINTER – SPRING
NOVEMBER – MAY

- General distribution
- Major concentrations
- Potential spawning regions (sandy shoals)

SUMMER – FALL

- Juveniles

Most adults in ocean

Figure 8-23. *General distribution of winter flounder in the Potomac estuary. Spawning regions are potential sites based on known habitat preferences. The only documented spawning has been at the mouth of the Potomac (Refs. 6, 7, 68); juveniles have been taken as far upstream as the Wicomico River (Refs. 3, 4).*

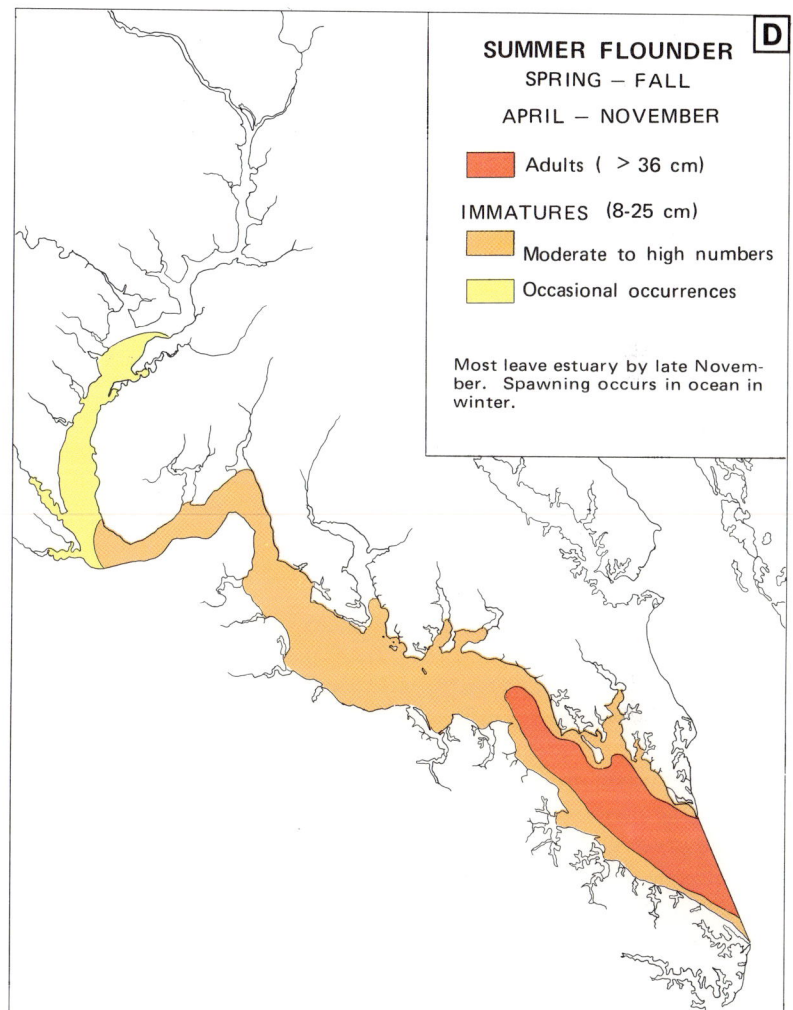

SUMMER FLOUNDER [D]
SPRING – FALL
APRIL – NOVEMBER

- Adults (> 36 cm)

IMMATURES (8-25 cm)

- Moderate to high numbers
- Occasional occurrences

Most leave estuary by late November. Spawning occurs in ocean in winter.

Figure 8-24. *General distribution of summer flounder in the Potomac estuary (Refs. 3, 31, 35, 66).*

larger of these fish consume small fish and crustaceans such as mysid shrimp.[31]

Silver perch are sometimes mistaken for white perch, but can be distinguished by their unforked tails. Silver perch are distributed primarily in the lower Chesapeake Bay and are not abundant in the Potomac. They are present in the Chesapeake Bay from April to November, and in the Potomac estuary from June to November. Greatest numbers appear in October. Spawning takes place in the lower Bay and offshore from May to early August.[8] Silver perch are bottom feeders on small crustaceans.[31]

Red drums and black drums occur sporadically in the Potomac. Both species are large and commonly reach 18 kg. Red drums, or channel bass, are identified by a prominent dark spot at the base of the caudal fin, black drums by a robust body and numerous barbels on the chin. Large specimens of both drums are seldom caught anywhere in the Potomac, except at the mouth. However, young may occasionally be found upstream or in the lower tributaries.[4] Both drums are generally found in the Potomac estuary from June to October, with young red drums remaining into late fall (Fig. 8-22).[65] Like others of the family, they feed largely on crustaceans (mysid shrimp and amphipods).[31] Black drums are also capable of crushing heavy mollusc shells and may do considerable damage to oyster or clam beds.

WINTER FLOUNDER AND SUMMER FLOUNDER

Winter flounder, like the estuarine hogchokers, are right-handed flatfish (that is, the dark-colored side is their right side). Summer flounder are left-handed. Winter flounder have small mouths and straight lateral lines (distinct lines of sensory organs seen along the sides of many fish), whereas summer flounder have large mouths reaching well behind their eyes and lateral lines that curve as they approach the head.

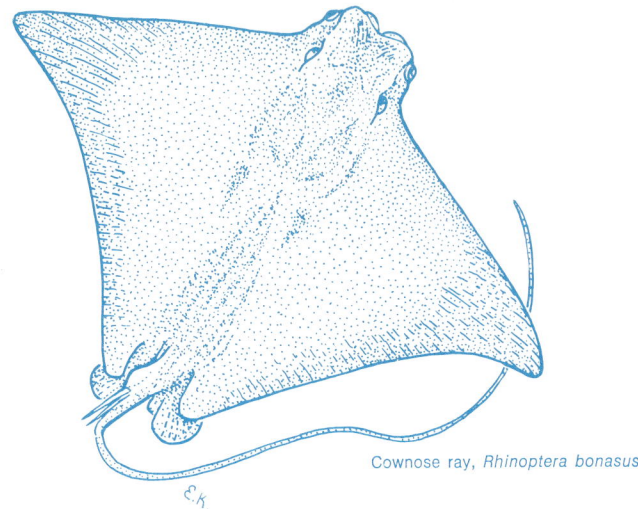

Cownose ray, *Rhinoptera bonasus*

Winter flounder are usually no more than about 48 cm long and about 1.5 kg in weight, while summer flounder may reach a length of over 90 cm and 11 kg.[66] However, in mid-Atlantic estuaries, summer flounder populations usually consist of younger and smaller fish weighing between 0.2 and 2.5 kg and occasionally up to about 4 kg.[66]

Both flounder species range along the mid-Atlantic coast. Winter flounder are distributed from Labrador to Georgia, while summer flounder have a slightly more southerly distribution from Maine to Florida.[31] Summer flounder typically follow a spring-summer pattern of appearance in the Potomac estuary. But winter flounder, unlike almost all other marine fish, appear in the Potomac during the cold season, from November to May (only stragglers occur there from June to October).[67] Spawning occurs both in shallow coastal waters and in Atlantic Coast estuaries. Winter flounder are sometimes found in tidal fresh and low brackish waters, but in the Potomac they appear mostly in the lower half of the estuary in mesohaline waters (Fig. 8-23). They have been recorded upstream to Nanjemoy Creek.[3]

Winter flounder prefer channel waters, but move towards the shallows and into the mouths of tributaries to spawn. Spawning occurs from February to April (Folio Map 7).[8] Eggs are laid on the bottom, loosely adhering to sandy substrates, and larvae apparently remain close to the spawning grounds. Larvae have been found in the Maryland portion of the Chesapeake Bay in salinities ranging from 0 to 22 ppt, but peak larval numbers occurred in salinities of 6 to 15 ppt.[45] Larvae are usually abundant for one month, from the last half of March through the first half of April. Spawning has been documented only at the mouth of the Potomac,[68] but sandy shoal areas of the estuary from Morgantown to the mouth are also potential spawning sites. As the adults move out of the estuary in the spring, the developing juveniles move into the shallow shore areas in the main stem and the tributaries where they spend their first summer. With the onset of winter, first-year fish (10 to 17 cm long) are found in the deeper channels.

Summer flounder enter the Potomac before the winter flounder have left. They usually arrive in April and leave by November, but, in years of high abundance they arrive earlier (March) and leave later (December).[53] Summer flounder that leave the Potomac and the Chesapeake Bay in the fall move toward overwintering grounds along the edge of the continental shelf. Spawning occurs from late September to December during this offshore migration.[66] Eggs and newly hatched larvae drift with ocean currents, moving shoreward towards coastal bays and estuaries. Eventually, young summer flounder move into the Potomac, generally arriving in mid-summer when they are approximately 5 cm in length.[31] They have been found as far upstream as Indian Head,[3] but most remain in the lower reaches of the estuary (Fig. 8-24).

Both summer and winter flounder are benthic feeders. Winter flounder, with their smaller mouths, tend to feed on small crustaceans (such as mysid shrimp), worms, and molluscs, whereas summer flounder may also eat larger shrimp, crabs, and fish.[31] Large populations of winter flounder are a boon to the commercial fishermen. Summer flounder are important as sportfish.

Summer Transients
COWNOSE RAYS

The cownose ray is the only species of the group of primitive cartilaginous fishes (sharks, skates, and rays) that commonly frequents Potomac waters. The presence of cownose rays is often signaled by the sight of their paired dark fins breaking the surface. They are large-winged, flat-nosed fish with long whip-like tails. Their mouths have grinding plates that are used to crush clams, oysters, and other molluscs.[69] As they root over clam beds in shallow waters, they stir up clouds of mud and sand from the bottom that are often visible from the surface. Schools of cownose rays can destroy a considerable area of clam beds, leaving large shallow pits in their wake.[70] Uprooting and destruction of rooted aquatics, eelgrass beds in particular, have been attributed to their grazing activities.[71] They are one of the few effective predators on large hard-shelled invertebrates.

Cownose rays are common along the Atlantic Coast from Massachusetts to Florida,[31] but large migrations of these rays into the Chesapeake Bay occur sporadically. It is speculated that they migrate into areas north of Cape Hatteras in spring and summer and south in fall and winter.[70]

Cownose rays do not migrate far into mesohaline regions[2] and thus do not move much beyond the lowermost reaches of the Potomac. They may, however, be far more common than scientific records imply because they avoid the nets and trawls used in conventional sampling methods.

NORTHERN PUFFERS

Northern puffers at one time were far more common than they have been in recent years. Puffers are interesting fish because of their ability to puff into a balloon shape when agitated. Their strong, beaklike mouths are used to crush the shells of the benthic invertebrates that they eat. A small fishery developed for them in the late 1960s and early 1970s, but the trade disappeared as their numbers decreased. Whether overfishing contributed to their decline is uncertain. They may be regularly found at salinities as low as 9 ppt and have been recorded upstream as far as the Rt. 301 bridge.[3] Most migrate into the ocean and northwards in winter.[2] Some spawning apparently occurs in the Chesapeake Bay, since larvae and juveniles have been found there in salinities of 12 to 21 ppt (see Folio Map 7).[45]

INSHORE LIZARDFISH AND NORTHERN SEAROBINS

Inshore lizardfish have never been common in the Chesapeake Bay, and in the Potomac, they generally appear only at the mouth. Their large pectoral fins with finger-like rays propel them across the bottom. Northern searobins at times may be numerous in the Potomac. Their wing-like pectoral fins are also adapted for crawling along the bottom. They are known to occur in waters with salinities as low as 5 ppt and have been taken in recent years at Ragged Point.[12] Historically, northern searobins were recorded as far upstream as Gunston Cove,[6] but they are far more common in the lower Chesapeake Bay than in the Maryland portion.[2]

BUTTERFISH AND HARVESTFISH

Butterfish and harvestfish are closely related, disc-shaped fish. Harvestfish are deep-bodied and generally no longer than 17.5 cm in the Chesapeake Bay.[31] Butterfish are more elongated, reaching approximately 20 cm.[72]

Butterfish are more abundant in the lower Chesapeake Bay than in the Potomac, although they are known to tolerate salinities down to 5 ppt.[2] The extent of their distribution in the Potomac is uncertain, as few surveys sample the habitat of this pelagic species. A year-round survey in the region of Morgantown did not collect either species.[23] However, butterfish are taken in commercial catches in zone 3 above Morgantown (see Table 10-2 in Chapter 10). Commercial fishery records show that butterfish populations vary widely in the Chesapeake Bay (e.g., 1.5 million pounds in 1961 and 126 thousand pounds in 1976),[53] but, even in years of high abundance, Potomac catches never run more than about 10,000 pounds (4,536 kg).

Butterfish arrive in the Chesapeake Bay in April, spawn in the lower Bay in June and July, and leave in November.

Harvestfish are more abundant in the Potomac than would be suggested by catches made during scientific surveys, because they are difficult to capture in standard sampling gear.[73] They are found in the Potomac at least to Cobb Island[32] and probably move farther upstream since it is known that they tolerate salinities down to 4 ppt.[2] They are present in both the main stem and the tributaries during summer and fall (July through October). Spawning occurs in June and July, but apparently only in the higher salinities of the lower Bay.[8]

Small harvestfish are often found living among the tentacles of sea nettles.[74] It is uncertain whether the harvestfish are immune to the stinging poison or whether they are particularly adroit in avoiding the tentacles. Sea nettles provide them with protection from predators and a ready food source from particles brought in by the pumping action of the medusae (harvestfish have also often been observed feeding on various parts of the sea nettle medusae).[74] Free-swimming harvestfish feed on both sea nettles and comb jellies.[74] In the Potomac estuary, harvestfish and butterfish are the only known predators of true jellyfishes (other than jellyfishes themselves).

HALFBEAKS

Halfbeaks look much like the Atlantic needlefish discussed below, but have short abbreviated upper jaws and extended lower jaws. There are three species of the halfbeak family found in the Chesapeake Bay, but only one, the halfbeak, is occasionally encountered in the Potomac estuary. The halfbeak is not an abundant species, and little data are available on its distribution in the Potomac. They tolerate salinities down to about 12 ppt.[2]

LINED SEAHORSES

Lined seahorses may be more common to the Chesapeake Bay than is apparent from the records. They tend to inhabit shallow areas with abundant rooted aquatic vegetation where most sampling gears are ineffective. Thus, they tend not to be taken in surveys, but are frequently captured by curious naturalists. In the Potomac, young lined seahorses have been captured in the Wicomico River[4] where salinities are substantially lower than the salinity tolerances listed in the literature (around 15 ppt).[2] Reproduction occurs in a similar manner as the pipefish (discussed above in the section on estuarine fishes).

STRIPED MULLETS

Striped mullets are schooling fish that swim close to the surface and often leap from the water[31] in a characteristic manner that quickly identifies them. They frequent tidal creeks and flats,[2] unlike most marine fishes which are more open-water-oriented. Of the two mullet species found in the Potomac estuary, striped mullet is more common than white mullet, although both are only occasional visitors. During some years, striped mullet are taken by commercial fishermen in small amounts (up to a few thousand pounds), but are absent from the catch in other years.[53] Although adult striped mullets have been taken from low saline waters of the upper Chesapeake Bay,[31] they are generally not very tolerant of anything below high mesohaline salinities,[2] and adult populations would be found only towards the mouth of the Potomac. Their young, however, have been found in the shore zone as far upstream as Rocky Point, just within the Wicomico River.[3] Spawning occurs in the ocean in winter and early spring, and schools of young mullets, 25 to 36 mm long, begin to enter the Chesapeake Bay in April.[31] Schools of increasingly larger fish enter through the summer and stay until November. The largest runs have been observed in October.[31] Striped mullets in the fall schools average 20 to 30 cm in length.[31]

ATLANTIC NEEDLEFISH

The Atlantic needlefish, as its name implies, is a long slender fish and has elongated scissor-like jaws. They are primarily inshore, shallow water species, which are usually seen schooling at the surface. Atlantic needlefish are known to frequent fresh water and have even become landlocked in coastal freshwater lakes.[75] They are probably the most euryhaline of all marine species in the Potomac estuary and are commonly found upstream as far as Washington, D.C. In Maryland waters, Atlantic needlefish generally are no longer than 50 cm, although they reach lengths of 120 cm in other regions.[17] They feed on shrimp and small fish, such as killifishes, but silversides are their principal prey.[17] [39]

Atlantic needlefish move into the Potomac in early April during the blueback herring runs[75] and remain until fall. They are found throughout the Potomac main stem and all of its tributaries.

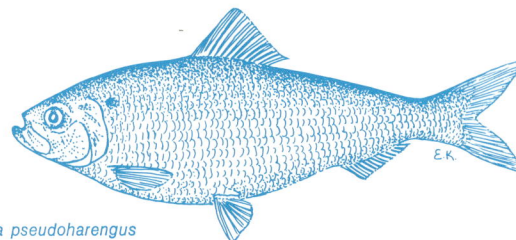

Alewife, *Alosa pseudoharengus*

Unlike most marine fish, Atlantic needlefish spawn within bays and estuaries in fresh and brackish waters.[39] Eggs are laid in shallows in submerged masses of algae.[39] Spawning in the Potomac usually occurs from late May to early June,[39] but may continue into August.[75]

BLACK SEA BASS

Black sea bass, occasional visitors to the Potomac, are found only near the mouth. They are relatively small fish in the Chesapeake Bay, reaching only about 22 cm in length.[31] They are easily recognized when they are fresh from the water by their distinctive coloration — bright blue streaks around their eyes and blue scales, each outlined with black.

In some years, a few are caught by commercial fishermen[53] and the sport catches at Cornfield Harbor (near Point Lookout, Maryland) are known to include black sea bass.[76]

ANADROMOUS AND SEMIANADROMOUS FISHES

Anadromous and semianadromous fishes, as stated previously, are fish that spend most of their lives in marine or estuarine waters but return to fresh water to spawn. This group of fishes includes some of the most abundant species found in the Potomac estuary, as well as some that are greatly prized by commercial and sportfishermen. Because of their economic importance, many scientific studies and field surveys have been conducted on these species. Consequently, more information is available on their distributions than on those of any other category of fish in the Potomac estuary.

The anadromous species of the Potomac (those that ascend to fresh water from the ocean) include herrings and shad, sea lamprey, and two species of sturgeon (Table 8-3). Species such as striped bass and white perch are termed semianadromous because many or all spawning adults ascend to fresh water from lower regions of the estuary. Yellow perch and gizzard shad are also considered semianadromous. Folio Map 8 shows the extent of spawning migrations of eight major species of anadromous and semianadromous fishes in the Potomac.

Table 8-3 — **Anadromous and Semianadromous Fishes of the Potomac Estuary**

Anadromous	Semianadromous
Alewife	Striped bass
Blueback herring	White perch
American shad	Gizzard shad
Hickory shad	Yellow Perch
Sea lamprey	
Atlantic sturgeon	
Shortnose sturgeon	

Anadromous Species

HERRINGS AND SHAD

Herrings and shad belong to the same family of fishes, Clupeidae.* Two anadromous herrings occur in the Potomac — alewife and blueback herring. Blueback herrings are generally 2 to 10 times more common than alewives.[77] There are also two anadromous shad species in the Potomac — American shad and hickory shad. American shad (or white shad as it is sometimes called) is by far the more common, although in recent years its numbers have declined sharply, not only in the Potomac, but elsewhere in the Chesapeake Bay.

Herrings and shad are remarkably alike in appearance, with laterally compressed "flat" bodies and silvery scales that easily flake off. Both shad species grow larger than the herrings. Female American shad are, on the average, about 0.7 kg heavier than the males, normally reaching lengths of about 50 cm and weights of 2 to 2.5 kg. Herrings rarely exceed about 38 cm and 0.2 kg, and average about 28 cm.[31]

Hickory shad may be distinguished from American shad by their more sharply projecting lower jaw and their smaller size.[78] The subtle differences between bluebacks and alewives are generally evident only to the more knowledgeable observer. Alewives are sometimes referred to as "big eyes" because their eyes are larger in proportion to their bodies. This is the main criterion for separating these two species without resorting to dissection and examination of the abdominal cavity lining (which is black in blueback herrings, pale in alewives).

Alewives and Blueback Herrings

Alewives and blueback herrings are discussed together since their distributions and movements are virtually the same. Fishermen seldom make distinctions between the two. Furthermore, since their egg and larval stages are seldom separated by species in scientific surveys, their spawning and early nursery grounds have only been defined collectively.

Both species spend the greater part of their lives along the northeast coast of North America from Nova Scotia to South Carolina. Alewives may extend southward to northern Florida.[78] Few details are known about their migratory habits along the coastline, but herring schools generally stay within 80 nautical miles of the coast in waters less than about 250 feet (76 m) deep. There are also thriving landlocked populations of alewives in the Great Lakes. Both species may be found in the Potomac main stem from the mouth to Little Falls; in the Anacostia River; and within all major tributaries to the estuary, from the Wicomico River upstream.

Alewives usually enter the Potomac several weeks earlier than bluebacks and well before the shad. They have usually arrived by the end of March and occasionally start arriving as early as February. Bluebacks begin arriving by early April and peak in the last part of the month, largely replacing alewives.[19]

Spring spawning runs of both herrings are primarily composed of 3 to 5 year olds.[79] Alewives spawn from late March through April, while bluebacks spawn from mid-April to mid-May (see Folio Map 7), both mostly in the tributaries. After spawning, adults gradually move back

---

*There are five other (nonanadromous) members of this family that occur in the Potomac. See Appendix Table 6.

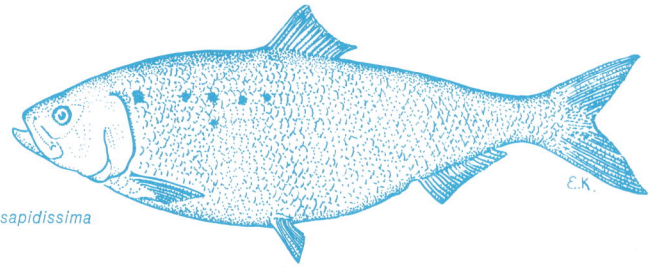

American shad, *Alosa sapidissima*

downstream and, by midsummer, have left the Potomac for the ocean. Commercial catches at the mouth of the estuary indicate that herrings leave the Potomac a few weeks behind the shad. This can be attributed to the greater distance that herrings must travel from their spawning grounds. By late summer, most adult herrings have returned to the ocean, and few are found in the Chesapeake Bay.

During spawning, the two herrings migrate into the tributaries to different extents (see Folio Map 8). Alewives penetrate the farthest, moving into small freshwater streams only a few feet wide and a few inches deep to spawn. In the Potomac, their progress is often inhibited or halted by man-made structures such as dams and water gaging stations, and by natural obstructions such as large fallen trees or beaver dams. Surveys of the upper reaches of Potomac tributaries have documented the loss of former spawning sites because of such obstructions.[20] Bluebacks tend to spawn in tidal fresh and low brackish waters farther downstream than alewives, but in many of the tributaries with obstructions, alewives and bluebacks spawn within the same region. Both herrings also spawn in the main stem Potomac[36] (Fig. 8-25). Their eggs are demersal, although some may be washed up into the water column by currents.

The predominance of tributary spawning is evidenced by the greater larval densities found at tributary sampling stations than at mainstem stations during surveys in 1974[4][36] (Fig. 8-26). Samples taken in the uppermost tributaries had the most larvae. There were no sampling stations on Chaptico Bay, but at least marginal spawning in this bay has been documented in other studies.[20]

Greatest larval densities in the main stem have been reported from Washington, D.C., down to Possum Point, with marginal spawning down to Maryland Point and perhaps below (Fig. 8-26).[47] A few larvae have been collected in the main stem down to the mouth of the Wicomico River, but no eggs have been taken.[4][36]

As with other anadromous species, herring larvae are transported by currents somewhat downstream, but apparently not a great distance. In 1975, peak clupeid larval densities were within 5 nautical miles downstream of centers of egg densities.[47]

A general herring nursery area stretches from about Maryland Point upstream to Washington, D.C., and includes natal tributary streams.[3][4][80] Annual surveys over a period of 15 years show that some juveniles remain in the nursery area even into September (Fig. 8-26);[3] however, by early winter, most have left the Chesapeake Bay entirely. A few stragglers may spend their first or even second winter within Virginia waters of the lower Bay.[31]

The youngest herring larvae feed on small organisms such as rotifers and copepod nauplii. Older larvae and young juveniles feed primarily on copepods. Mysid shrimp become the most important component in the diets of late juveniles. Herrings reach lengths of 5 to 9 cm by November of their first year.[31]

The small and bony herrings are caught in great numbers by commercial netters who sell them to fish processing plants primarily to be made into fertilizers. They are also highly prized by many sportsmen who line up along the spawning streams with dip nets to harvest the runs.

American Shad*

American shad are distributed from the Gulf of St. Lawrence to Florida, with the center of abundance between North Carolina and Connecticut.[81] There are distinct seasonal patterns of movement through marine, estuarine, and fresh waters for each life history stage of American shad. The mature fish enter the Chesapeake Bay from the Atlantic Ocean in spring, exhibiting a strong homing instinct for their natal stream. These fish have spent the previous 3 to 5 years along the Atlantic Coast as immatures, moving northward in fall and winter, southward in spring. There is strong evidence that the triggering mechanism for movement of shad into their natal estuary is the ambient ocean water temperatures at the mouth of these estuaries.[82] In the Chesapeake Bay, the main body of the American shad population ascends the rivers when water temperatures are between about 13 and 19° C.[81] In March (occasionally late February), earliest arrivals may be seen at or close to the spawning grounds, but the bulk of the run arrives in mid-April with males preceding females.[52][83][84] Peak spawning occurs from mid-April to mid-May (see Folio Map 7).[19]

After spawning, adult American shad migrate downstream. They remain in the Potomac a relatively short time, usually leaving by the end of June.

Spawning grounds are in tidal fresh waters over shallow flats in the main stem between Mattawoman and Piscataway creeks and sometimes upstream to Broad Creek (Fig. 8-25). Primary sites are off Gunston Cove and Occoquan Bay.[36] Marginal spawning may occur as far downstream as Maryland Point.[22] There is no documentation of shad spawning in the tributaries or in low-brackish waters,[4][20] as has been observed for herrings. From historical references at the turn of the century, it is known that peak spawning once occurred as far upstream as Great Falls.[6] The displacement of spawning downstream from previously favored sites has accompanied increasingly degraded water quality in those locations.

American shad eggs are large and demersal and settle to the bottom where they sometimes become buried in silt or sand. Rarely are currents sufficient to keep the eggs suspended where they would be susceptible to capture by plankton nets. Consequently, information on shad egg distributions is sparse. Furthermore, larval stages are difficult to distinguish from those of the other herrings. Thus, a quantitative description of spawning and nursery areas in the Potomac has not been made to date.

Juveniles remain, for the most part, in the same general vicinity or slightly downstream from the spawning grounds throughout the summer (Fig. 8-26).[3] In this region, salinities increase during the

---

*Little is known about specific distributions of hickory shad. Distribution and life history cycles approximate those of the American shad (see Folio Maps 7 and 8).

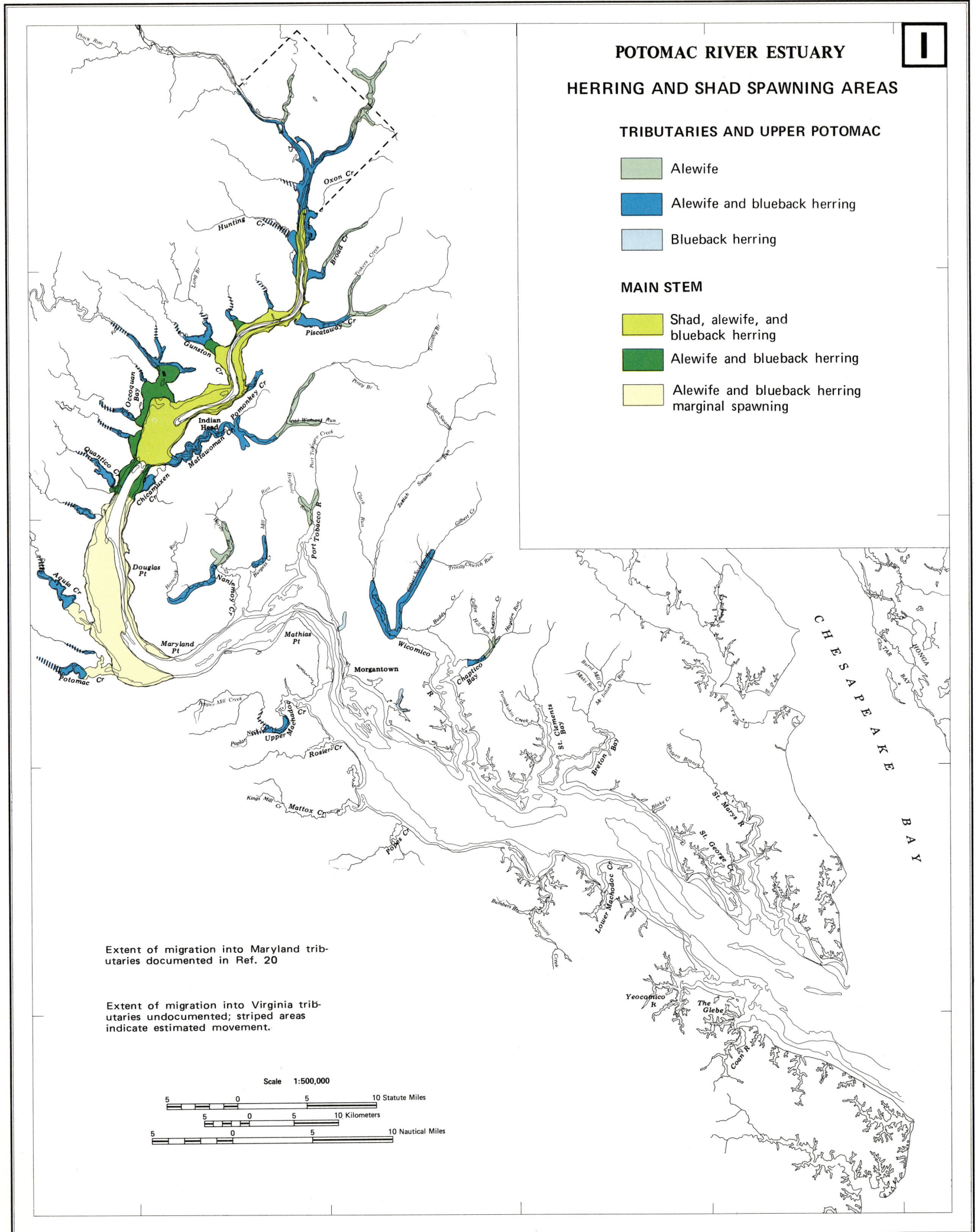

POTOMAC RIVER ESTUARY

I

HERRING AND SHAD SPAWNING AREAS

TRIBUTARIES AND UPPER POTOMAC

Alewife

Alewife and blueback herring

Blueback herring

MAIN STEM

Shad, alewife, and blueback herring

Alewife and blueback herring

Alewife and blueback herring marginal spawning

Extent of migration into Maryland tributaries documented in Ref. 20

Extent of migration into Virginia tributaries undocumented; striped areas indicate estimated movement.

Scale 1:500,000

5   0   5   10 Statute Miles

5   0   5   10 Kilometers

5   0   5   10 Nautical Miles

Figure 8-25. *Herring and shad spawning areas in the tributaries (Refs. 1, 4, 20, 80) and main stem (Refs. 4, 5, 36, 80) of the Potomac estuary.*

POTOMAC RIVER ESTUARY

M

ALEWIFE, BLUEBACK HERRING, AND AMERICAN SHAD

HERRING LARVAE *

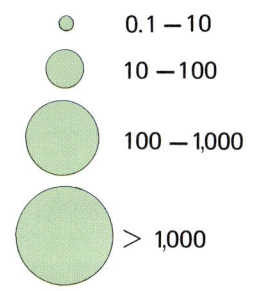

0.1 — 10

10 — 100

100 — 1,000

> 1,000

Average density of larvae per 1,000 m^3
(April 3 to June 14, 1974)

• Black dot is sampling station

* No species distinctions made between
alewife and blueback herring larvae

HERRING AND SHAD YOUNG

Primary nursery for American shad
and herring young (July – September)

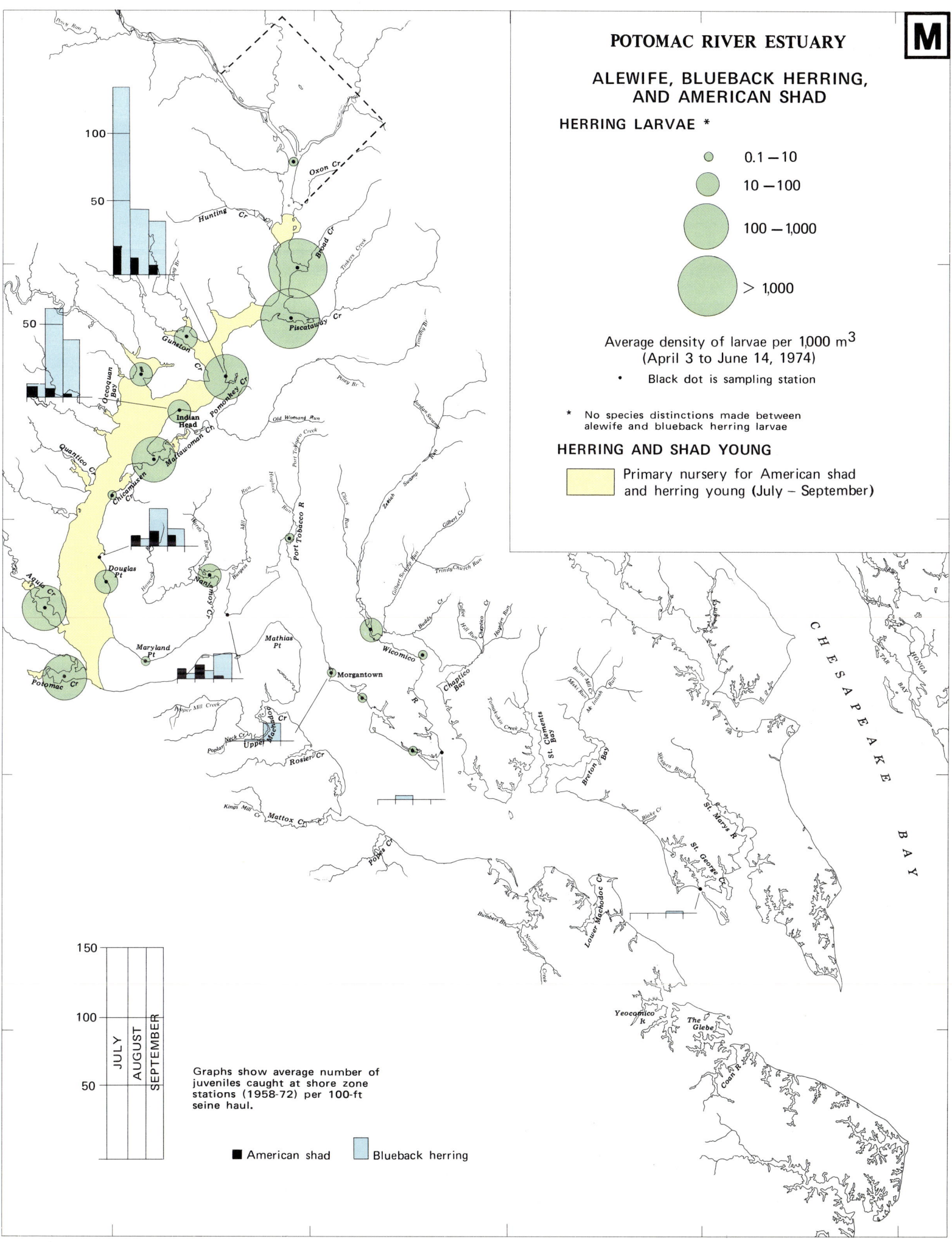

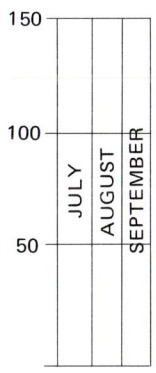

100

50

50

50

Pina Run

Oxon Cr.

Hunting Cr.

Broad Cr.

Tinkers Creek

Piscataway Cr.

Gunston Cr.

Occoquan Bay

Piney Br.

Pomonkey Cr.

Indian Head

Mattawoman Ch.

Old Womans Run

Port Tobacco R.

Quantico Cr.

Chicamuxen Cr.

Nanjemoy Cr.

Douglas Pt

Aquia Cr.

Maryland Pt

Mathias Pt

Wicomico R.

Morgantown

Chaptico Bay

Potomac Cr.

Upper Machodoc Cr.

Rosier Cr.

Mattox Cr.

Popes Cr.

St. Clements Bay

Breton Bay

St. Marys R.

St. George Cr.

Lower Machodoc Cr.

Yeocomico R.

The Glebe

Coan R.

C H E S A P E A K E   B A Y

150

100

50

JULY
AUGUST
SEPTEMBER

Graphs show average number of
juveniles caught at shore zone
stations (1958-72) per 100-ft
seine haul.

■ American shad   ▨ Blueback herring

Figure 8-26. *Densities of early life stages of herring and American shad observed at various locations in the Potomac estuary; herring larval data are from a 1974 survey (Ref. 4); juvenile herring and American shad data are from shore seining surveys from 1958 to 1972 (Ref. 3).*

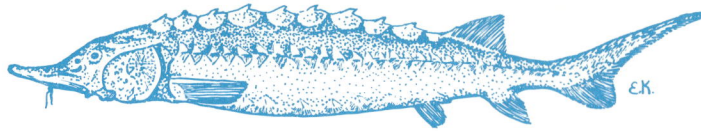

Atlantic sturgeon, *Acipenser oxyrhynchus*

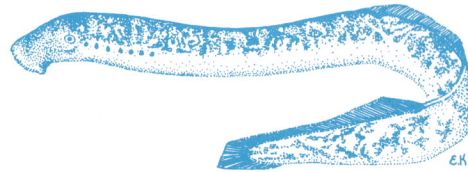

Sea lamprey, *Petromyzon marinus*

168　summer months until they become low brackish (1 to 4 ppt). Young American shad utilize the same low salinity nursery areas as striped bass, white perch, and many other estuarine-dependent species. The young grow rapidly in their first summer, reaching about 5 cm by July and about 8 to 12 cm by October.[77] Under particularly favorable environmental conditions, they may reach 14 to 17 cm by fall.[85]

Some juveniles begin their seaward migration early, but the majority are still found upstream in September and October. By late fall, however, they leave the Potomac, moving into the Chesapeake Bay and towards the ocean. Some remain in the lower Chesapeake Bay throughout their first winter, but most of the immature fish return to the Atlantic where they remain for the next 3 to 4 years.[81]

Juvenile shad feed mainly on copepods, but as they grow, mysid shrimp become an important component of their diet. Adults are known to feed on copepods as well as ostracods, amphipods, mysids, isopods, insects, algae, and even small fish.[31]

STURGEONS

Two species of sturgeon, Atlantic sturgeon and shortnose sturgeon, occur in the Potomac estuary. The shortnose is considered an endangered species along the entire Atlantic Coast.[86] These large, unusual, and primitive fish have large bony plates or shields along their backs and sides, and mouths extending beneath long snouts. As their name implies, shortnose sturgeons have shorter snouts than Atlantic sturgeons. Atlantic sturgeons up to 5.5 m long have been recorded, but one 2 to 2.5 m long would be considered very large.[16] Shortnose sturgeons are smaller, growing to no more than 1 m.[16]

Adults of both sturgeon species range along the northern Atlantic Coast.[87] Atlantic sturgeons are closely associated with estuaries but sometimes wander across the continental shelf to offshore banks where they are often caught by commercial fishermen.[19] Little specific information is available on shortnose sturgeons. Presumably, their seasonal migration patterns and behavior are similar to those of the Atlantic sturgeon.

In spring, sturgeons move into estuaries towards fresh waters to spawn. They arrive in the Potomac in April, moving to low brackish and tidal fresh waters. Little documentation is available on their exact distributions within the Potomac, except for their presence in the main stem at least to the vicinity of Washington, D.C.[6]

SEA LAMPREYS

The eel-like sea lamprey is a parasitic species belonging to a primitive group of jawless fishes that have cartilaginous skeletons. It has a circular disc-like mouth opening (lined with rings of rasping teeth), which it attaches to the flesh of other fishes to feed on blood and body fluids. Most other lampreys are freshwater species, but the sea lamprey is an anadromous species and spends its mature years in the ocean.

Sea lampreys range along the entire north Atlantic Coast, apparently staying close to land, although some stray into deeper continental waters.[88] They may enter mid-Atlantic estuaries as early as the end of March, but they do so mainly in April, coinciding with the shad runs.[15]

After sea lampreys invaded the Great Lakes in the 1950s, following completion of the St. Lawrence Seaway, the landlocked populations proliferated to such an extent that they ravaged commercial trout and whitefish fisheries. Their populations are now controlled. Sea lampreys, although common in the Potomac estuary and the Chesapeake Bay, have never been abundant in this region. Consequently, they have had no detrimental impact on the fisheries here.

Sea lampreys ascend the Potomac at least to Washington, D.C.,[6] and presumably enter most major tributaries as well. Their occurrence has been documented in the nontidal freshwater stretches of Mattawoman and Piscataway creeks, Port Tobacco River, and St. Clements Bay.[2] [13]

Although they prefer to spawn in nontidal waters, sea lampreys will spawn in tidal water when their passage is blocked. Preferred spawning sites are rapidly flowing water over gravel bottoms.[88] Spawning begins at 11°C, peaks at 14 to 15°C, and is completed by the time water temperatures reach 24°C.[16] After spawning, adults die.

The larval or worm-like "ammocoetes" are unlike the adults in appearance. Not yet parasitic, they do not have the adults' sucking discs, but instead have broad, hooded upper lips and no teeth. Ammocoetes remain in their natal freshwater streams, burrowing under stones or in the mud and feeding on planktonic organisms until they are 3 or 4 years of age.[15] At that time, they are 10 to 15 cm long and have assumed the adult form. They then migrate from the streams to the ocean. The seaward migration begins in summer, and they usually reach salt water by late fall or early winter. The period of time spent in the ocean is unknown.[88]

Semianadromous Species
STRIPED BASS

The presence of the striped bass on the Great Seal of Maryland indicates the high esteem in which Marylanders hold the fish, which are colloquially known as rock or rockfish. Avidly sought after by sportfishermen, they are also of great economic value to commercial fishermen. Striped bass can be found along the entire east coast of the Atlantic Ocean from Florida to Canada. However, spawning occurs in only a few areas, with the Hudson River and the Chesapeake Bay regions accounting for nearly all East Coast stocks. Recent studies have suggested that the Chesapeake Bay may contribute as much as 90 percent of the striped bass present along the coast.[89]

The high degree of localization of spawning makes striped bass populations vulnerable to man-made environmental modifications. This vulnerability, combined with the fish's great popularity, has made the welfare of the striped bass a major concern. Legal controversies have arisen concerning construction or operation of electric generating sta-

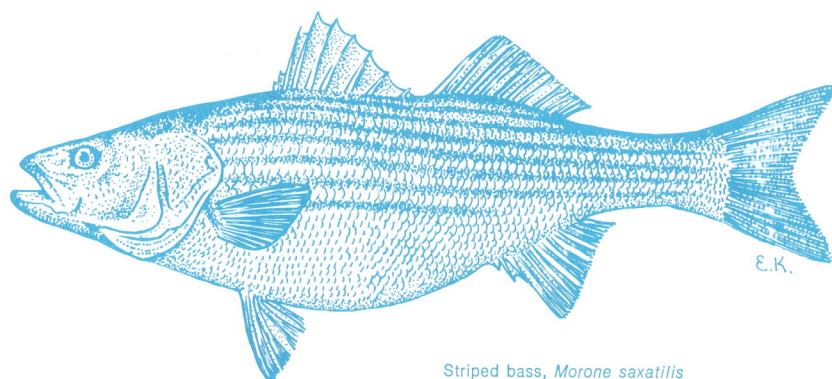

Striped bass, *Morone saxatilis*

tions adjacent to or within striped bass spawning and nursery areas. In part, the challenges have led to the studies responsible for much of the detailed information on striped bass spawning and population dynamics available today.

Striped bass are voracious predators and, as a result, grow rapidly. In their second year, they are already pan-sized fish[90] (Table 8-4). In the first few years of life, they tend to be pelagic schooling fish and are found pursuing forage fish in open waters of the Chesapeake Bay. As they grow older, they tend to be more solitary. Striped bass up to 30 years old have been caught, but most fish collected are less than 8 years old.[90]

Table 8-4 — Average Total Length and Weight of Striped Bass at Different Ages

Age (years)	Total Length (cm)	Weight (kg)
1	12.7	0.02
2	22.9	0.23
3	35.6	0.68
4	44.4	1.25
5	53.3	2.15
6	61.0	3.06
7	68.6	4.42
8	73.7	5.67
9	81.3	7.26

Source: Ref. 90

Tagging studies have suggested that striped bass spawn in the same area year after year, probably returning to the locale where they themselves were spawned.[90-92] Results of recent studies have indicated that the striped bass populations of various Chesapeake Bay tributaries, including the Potomac, are discrete subdivisions of the Chesapeake Bay stock.[93]

Striped bass exhibit a very complex pattern of seasonal and spawning migratory movements during their life cycle. Distinctive migratory differences are evident between the two sexes; as much as half of the population of females over 3 years old leaves the Chesapeake Bay to enter coastal waters, while most of the males less than 6 years of age remain year-round in the bay.[94] The differences produce patterns that can be generally described by season and life stage.

Immature and Adult Seasonal Distributions

In summer, after spawning in the Potomac, adults move downstream towards the shoals and surface waters of the lower two-thirds of the estuary (Fig. 8-27). There is a general dispersion of adults and immatures, with greater concentrations appearing towards the mouth.[91] Few striped bass inhabit the deep channel habitats at this time. There is also some interchange of fish between the Potomac estuary and other Chesapeake Bay tributaries (Fig. 8-27).[90]

In fall (usually October), movements of both adults and young striped bass in the Potomac are downstream or towards deeper waters (Fig. 8-28). In this season, some of the schooling fish in open waters

support an active fall sportfishery. At the same time, many ocean migrants return to the Potomac to overwinter near the mouth of the estuary (Fig. 8-28).[90-92]

In winter, during the coldest weather, striped bass concentrate in waters greater than 30 feet (9 m) deep (Fig. 8-29) where temperatures are somewhat warmer than those at the surface. During warmer periods, the overwintering fish, still actively feeding, often move out of the deep waters in search of food. Although some striped bass are found in winter in the mid-region of the estuary from Aquia Creek to below Morgantown,[52] the largest numbers are concentrated nearer the mouth of the Potomac. Towards the end of winter, some sexually mature fish leave to join spawning runs in other Chesapeake Bay tributaries.[91]

In spring, many sexually immature striped bass move out of the Potomac and the Chesapeake Bay and migrate northward along the Atlantic Coast (Fig. 8-30). At this time, immature 3-year-old females often leave to participate in this northern coastal movement,[91] while other immature females begin to move towards shoals. Some immatures of both sexes disperse somewhat upstream in the same direction as spawning adults.[83][95]

By the end of March, sexually mature striped bass begin to leave their overwintering sites and migrate upstream to tidal fresh or low brackish waters to spawn. A few mature fish that have overwintered in the lower Potomac leave the estuary in spring to spawn in other upper Chesapeake Bay tributaries, such as the Nanticoke and Choptank rivers. At the same time, some older fish that have previously left the Potomac for coastal waters return to the estuary to spawn.

Males, which generally outnumber females, usually arrive first on the spawning grounds; females arrive a week or two later. In a year of relatively normal runs, such as 1975, the ratio of males to females may be as high as 9 to 1 at the peak of the spawning season.[52] In years of poor runs (e.g., 1977), the ratio is approximately the inverse.[95]

Spawning populations consist of males 2 years old or older and females 4 years old or older. A survey of the Potomac estuary striped bass spawning population in 1975 showed that 97 percent of the males were 2 and 3 years old, while 99 percent of the females were 4 and 5 years old.[52]

Few adults migrate as far upstream as Washington, D.C. In earlier years, the riffles and strong running waters of this portion of the estuary were centers of striped bass spawning, but with the spread of urban development and the accompanying change in the Potomac estuarine environment, only a small number of striped bass now reach this far in their spawning migrations.

Egg and Larval Distributions

Delineation of the seasonal and temporal boundaries of striped bass spawning in the Potomac estuary has been the focus of much scientific effort in the past few years. These studies have shown that the major spawning area is from Maryland Point upstream to Possum Point. Moderate spawning activity spreads upstream to just above Indian

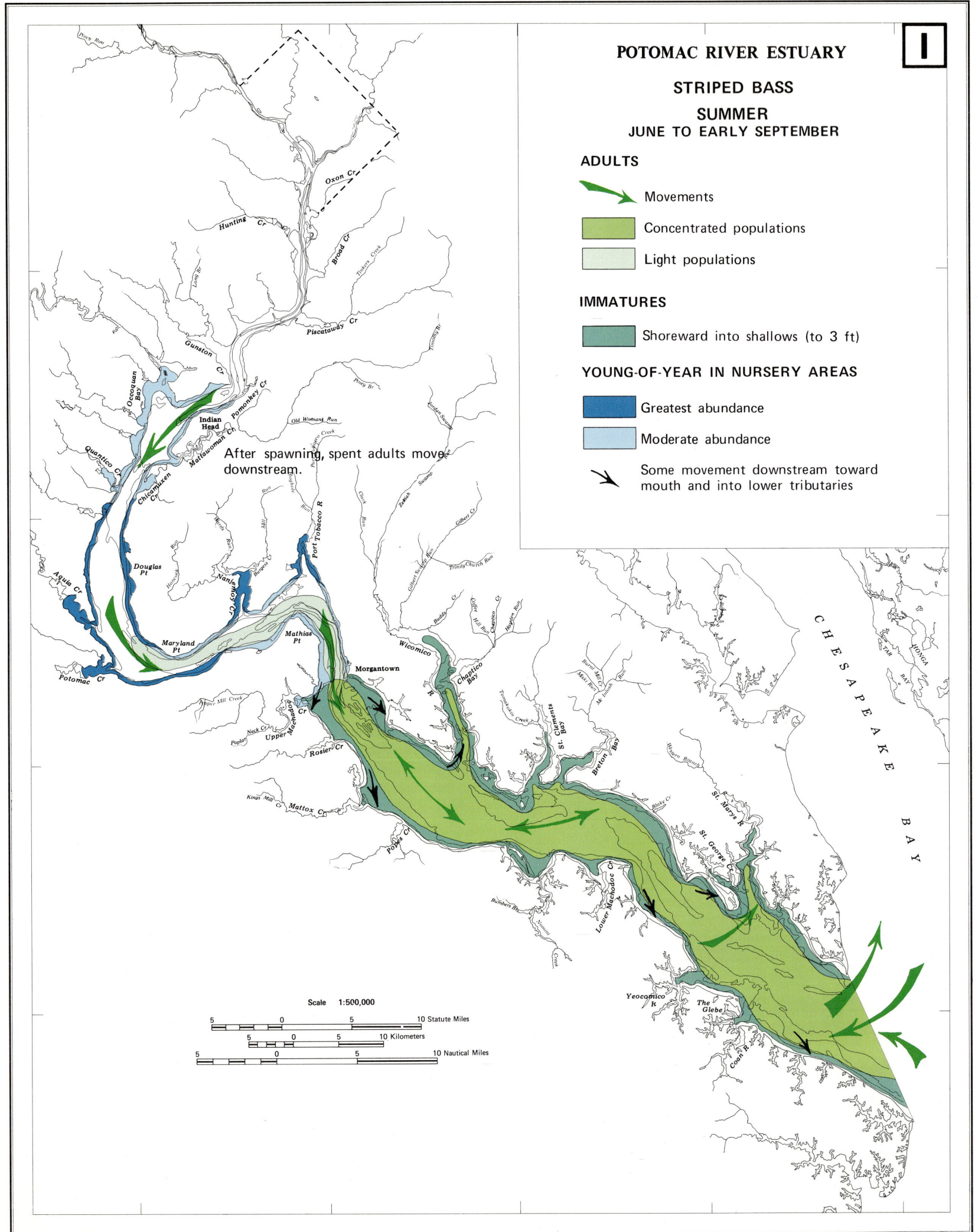

After spawning, spent adults move downstream.

POTOMAC RIVER ESTUARY | **I**

STRIPED BASS

SUMMER
JUNE TO EARLY SEPTEMBER

ADULTS

→ Movements

Concentrated populations

Light populations

IMMATURES

Shoreward into shallows (to 3 ft)

YOUNG-OF-YEAR IN NURSERY AREAS

Greatest abundance

Moderate abundance

↘ Some movement downstream toward mouth and into lower tributaries

Figure 8-27. *Summer distributions and movements of adult and immature striped bass in the Potomac estuary (Refs. 36, 47, 90, 91).*

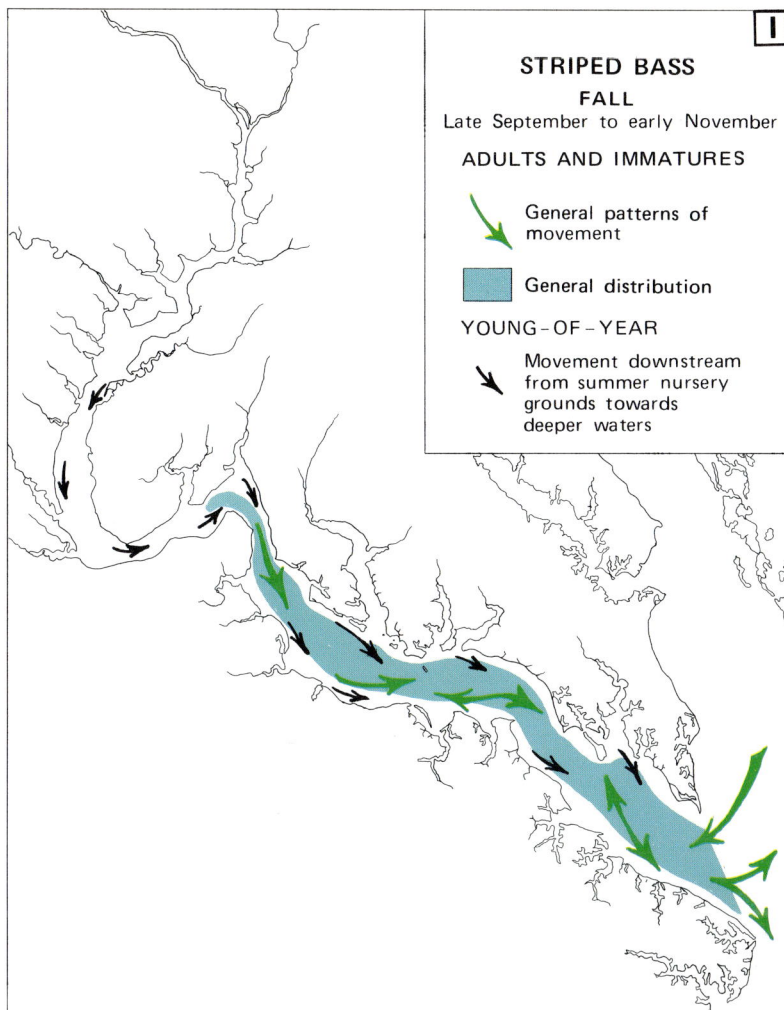

Figure 8-28. *Fall distributions and movements of adult and immature striped bass in the Potomac estuary (Refs. 90–92).*

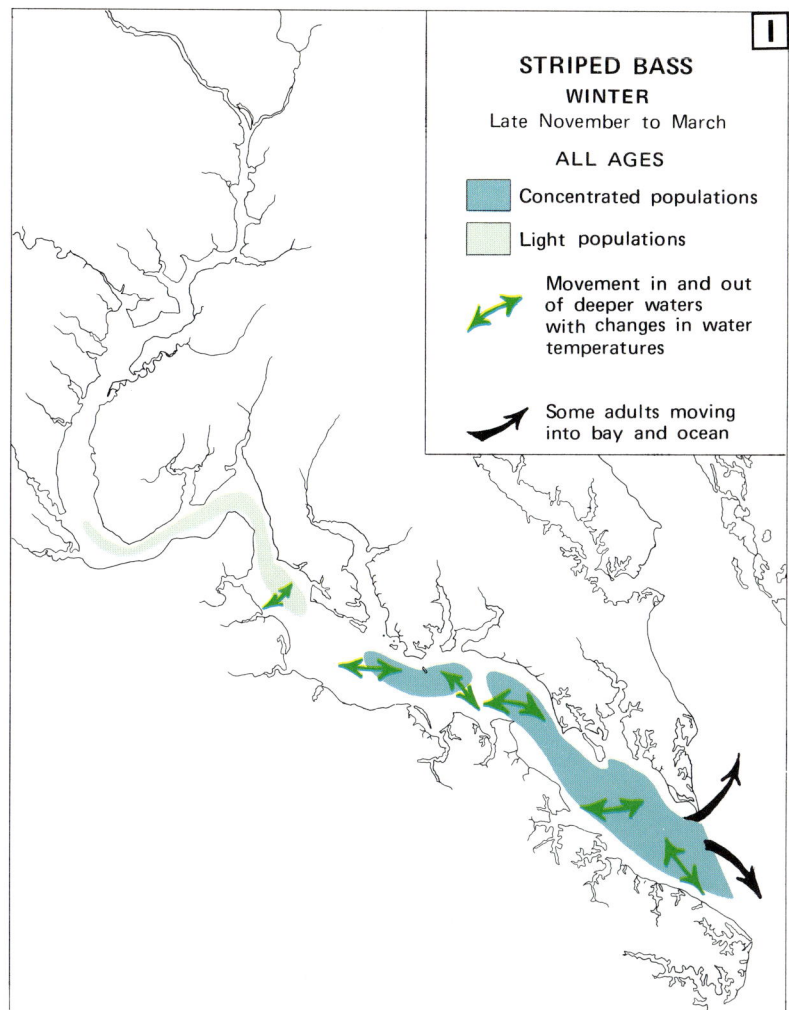

Figure 8-29. *Winter distributions and movements of adult and immature striped bass in the Potomac estuary (Refs. 52, 91).*

Head and downstream to Nanjemoy Creek, and only sparse spawning activity is detected above Indian Head or below Maryland Point. In recent years, only marginal spawning has been documented in the upper 20 miles of the estuary.

Timing and location of spawning are controlled by water temperature and salinity, and probably by a number of other unidentified factors. Striped bass spawn in temperatures from 10 to 23° C, with peak spawning between 14 and 15° C[96] (see Folio Map 7). In the Potomac, waters usually warm to 10° C by early April, and some initial spawning occurs at that time. However, peak spawning does not usually occur until mid-to-late April, when the optimum spawning temperature range is approached. There is some variation, as in 1976, when an abnormally warm February caused an early rise in water temperatures and triggered a spawn in late March. Temperatures in the Potomac usually reach the observed upper limit for spawning by the end of May, and eggs spawned after that time have a poor survival rate in the warming waters. Laboratory experiments have indicated that optimum development of eggs occurs between 16 and 20° C.[97–99]

Peak spawning occurs in the Potomac mostly in tidal fresh waters (< 0.5 ppt) in the region 15 to 20 nautical miles above Maryland Point (Fig. 8-30).[96] This spawning region may shift upstream or downstream depending on freshwater river flow. Incubation of eggs is rapid (3 days at 16° C).[98]

The fragile, newly-hatched larvae are planktonic and thus are dispersed and displaced by estuarine currents. In general, since both eggs and larvae would be expected to be transported downstream with time, highest abundances of late stages should be found downstream of regions with peak egg densities. But detailed distribution and abundance data on separate developmental stages of striped bass collected in 1974, 1975, and 1976 (Fig. 8-31) do not show evidence of displacement of eggs and larvae. Larval distributions always declined downstream of the freshwater-saltwater interface near Maryland Point.

It may be that larval mortality was high near the downstream end of the spawning area in these years, with complex hydrodynamic and behavioral effects also playing a role.[100] It is also evident that ichthyoplankton distributional patterns can vary widely from year to year. Such differences also contribute to the great variability of annual spawning success of striped bass populations.

The large fluctuations in abundance of striped bass over periods of years are a result of this variation in spawning success. Analyses of historical catch records show that populations are dominated by single yearclasses (i.e., fish produced in a single year), with a dominant yearclass appearing irregularly.[101] Since it has been shown that abundance of juveniles is a good indicator of the size of the yearclass, it appears that the success of a particular year's spawning is largely determined by the extent of survival between egg and juvenile stages, although size and fecundity of the parent stock also play a role.[102]

One factor critical for the survival of larvae is food density when feeding begins, generally about 9 days after egg fertilization. Laboratory studies have shown that 1,800 to 15,000 larval copepods (nauplii) per liter of water are required to ensure survival of normal densities of striped bass larvae.[103] [104] Presumably, food densities could be responsible for poor or dominant yearclasses, although this hypothesis remains controversial. The high zooplankton densities required by the striped bass larvae may be produced in response to nutrient influxes in years of cold winters and high spring freshwater runoff.[105]

Juvenile Distributions

During early summer, juveniles concentrate along the shores in waters less than 6 feet (1.8 m) deep in an area between Quantico and Mathias Point (Fig. 8-27).[4] [5] Apparently, some move into major tributaries of this stretch of the estuary, such as Aquia and Potomac creeks in Virginia and Nanjemoy Creek and Port Tobacco River in Maryland. Heavy concentrations of these small juveniles (less than 6 cm long) were

171

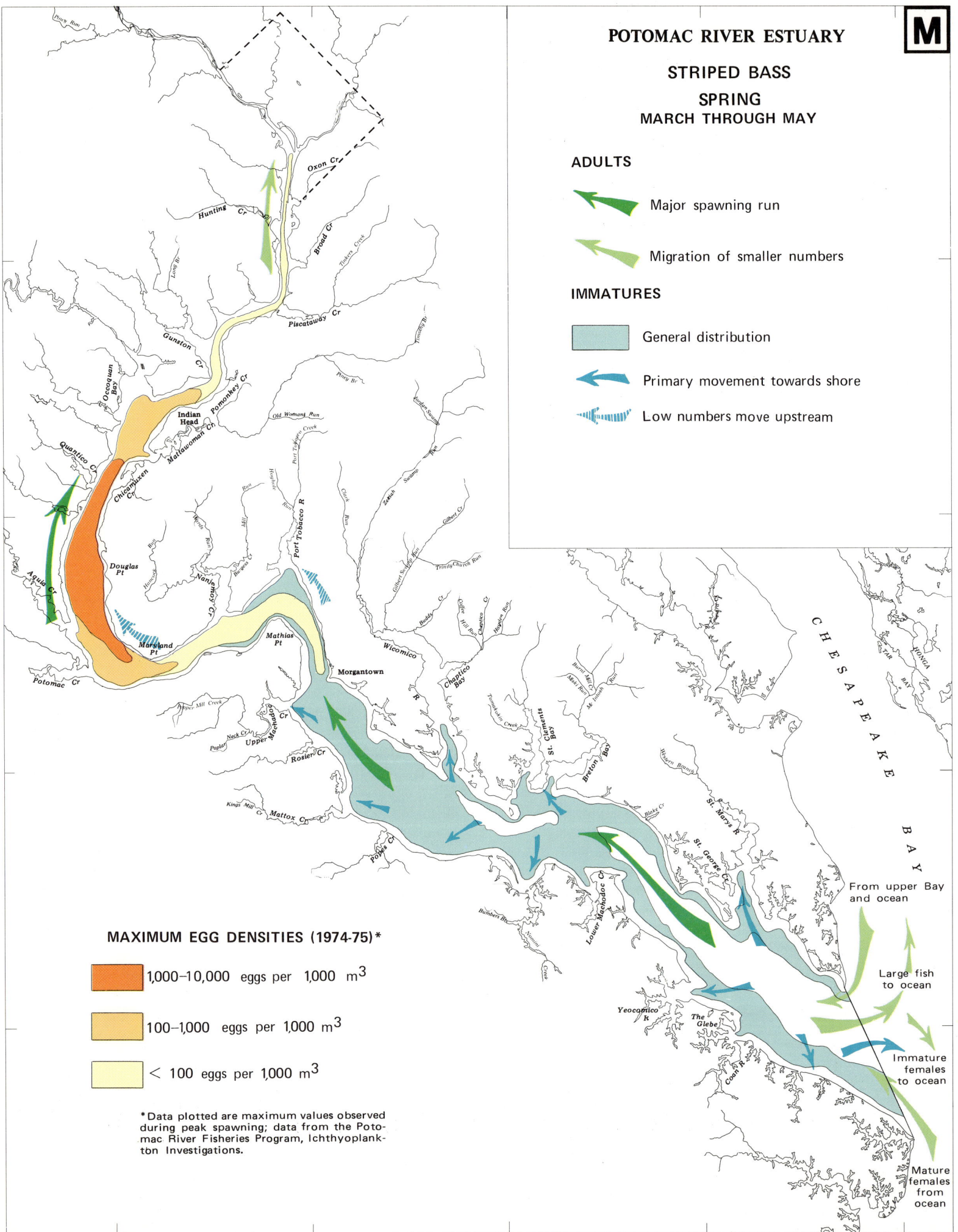

Figure 8-30. *Spring distributions and movements of adult and immature striped bass in the Potomac estuary (Refs. 36, 47, 52, 90, 91); distribution of spawning activity is based on maximum egg density data collected in 1974 and 1975 during peak spawning (Refs. 36, 47).*

STRIPED BASS

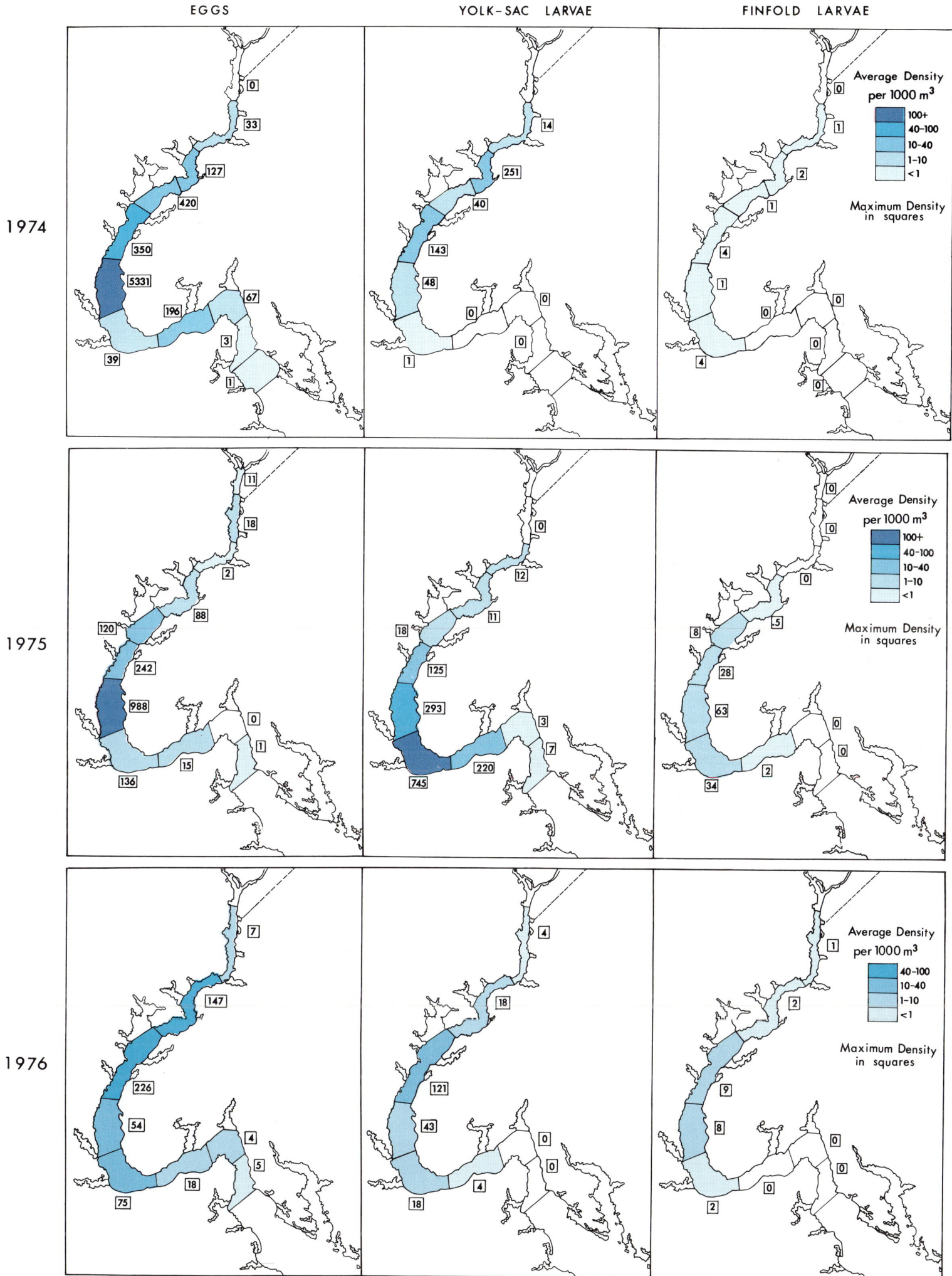

Figure 8-31. *Average densities of eggs and larvae of striped bass collected within each sampled segment of the Potomac estuary during ichthyoplankton surveys in 1974, 1975, and 1976 (Ref. 36). Maximum density of eggs or larvae in each segment for each year shown in squares.*

173

found in the Port Tobacco River in a 1975 survey.[47] The other tributaries were not sampled, but juveniles presumably moved into them as well. They tend to migrate downstream gradually as they grow, although some move randomly upstream. As early as July, juveniles are found as far downstream as St. George Island (Fig. 8-32).[3] As summer progresses and the young striped bass reach a length of 6 cm, they begin to move out of the postlarval nursery area and towards the vicinity of the overwintering grounds (Fig. 8-28). The downstream displacements of juvenile striped bass from the major nursery areas are variable from year to year.

Feeding Habits

Food preference of striped bass varies with age (see Fig. 6-3 in Chapter 6). Larvae just starting to feed concentrate on rotifers and copepod nauplii. By the time the larvae are 10 to 15 mm long, they are feeding on cladocerans and copepodites and, at 20 to 30 mm, on adult copepods, mysid shrimp, and amphipods. In a very short time (by the time they reach 50 to 70 mm in length), small fish (anchovies, silversides, and juvenile menhaden) become an important part of the young striped bass diet. The species consumed by the developing larvae in the Potomac estuary during late spring and early summer are primarily the copepods, *Acartia tonsa* and *Eurytemora affinis;* the cladocerans, *Bosmina longirostris* and *Daphnia* spp.; and the opossum shrimp.

Adult striped bass are voracious predators, and are near the top of the food web in the ecosystem of the Potomac estuary. Major fish species consumed are the ubiquitous forage species: bay anchovies, silversides, and Atlantic menhaden. Although striped bass also feed on benthic macroinvertebrates (primarily isopods, amphipods, and other crustaceans such as crabs and shrimps), the forage fish schools are their most important food source.[31] Feeding is apparently reduced during the spawning season, but increases to normal levels immediately afterwards and continues through winter.

WHITE PERCH

White perch are found throughout tidal waters of the Potomac. Frequently, thousands are hauled on board in a single trawl during scientific surveys. In a year-long study of the region between Maryland Point and Indian Head, white perch were the most numerous species caught, both in the bottom trawls and the shore seine hauls. They were outnumbered only by two other species (bay anchovy and Atlantic menhaden) in mid-water trawl samples.[5] White perch tend to inhabit open waters both close to shore and in deeper channel areas, but they also frequent quiet streams well up into the tributaries. This species is bottom-oriented, generally meandering close to the bottom in search of food.[106]

Although closely related to striped bass (they are in the same family, Percichthyidae, and the same genus, *Morone*), adult white perch are quite distinct in appearance, behavior, and seasonal distribution. They are smaller than striped bass and lack the distinctive stripes of the

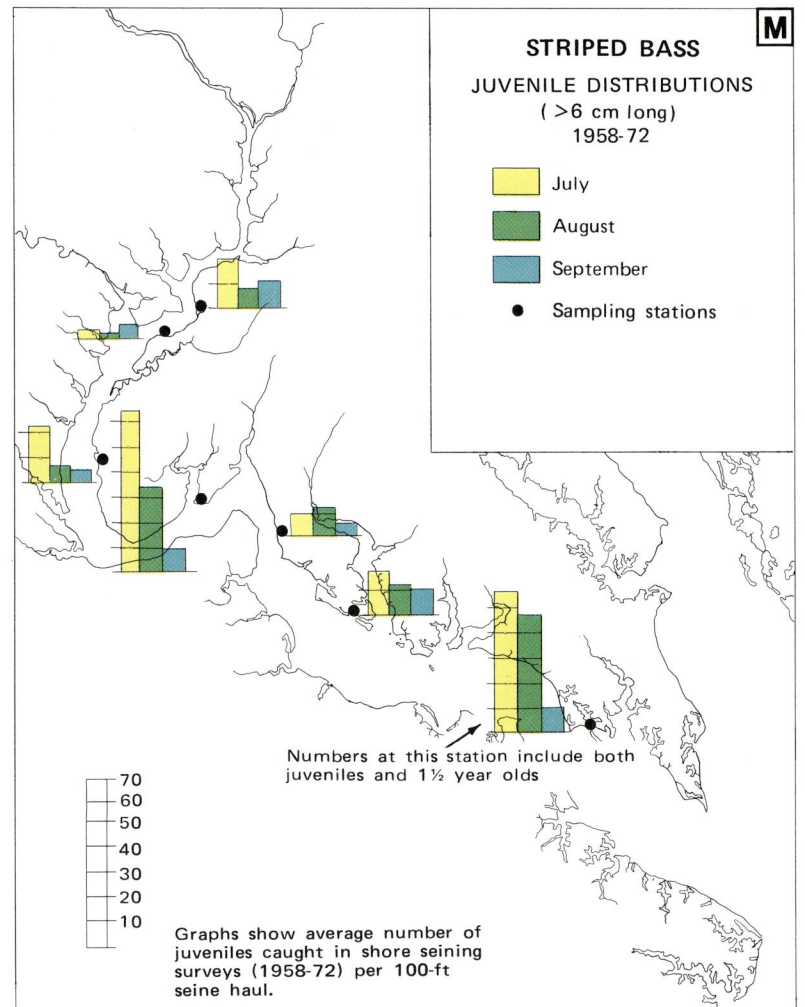

Figure 8-32. *Average densities of juvenile striped bass recorded from shore seining surveys from 1958 to 1972 (Ref. 3).*

latter. The average length of a mature adult is usually in the range of 16.5 to 19 cm, although 30 cm fish are sometimes encountered.[106] Their weight rarely exceeds 0.5 kg (Table 8-5). Most males mature by 2 years of age, most females by 3 years, and few of either sex live beyond 9 or 10 years. White perch populations have been consistently large over the past 10 years, and competition for food has apparently resulted in large numbers of small fish.[107]

The white perch population in the Potomac is apparently indigenous and distinct from populations in other regions of the Chesapeake Bay.[108] They are distributed throughout the Potomac tidewaters,[1 4 5 36 47] but based on distributions in other estuaries, their relative abundance probably decreases downstream in waters over about 5 ppt salinity.[109] Older, larger fish are more oriented to higher salinities than are younger individuals. All ages of fish exhibit extensive seasonal movements within the Potomac.

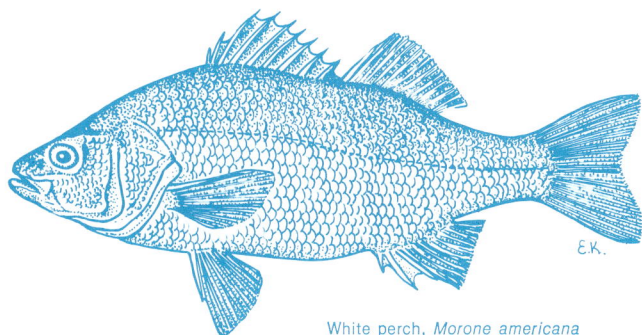

White perch, *Morone americana*

Table 8-5 — Average Total Length and Weight of White Perch by Age and Sex [a]

Maximum Age (years)	Males		Females	
	Total Length (cm)	Weight (g)	Total Length (cm)	Weight (g)
2	8.4	20	10.0	23
3	13.0	43	13.5	51
4	15.2	74	16.0	105
5	17.0	108	18.3	125
6	19.0	139	20.0	156
7	20.3	156	22.0	193
8	22.6	187	23.4	238
9	23.4	252	25.0	320
10	24.6	352	26.4	445
11	—	—	27.7	468

(a) Derived from Fig. 10 in Ref. 106

Seasonal Distribution

In the spring, white perch spawn in the main stem and the tributaries[36] where they enter even the tiniest streams of the Potomac (Fig. 8-33).[1] Movement towards spawning grounds begins by the end of March, preceding the spring migration of striped bass by a few weeks (see Folio Maps 7 and 8). By the first week of April, most ripe adults are on the spawning grounds, and large concentrations remain through May. Primary spawning is in tidal fresh water, but some spawning also occurs in low brackish waters in salinities up to about 2 ppt. The most intensive spawning is in the tributaries, principally those between Nanjemoy and Broad creeks,[1 4 20] but spawning fish move as far upstream as the District of Columbia and even into Rock Creek and the Anacostia River.[1] Spawning in these regions, however, is marginal. Likewise, less spawning occurs in tributaries below Nanjemoy Creek; spawning in the main stem nearly ceases below Maryland Point.[36]

Spawning begins when water temperatures reach 8 to 10°C, and is inhibited at temperatures above 15°C.[106] Optimum spawning temperatures (12 to 14°C)[106] are ordinarily encountered in the Potomac from the first part of April to the end of May, but sporadic spawning has been observed at the end of March and into the first week of June. Preferred locations are shallow waters along the shores, often under overhanging banks. White perch lay demersal eggs with thick capsules that stick to firm substrates such as branches or stones.[8] However, their eggs are easily dislodged and may become pelagic. Floating eggs that have previously been attached retain distinct adhesive discs on their outer membranes.

In 1974 and 1975, quantitative studies designed to delineate striped bass spawning and nursery areas provided simultaneous information on white perch egg and larval densities (Fig. 8-34).[36] However, the magnitude of spawning in tributaries relative to that in the main stem could not be determined because tributaries were not sampled intensively.

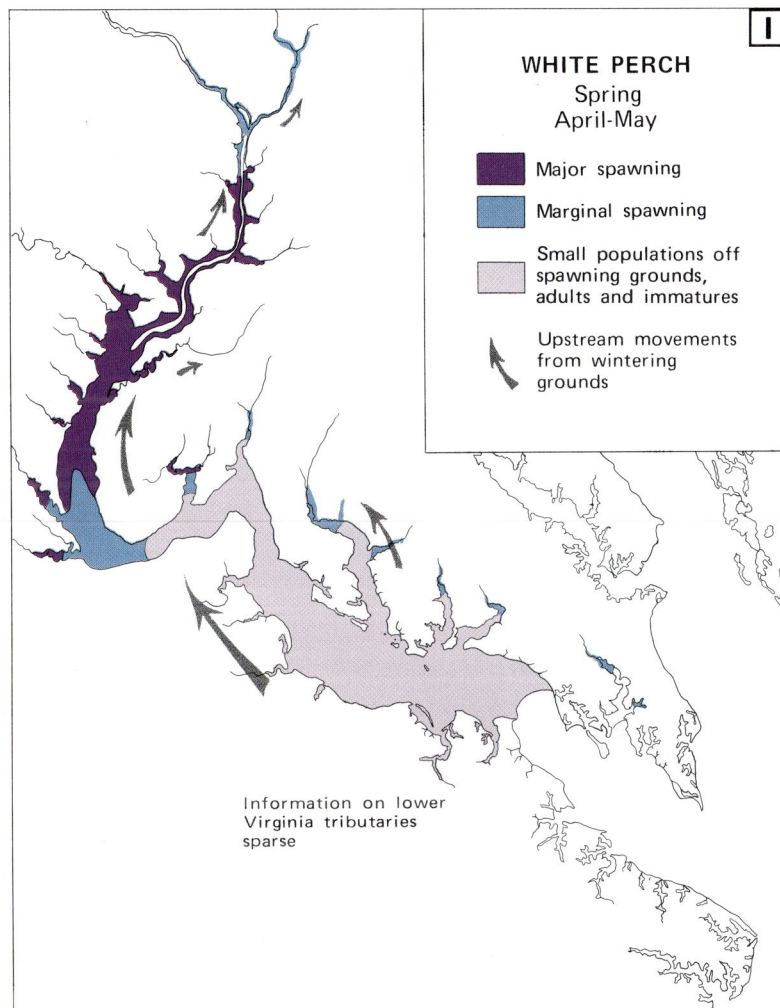

Figure 8-33. *Spring movements and spawning distributions of white perch in the Potomac estuary (Refs. 1, 4, 5, 20, 36, 52).*

In both study years, mainstem spawning seemed to be concentrated in two areas: one from Indian Head upstream to Broad Creek, and another in the bend of the estuary around Maryland Point.[5 36 47] However, there is a good possibility that high egg densities found near Maryland Point may be due to eggs spawned in Aquia and Potomac creeks that were dislodged and transported out into the main stem. There appeared to be a general downstream movement of larvae hatched in the uppermost spawning area, and the entire stretch of the main stem from Maryland Point upstream to Broad Creek was utilized as an early nursery area. High densities of white perch larvae at the mouths of major tributaries could be attributed to downstream transport of larvae.

During the spawning season, large numbers of immature white perch are found along with the mature adults in both the main stem and the tributaries above Nanjemoy Creek.

WHITE PERCH

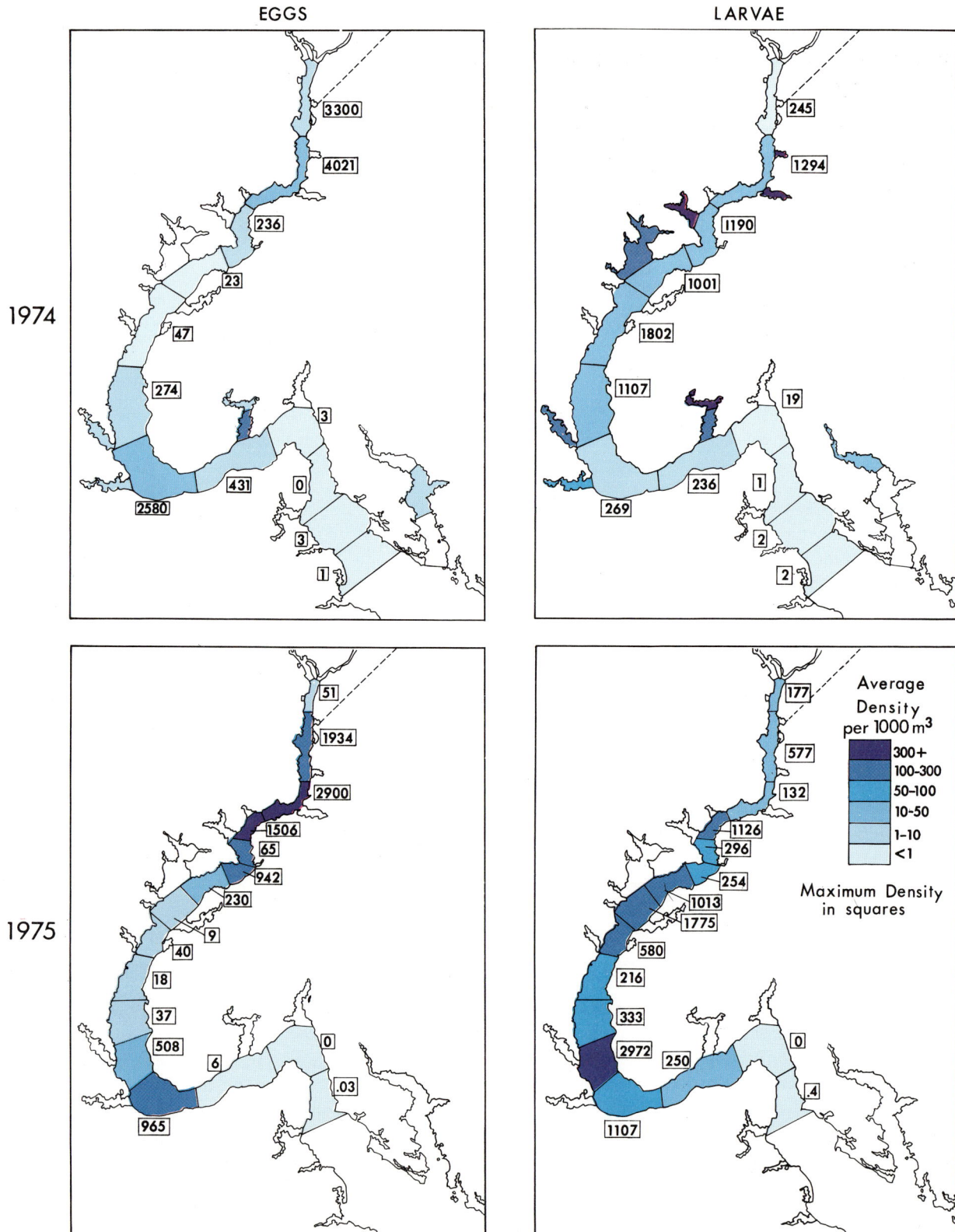

Figure 8-34. *Average densities of eggs and larvae of white perch collected within each sampled segment of the Potomac estuary during ichthyoplankton surveys in 1974 and 1975 (Ref. 36). Maximum density of eggs and larvae in each segment for each year shown in squares. Tributary data for 1974 from Ref. 4.*

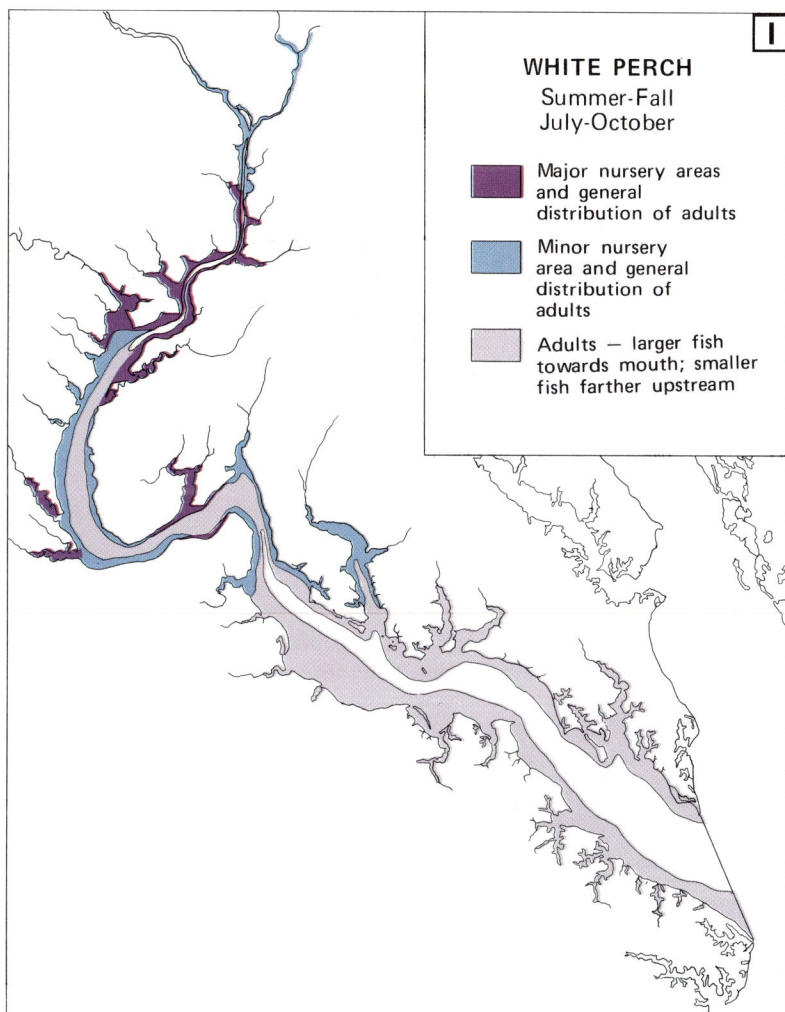

Figure 8-35. *General summer and fall distributions of adult and juvenile white perch (Refs. 3, 4, 5, 23, 36).*

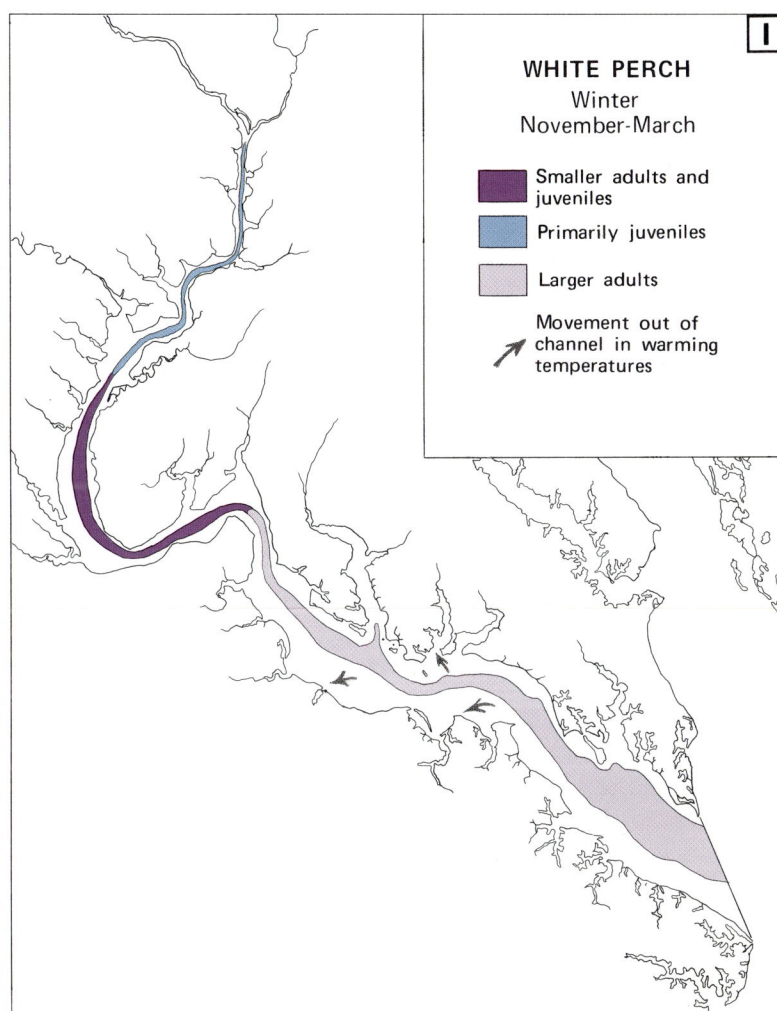

Figure 8-36. *General winter distributions and movements of adult and juvenile white perch (Refs. 5, 52, 106).*

In the summer to fall season, post-spawning adults move from the smaller streams into the lower portions of the tributaries and into the main stem (Fig. 8-35). Here the adults randomly disperse throughout summer and fall with greatest concentrations above Mathias Point.[3 5 23]

During the summer feeding season, white perch are concentrated towards shoals, although some are also found over deeper channel waters. Young-of-the-year and immature 1-year-olds are more strongly oriented towards the shallows than are adults. Heaviest concentrations of these juveniles occur in two regions: the main stem from Piscataway Creek down to the mouth of Mattawoman Creek, and from Maryland Point to Mathias Point.[36] The latter concentrations in the lower estuary may partly arise from migration of young out of the major spawning tributaries of Nanjemoy, Aquia, and Potomac creeks.[1 4] Smaller populations of juveniles are present in the summer along all shores of the main stem and throughout most tributaries.

In the winter (November-March), white perch move toward deeper waters as colder weather approaches (Fig. 8-36). However, concentrations in channels appear only during coldest periods. During milder weather, the fish move into shoal waters, although seldom close to shore.[22 106]

Feeding Habits

Young white perch are predators, feeding primarily on copepods, cladocerans, rotifers, amphipods, and insect larvae. As in the case of striped bass, their diet changes with growth. A study of their food habits determined that the smallest larvae (about 0.5 cm) preferentially feed on rotifers, but, by the time they reach 0.7 or 0.8 cm, they switch to cladocerans and copepods.[4] The study also revealed strong correlations between the availability of rotifers and successful development of yolk-sac larvae. The diet of juveniles (4.6 to 7.8 cm long) was found to consist of 80 percent copepods and 15 percent cladocerans. Insect larvae accounted for the rest, and no rotifers appeared in their stomachs.[4] Adult white perch are rapacious feeders on all forage species. Because of their abundance, they can cause significant cropping of populations of these small species. Because white perch are a desirable food fish, they are popular with both sport and commercial fishermen.

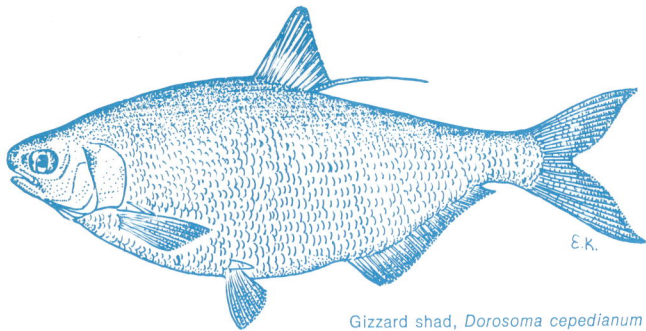

Gizzard shad, *Dorosoma cepedianum*

GIZZARD SHAD

Gizzard shad are members of the herring and shad family (Clupeidae) that never venture into oceanic waters. Although similar in appearance to other clupeids, gizzard shad differ in their feeding habits, ingesting bottom muds and associated fauna rather than pelagic invertebrates. Their common name comes from their thick stomach walls.

Gizzard shad are widely distributed throughout the upper tidewaters of the Chesapeake Bay, but are less common in higher salinity areas. In the Potomac, they occur along the entire main stem and in most of the larger tributaries from the Wicomico River upstream (Fig. 8-37). They penetrate upstream to fresh water and have been found in the Potomac to Rock Creek and in the Anacostia to just above the confluence of the Northeast and Northwest Branches.[1][20]

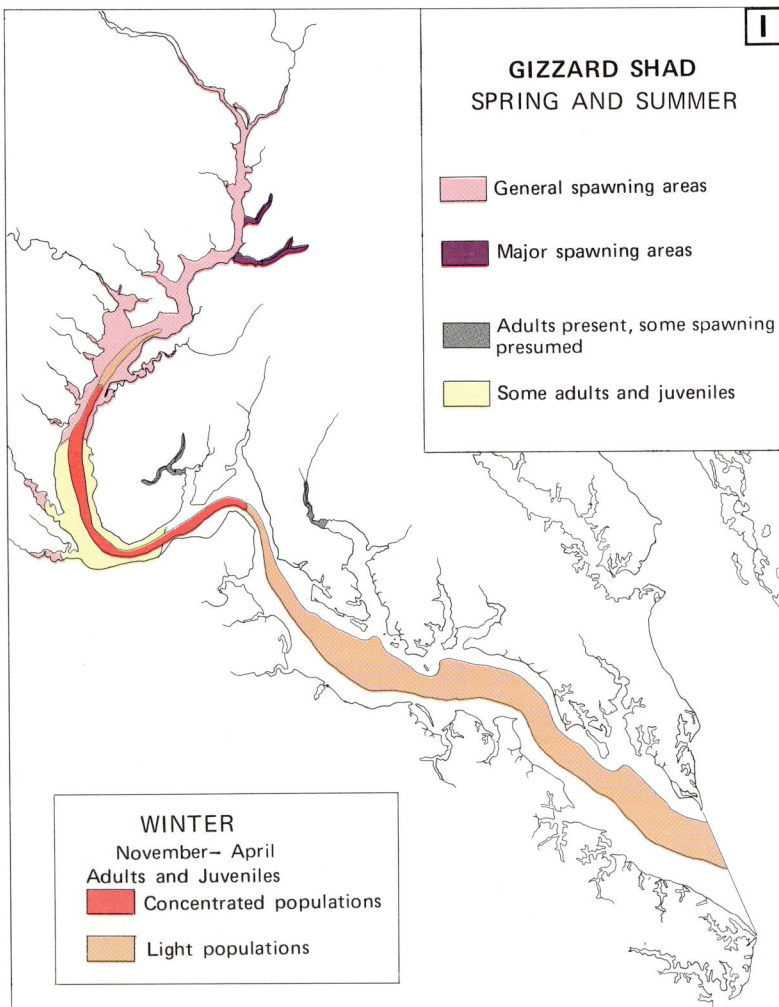

GIZZARD SHAD
SPRING AND SUMMER

General spawning areas

Major spawning areas

Adults present, some spawning presumed

Some adults and juveniles

WINTER
November– April
Adults and Juveniles

Concentrated populations

Light populations

Figure 8-37. *General seasonal distribution of gizzard shad in the Potomac estuary (Refs. 1, 4, 5, 36, 52); spawning distributions are based on 1974 data (Ref. 4).*

Seasonal migrations are marked by spring runs into tidal freshwater spawning areas (see Folio Map 8) and by fall runs back downstream to deeper estuarine waters. In the Potomac, overwintering populations are present downstream from Morgantown to the mouth. Recent studies of gill net catches[52][83] show that the gizzard shad is a relatively abundant species in the Potomac main stem. In February and March, they ranked fourth in abundance in the lower estuary off the mouth of St. Clements Bay (after striped bass, Atlantic menhaden, and white perch). At the same time, they ranked second at Maryland Point and first at Possum Point and above. In year-round sampling of both channel and shore regions between Maryland Point and Mattawoman Creek, they ranked either fourth or fifth.[5]

Gizzard shad spawn primarily in tidal fresh waters from April to June, with peak spawning in May. Their spawning activity occurs somewhat later than that of the anadromous species and is more prolonged. Their spawning areas in the Potomac include most of the large tributaries above Maryland Point and the upper freshwater region of the main stem down to approximately Douglas Point (Folio Map 8).[4][5] There is no documentation of spawning in any of the tributaries below Maryland Point, but adult gizzard shad have been collected in Allens Fresh in the upper Wicomico River and in the upper portions of Nanjemoy Creek.[1] Presumably, some spawning occurs in these tributaries. Mattawoman and Piscataway creeks are apparently the locations of the most intensive spawning.[4]

After spawning and throughout the summer, adults and juveniles may be found both in tidal fresh and low brackish waters in shallow shore regions close to the bottom.

Although abundant in the regions of the estuary where gill nets are regularly set by commercial fishermen, gizzard shad are considered trash fish, and only a small poundage is landed or sold. They are forage for larger species and serve as a useful ecological link by converting bottom mud-dwelling organisms into a form usable by more valuable, larger fish.[31]

YELLOW PERCH

Yellow perch are similar in size to white perch, but are more distinctively marked. They have six to eight vertical black bands contrasting sharply with their bright yellow and gold body color. During spawning season, the lower fins of the males turn from yellow to a brilliant orange-red color. Females grow faster than males, reaching 8 inches (about 20 cm) — the legal size for commercial catch — at 3 years (Table 8-6); males reach legal size at about 4 to 5 years. Males mature at the age of 1, when they are only about 10 cm long; females are older, aged 2 or 3 years and 17 to 20 cm long, before they spawn.[110]

Yellow perch, which are strictly freshwater fish in many areas of North America, have adapted exceptionally well to estuarine conditions. Fish in estuarine populations grow faster and live longer than those in freshwater populations (maximum 10 to 12 years vs 5 to 7 years in fresh waters), and females attain maturity later.[110]

Yellow perch, *Perca flavescens*

Table 8-6 — Average Total Length and Weight of Various Age Groups of Male Yellow Perch and Lengths of Female Yellow Perch [a]

Maximum Age (years)	Males		Females
	Total Length (cm)	Weight (g)	Total Length (cm)
2	10.1	11.3	11.4
3	15.7	39.7	17.5
4	19.0	73.7	21.3
5	21.6	121.9	24.3
6	23.6	164.4	27.0
7	25.1	184.3	29.0
8	26.2	229.6	30.5
9	27.0	246.6	31.1
10	27.7	266.5	32.5
11	28.0	272.2	33.2
12	28.4	297.7	34.8
13	28.4	297.7	—

(a) Derived from Fig. 7 in Ref. 110.
Female weights are not included because the weight of roe carried in spring months distorts the comparisons between the sexes.

Yellow perch are gregarious, sometimes loosely schooling, and may be found both in open waters or shallow, vegetated areas. Adults are distributed primarily in tidal waters up to 11 or 12 ppt salinity, except during the spawning season. There is strong evidence that yellow perch populations are indigenous to their own river system — if they are transported from their home stream into another area, they will leave the new stream almost immediately.[111] In the Potomac, it is uncertain whether there is a homogeneous estuary-wide population or whether each tributary supports distinct subpopulations.

Seasonal Distributions

In late winter to early spring (towards the end of February), when temperatures are only 4 to 5°C, yellow perch move from lower regions of the tributaries to the spawning grounds (Fig. 8-38). They are one of the earliest of the anadromous spawners, frequently appearing while snow and ice still remain. Males arrive first, followed a week later by females. Spawning is generally confined to areas from just below the freshwater-saltwater interface (salinities from about 2 ppt and below) to just a short distance beyond the extent of tidal fresh water (see Folio Map 7). Migration into nontidal fresh water is limited, and perch are only rarely encountered in the headwaters of the tributaries. In the Potomac near Washington, D.C., and in the Anacostia River, yellow perch populations are small, and little spawning apparently takes place. At the turn of the century, however, yellow perch were an abundant and prized sport and food fish in these regions.[6]

Spawning activity usually occurs in temperatures between 5 and 7°C (see Folio Map 7) in narrow, slow-moving streams a few feet in depth. The eggs are laid in long, accordion-shaped, jelly-like masses, which may stretch along the stream for 30 or 40 feet (9 or 12 m) and may become entangled in submerged bushes and tree limbs.[112] Aquia Creek,

with its dense larval concentrations, is a major spawning center. Potomac, Occoquan, and Nanjemoy creeks, and the Wicomico River are also active spawning areas,[4] and tributaries below the Wicomico River appear to be marginal spawning areas. After spawning, females leave almost immediately, while males remain throughout the 2 to 3 week spawning period.

The low temperatures during spawning cause an extended incubation period of 2 weeks — relatively long when compared to the 2- to 3-day hatching time for striped bass. Because of their long incubation, hatching young are in a much more advanced developmental stage than other anadromous or semianadromous species and have well-developed mouths and pectoral fins.[8] After hatching, larvae move downstream, and many are transported to the mouths of the tributaries and out into the main stem. Yellow perch young are voracious feeders and grow quickly; by their first fall, they are about 8 cm long. Like many other juvenile fish, they are concentrated in the shore zone. Greatest abundances have been found in tributaries between Nanjemoy and Piscataway creeks.

In summer to fall, after the spawning season, adults spread randomly throughout the estuarine system to salinities as high as 10 to 12 ppt (Fig. 8-38). Greatest concentrations are between 5 to 7 ppt. At this season, they are mostly nonschooling.

In winter in the Potomac main stem, there is some suggestion that yellow perch move downstream, but their overwintering locations are not known. In the Magothy River near Annapolis, there is upstream movement during winter away from the mouth (salinity 10 to 12 ppt) and towards a midestuary region (salinities of 2 to 5 ppt).[113] In the Potomac estuary, yellow perch have been found at least to Morgantown where winter salinities average 7 to 9 ppt.[23]

Yellow perch enter the commercial and sport fishery primarily in the spring when they are concentrated in the spawning areas. Since many adults spawn 1 or 2 years prior to reaching the legal commercial catch size of 8 inches (about 20 cm), populations are not expected to be endangered by intensive seasonal fishing.

Juvenile yellow perch feed primarily on small crustaceans, insects, worms, and molluscs, while adults prey on small fish — mostly bay anchovies, silversides, and killifishes.[31] Feeding habits of the young are similar to those of white perch and striped bass.

CATADROMOUS FISHES

AMERICAN EELS

The American eels are catadromous fish, that is, those that spend most of their lives in fresh or estuarine waters but return to the ocean to spawn — a behavior pattern the reverse of that displayed by anadromous fishes. American eels are found in rivers and estuaries from the Gulf of Mexico to Labrador,[58] and they are ubiquitous in the Potomac.

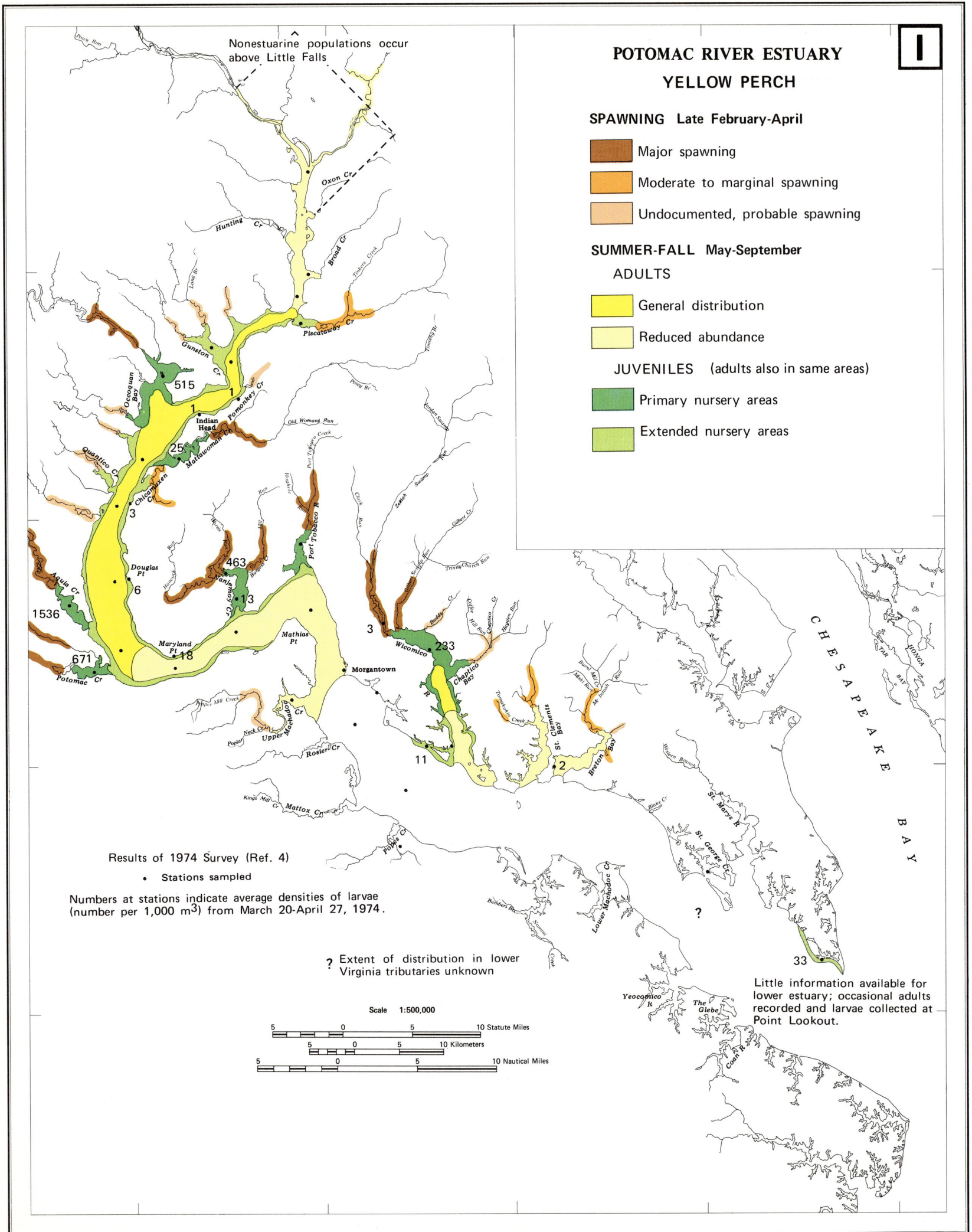

POTOMAC RIVER ESTUARY

YELLOW PERCH

I

SPAWNING Late February–April

- Major spawning
- Moderate to marginal spawning
- Undocumented, probable spawning

SUMMER-FALL May–September

ADULTS

- General distribution
- Reduced abundance

JUVENILES (adults also in same areas)

- Primary nursery areas
- Extended nursery areas

Nonestuarine populations occur above Little Falls

Results of 1974 Survey (Ref. 4)

- Stations sampled

Numbers at stations indicate average densities of larvae (number per 1,000 m^3) from March 20–April 27, 1974.

? Extent of distribution in lower Virginia tributaries unknown

Scale 1:500,000

Little information available for lower estuary; occasional adults recorded and larvae collected at Point Lookout.

Figure 8-38. *General seasonal distributions of adult and juvenile yellow perch, including areas of spawning in the Potomac estuary (Refs. 1, 3-5, 20, 36, 112); densities of larvae are based on 1974 data (Ref. 4).*

American eel, *Anguilla rostrata*

The American eel has an unmistakable, serpentine body and a long continuous fin that runs along its back around the top of its tail and forward again. They may grow to 1.2 m, with the largest individuals being females, which mature when they reach a length of about 45 cm.[15] These fish are nocturnal, often burying themselves in bottom muds during daylight hours.

In one of the most dramatic migrations of all fish species, sexually mature American eels (5 to 10 years of age) migrate from coastal rivers and estuaries into the Atlantic, leaving the Potomac from late August to mid-November. They converge at spawning grounds located in a broad region of the Sargasso Sea in the north Atlantic Ocean. After spawning, the adults die. The newly hatched eels are transparent, leaf-shaped creatures (called leptocephali) that are swept towards the coast by ocean currents in their first year of life.[114] By the time they approach the coast, they are 6 to 9 cm long and are metamorphosing into small, transparent "glass" eels. As they gradually become pigmented and transformed into "elvers," they begin moving into the estuaries. These elvers arrive in the Potomac estuary from mid-April to mid-May in runs that last a few days to a month.[15] The young eels move along the banks in shallow waters, migrating primarily at night and hiding in crevices and behind rocks during the day. Only a portion of the runs reach strictly fresh water; the rest remain within the estuary. Apparently, females migrate farther upstream than males.[13][88]

In a survey of nontidal freshwater regions of three tributaries — Piscataway and Mattawoman creeks and the Port Tobacco River — elvers 6 to 14 cm made up 30 to 64 percent of the total fish caught. Few American eels over 25 cm were captured.[13] American eels were found to be far more abundant in the estuarine portion of the Port Tobacco River than in its freshwater portion.[13]

Larger adult eels are distributed widely throughout the Potomac estuary, where they remain until sexually mature (eels up to 20 years old have been recorded[15]). Numbers upstream decline in October and remain low throughout the winter. The location of overwintering grounds is uncertain, but a general downstream movement to deeper, warmer waters probably occurs.

American eels are omnivores, feeding on crustaceans, molluscs, annelid worms, submerged vegetation, and fish.

Chapter 8 — References

1. Maryland Department of Natural Resources, 1969-1971
2. Musick, 1972
3. Scott, R.F., and Boone, 1973
4. Loos, 1975
5. Ecological Analysts, Inc., 1974
6. Smith and Beane, 1899
7. Elser and R.J. Mansueti, 1961
8. Lippson, A.J., and Moran, 1974
9. Howden and R.J. Mansueti, 1951
10. Wiley, 1970
11. Giraldi and Dietemann, 1974
12. Lippson, R.L., 1969-1971
13. Dintaman, 1975
14. Jenkins and Zorach, 1970
15. Schwartz, 1961a
16. Scott, W.B., and Crossman, 1973
17. Schwartz, 1962
18. Lee, Norden, Gilbert, and Franz, 1976
19. Mansueti, A.J., and Hardy, 1967
20. O'Dell, King, and Gabor, 1973
21. Schwartz, 1960
22. O'Dell, Dintaman, and Gabor, 1976
23. Academy of Natural Sciences of Philadelphia, 1970b
24. Richards and Castagna, 1970
25. Schwartz, 1963
26. Allee, Kelso, and Rosenblatt, 1976
27. Breder and Crawford, 1922
28. Schwartz, 1961b
29. Kendall and Schwartz, 1968
30. Schwartz, 1968
31. Hildebrand and Schroeder, 1928
32. Smith, 1892
33. Anjard, 1974
34. Cole, 1967
35. Rohde, 1974
36. Maryland Department of Natural Resources, 1974-1976
37. Mansueti, R.J., and Sheltema, 1953
38. McAtee and Weed, 1915
39. Foster, 1974
40. Foster, 1967
41. Richards and Bailey, 1967
42. Fritz, Meredith, and Lotrich, 1975
43. Tyler, 1963
44. Massmann, Ladd, and McCutcheon, 1952
45. Dovel, 1971
46. Massmann, 1954
47. Mihursky et al., 1976
48. Dovel, Mihursky, and McErlean, 1969
49. Schwartz, 1974
50. Schwartz, 1971
51. Richardson and Joseph, 1975
52. Wilson et al., 1975
53. U.S. Department of Commerce, 1960-1976
54. June and Nicholson, 1964
55. Nicholson, 1972
56. Henry et al., 1965
57. Walford et al., 1978
58. Dahlberg, 1975
59. Wiley and Boone, 1973
60. Dovel and A.J. Lippson, 1973
61. Pacheco, 1962
62. Young, 1953
63. Joseph, 1972
64. Dovel, 1968
65. Mansueti, R.J., 1960a
66. Lux, Hamer, and Poole, 1966
67. Koo, 1973
68. Koo, 1967
69. Bigelow and Schroeder, 1953
70. Mansueti, R.J., 1957
71. Orth, 1975
72. DuPaul and McEachran, 1973
73. Lippson, R.L., 1978
74. Mansueti, R.J., 1963
75. Hardy, J.D., 1978
76. Alexandria Drafting Company, 1973
77. Merriner, 1976
78. Hildebrand, 1963
79. Marcy, 1969
80. Warinner, Miller, and Davis, 1970
81. Mansueti, R.J., 1955a
82. Leggett and Whitney, 1972
83. Wilson et al., 1974
84. Wilson et al., 1976
85. Mansueti, R.J., and Kolb, 1953
86. Maryland Department of Natural Resources, 1977a
87. Breder, 1948
88. Bigelow and Schroeder, 1948
89. Kumar and Van Winkle, 1978
90. Mansueti, R.J., 1961a
91. Nichols and Miller, 1967
92. Massmann and Pacheco, 1961
93. Morgan, Koo, and Krantz, 1973
94. Kohlenstein, in press
95. Jones et al., 1977
96. Boynton et al., 1977
97. Mansueti, R.J., 1955b
98. Morgan and Rasin, 1973
99. Morgan, 1973
100. Polgar et al., 1976
101. Koo, 1970
102. Polgar, 1975
103. Miller, 1976
104. Doroshev, 1970
105. Heinle, Flemer, and Ustach, 1977
106. Mansueti, R.J., 1961b
107. Academy of Natural Sciences of Philadelphia, 1977
108. Ritchie, King, and A.J. Lippson, 1973
109. Ritchie, 1977
110. Muncy, 1962
111. Mansueti, R.J., 1960b
112. Mansueti, A.J., 1964
113. Dovel, 1967
114. Lippson, R.L., 1973

Birds and Mammals

Birds and Mammals

The diverse habitats and abundant food supplies found in and adjacent to estuaries frequently support large populations of water-oriented birds and mammals. Some of these animals are at the highest levels of the estuarine food web and form a link between the truly aquatic component and the terrestrial- or land-associated systems. For man, they provide aesthetic and recreational enjoyment as well as a source of income. The information given here on the birds and mammals commonly found in the Potomac estuary is less detailed than that for fish, plankton, and benthic invertebrates, because some of the species have not been studied as intensively as the animals that spend their entire life cycle in the waters of the Potomac.

BIRDS

Many species of birds inhabit the shores of the Potomac estuary, either as native residents that live and breed in its marshes and swamps and along its open shorelines, or as occasional visitors and seasonal migrants that frequent the wetlands to rest and feed. Among the birds found along the Potomac are waterfowl (ducks, geese, and swans), rails, waders (herons, egrets, and their allies), birds of prey (eagles, owls, osprey, and hawks), shorebirds (e.g., sandpipers, willets, and plovers), and the gulls, terns, and their allies. Also, most songbirds and perching birds common to the East Coast may be seen at one time or another in and about the edge of the estuary, with many dwelling in marshes or swamps[1] (Appendix Table 7). The species discussed here are among the most conspicuous or familiar water-oriented birds frequenting the estuary.

Waterfowl

Each fall and winter, following the completion of their summer breeding season in the northern United States and much of Canada, thousands of waterfowl begin their annual migrations south along the Atlantic flyway toward milder climates for overwintering. The Chesapeake Bay and its tributaries have always been an important winter habitat for many of these birds. The Potomac estuary is not as well endowed with wetland habitat as other areas of the Chesapeake system (e.g., the Eastern Shore of the Chesapeake), but its shoal waters, marshes, and swamps still attract and support substantial numbers of waterfowl that return year after year to rest and feed.

Most of the ducks and geese that are sought by hunters are herbivorous grazers that feed on the seeds, tubers, rootstocks, and foliage of submerged and emergent rooted aquatic vegetation. It has been estimated that rooted aquatic plants constitute nearly one half of the summer diet of most waterfowl species, with animals making up the remainder.[2] Recent deterioration of once-extensive submerged plant beds along much of the Potomac shoreline may have forced more ducks, geese, and swans to feed on inland grain fields. Many species of ducks are also carnivorous, consuming small animals that live in wetlands — including young *Macoma* clams (*Macoma balthica*), marsh snails and other small molluscs, amphipods, mud crabs, and insects.

The four major types of waterfowl inhabiting the Potomac are diving ducks, dabbling or puddle ducks, geese, and swans.

Data presented on the waterfowl maps in this chapter came from two sets of surveys conducted by the Maryland Department of Natural Resources. No data on Virginia tributaries were obtained. In the Wetland Habitat Inventory,[3] individual marshes and wetlands were visited during the year 1967 to 1968, and the presence and relative abundances of waterfowl were noted. Although these are the only data available for many of the smaller wetland areas, they may not be representative of long-term patterns of distribution because they were collected over a single, 1-year period. The Marsh and Waterfowl Investigations,[4][5] on the other hand, have been carried out annually since 1956. In these aerial surveys, abundances of waterfowl species in various survey zones were estimated, and data were provided as either 3-year[4] or 5-year averages.[5]

DIVING DUCKS

Diving ducks are so called for their excellent diving ability, which enables them to feed on submerged plants, fish, and benthic invertebrates. Diving ducks commonly found on the Potomac estuary include canvasbacks, greater and lesser scaup, mergansers, redheads, buffleheads, goldeneyes, ruddy ducks, and sea ducks (oldsquaw and several species of scoters). They are distinguished from dabbling or puddle ducks by their lobed hind toe; a smaller wing surface in relation to body size; red, yellow, or white eyes; and noniridescent wing colors. Their legs are positioned far back on their bodies, which makes walking awkward for them. Unlike puddle ducks, diving ducks cannot spring directly from the water, but must patter some distance along the surface before taking flight.

The majority of all ducks occurring seasonally along the Potomac are diving ducks. Often, hundreds of thousands visit during fall migrations. Although large numbers of divers may be found throughout the estuary, the greatest concentrations are generally along the expansive open waters of the lower estuary (Fig. 9-1). These concentrations consist mainly of sea ducks, which prefer the deeper open waters of the lower Potomac estuary, the Chesapeake Bay, and the Atlantic Ocean.

Canvasbacks

Probably the most well-known of all diving ducks, the canvasbacks have long been considered a thrill to sportsmen and epicureans alike. The large size, dark head, and white back of the males make them easily distinguishable at long range. Canvasbacks, like the Canada geese, fly in V-formation during their long migrations. This hardy species is one of the last to leave its summer breeding grounds in central and northwestern Canada, moving south only after the first hard freeze. It is also one of the first species to return north when the ice first breaks in February or March.[6] Canvasbacks are sighted along the Potomac from late fall to spring, but are most abundant during the winter.[7] Only a few stragglers are seen in summer.

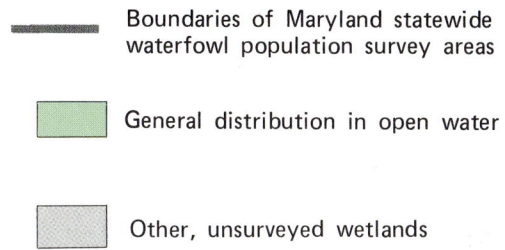

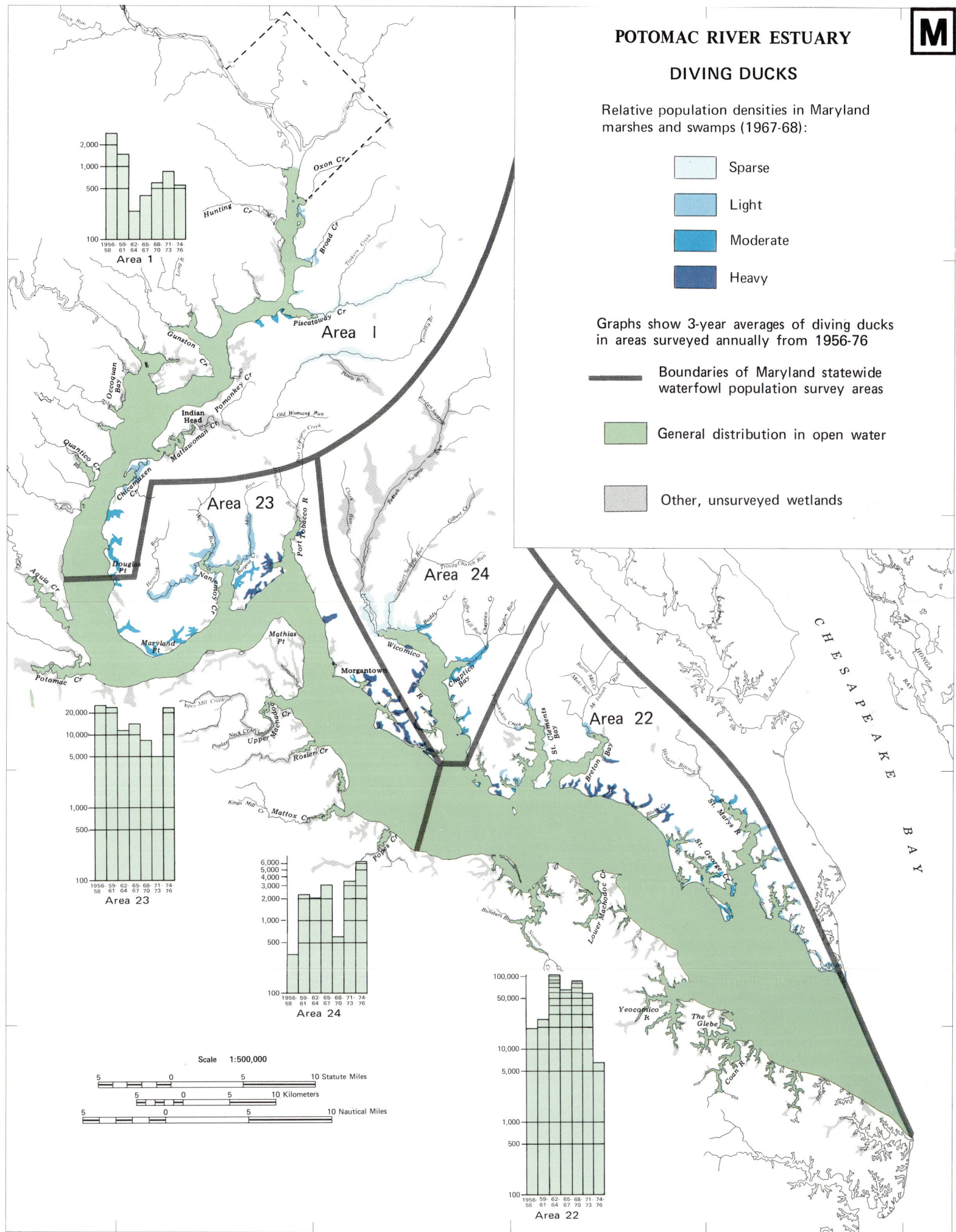

Figure 9-1. *Distribution of diving ducks in the Potomac estuary. Relative population densities in Maryland wetlands (1967-68) derived from Ref. 3. Graphs derived from 1956-76 survey data in Ref. 4.*

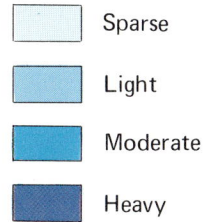

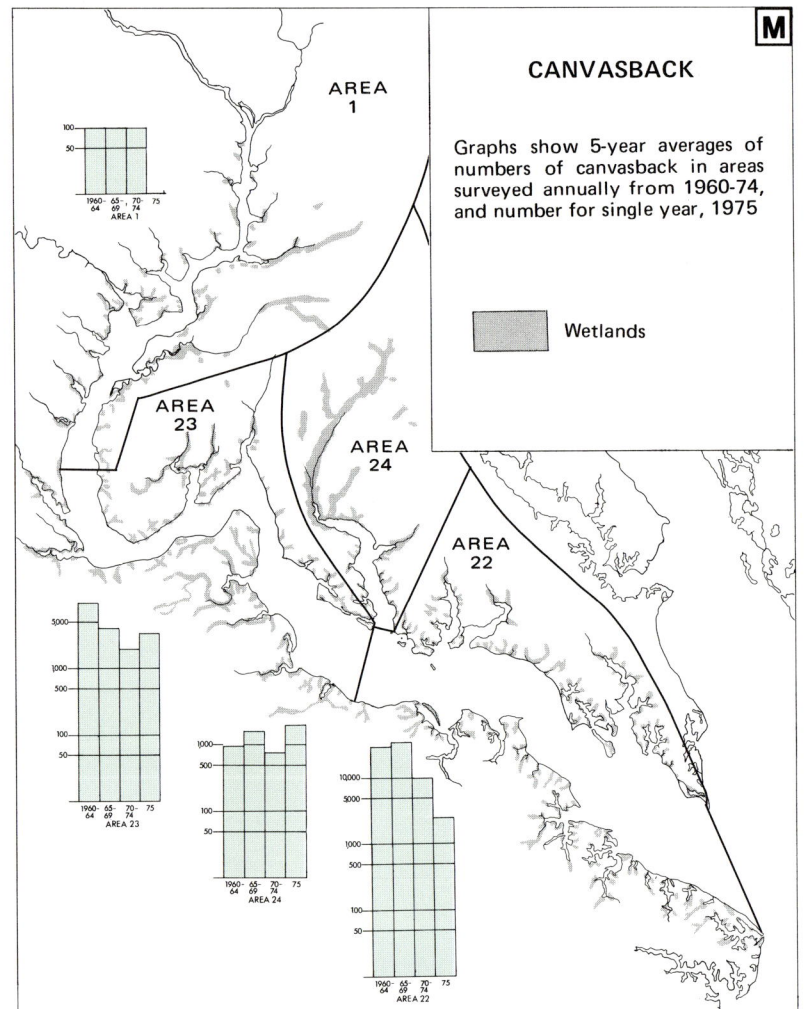

Canvasbacks were once very abundant, but their numbers have declined drastically since the turn of the century due to loss of breeding habitat and possibly overhunting in the early 1900s. Federal regulations instituted as a result of that decline currently prohibit the hunting of canvasbacks in much of the Chesapeake Bay region. Most of the visiting canvasbacks inhabit the open water and deeper shoreline areas of the central and lower Potomac estuary (Fig. 9-2).

In freshwater embayments, canvasbacks feed primarily on submerged rooted aquatic plants. In brackish waters, they may also eat molluscs, insects, and crustaceans.[7] Significant long-term fluctuations in canvasback populations and a general decline in numbers on the Potomac since the 1950s may reflect fluctuations in the availability of some principal food items — the small-sized, brackish-water clam (*Rangia cuneata*), the *Macoma* clam (*Macoma balthica*), wild celery, and pondweeds.[8]

Greater Scaup and Lesser Scaup

The scaup, or "blue bills" as they are sometimes called locally because of the bluish-gray cast of their bills, have historically been important game birds in the Chesapeake Bay region. Although the greater and lesser scaup look very much alike, they have different habits and distributions.

In winter, when greater scaup are distributed along the Atlantic coastline, their preferred habitat is in broad saltwater bays. Most of the greater scaup visiting the Potomac in the winter are found on the open waters and within coastal marshes in the lower half of the estuary.[6] Lesser scaup, on the other hand, tend to winter farther south than the greater scaup and prefer fresher waters.[6]

Scaup have been recorded all along the Potomac, but are primarily found from the mid to lower estuary (Fig. 9-3). Their large, dense flocks, which form on open water areas during their winter migrations, have earned them the nickname "raft ducks."

The diets of both species consist of vegetable and animal matter, although during fall and winter migrations, they probably consume more invertebrates than submerged aquatic plants.[9] Both duck species are good divers and are known for their ability to remain under water for long periods of time.

Mergansers

Long and narrow bills with sharp serrations or "teeth" distinguish mergansers, or fish ducks, from other diving ducks. This morphological adaptation makes mergansers very effective in catching and eating fish. The males of the hooded and American species are black and white, while male red-breasted mergansers have a more reddish or cinnamon color.[6] The colors of the females of the three merganser species are more subdued. All three species, the hooded, American, and red-breasted, can be expected to winter in some portion of the Potomac estuary. They are less abundant than the more common diving ducks such as scaup. The hooded mergansers prefer inland wooded freshwater streams and ponds and are the only mergansers that might breed in the Potomac area. The American mergansers overwinter along fresh-

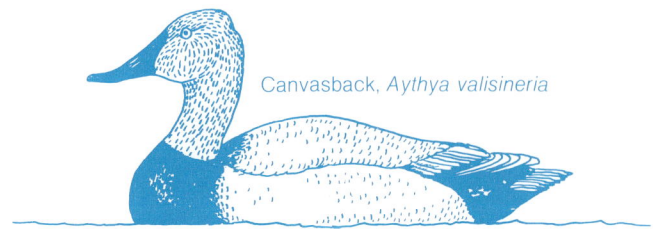

Figure 9-2. *Distribution of canvasback in the Potomac estuary. Graphs derived from 1960-75 survey data in Ref. 5.*

water streams and clear ponds with sandy or rocky bottoms. Although the red-breasted mergansers may travel to inland freshwater areas to feed, they prefer the higher salinity portions of the estuary where they prey on the more abundant and larger fish. The mergansers start their spring migration in March and April, although some red-breasted mergansers may stay on the wintering grounds until May.

Redhead Ducks

Redhead ducks are named for the distinctive red heads of the males. They are fast flyers and often drop suddenly to the water surface from great heights before diving into deeper waters to feed on vegetation. These ducks also feed on insects, small clams, and snails in shallower waters.

Redheads make their fall migrations through the Potomac region between early October and late December, with their peak concentra-

Lesser scaup, *Aythya affinis*

Redhead, *Aythya americana*

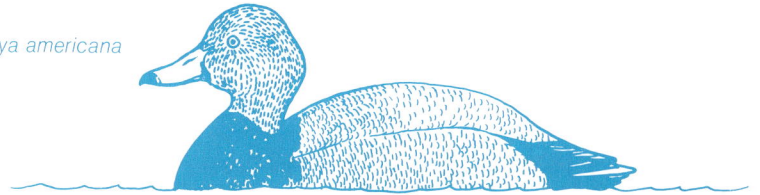

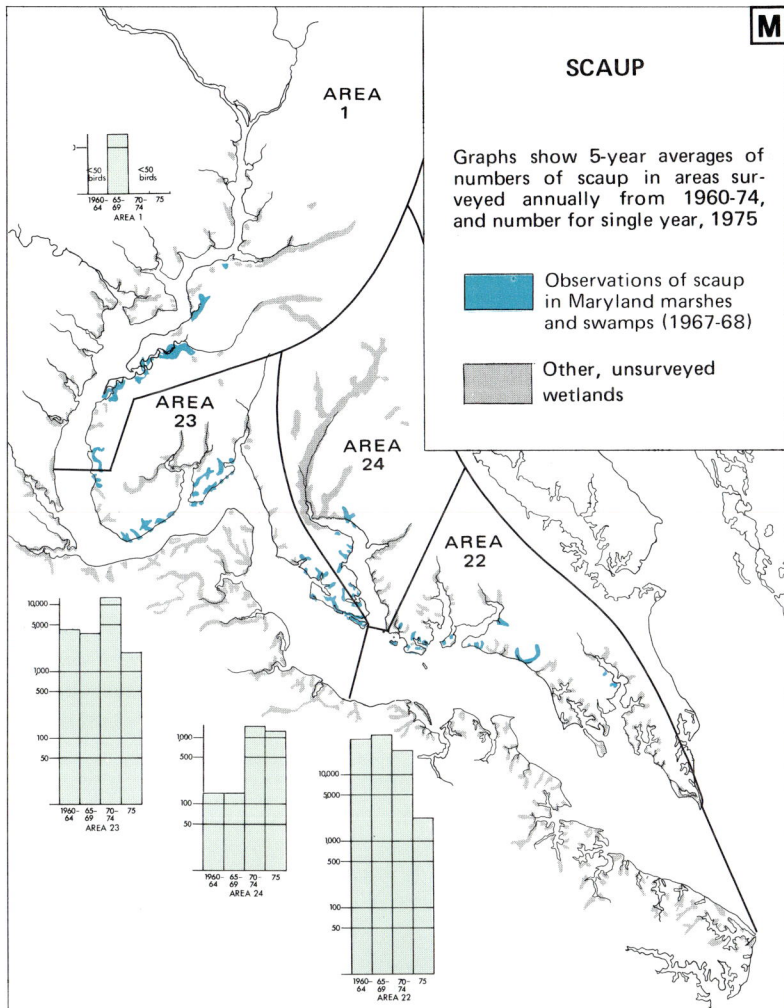

SCAUP

Graphs show 5-year averages of numbers of scaup in areas surveyed annually from 1960-74, and number for single year, 1975

■ Observations of scaup in Maryland marshes and swamps (1967-68)

□ Other, unsurveyed wetlands

Figure 9-3. *Distribution of lesser and greater scaup in the Potomac estuary. Observations of both scaup species in Maryland wetlands in 1967-68 from data reported in Ref. 3. Graphs derived from 1960-75 survey data in Ref. 5.*

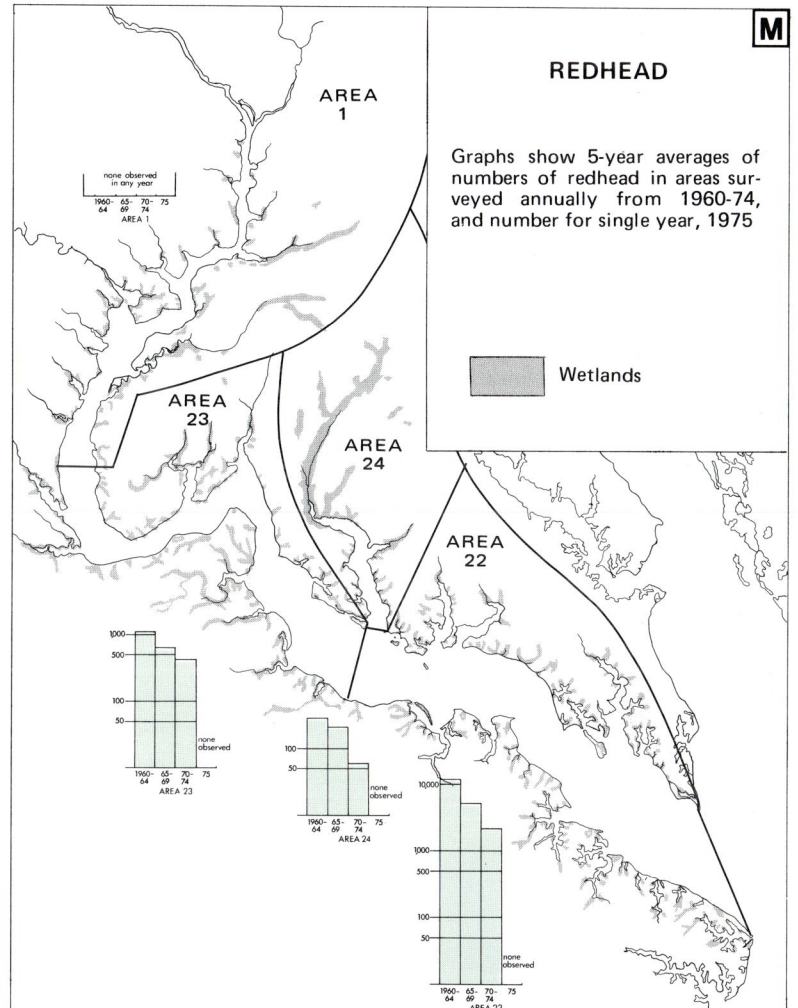

REDHEAD

Graphs show 5-year averages of numbers of redhead in areas surveyed annually from 1960-74, and number for single year, 1975

□ Wetlands

Figure 9-4. *Distribution of redhead in the Potomac estuary. Graphs derived from 1960-75 survey data in Ref. 5.*

tions occurring from mid-November to mid-December. They are generally most numerous along the open coastal areas of the lower estuary (Fig. 9-4) where they feed on redhead grass, widgeon grass, and eelgrass. When they return north, from mid-March through mid-May, redheads visit the slightly less saline waters of the middle estuary, where they feed primarily on wild celery and naiads.[7]

Formerly very common winter visitors to the Potomac area, redhead ducks have rapidly declined in numbers since the 1960s (Fig. 9-4). Few are now found in Maryland, although their breeding numbers on the North American continent are not severely depressed.[10] These ducks have become less abundant throughout their range due to destruction of their breeding grounds in central Canada and the northcentral United States.[10] Their decline in the Potomac in recent years may be related to loss of submerged aquatics on which they are very dependent.[10]

Sea Ducks — Oldsquaw and Scoters

Oldsquaw and several species of scoters are frequently seen transients and winter residents throughout the Chesapeake Bay system. Adult male oldsquaw have distinctive long tails and black and white coloring. The females have short tails and gray and white bodies. Oldsquaw, known for their raucous chattering, are most numerous on the deeper portions of the lower Chesapeake Bay and on the ocean shoreline during their fall and winter migrations. Substantial numbers are also observed in winter on the deeper, offshore waters of the lower Potomac main stem (Fig. 9-5). Their fall migrations extend from late October to mid-December, but the greatest numbers enter the Potomac area between early November and early December.[7] Oldsquaw usually fly in small flocks, but they occasionally form larger flocks in the lower estuary during fall migrations. Only a few individuals visit the upper portion of the estuary.[9] Few oldsquaw are taken by hunters, since their

Oldsquaw, *Clangula hyemalis*

Surf scoter, *Melanitta perspicillata*

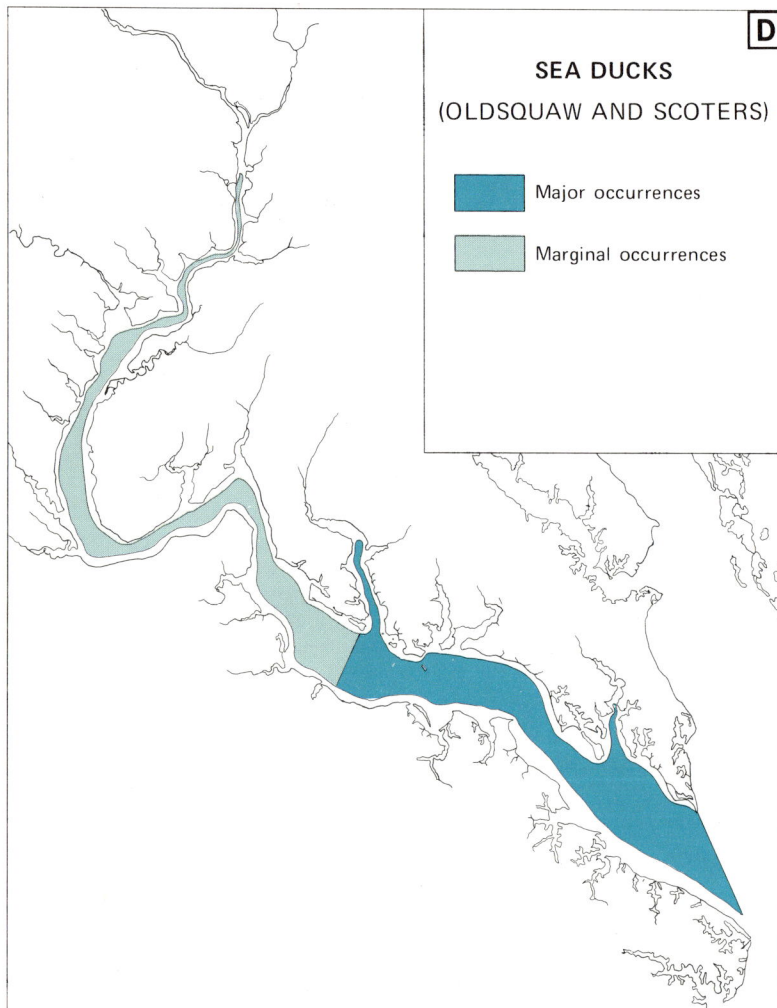

SEA DUCKS

(OLDSQUAW AND SCOTERS)

Major occurrences

Marginal occurrences

Figure 9-5. *General distribution of sea ducks (oldsquaw and scoters) in the Potomac estuary. (Source: Refs. 7, 9)*

meat has a peculiar flavor caused by their diet of clams, worms, crustaceans, and small fish, and since they tend to inhabit offshore waters seldom frequented by hunters.

Scoters are large black birds with large heads. They can often be observed flying just above the water surface, in straight-line formations that undulate with large surface waves. The white-winged scoter is the most commonly encountered species in the Potomac estuary. Like oldsquaw, scoters dive deeply for small aquatic invertebrates. They breed in freshwater areas of northern Canada and Alaska in summer and migrate to the mid-Atlantic coast for winter. Although scoters prefer high salinity waters of the ocean and the lower Chesapeake Bay, moderate numbers are found in the lower mesohaline region of the estuary, and smaller flocks regularly appear in the upper estuary.[7] These birds are occasionally spotted as far upstream as Washington, D.C.[9]

DABBLING (PUDDLE) DUCKS

Dabbling ducks are distinguished from divers by the central position of their legs, which allows them to walk more easily on land. They have an unlobed hind toe, a large wing surface area relative to their body size, an iridescent patch on their wings, and usually, brown eyes. Dabblers feed by tipping "bottoms-up" in shallow waters. Unlike the divers, they take flight with a single bound from the water.

Although dabblers are not as numerous as diving ducks in the Potomac estuary, they are prized by hunters for their excellent flavor. They are generally found close to protected shorelines and along the waterways of freshwater tributary marshes (Fig. 9-6). Estuary-wide distributions appear to have changed over the 1956 to 1976 period, with the highest concentrations being found in the lower estuary during the beginning of the period, but occurring in the upper estuary in more recent years. However, the design of the Marsh and Waterfowl Investigation surveys changed in the early 1960s, and the apparent shifts in the distributions along the Potomac may be related to the change in survey design.[10]

Dabblers venture farther inland than the open-water-oriented diving ducks. Some dabblers eat small animals, but most prefer feeding on rooted aquatic vegetation and plant seeds in shallow waters. Certain species began frequenting grainfields more often to feed when once-plentiful rooted aquatic vegetation began to disappear.

Mallards

A colorful bird, the male mallard is recognized by its characteristic green head, yellow bill, and white neck ring. The female mallard, in contrast, is a comparatively drab duck with a mottled chestnut colored body and darker colored bill and legs.[6] Mallards are relatively abundant transients during spring and fall migrations, and are common winter residents in the Potomac estuary.[7] A portion of the mallard population is nonmigratory, breeding in the Potomac in spring and remaining throughout the summer. The number of mallards found in the Potomac drainage area fluctuates somewhat erratically from year to year. Mates are selected on the wintering grounds or during the migrations to the summer breeding territory north of the Chesapeake Bay.

Shallow fresh and brackish water areas, especially those near large grainfields, are the preferred habitats of mallards (Fig. 9-7). In recent years, highest concentrations have been found in the upper estuary where mallards feed and roost in the many swamps, marshes, and freshwater impoundments. Birds inhabiting freshwater marshes and swamps feed on the seeds of rice cutgrass, Walter millet, smartweeds, bulrushes, common threesquare, arrow arum, and oaks. In open or more brackish waters, they eat the seeds, leaves, stems, or rootstalks of widgeon grass and pondweeds, and the seeds of threesquares, rushes, and a number of grasses.[7] They also feed in grain fields adjoining the marshes. In contrast to many other waterfowl species, mallards appear to have become more abundant over the last 15 years (Fig. 9-7); however, part of the apparent increase may be a reflection of the change in survey design.[10]

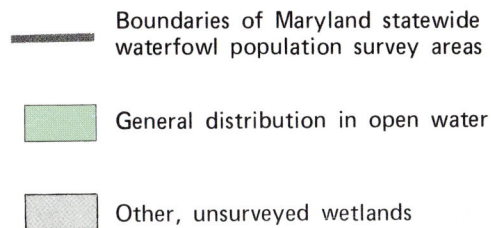

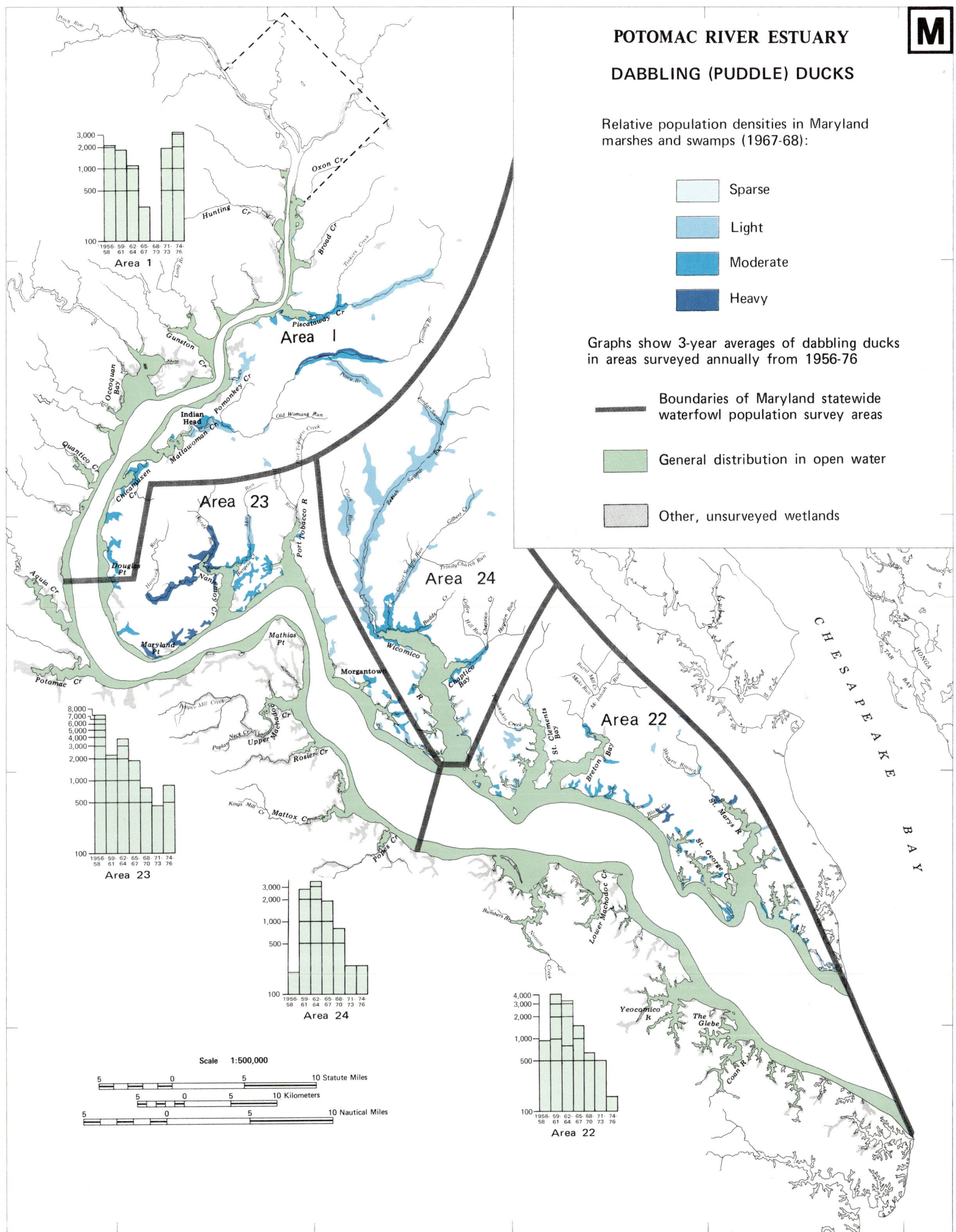

Figure 9-6. *Distribution of dabbling ducks in the Potomac estuary. Relative population densities in Maryland wetlands (1967-68) derived from Ref. 3. Graphs derived from 1956-76 survey data in Ref. 4.*

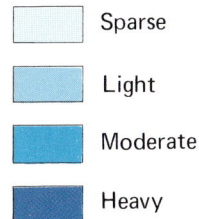

Mallard, *Anas platyrhynchos*

Black duck, *Anas rubripes*

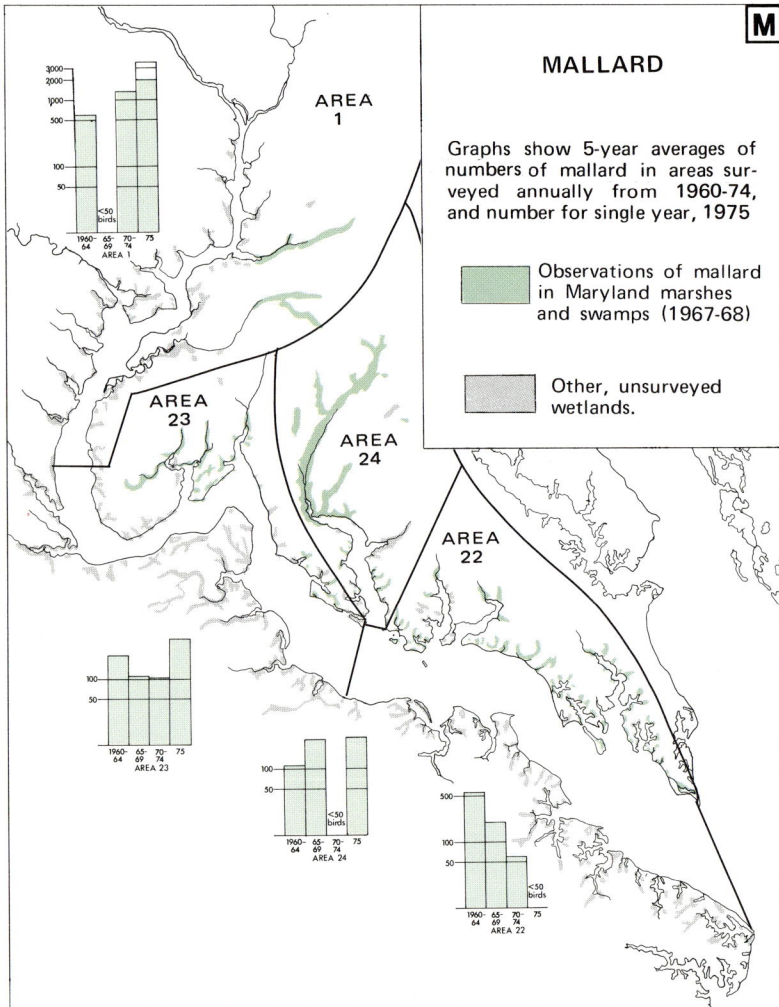

Figure 9-7. *Distribution of mallard in the Potomac estuary. Observations of mallard in Maryland wetlands in 1967-68 from data reported in Ref. 3. Graphs derived from 1960-75 survey data in Ref. 5.*

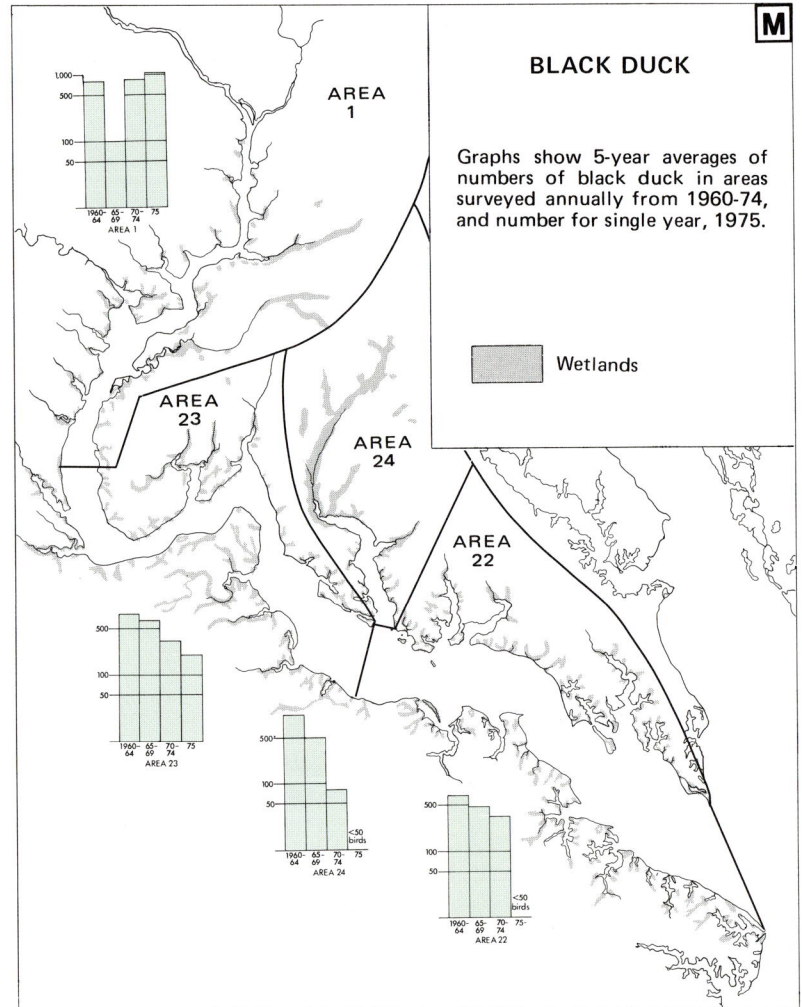

Figure 9-8. *Distribution of black duck in the Potomac estuary. Graphs derived from 1960-75 survey data in Ref. 5.*

Black Ducks

Close relatives of the mallards, black ducks are also commonly found in the Potomac estuary throughout the year. They are both winter and summer residents and spring and fall transients.[7] During their fall southward migrations, these dark colored ducks have been estimated to outnumber all other dabblers combined. Those black ducks that do breed along the Potomac, nest between late March and mid-June. Black ducks usually build their nests in brushy borders of tidal creeks, along marginal thickets of estuarine bays, in marshes, wooded bottomland and swamps, cultivated fields, and freshwater impoundments. They feed primarily on small molluscs, insects, and submerged rooted aquatic plants. Their distribution throughout the estuary has varied, but since 1970, highest concentrations have been observed in the upper estuary (Fig. 9-8). Again, the change in survey design might be a factor influencing reported numbers.[10]

Wood Ducks

The male wood duck is one of the most handsome of all waterfowl. It has a chestnut colored breast with a vertical white stripe along the sides, a white chin, and white cheek stripes. Both males and females have colorful crested heads. They are valued, not only by hunters for their flavor and as trophies, but also by fishermen who use their plumage to make trout flies.

Wood ducks are rare visitors to coastal areas, preferring the seclusion of isolated inland wooded swamps and forested streams where they forage for acorns and berries. They are skillful flyers that can twist and dodge agilely through even the densest stands of trees. Because wood ducks have regularly used artificial nesting boxes provided by man since the 1930s, populations along the Potomac estuary have been enhanced.[2] The most dense populations are commonly found among the freshwater swamps of the upper drainage basin

Baldpate, *Anas americana*

Wood duck, *Aix sponsa*

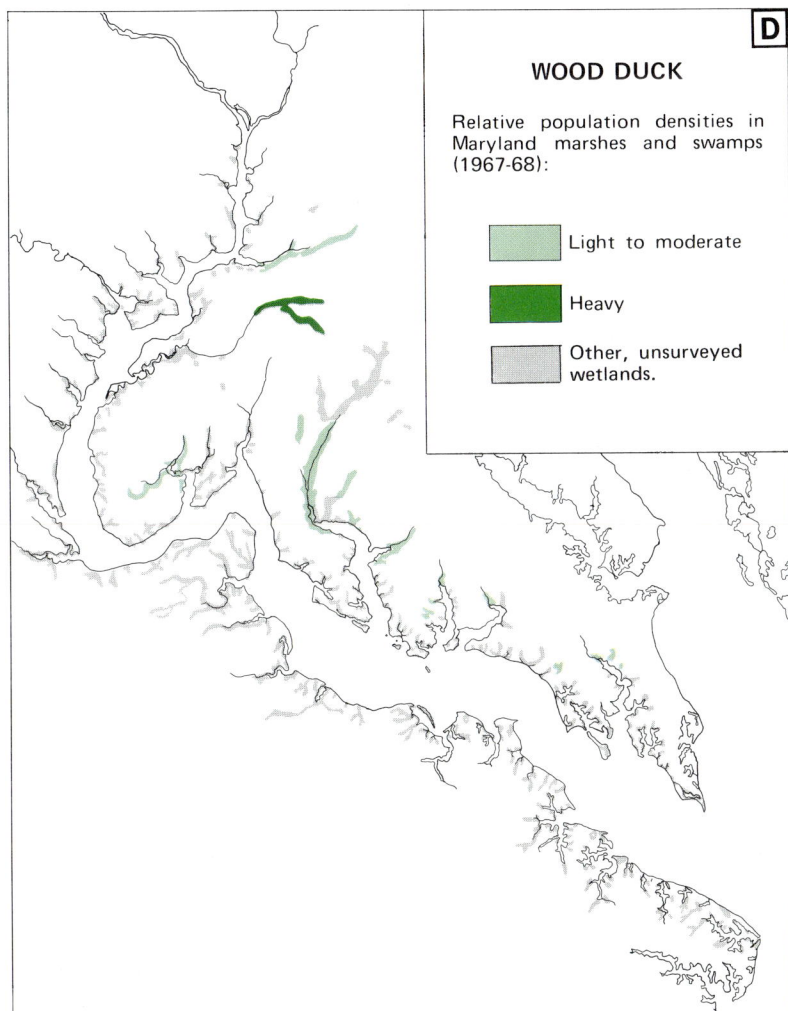

WOOD DUCK

Relative population densities in Maryland marshes and swamps (1967-68):

Light to moderate

Heavy

Other, unsurveyed wetlands.

Figure 9-9. *Distribution of wood duck in the Potomac estuary. Relative population densities in Maryland wetlands (1967-68) derived from Ref. 3.*

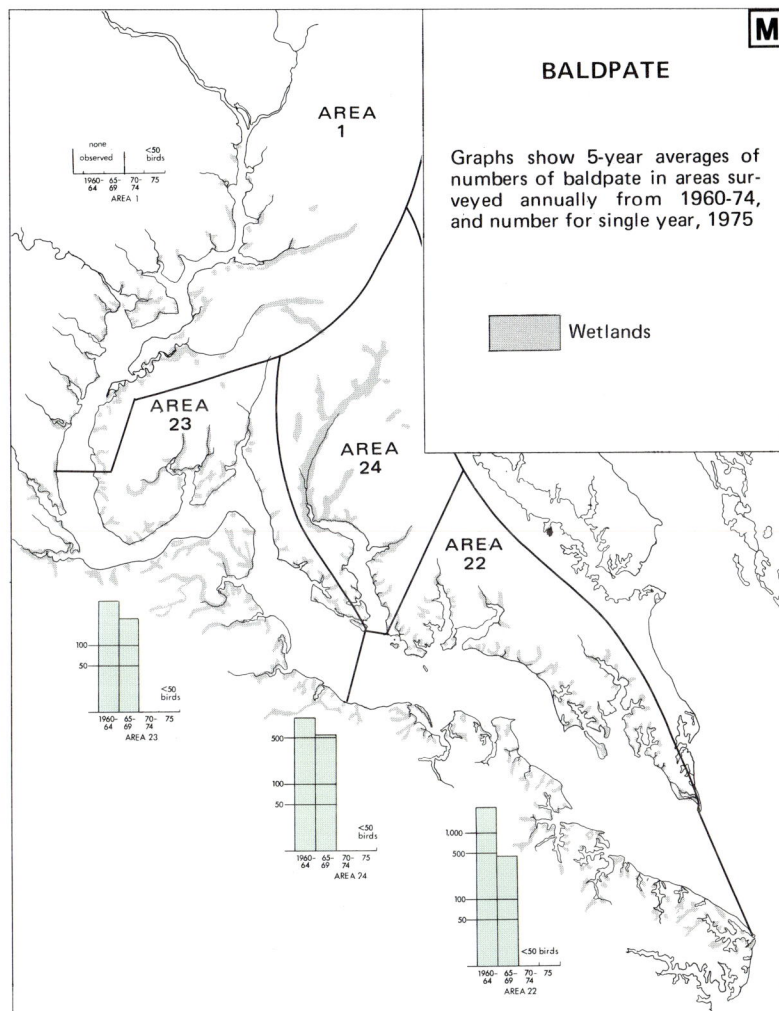

BALDPATE

Graphs show 5-year averages of numbers of baldpate in areas surveyed annually from 1960-74, and number for single year, 1975

Wetlands

Figure 9-10. *Distribution of baldpate in the Potomac estuary. Graphs derived from 1960-75 survey data in Ref. 5.*

during the summer[3] (Fig. 9-9). The majority of wood ducks migrate south of the Chesapeake system for overwintering by early November and start returning in March for nesting.

Baldpates

The baldpate is a small duck with a small head and short bill. The male has a light-colored, almost white, head with a dark green patch extending back from the eye. The wing patches are green; the breast and under parts are snow white. The female is not as conspicuously marked as the male. The head is gray in color, and the underparts are white.[6] Baldpates breed in the central western part of the North American continent and overwinter from the Chesapeake Bay and its tributaries southward.

Baldpates generally arrive in the Potomac estuary from early September to mid-December.[7] In the Potomac, they are most abundant along the shallow oligohaline portions of the estuary and the freshwater tributary marshes and streams (Fig. 9-10), where they feed primarily on wild celery, naiads, common waterweed, sago pondweed, and wild rice. In the saltier waters of the lowermost portions of the estuary, redhead grass and widgeon grass become more important components of their diets.[7] Baldpates sometimes steal deeper-water rooted plants from diving ducks emerging at the surface, and therefore, are often called "robber ducks." Numbers of baldpate in the Potomac have been very low in recent years (Fig. 9-10).

GEESE AND SWANS

Members of the goose family that may be found in the Potomac estuary include the Canada goose and the brant. Swans, the largest of all waterfowl, are represented in the Potomac by only one species, the whistling swan.

POTOMAC RIVER ESTUARY

CANADA GOOSE

M

General open-water distribution

Recent landward extension into grain-producing farmlands.

Canada geese observed along marsh creeks during 1967-68

Graphs show 5-year averages of numbers of geese in areas surveyed annually from 1960-74, and number per single year, 1975

Statewide waterfowl population survey areas

Wetlands

Area 1

Area 23

Area 24

Area 22

CHESAPEAKE BAY

Piscataway Cr

Indian Head

Mattawoman Cr

Chicamuxen

Douglas Pt

Nanjemoy

Maryland Pt

Mathias Pt

Morgantown

Wicomico R

Chaptico Bay

St. Clements Bay

Breton Bay

St. Mary's R

St. George

Port Tobacco R

Gunston Cr

Pomonkey Cr

Occoquan Bay

Quantico Cr

Aquia Cr

Potomac Cr

Upper Machodoc

Rosier Cr

Mattox Cr

Kings Mill Cr

Popes Cr

Neck Cr

Lower Machodoc Cr

Nomini Creek

Bumbers Br

Yeocomico R

The Glebe

Coan R

Old Woman's Run

Oxon Cr

Broad Cr

Hunting Cr

AREA 1
1960-64 | 65-69 | 70-74 | 75
500 / 100 / 50

AREA 23
1960-64 | 65-69 | 70-74 | 75
3000 / 2000 / 1000 / 500 / 100 / 50

AREA 24
1960-64 | 65-69 | 70-74 | 75
2000 / 1000 / 500 / 100 / 50

AREA 22
1960-64 | 65-69 | 70-74 | 75
2000 / 1000 / 500 / 100 / 50

Scale 1:500,000

5 0 5 10 Statute Miles
5 0 5 10 Kilometers
5 0 5 10 Nautical Miles

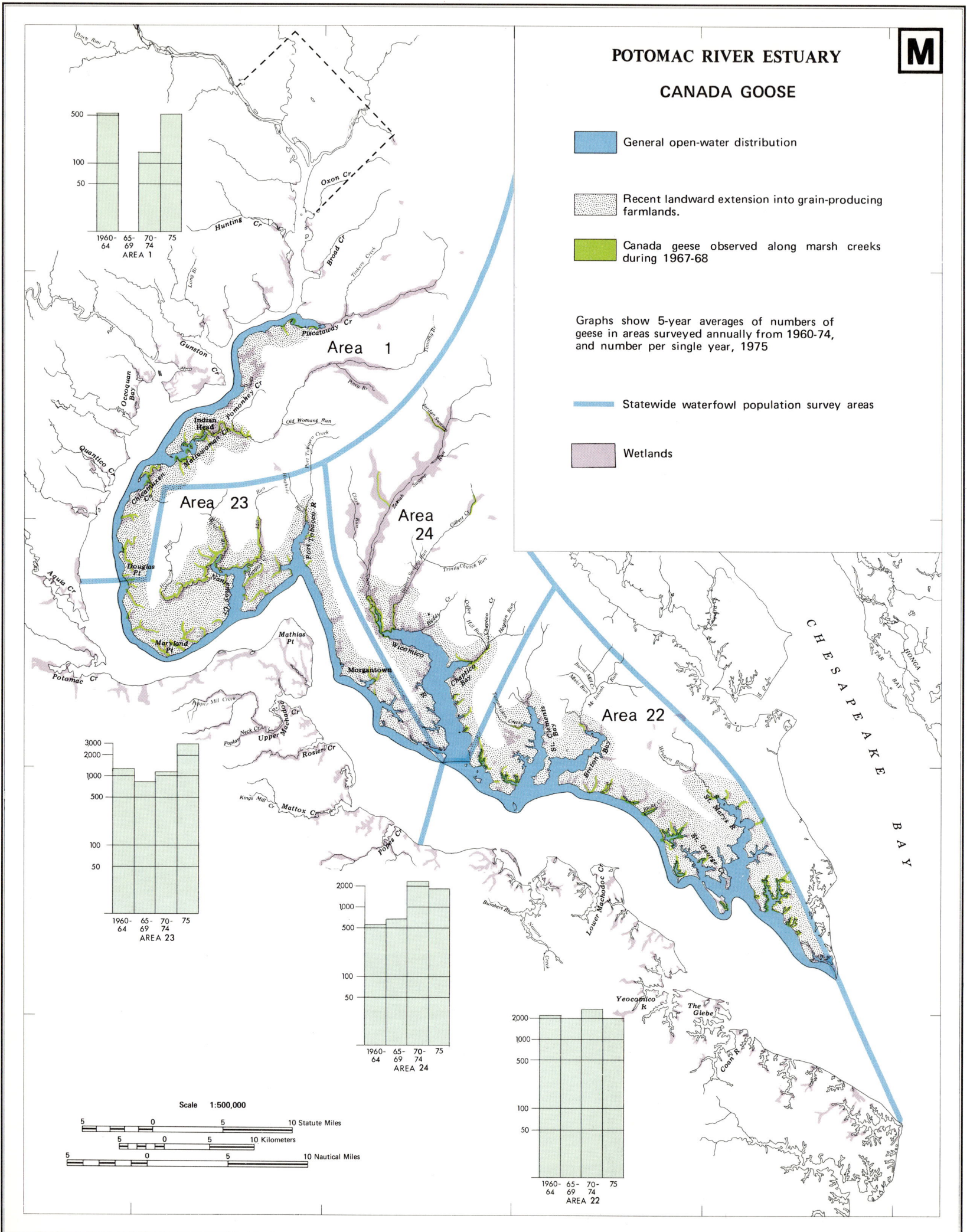

Figure 9-11. *Distribution of Canada goose in the Potomac estuary. Observations of Canada goose in Maryland wetlands in 1967-68 from data reported in Ref. 3. Graphs derived from 1960-75 survey data in Ref. 5.*

Canada goose, *Branta canadensis*

Canada Geese

The large size, conspicuous white sides of their heads, and long black necks of Canada geese distinguish them immediately from all other waterfowl inhabiting the Potomac. Their V-shaped flocks passing noisily overhead herald the beginning of fall and the return of spring all along the Atlantic flyway.

Although most Canada geese within the Chesapeake system congregate along the expansive and well-protected shallow wetlands of Maryland's Eastern Shore, many settle along the shallow shorelines of the mid to lower Potomac estuary (Fig. 9-11). They arrive during fall migrations from their breeding grounds in eastern Canada between September and November. Some continue as far south as South Carolina, stopping only briefly along the Potomac to feed and rest. A small percentage remain in the area until February or March.[7] Canada geese usually mate for life. The male, female, and brood remain together through the winter and return to their northern breeding grounds from late February through April. In the coldest years, some Canada geese remain in the Potomac estuary until late April. Over the estuary as a whole, the Canada goose population has increased since 1960 (Fig. 9-11), with the major increases appearing in the midestuary survey areas.

Both visiting and overwintering birds graze on algae, submerged rooted aquatics (especially widgeon grass), and emergent wetland plants (threesquares in the fresh and slightly brackish marshes and saltmarsh cordgrass and saltgrass in the brackish and salt marshes). Canada geese have adapted well to feeding on grain and agricultural green crops whenever suitable aquatic vegetation is absent or depleted. In recent years, geese have taken to feeding in agricultural fields where automatic harvesters leave corn and soy beans in ample supply. Often, young sprouts of rye, barley, wheat, and hay crops are preferred over corn and soybeans.[7] The ability of geese to adapt may account for their continued abundance within the Potomac area, in contrast to certain species of ducks whose dwindling numbers may reflect the continuing loss of naturally occurring beds of aquatic vegetation necessary in their diets.

Brants

Brants closely resemble their larger and more well-known cousins, the Canada geese. Brants are found only on salt water and almost exclusively inhabit open coastal bayfronts and ocean bays where their chief food items (eelgrass, sea lettuce, and widgeon grass) are present. Because of the limited suitable habitat, brants are rare in the Potomac and only occasionally visit the estuary during their spring and fall migrations.

Whistling Swans

Whistling swans are large, long-necked, pure white relatives of the Canada geese. They are best known for the distinctive high-pitched, musical "whoops" they make as they fly overhead in wedge formation during their migrations. They generally migrate from mid-October through late November to their southern wintering grounds along the east coast of Maryland, Virginia, and North Carolina.[11] Spring migrations northward to the Arctic Islands and Alaska usually occur from the first weeks in March to the last week of April, with the majority of birds heading north from mid-March to the first week in April.[7]

Because they are more awkward on land than are Canada geese, most whistling swans settle on open shoal areas of the estuary to feed. Before taking flight, they must patter quite a distance over the water. Once in the air, however, they are strong and graceful fliers.

During their fall and spring migrations, whistling swans frequent the open shallows of tidal fresh and slightly brackish embayments where submerged vegetation is most abundant. Small flocks or scattered individuals may also settle on interior impoundments and reservoirs. In most freshwater areas the submerged rooted aquatic, wild celery, is their preferred food item. In brackish waters widgeon grass and sago pondweed are eaten, along with small, thin-shelled *Macoma* clams and soft-shell clams. During the extremely harsh winters, when vegetated shoal waters remain covered with ice for long periods of time, whistling swans frequently feed on waste corn and other grains in nearby fields.

Although the majority of overwintering whistling swans in the Chesapeake system settle along the central portion of the Eastern Shore of Maryland, hundreds return every year to the shoal-water areas of the middle and lower Potomac estuary (Fig. 9-12). These flocks represent approximately 5 to 10 percent of the total winter populations in the upper Chesapeake Bay. Very few whistling swans settle in the more heavily populated portions of the upper Potomac, while numbers found in the lower portion of the estuary have generally been high, particularly in recent years (Fig. 9-12).

Other Water-Oriented Birds

American Coots

The pure white bills and distinctive white borders along the inner wing edges of these slate-black members of the rail family make American coots easily distinguishable when they are flying. Although American

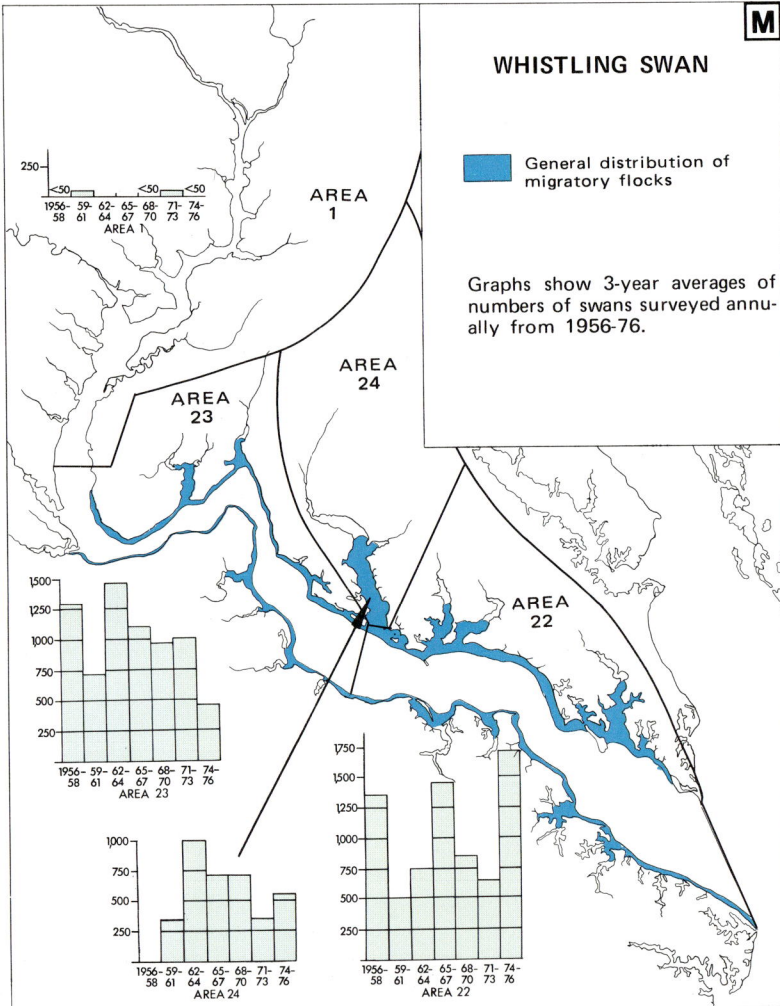

194

M

WHISTLING SWAN

General distribution of migratory flocks

Graphs show 3-year averages of numbers of swans surveyed annually from 1956-76.

Figure 9-12. *Distribution of whistling swan in the Potomac estuary. Graph derived from 1956-76 survey data in Ref. 4.*

coots are not taxonomically true waterfowl (which are of the Order Anseriformes), they are often grouped with ducks, geese, and swans because of their similar seasonal migrations and their close associations with waterfowl populations. American coots are the most aquatic of the rail family (the Rallidae), which also includes gallinules and rails. Although they do not have webbed feet, their toes have expanded pads, or lobes, which aid them in swimming. Coots swim buoyantly and on an even keel, nodding their heads rhythmically as they go. They are sluggish flyers, however, and when taking flight, must run quite a long distance over the water before getting aloft. Although they usually are among the more plentiful of all waterbirds, their very wild flavor makes these birds unpopular with most hunters.

American coots are found in the Potomac estuary as either transients during their early fall and spring migrations or as winter residents. Their southward migrations usually extend from late September to late December. Peak numbers of American coots arrive in the Potomac between October and the end of November.[7] At one time, the central Potomac estuary provided an important overwintering habitat for large numbers of these birds, while few were found as far inland as Washington, D.C.[7][9] Since the late 1960s, however, numbers of American coots visiting the Potomac estuary have declined dramatically. Very few birds were sighted during winter aerial surveys from 1971-76.[10]

American coots feed on submerged rooted aquatic plants, but are also fond of algae. In the oligohaline and lower mesohaline waters of the middle Potomac, coot feed chiefly on the leaves, stems, and rootstalks of redhead grass and widgeon grass,[7] as well as the leaves of Eurasian water milfoil. In other areas, they eat the foliage and rootstalks of saltgrass; the seeds of arrow arum, swamp rose, and greenbrier; and various insects.[7]

Great Blue Herons

Less common, but among the most conspicuous of all birds frequenting the Potomac wetlands, are the graceful wading birds: the herons, egrets, and bitterns. The largest of these birds is the great blue heron, usually seen standing motionless in, or wading deliberately through, quiet shallow waters in search of prey. When disturbed, this distinctive four-foot high bird escapes pursuit with a few slow powerful beats of its large wings. The great blue heron is impressive in flight, with its neck folded back on its shoulders, and its long legs stretched out behind.

Great blue herons are predators, eating small snakes, insects, mice, frogs, salamanders, young fish, crabs, and at times, wetland birds.[12] They fish by waiting patiently for prey to come within striking range or by walking slowly through the water. Occasionally, these birds will drop down with outstretched wings onto the surface of deeper open waters and then feed as they float. Blue herons are adaptable birds. They are equally at home on small streams, upland meadows, crop fields, salt marshes, mud flats, and shallow coastal bays. They can be found in almost all regions of the Potomac estuary where quiet shallows exist.

Nesting occurs in mid-March to mid-July in colonies ranging in number from a few to many pairs of birds.[12] Their large flat nests, averaging 3 feet in diameter, are constructed of large sticks and lined with soft grasses and other materials. They are usually built in isolated patches of woodland swamps, on islands, in the tops of tall trees, or on other structures at the water's edge. An exceptionally large heron rookery, holding 750 nests, is located on Nanjemoy Creek.[13] Every spring these majestic birds return to this site to nest, then depart to migrate southward in midsummer with their broods. A few overwinter on the Potomac, and a small number even wander northward.

Great blue herons are legally protected by the Federal Migratory Bird Treaty, but their large size makes them easy and tempting targets for thoughtless shooters. In addition, nesting colonies are often broken up by intruders.

Bald eagle, *Haliaeetus leucocephalus*

Birds of Prey

Bald Eagles

The bald eagle is primarily found along estuaries of the Atlantic Coast from Maine to Florida. It is one of six officially designated endangered species in the Potomac estuarine region.[14] In the Potomac region, bald eagle nests have been found along marshes and secluded open water areas from Gunston Cove to the mouth. Figure 9-13 presents distributional data collected during several different surveys over the period 1967 to 1976. Because a number of different observers were involved and the surveys covered different periods of years, all the data are not consistent. However, in aggregate, they do represent general distributions.

Most of the eagles inhabiting Potomac wetlands are permanent residents that breed along the estuary. Small numbers, however, may wander northward in summer after nesting. Bald eagle populations declined in recent times, primarily in response to the loss of suitable, undisturbed nesting habitat and the deleterious effect of derivatives of the insecticide DDT on reproductive success. However, their productivity in the Potomac estuary since 1970 has shown improvement. In 1970, only 4 young were known to have been hatched along the Maryland shore of the Potomac estuary; 9 were reported hatched in 1976.[15] While this improvement is significant, the instability of production from 1970 to 1976 (ranging from 2 to 12 young hatched per year) indicates that the population has not fully recovered from factors causing the low levels observed in the late 1960s. However, the increase in productivity does give some reason for cautious optimism for population recovery.[15] Federal law prohibits the destruction of bald eagles or their nests.

Bald eagles primarily eat fish, but also consume injured water birds and small aquatic animals such as turtles. Eagles are also scavengers and feed on carcasses, especially those of road-killed animals.

Ospreys

The osprey, or "fish hawk" as it is often called, uses estuarine areas for feeding and nesting. As its name suggests, the fish hawk's principal food item is fish (especially eel, catfish, and menhaden) which it captures with its sharp-taloned feet.

Breeding pairs of ospreys build large woody nests, similar in structure to those of the bald eagle, along coastal marshes and open water embayments. Tall dead or dying trees bordering estuarine shorelines are the favored nesting sites. In recent years, increasing numbers of ospreys have nested on offshore duck blinds, on channel markers, and on other man-made structures.[16] [17]

The osprey is an abundant summer resident of Potomac wetlands. Breeding pairs nest along the mid to lower reaches during the summer (Fig. 9-13), but move south for the winter.[18] Over 200 active nests have been observed between 1970 and 1973 along both the Maryland and Virginia sides of the Potomac from near Maryland Point to the mouth.[15] [17] [18] Primary concentrations occurred along the Wicomico River and St. Clements and Breton bays. Like the bald eagles, the ospreys have suffered from a lack of suitable nesting sites, but are adapting to alternate sites such as duck blinds and navigational markers. The cumulative effects of pesticides and other chemicals such as DDT and PCB have had a subtle but definite impact on osprey populations.[9]

Osprey, *Pandion haliaetus*

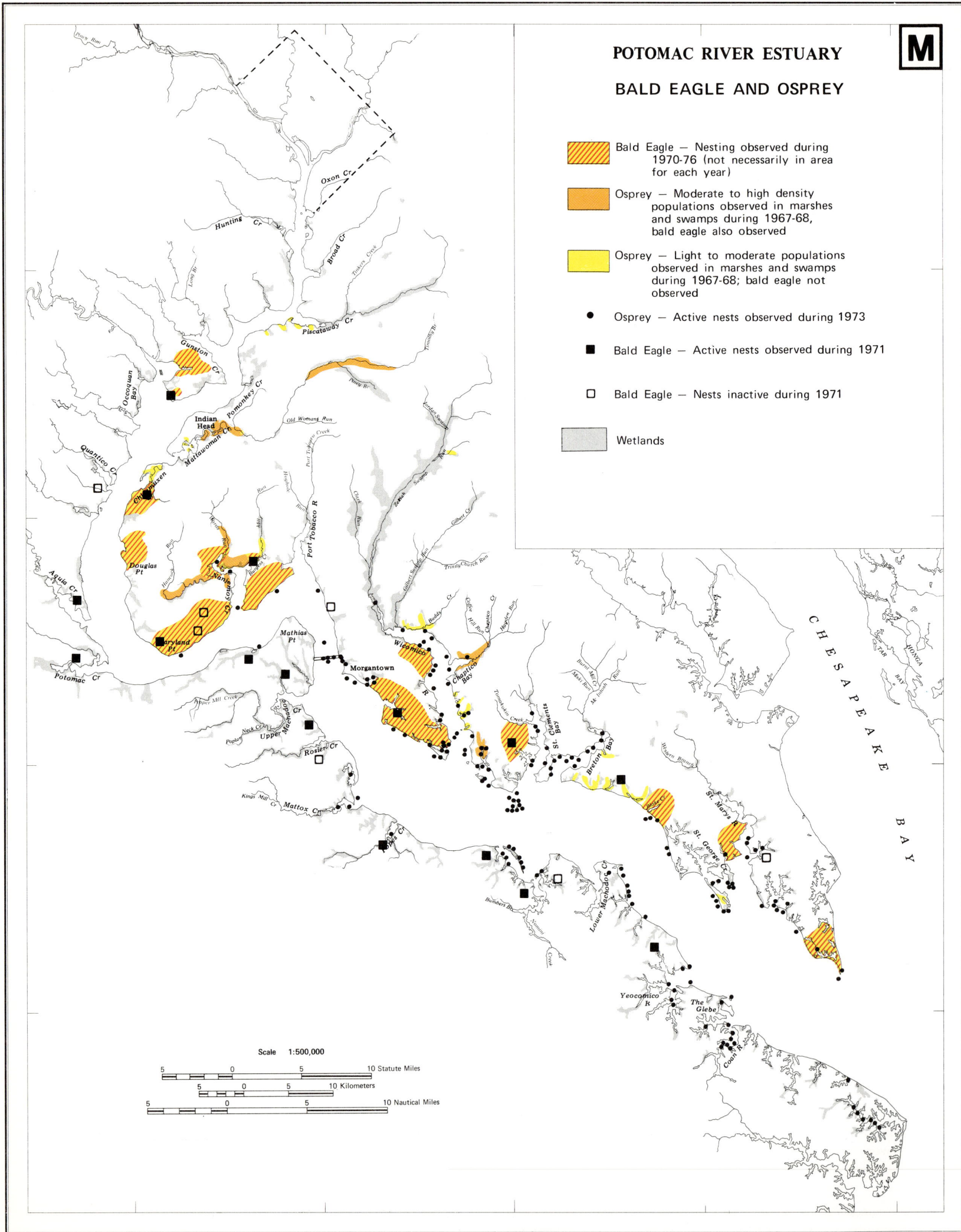

POTOMAC RIVER ESTUARY

BALD EAGLE AND OSPREY

M

Bald Eagle — Nesting observed during 1970-76 (not necessarily in area for each year)

Osprey — Moderate to high density populations observed in marshes and swamps during 1967-68, bald eagle also observed

Osprey — Light to moderate populations observed in marshes and swamps during 1967-68; bald eagle not observed

● Osprey — Active nests observed during 1973

■ Bald Eagle — Active nests observed during 1971

□ Bald Eagle — Nests inactive during 1971

Wetlands

Scale 1:500,000

Figure 9-13. *Distribution of bald eagle and osprey and their nesting sites in the Potomac estuary. (Sources: Refs. 3, 15)*

MAMMALS

Mammals are a significant, although sometimes overlooked natural resource of the Potomac estuary. Many of the well-known mammals in and along the Potomac are only visitors to its wetlands, attracted there by an abundant food supply and an environment that is relatively undisturbed by man. These include big game (primarily white-tailed deer), upland game (e.g., rabbits), a number of furbearers (e.g., the gray and red fox, raccoons, and mink), small rodents (e.g., mice, rats, voles, and lemmings), and small insectivores (e.g., shrews and moles) (see Appendix Table 8).[1][3][9] However, four common species of furbearers are truly aquatic, living along the banks or in the wetlands of the Potomac estuary. They are the muskrat, nutria, beaver, and river otter. All four are trapped and sold in Maryland for their pelts and occasionally meat.[1]

Muskrat, *Ondatra zibethicus*

Muskrats

Muskrats are probably the most ubiquitous and abundant of the aquatic furbearers inhabiting the Potomac estuary basin. Muskrat pelts alone account for over one half of the total income from trapping in Maryland.[19] They are found in almost all wetland areas (Fig. 9-14), including the extensive brackish marshes of the middle and lower estuary.[9] These excellent swimmers feed primarily on marsh vegetation such as common threesquare, cattail, and arrowhead.[20] At times, they will also consume filamentous algae,[20] small invertebrates, insects, birds, and fish. Muskrats build large, above-water houses of plant stems and other marsh vegetation and often dig burrows deep into creek and stream banks.

Muskrats are prolific rodents that seem to have an ability to withstand intense trapping pressure.[6] In Maryland, during their extremely long breeding season from late January through October, they may have one litter every 120 days with an average of 4 to 5 young per litter.[20] The young are weaned quickly and begin to roam and feed on their own when they are only two to three weeks old.[20]

Nutria

Nutria, large aquatic rodents that somewhat resemble guinea pigs, were imported into Maryland from South America in the 1940s for fur farming. During this period, as some were turned loose or escaped, they became established in wetland areas. Since nutria are not well adapted to the cold, it was thought at first that their populations were kept in check during severe Maryland winters.[21] Since the mid-1960s, however, they have increased in some areas of the Eastern Shore to nuisance levels. They are prolific breeders, producing two and sometimes three litters per year, each with seven to twelve young. The young can travel

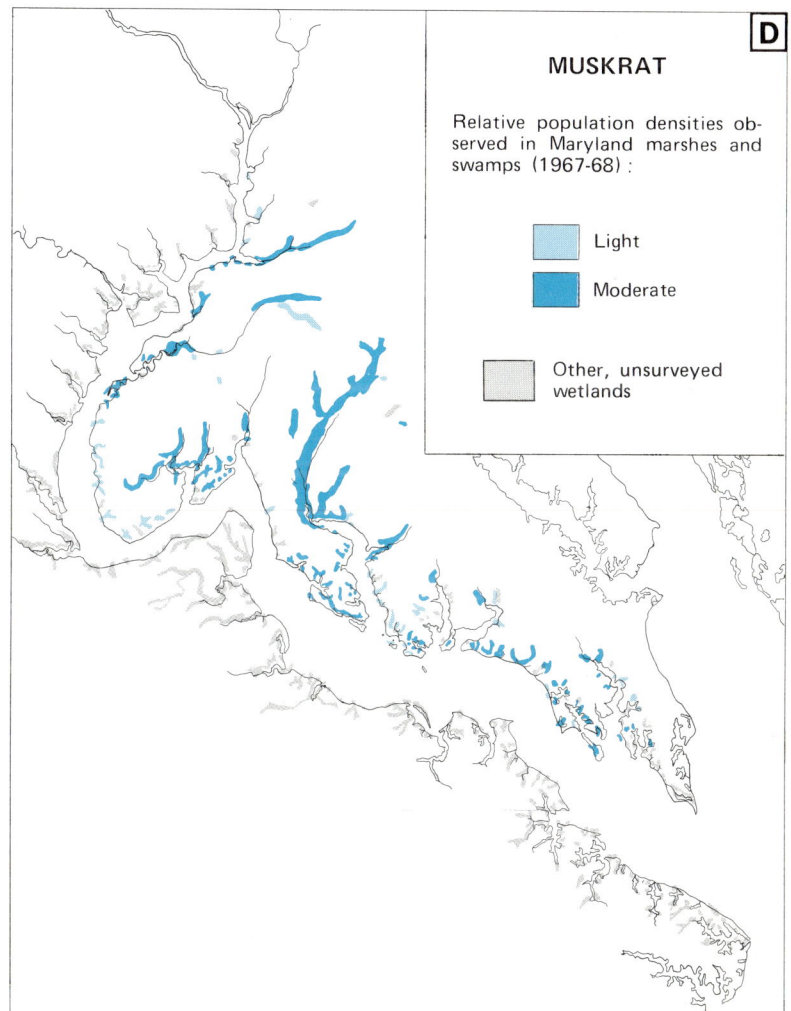

Figure 9-14. *Relative population densities of muskrat in Maryland wetlands of the Potomac estuary (1967-68), derived from survey data in Ref. 3.*

with the mother from birth, attached to nipples conveniently located on her back so they can suckle while she swims. They become sexually mature when they are only five months old. Nutria compete with some native furbearers (muskrat and beaver) for food and habitat, and they may damage dikes, marshes, farm ponds, and commercial croplands (especially corn). In 1970, the Maryland State Legislature asked the Fish and Wildlife Administration to conduct studies of their life history for future management.[21]

Nutria are not currently a nuisance along the Potomac estuary. Only small populations have become established along the fresh and brackish coastal marshes of the middle estuary (Fig. 9-15). They live in banks along the river and subsist on reeds, grasses, rushes, cattails, sedges, and other aquatic plants.

Nutria, *Myocastor coypu*

Beaver, *Castor canadensis*

198

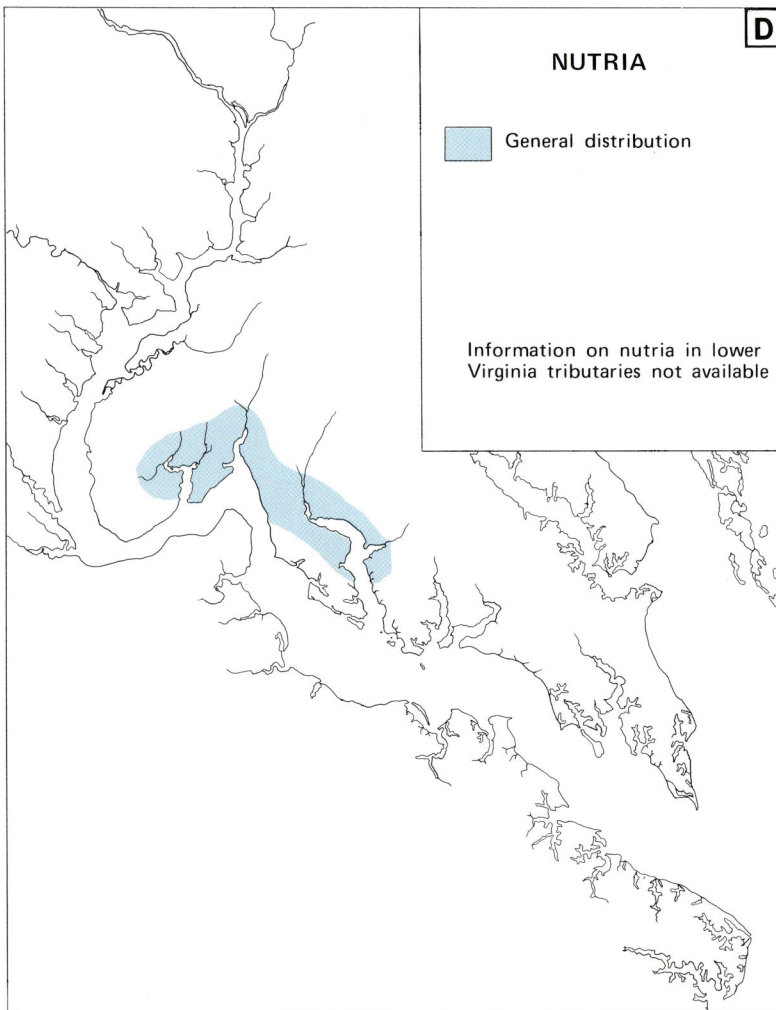

Figure 9-15. *Current distribution of nutria in the Potomac estuary.* (Source: Ref. 22)

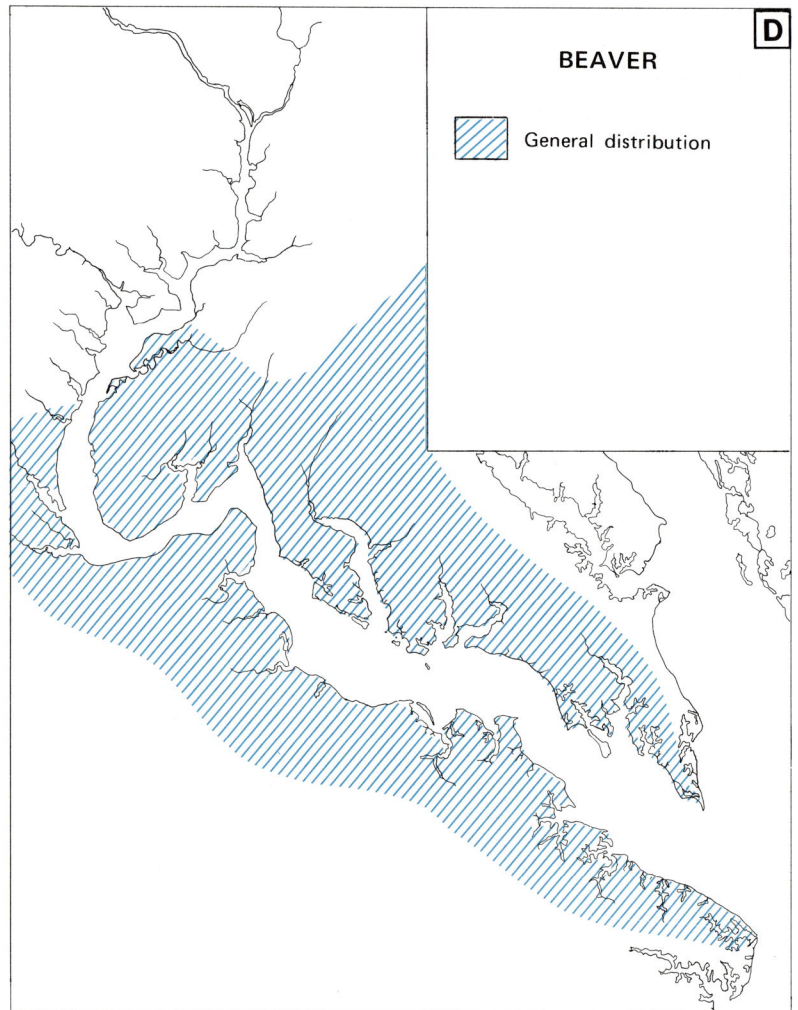

Figure 9-16. *Current distribution of beaver in the Potomac estuary.* (Source: Ref. 21)

Beaver

The largest rodents in North America, beaver were once plentiful in the Potomac estuary. However, they virtually disappeared from Maryland by 1800 because of heavy trapping and land-use changes.[23] Today, due to migration from adjacent states, restocking programs, and protective laws, beaver populations are steadily growing. Current trapping levels yield a small income for Maryland trappers.

Beaver live in small colonies of about six animals each along most of the Potomac from Indian Head to the mouth[20] and prefer unpopulated areas (Fig. 9-16). They build lodges in the slow-moving backwater areas near the inland heads of creeks and streams. In these areas, their principal food items — the bark, twigs, and leaves of birch, willow, and poplar trees — are most abundant. This strictly vegetarian diet also includes small amounts of cattails, grasses, sedge roots, and

some other freshwater aquatic plants. During the fall, beaver stockpile young green tree branches to subsist on throughout the winter.[20]

Beaver are believed to mate for life. Their breeding season is short, from mid-January through February. Following a gestation period of approximately 120 days, a female beaver produces its only offspring of the year — a litter of one to eight young.[20] When a beaver is two years old, it is driven from the lodge in spring to fend for itself.

Considered the "gypsies" of furbearing mammals, beaver move from stream to stream as the local supply of food is reduced. The average beaver fells 200 to 300 trees a year for food and shelter. It is debatable whether beaver are a benefit or a liability to some areas. Although the numerous lodges and dams found along the Potomac may in some cases effectively block the upper end of the spawning pathways of certain anadromous and semianadromous fishes,[24] the back-water

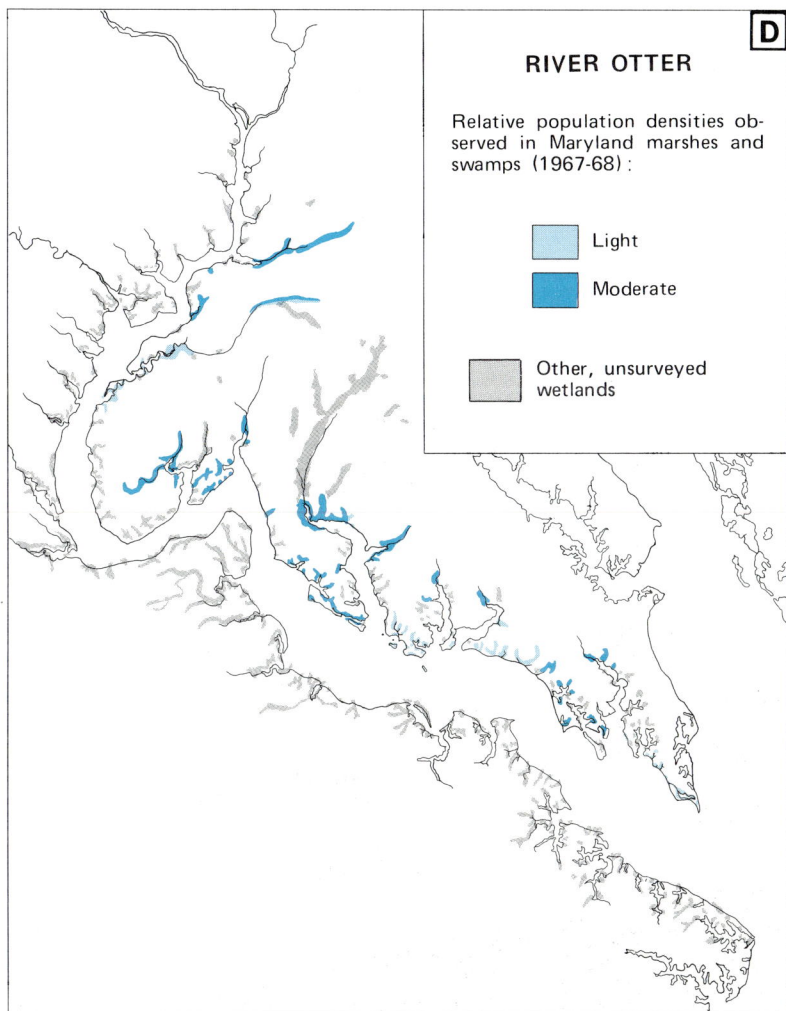

Figure 9-17. *Relative population densities of river otter in Maryland wetlands of the Potomac estuary (1967-68), derived from survey data in Ref. 3.*

They breed in early spring, producing only one litter of one to four young per year. The number of otter harvested from marshes along the Maryland portion of the Potomac estuary is second only to that harvested from marshes bordering the main stem of the Chesapeake Bay. In fact, St. Marys County produces the second greatest number of otter pelts of all the counties in Maryland.[25]

River otter, *Lutra canadensis*

ponds created by these structures provide acres of new habitat suitable for certain waterfowl and fishes.

River Otters

River otters are large aquatic furbearers whose pelts are quite valuable. They occur in low to moderate numbers along the tidal freshwater streams, rivers, and marshes of the Potomac estuary (Fig. 9-17). Otter tend to be shy and inconspicuous, utilizing cover along stream banks and raising their young in dens. They frequently take over muskrat lodges for shelter.

River otters are carnivorous, feeding primarily on fish, shellfish, frogs, turtles, and other small aquatic animals. They occasionally eat ducks and small mammals. Otter are deft swimmers and often make slides down steep mud banks for play or for quickly entering the water.

Chapter 9 — References

1. Metzgar, 1973
2. Studholme et al., 1965
3. Maryland Department of Natural Resources, 1967-1968
4. Maryland Department of Natural Resources, 1956-1976
5. Maryland Department of Natural Resources, 1960-1975
6. Trippensee, 1953
7. Stewart, 1962
8. Perry, Andrews, and Beaman, 1976
9. Wass, 1972
10. Stotts, 1978
11. Pough, 1951
12. Caulk, 1965
13. Corddry, 1978
14. Maryland Department of Natural Resources, 1977a
15. Taylor, G. J., 1977
16. Tyrrell, 1936
17. Henny, Smith, and Stotts, 1974
18. Wiemeyer, 1971
19. Maryland Department of Natural Resources, 1978
20. U.S. Department of the Army, 1976
21. Goldsberry, 1971
22. Willner, Chapman, and Pursley, in press
23. Goldsberry, 1973
24. O'Dell, King, and Gabor, 1973
25. Mowbray, Chapman, and Goldsberry, 1976

Chapter 10

The Fishery

The Fishery

From earliest times, people living along the Potomac estuary have reaped great harvests from its waters. Early settlers and the Indians before them found the Potomac and its tributaries teeming with herrings and shad in the spring and inhabited by many other varieties of finfish and shellfish year-round. Today, the Potomac contributes substantially to the seafood harvest of the Chesapeake Bay area. From 1960 to 1976, striped bass landings in the Potomac estuary accounted for 20 to 25 percent of the total Chesapeake Bay landings by weight. The Potomac oyster catch made up 10 to 25 percent, and alewife landings accounted for 20 to 40 percent of the Bay catch during the same period.[1]

Seafood has been commercially harvested in the Potomac estuary from colonial times to the present. This commercial harvest has contributed to the decline of some species that formerly were collected in huge numbers. The Atlantic sturgeon, for example, was once important in the fishery of the Potomac estuary but has virtually disappeared, chiefly due to overexploitation.[2] Sportfishing, although also pursued in colonial times, has only within the last decade reached a level of intensity where the sports harvest of some species may equal or exceed commercial catches.[3] The possible deleterious effects of overexploitation are being countered by fisheries management programs that regulate fishing methods and catch size.

Commercial fishing effort and success are strongly influenced by fluctuations in target populations that are caused by natural phenomena as well as by man's activities. Commercial catches may be dominated by different species at different times as some populations decline and others prosper. However, the use of commercial landings to determine population sizes of different species is complicated by the fact that fishing intensity is very responsive to market demand. The same amount of a highly prized fish might be landed in two different years despite large changes in natural abundance, because high prices make it profitable to expend extra effort to catch the species in years when populations are smaller. Thus, commercial catch records may give some indication of changes in abundance of sought-after species, but the catches are also influenced by economic factors.

Because commercial fishing has been, until recent times, the principal means of harvesting valuable species, it has been regulated to a greater degree than sportfishing. Proper fisheries management has required the keeping of accurate catch records. For many regions of the United States, such records extend back many decades. However, separate records for the Potomac estuary extend back only to 1960.* These records are the basis for the following discussion of commercial fishing in the Potomac, which includes the locations where various species are fished, the seasonal changes in catch of the species, and the long-term changes in their abundance. The discussion of the sport harvest is much briefer than that of the commercial fishery because sportfishing is largely unmanaged and unmonitored, and data are much more limited.

FISHERIES MANAGEMENT

Although Maryland legally owns the Potomac to the low tide line along the Virginia shore (see Folio Map 1 for state boundaries), a compact drawn up between Maryland and Virginia in 1958 decrees that the commercial fisheries in the main stem of the Potomac estuary will be regulated through the Potomac River Fisheries Commission (PRFC) with members appointed from Virginia and Maryland. The PRFC is responsible for the management of oyster and clam beds, the issuance of licenses for nonstationary fishing gear, and the issuance of permits for fixed fishing devices such as anchor gill nets, pound nets, and crab pots. Commercial fishermen harvesting either finfish or shellfish within the jurisdictional boundaries set up by law are required to report their monthly catches to the PRFC offices in Colonial Beach, Virginia.[4] Since 1960, these reports have been forwarded to the National Marine Fisheries Service of the National Oceanic and Atmospheric Administration (NOAA), and the data appear as separate Potomac estuary catches in monthly and annual reports published by NOAA. Prior to 1960, Potomac landings of most species were included in Chesapeake Bay totals for Maryland and Virginia. Reports and summaries of landings are still separated into those from the Maryland side and those from the Virginia side. Jurisdiction over the fisheries and reporting of the catches within the tributaries of the Potomac are maintained in the respective states.

Boundaries of the PRFC jurisdiction are shown in Fig. 10-1 and Folio Map 6. For reporting purposes, the Potomac main stem is further divided into four commercial zones, which are also shown in Fig. 10-1. Monthly catch records by zones provide a basis for examining seasonal changes in the patterns of commercial harvest in the Potomac, and annual records provide some insight into long-term trends in population abundance.

Sportfishing in the Potomac estuary main stem is also under PRFC jurisdiction. However, as noted earlier, it is not as rigidly controlled as commercial fishing. The majority of existing sportfishing regulations for tidal waters deal with size limits. Notable exceptions are regulations that control the striped bass sport catch and a daily limit on the numbers of crabs that may be taken by sportfishermen.[4]

FISHING METHODS AND FISHING GROUNDS
Commercial Finfishing

Historically, many methods for commercially harvesting finfish have been used in the Potomac. Drift, stake, and anchor gill nets; haul seines; fyke nets and hoop nets; pound nets; fish trot lines; and fish or eel pots have all been employed. In more recent years, however, the principal fishing devices used in the main stem have been stake and anchor gill nets, pound nets, and eel pots. Stake gill nets are used mostly in the upper half of the estuary above Mathias Point (Fig. 10-1). Nets are put out in early spring (March) to catch the ascending schools of

---

* Records of the catch of a few species extend back before 1960.

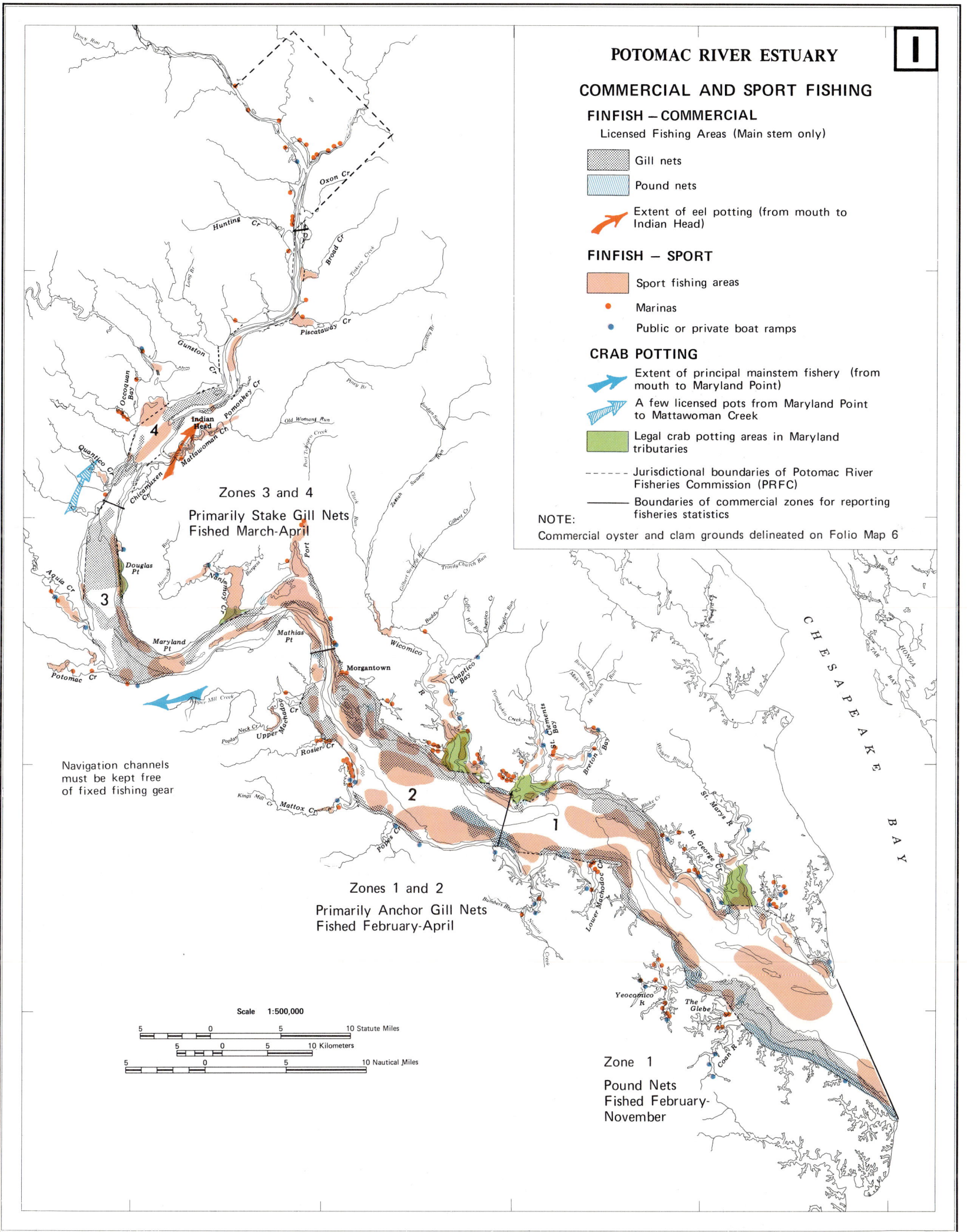

POTOMAC RIVER ESTUARY

I

COMMERCIAL AND SPORT FISHING

FINFISH — COMMERCIAL

Licensed Fishing Areas (Main stem only)

Gill nets

Pound nets

Extent of eel potting (from mouth to Indian Head)

FINFISH — SPORT

Sport fishing areas

● Marinas

● Public or private boat ramps

CRAB POTTING

Extent of principal mainstem fishery (from mouth to Maryland Point)

A few licensed pots from Maryland Point to Mattawoman Creek

Legal crab potting areas in Maryland tributaries

- - - - Jurisdictional boundaries of Potomac River Fisheries Commission (PRFC)

——— Boundaries of commercial zones for reporting fisheries statistics

NOTE:

Commercial oyster and clam grounds delineated on Folio Map 6

Zones 3 and 4

Primarily Stake Gill Nets
Fished March-April

Navigation channels must be kept free of fixed fishing gear

Zones 1 and 2

Primarily Anchor Gill Nets
Fished February-April

Zone 1

Pound Nets
Fished February-November

Scale   1:500,000

5   0   5   10 Statute Miles
5   0   5   10 Kilometers
5   0   5   10 Nautical Miles

Figure 10-1. *Commercial and sport fishing zones, gear locations, and fishing areas. (Sources: Refs. 5, 6, 7, 8)*

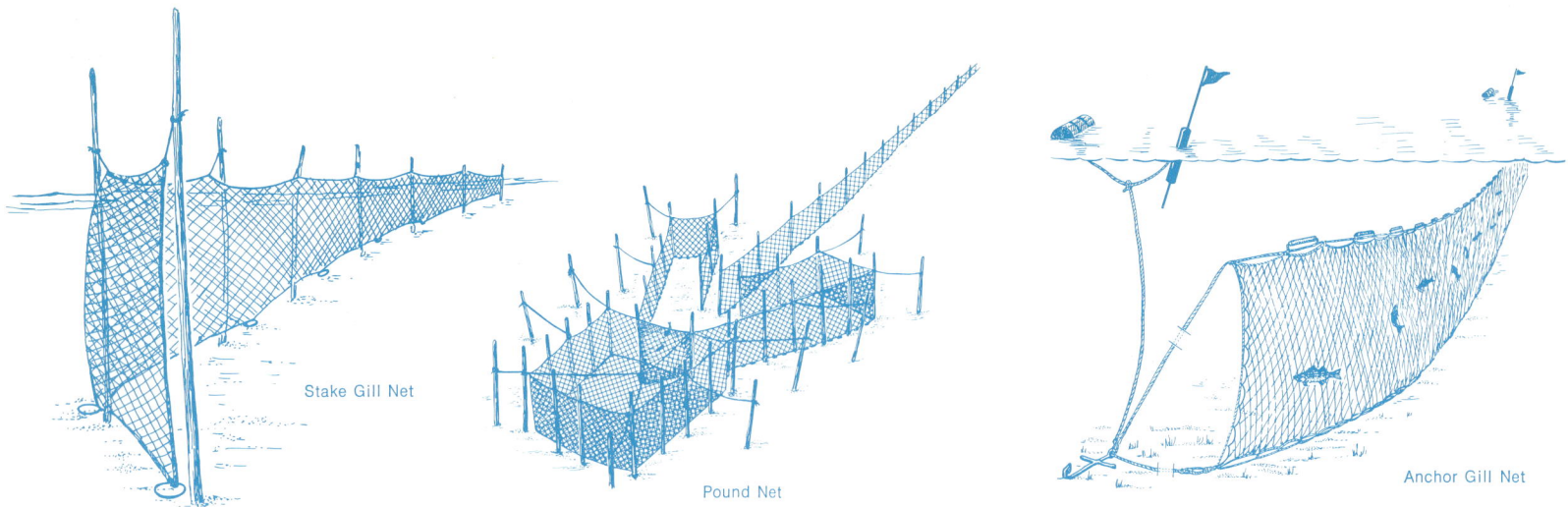

Stake Gill Net

Pound Net

Anchor Gill Net

anadromous fish (primarily striped bass and shad) and are usually dismantled by the end of April. In the lower half of the estuary, anchor gill nets are the primary fishing devices used. They are also set to coincide with the spring anadromous fish runs, but are fished earlier (starting in late February) to catch the schools as they first enter the estuary. They remain set until the end of April to catch anadromous fish as they leave after spawning.

Pound nets are primarily placed in the lower estuary (see Fig. 10-1), most being set along the Virginia shore off Nomini Bay and the Yeocomico River, and from The Glebe to the Chesapeake Bay. Pound nets and stake gill nets, which cannot be set efficiently in deep waters, are generally found in shoal areas that are less than 18 feet [5.5 meters (m)] deep. Pound nets are fished most of the year except in winter during bad weather. They are less selective than gill nets, and the catches from them are far more varied. In spring, they trap the anadromous species; in summer, they trap mostly Atlantic menhaden, Atlantic croakers, spot, summer flounder, and bluefish, as well as an assortment of less numerous fishes. In fall and winter, they mainly catch striped bass and winter flounder.

The eel fishery employs pots rather than nets. Eel pots must be licensed, but not for precise locations as is required for fixed nets. The Indian Head area is approximately the upstream limit of commercial eel potting.

Sportfishing

The most frequented sportfishing areas in the Potomac estuary are illustrated in Fig. 10-1.[7] The intensity of sportfishing in various parts of the estuary varies seasonally as certain sought-after species move in and out of their spawning and feeding grounds. Above Maryland Point, sportfishermen concentrate on spring anadromous fish runs. In March, fishermen may frequent streams known to harbor schools of spawning yellow perch. Soon after that, in April and May, they use drop nets and seines to intercept the herrings as they move up the tributaries to spawn. The spring sportfishery is primarily in the tributaries, and anglers may go after American shad in the main stem. The sport species that are most in demand and that inhabit the region above Maryland Point in relatively large numbers during the summer are white and yellow perch in the main stem and catfishes, black crappie, largemouth bass, and chain pickerel in the tributaries.[7]

A greater variety of sportfish are present and more sportfishing is done in the lower half of the estuary than in the upper portion. Most species are present from late spring until fall, and the sportfishing season generally extends from May to October.[10] Migrants from the ocean, such as spot, Atlantic croaker, weakfish, and bluefish, make their way into this part of the estuary each summer. Schools of large striped bass and the ever-present white perch are also found in the lower estuary. In the regions closest to the mouth, the less abundant ocean sportfish such as summer flounder, red drum, channel bass, black drum, black sea bass, and cobia are found. Fishermen may also move up towards

the limit of tidal waters in the lower tributaries to catch white perch and freshwater species.[7]

Most fishing is conducted from boats,[3][11] although, where docks and piers are conveniently located, shore fishing efforts may nearly equal those from boats.[10] Still-fishing from boats is employed for bottom species such as white perch, spot, and Atlantic croaker, but for the pelagic bluefish and striped bass, trolling (towing lures from a moving boat) is the most popular method.

For reasons that are not well known, certain areas consistently harbor concentrated populations of fishes such as striped bass and bluefish. Other than seeking such traditional "fishing holes," fishermen in the Chesapeake Bay usually concentrate in areas close to population centers or close to docking and service facilities. Since there are no large population centers in the lower half of the Potomac, the locations of marinas, boat ramps, and fishing piers have a strong influence on the sportfishing activity there.[3] Table 10-1 shows the number of boat launchings from different areas along the estuary during the 1974 boating season. The number of launchings made expressly for fishing is not known, but may have been a substantial portion of the total. The data suggest that there was greater boating activity (presumably for fishing) in the portion of the estuary from Nanjemoy Creek downstream than in the upper portion.

Table 10-1 — **Estimated Total Recreational Boat Launchings on the Potomac Estuary and Its Tributaries During the 1974 Boating Season** [a]

Potomac, Washington, D.C. area	230
Upper Potomac, Mattawoman Creek, Nanjemoy Creek	7,964
Port Tobacco River	15,315
Wicomico River	5,973
Breton Bay, St. Catherine Sound, St. Clements Bay	16,157
St. George Creek, St. Marys River	21,594
TOTAL - Potomac Estuary and Tributaries	67,233

(a) Source: Ref. 6.

The intensity of sportfishing pressure in certain areas of the Potomac can be high, particularly close to adequate marina facilities. For example, a survey conducted during the sportfishing season (May through October 1970) found an average of 39 fishermen per day in a 4-nautical-mile segment on the eastern side of the Potomac estuary near the Rt. 301 bridge — an area with easy access to a marina and boat launching ramp.[10] The survey showed that the greatest numbers of fishermen were out on the estuary during July and August when 60 percent of the total catch was collected.

Crabbing

Blue crabs are caught commercially in the Potomac estuary with crab pots and trot lines. Since crab pots are fixed fishing devices, those set in the main stem must be licensed by the PRFC.[4] Figure 10-1 shows the general extent of the commercial crab fishery in the Potomac. Most potting is done below Maryland Point. In the tributaries, crab potting is allowed in only a few areas (see Fig. 10-1).[8] Trot lines may be set

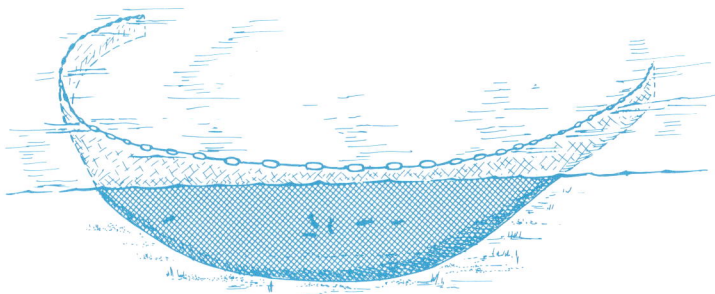
Haul Seine

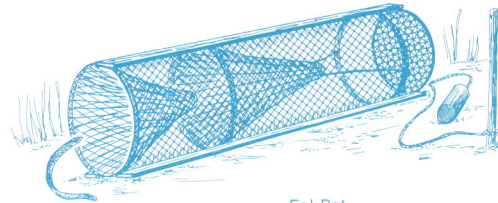
Eel Pot

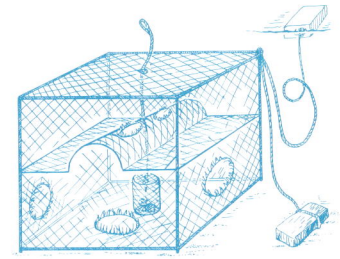
Crab Pot

anywhere in the main stem and tributaries, but commercial crabbers must have a license to operate them.[4][8]

Over the years, recreational crabbing has become as popular as sportfishing in the Chesapeake Bay region. In the Potomac estuary, most recreational crabbing occurs in the main stem and tributaries below Morgantown. The noncommercial crabber relies on a variety of capture methods. Many use trot lines that are similar to those used by commercial watermen, but the length of their lines is limited to 100 yards (91 m) by law.[4] Up to five collapsible crab traps may be set by one person at any one time. Unlicensed crab potting is permitted, with each individual limited to one pot.[5] In one favorite crabbing technique, lines baited with chicken necks are lowered to the bottom and then raised slowly to the surface where the crabs that hold onto the bait are netted. Soft crabbing is also pursued recreationally. Poling along the shore or wading knee deep, a practiced soft-crabber can spot recently shedded crabs that are barely visible on the bottom. As stated in PRFC regulations, each sport crabber may harvest only one bushel of hard crabs per day, or three dozen soft or peeler crabs.[4]

Oystering

In most areas of the Potomac main stem that are under PRFC jurisdiction, hand tongs are the sole legal method for gathering oysters.[4] With certain restrictions, hand scraping using an oyster dredge is permitted downstream from a line extending from Herring Creek, Maryland, to Bonum Creek, Virginia. Patent tonging from sailboats without engines is legal in some tributaries. Specific restrictions for Maryland and Virginia tributaries are listed in the fisheries rules and regulations of the respective states.[8][12] The season for taking oysters from public oyster grounds is September 1 through April 30, but specific restrictions within this period may also be imposed.

There are privately leased oyster bars in Maryland in the Wicomico River, St. Clements Bay, and Breton Bay; and in Virginia in Nomini Creek, Lower Machodoc Creek, and the Yeocomico and Coan rivers (see Folio Map 6). These bars are usually subject to less stringent fishing regulations than public bars, and oysters may be taken from them all year.

Recently, some of the most productive bars in the main stem have been: Heron Island Bar and Sheepshead Bay Bar, both off the mouth of St. Clements Bay (Maryland Bar Nos. 15 and 13, respectively, on Folio Map 6); Ragged Point Bar (Virginia Bar No. 22); Jones Shore Bar near the mouth of the river (Maryland Bar No. 26); Cornfield Harbor Bar along Maryland's shore (No. 27); and Hog Island Bar off Virginia (No. 28).[13] These bars, along with other oyster grounds that have historically been productive in the Potomac, are now producing far less than they did during the late 1960s. As a result of low salinities caused by Hurricane Agnes in 1972, oysters on bars in the main stem above Swan Point suffered 100 percent mortality, while, from Cobb Island downstream, mortalities ranged from 50 to 5 percent.[13] No oyster set

occurred in that year. Because of these devastating losses, many bars were closed to oystering by the PRFC until 1976. Cultch and oyster seed planting efforts have been made in the last few years in attempts to rehabilitate many formerly reliable oyster bars. Most of the Potomac oyster production has been from the tributaries rather than from the main stem in the post-Agnes period.[1] Folio Map 6 shows the recent status of Potomac oyster bars.

Clamming

The soft-shell clam industry, which currently is at a low ebb in the Potomac, began in Maryland in the early 1950s with the invention of the hydraulic clam dredge, which permitted efficient harvesting of submerged clam beds. The development of this new industry coincided favorably with an increasing market for clams in New England, where the green crab destroyed the soft-shell clam populations on many mud flats.[14] Most of the clams harvested from the Chesapeake Bay region in the 1950s were shipped to New England.[15] In later years, the demand in Maryland increased, and "maninose" are now a highly prized seafood item in the Chesapeake Bay area.

Soft-shell clams may be found in the Potomac as far upstream as Mathias Point, but commercial clamming is not allowed above the Charles County line. There are relatively few areas of the lower Potomac where soft-shell clam populations reach densities which make clamming commercially feasible. Many areas are restricted because it is illegal to clam within 150 feet of either natural or leased oyster grounds in Maryland. The status of soft-shell clam beds is indicated on Folio Map 6. All beds were closed by the Maryland Department of Health in 1972 following Hurricane Agnes and were kept closed between 1973 and 1976 by the PRFC in hopes that harvestable populations would become reestablished.[5] All beds were reopened in 1976, but because of continued low densities, very little commercial clamming has been done.

CATCH DISTRIBUTIONS OF COMMERCIAL FINFISH

Of the more than 100 fish species known to occur in the Potomac estuary (see Appendix Table 6), about 20 finfish species* appear regularly in commercial catch records (Table 10-2). For eight of the twenty species, average annual landings in the Potomac from 1964-71** exceeded 100,000 pounds [46,000 kilograms (kg)]. They are: alewife, Atlantic menhaden, striped bass, spot, American shad, American eel, white perch, and the catfishes. The monthly catches of each species show characteristic seasonal and geographic patterns within the

---

* Four of the entries — alewife, catfishes, crappies, and flounders — may include more than one species.
** Although commercial catch records are reported by river zone, tabulations by zone were available only for this period; catch data for 1960-63 and the years since 1971 are available only for the entire estuary, with all zones combined.

Drift Gill Net

Table 10-2 — Average Annual Commercial Harvest (Average Pounds Landed) in the Potomac Estuary in the Years 1964-71, Presented According to Commercial Reporting Zone [a]

TOTAL ESTUARY			ZONE 1			ZONE 2		
Rank	Species	Landings	Rank	Species	Landings	Rank	Species	Landings
1.	Alewife [b]	7,044,637	1.	Alewife [b]	6,088,915	1.	Alewife [b]	942,332
2.	Atlantic menhaden	3,952,136	2.	Atlantic menhaden	3,512,670	2.	Atlantic menhaden	398,150
3.	Striped bass	1,117,248	3.	Striped bass	327,930	3.	Striped bass	351,261
4.	Spot	422,691	4.	Spot	294,210	4.	Spot	127,395
5.	American shad	366,495	5.	American eel	236,563	5.	American eel	34,150
6.	American eel	340,738	6.	American shad	144,640	6.	White perch	32,629
7.	White perch	191,327	7.	White perch	86,231	7.	American shad	23,370
8.	Catfishes [c]	161,088	8.	Flounders [d]	46,316	8.	Catfishes [c]	4,111
9.	Flounders [d]	47,309	9.	Bluefish	43,078	9.	Carp	1,903
10.	Bluefish	44,356	10.	Weakfish (gray trout)	25,415	10.	Yellow perch	1,218
11.	Weakfish (gray trout)	27,380	11.	Butterfish	13,775	11.	Bluefish	1,115
12.	Carp	27,313	12.	Black drum	7,754	12.	Weakfish (gray trout)	1,072
13.	Gizzard shad	20,597	13.	Carp	5,861	13.	Flounders [d]	878
14.	Butterfish	13,943	14.	Atlantic croaker	4,592	14.	Gizzard shad	705
15.	Black drum	7,766	15.	Catfishes [c]	2,352	15.	Hickory shad	572
16.	Atlantic croaker	5,059	16.	Yellow perch	1,672	16.	Atlantic croaker	315
17.	Yellow perch	3,768	17.	Hickory shad	424	17.	Butterfish	141
18.	Hickory shad	1,243	18.	Gizzard shad	393	18.	Black drum	12
19.	Red drum	370	19.	Red drum	370			
20.	Crappies [e]	187						

ZONE 3			ZONE 4		
Rank	Species	Landings	Rank	Species	Landings
1.	Striped bass	403,841	1.	Catfishes [c]	138,872
2.	American shad	180,820	2.	Striped bass	34,211
3.	White perch	67,016	3.	American eel	28,028
4.	American eel	41,995	4.	American shad	18,203
5.	Atlantic menhaden	41,316	5.	White perch	5,449
6.	Gizzard shad	19,496	6.	Carp	5,064
7.	Catfishes [c]	15,752	7.	Alewife [b]	1,121
8.	Carp	14,469	8.	Yellow perch	754
9.	Alewife [b]	12,628	9.	Crappies [e]	187
10.	Spot	1,085	10.	Hickory shad	22
11.	Weakfish (gray trout)	892			
12.	Hickory shad	225			
13.	Bluefish	163			
14.	Yellow perch	123			
15.	Butterfish	27			
16.	Atlantic croaker	11			

(a) See Fig. 10-1 for zones. Data tabulated by zone available only for the period 1964 to 1971, Ref. 9.
(b) Alewife landings include blueback herrings.
(c) May include several species.
(d) Presumed to include both winter and summer flounder.
(e) Presumed to include black and white crappies.

Potomac, which reflect the migration of species (see Chapter 8), as well as the responses of commercial fishermen to the changes in prices for the various species during the year. The patterns of catches of the eight species that dominated the total landings from 1964 to 1971 are illustrated in Fig. 10-2.

The group of fish listed as alewives in commercial fisheries records accounts for a major portion of commercial landings each year (Table 10-2). This group actually consists of two species — the blueback herring and the true alewife, in a ratio of approximately 4 to 1.[16] The river

herrings, as both species collectively are often called, are captured primarily in zones 1 and 2 during their upstream spawning run in the spring (Fig. 10-2). Catches decline farther upstream as populations leave the main stem to enter the tributaries where spawning occurs. Large numbers are caught again in June when they migrate seaward after spawning. Most of the river herring catch is taken by pound netters.

Atlantic menhaden, which by weight have dominated annual landings in recent years (see discussion below), are taken primarily

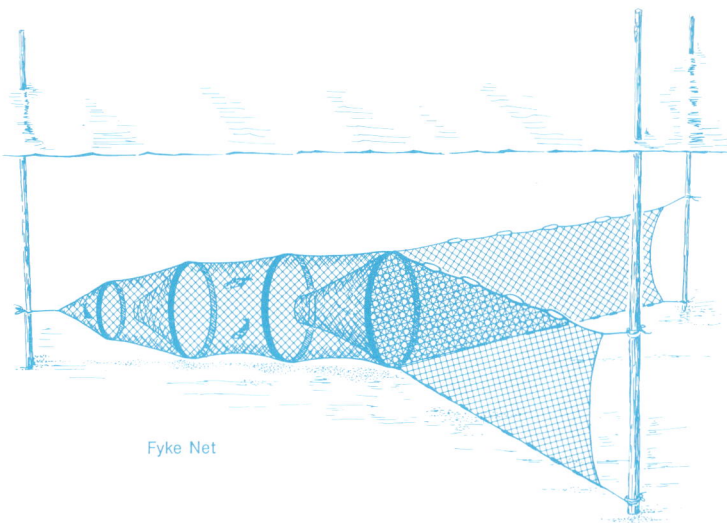

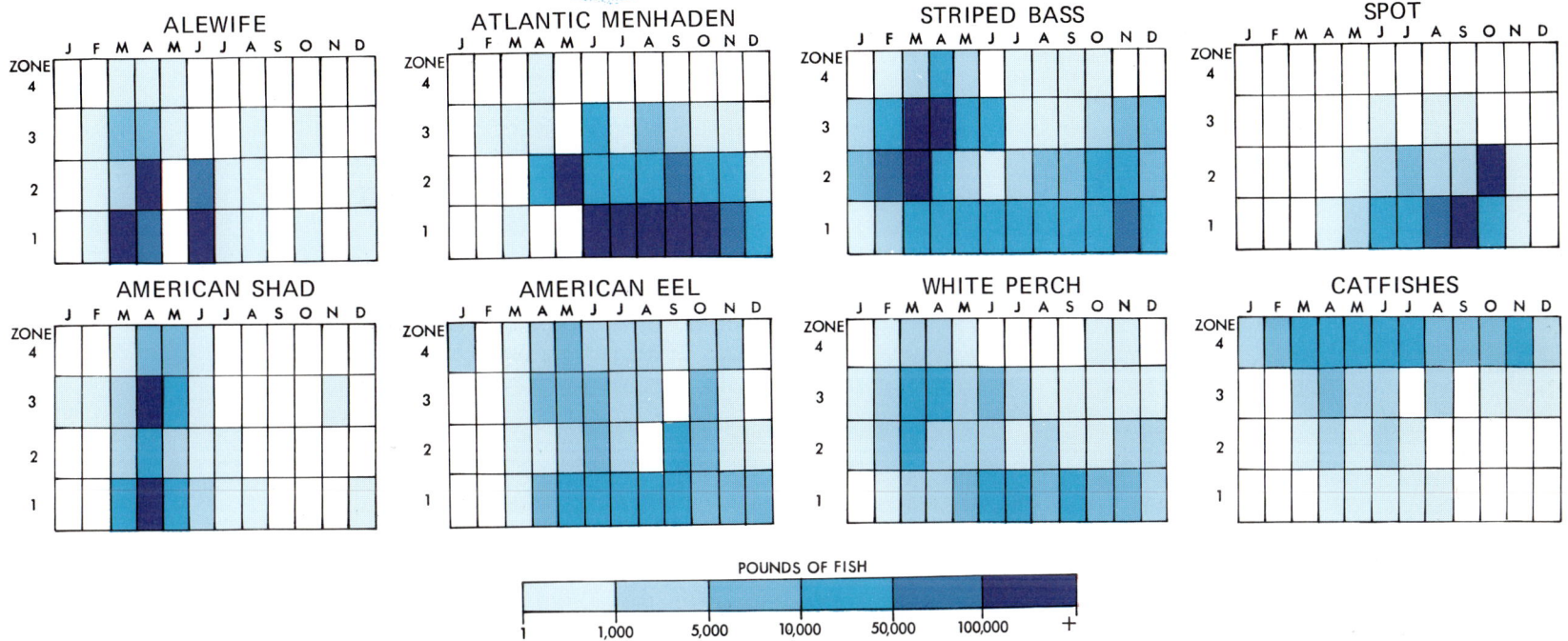

Figure 10-2. *Average seasonal distribution of catches of eight major species by zone in the Potomac estuary, 1964-71. (See Fig. 10-1 for zone boundaries.) (Source: Ref. 9)*

during summer and fall. Most of the catch comes from the high salinity areas in zone 1 (Fig. 10-2), and most of the menhaden are taken in pound nets, which are concentrated in zones 1 and 2 (Fig. 10-1).

Striped bass have consistently been the most valuable finfish species taken in the Potomac each year, bringing in the most dollars while generally ranking second or third in pounds landed.[1] Greatest monthly striped bass catches have traditionally been made during the spring spawning run (March and April) on, or just downstream of, the spawning grounds (zones 2 and 3, Fig. 10-2) where gill nets are the primary fishing gear. Moderate catches are made nearly year-round in zone 1 with gill or pound nets. This area is used by both immature and adult fish as a feeding and overwintering area (see Chapter 8). Annual average landings for zones 1, 2, and 3 are nearly equal despite the difference in the seasonal catch pattern.

Spot is one of the common marine species that frequents the Potomac during summer and fall and overwinters in oceanic waters. Since the harvestable-sized adult spot prefer higher salinity waters, most spot landings come from pound nets set in zones 1 and 2 (Fig. 10-2).

American shad is another anadromous species taken in relatively large numbers. Major landings are made during the spring spawning runs in zone 1 as the shad enter the estuary and in zone 3 just below the primary spawning grounds. Most of the catch is made by gill netters.

American eels are found throughout the Potomac estuary and its tributaries during all seasons, as evidenced by the widespread

distribution of the eel catch (Fig. 10-2). The larger catches in zone 1 may reflect the results of greater fishing intensity by eel potters in that area, combined with some landings by pound netters. Low winter catches are a result of the decreased activity of American eels during the cold months.

White perch is another ubiquitous species that is taken throughout the year in most of the estuary. Some of the heaviest catches are made during the spawning season (March and April) in the spawning area in the main stem (zone 3). Perch are taken with nearly all types of fishing gear.

The catfish group includes several species (see Chapter 8). They are the only freshwater fishes taken commercially in significant numbers. Nearly 90 percent of the annual average catch is taken in zone 4. Since they are resident in the Potomac, catfishes are taken year-round.

TRENDS IN CATCHES
Commercial Finfish

The catch of any species is affected by long-term trends that are evident when annual catch records for many consecutive years are examined in series. These trends are a function of both biological and socio-economic factors. Environmental or climatic conditions may result in production of a dominant yearclass that supports the fishery for many years, or they may cause poor spawning success and a yearclass that barely shows up in the catch records. Degradation of waters in spawning areas may cause a gradual decline in populations of some

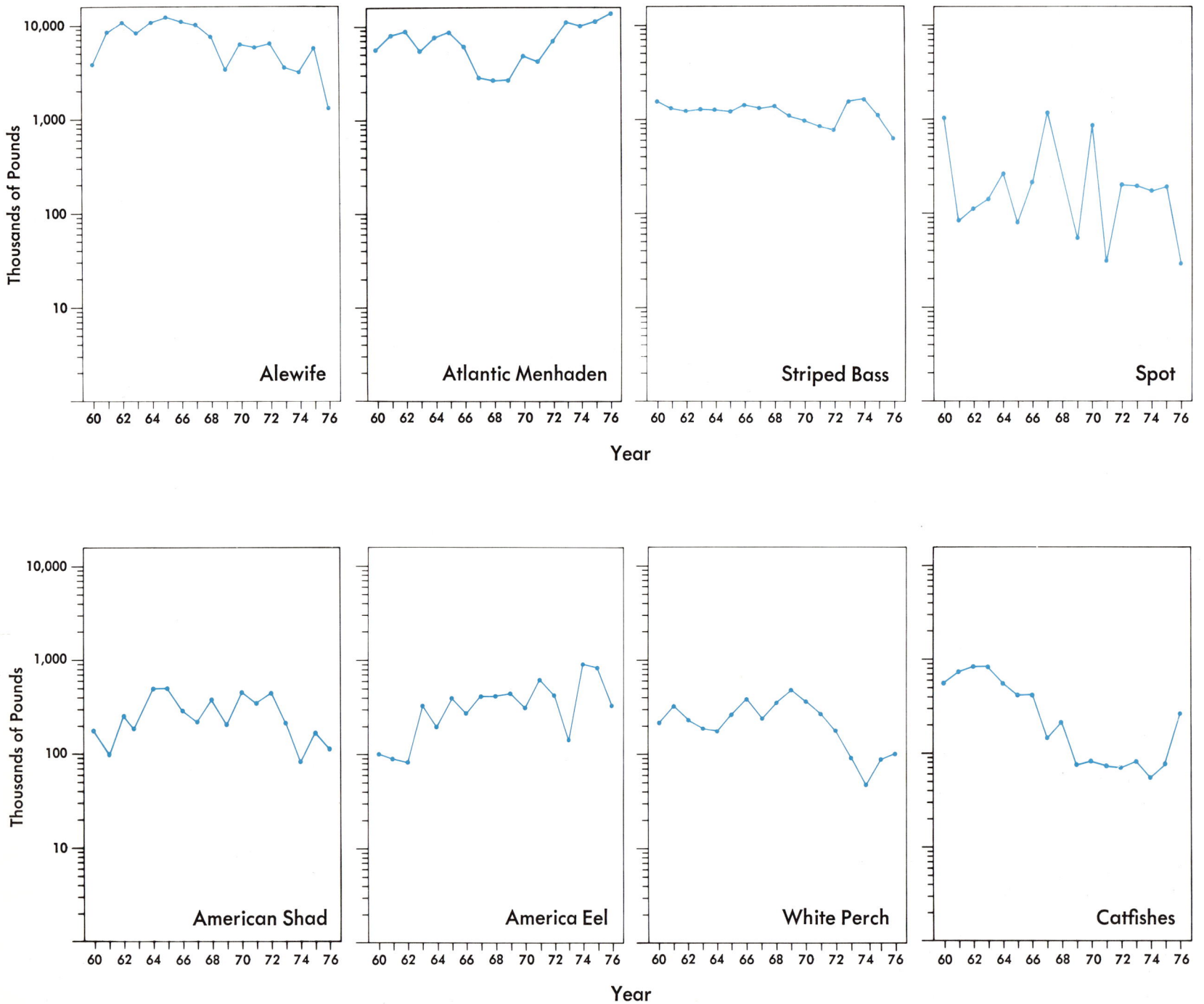

Figure 10-3.   *Annual commercial landings of eight major species from the Potomac main stem and tributaries, 1960-76. (Source: Ref. 1)*

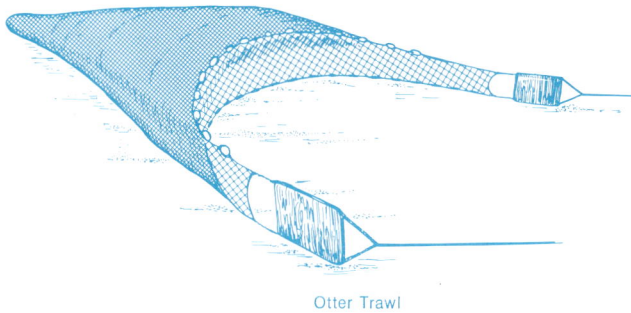

Otter Trawl

species. Occasional catastrophic natural events, such as Hurricane Agnes in 1972, may disrupt normal life cycles of many species, causing temporary population declines. Changes in fisheries regulations and fluctuating market demands and prices are two of the socio-economic factors that influence the amount of fishing pressure exerted by commercial fishermen. In response to such factors, catches of the important commercial species in the Potomac have often changed substantially over periods of several years.

River herrings (alewives) have generally ranked first in total landings in the Potomac. However, catches have shown a downward trend since the 1960s, a decline observed along the entire Atlantic Coast (Fig. 10-3).[16] It has been suggested that foreign fleets overfishing this species during the oceanic phase of its life cycle may be a major factor in the declines.[16] Some recovery was seen in 1975, but 1976 landings were again low.

Atlantic menhaden have long been taken in large numbers, but annual catches since 1973 have been the highest ever reported for the Potomac estuary (Fig. 10-3). Nearly all Atlantic menhaden landed are taken in zones 1 and 2 (Fig. 10-2). The increased catch since the late sixties probably reflects an increase in the Atlantic menhaden population in the Chesapeake Bay. However, increased fishing effort may also be a contributing factor. Atlantic menhaden have historically been sold to fish rendering plants, which processed them for fertilizer, chicken feed, or chemicals. Recently, however, crab potters have turned to menhaden as an inexpensive bait. Currently, more than half of the landings are sold as bait.[5] The increase in demand for menhaden has raised prices and created new interest in pound netting, the primary means of catching this species in the Potomac. In 1977, the PRFC issued 137 new pound net licenses. However, the outlook for the menhaden catch in the Chesapeake Bay is uncertain because the catch since 1972 has been composed primarily of fish less than 2 years old, and a series of poor yearclasses could endanger the fishery.[17]

Striped bass annual landings along the Atlantic Coast and in the Chesapeake Bay have fluctuated widely during the 20th century. Low catches were prevalent through the early 1900s, reaching the lowest recorded point in 1934 when the Chesapeake Bay total was only 642,000 pounds (291,000 kg).[18] In the same year, a mere 362 pounds (164 kg) were taken along the southern Atlantic Coast, and no striped bass were listed in the commercial statistics for the middle Atlantic or New England areas.[18] From this low year, the East Coast fishery increased steadily to a high plateau which was maintained through the 1960s.[18] Trends in the Potomac estuary landings prior to 1960 are assumed to parallel those exhibited by these East Coast landings, since they do so from 1960 to the present. Individual "dominant" yearclasses of striped bass (see Chapter 8) may support a large fishery for many years. It was a number of such yearclasses in the Chesapeake Bay, including the Potomac, which sustained high annual landings along the East Coast, in the Chesapeake Bay, and in the Potomac during the 1960s (Fig. 10-3).[18] The last dominant yearclass to appear in the Potomac was spawned in 1970. As this yearclass entered

the fishery at ages 2 to 4 years (1972 to 1974), Potomac landings increased. However, because members of the same age group were cropped year after year, annual landings have declined. Landings in 1976 were the lowest reported since separate Potomac records have been maintained.

Spot catches in the Potomac have fluctuated widely, but show a recent decline despite evidence that overall populations in the Chesapeake Bay have increased dramatically during the 1970s (Fig. 10-3).[19] One explanation for this phenomenon is that the majority of spot found recently in the Potomac and the Maryland portion of the Chesapeake Bay are juveniles that are too small to be harvested commercially.

The American shad was the most valued fish in the Potomac in the 19th century, when 2.5 to 3 million pounds (1.1 to 1.4 million kg) were landed annually.[16] Populations and catches have steadily declined since then. During the late 1960s, annual landings were on the order of 300,000 to 500,000 pounds (135,000 to 225,000 kg), with catches from 1974 to 1976 dropping to less than 200,000 pounds (91,000 kg) (Fig. 10-3).[1]

As in the case of spot catches, annual American eel landings have fluctuated widely. The 1974 landings were the largest ever recorded (Fig. 10-3). Recently, new markets have been established for this species, which traditionally was caught for use as crab bait. Strong demand for eels in Europe and Japan has created an export market that has caused prices to rise dramatically in recent years;[17] thus, fishing intensity has increased.

White perch landings have declined considerably since a peak catch in 1969 (Fig. 10-3). Potomac populations have been very large in recent years, but the mean size of the fish taken has tended to be small.[20] The decline in catches since 1969 may be a function of fishermen shifting emphasis to more profitable species.

Catfish landings declined from the early 1960s until 1974 (Fig. 10-3). Landings increased markedly in 1976. Whether these fluctuations reflect changes in population abundance or changes in market demand is unknown.

Landings of several other important commercial species have shown major changes in recent years. Winter flounder landings increased in the late 1960s, but have subsequently declined to previous levels (Fig. 10-4A). Summer flounder landings also increased in the 1960s, but have remained at about 50,000 pounds for the last 4 years (Fig. 10-4A). Bluefish annual landings have increased steadily since the late 1960s, leveling off since 1974 at about 500,000 pounds per year (Fig. 10-4B). This increase reflects a general increase in abundance of bluefish along the entire Atlantic Coast.[21]

Sport Finfish

The only comprehensive sportfish survey of the Potomac estuary was conducted over two fishing seasons, in 1960 and 1961.[11] Since that time, significant changes have probably occurred in the sportfish catch,

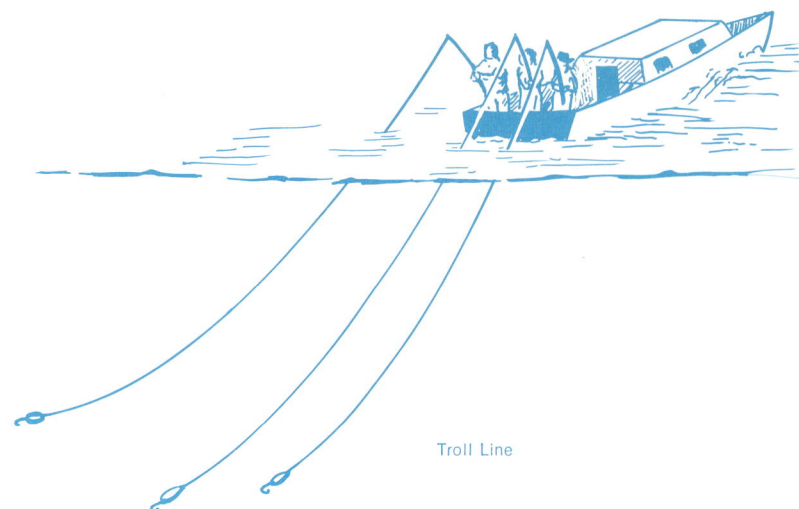

Troll Line

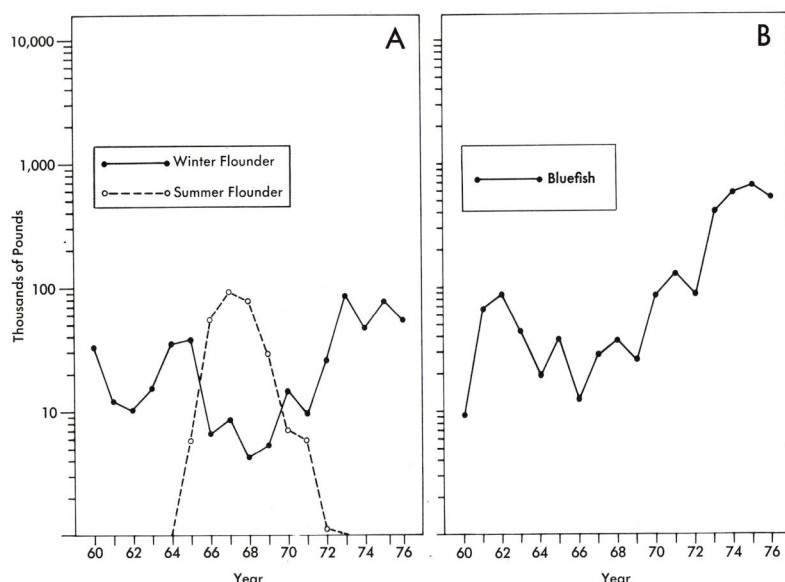

Figure 10-4. *Annual commercial landings of A—summer and winter flounder and B—bluefish from the Potomac main stem and tributaries, 1960-76. (Source: Ref. 1)*

because the abundance of certain species has declined or increased. Data from two sportfishing surveys of the portion of the Chesapeake Bay extending from Pooles Island (just north of Baltimore Harbor) down to the mouth of the Choptank River may serve as an indicator of sportcatch trends in the lower half of the Potomac estuary. The first was conducted in 1962 by the Maryland Department of Natural Resources;[22] the latter in 1976 by the Fisheries Administration of Maryland.[3] Sportcatch records from both the Potomac and the Chesapeake Bay surveys are shown in Table 10-3.

During the 1960 sportfishing season in the Potomac, striped bass ranked first by weight, white perch second, and spot third. The high rank of striped bass was created by the dominant yearclass in 1958, which entered the sportfishery as 2-year-old fish. In 1961, spot dominated, and the striped bass catch dropped to second (still supported by the 1958 yearclass). White perch catches remained at the same level, but dropped to third in rank. Bluefish did not appear in the survey in 1960 and were taken in small numbers in 1961.

In the Chesapeake Bay in 1962, striped bass, white perch, bluefish, and spot were the only four species that were reported as being taken in significant numbers; the same species dominated the 1961 catches in the Potomac. A clear change has occurred in the Chesapeake Bay

Table 10-3 — Results of Sport Fishing Surveys Conducted on the Potomac Estuary and the Upper Chesapeake Bay [a]

| | Potomac Estuary Main Stem | | | | Chesapeake Bay Segment (Pooles Island to Sharp Island) | | | |
| | 1960 Landings | | 1961 Landings | | 1962 Landings | | 1976 Landings | |
	Number (thousands)	Pounds (thousands)	Number (thousands)	Pounds (thousands)	Number (thousands)	Pounds (thousands)	Number (thousands)	Pounds (thousands)
Striped bass	173	357	96	191	464	864	48	419
Bluefish	—	—	13	18	150	112	485	2,216
White perch	236	63	206	62	882	678	468	176
Spot	85	41	803	321	19	8	54	16
Weakfish (gray trout)	—	—	72	43			—	—
Yellow perch	0.2	0.7	0.6	0.2	——————— not in area ———————			—
White catfish	0.6	0.8	0.08	0.1	——————— not in area ———————			—
American eel	3	3	0.2	0.2	———— not included ————		1.6	1.4
TOTAL FINFISH [b]	501	468	1,197	642	1,582	1,709	1,069	2,852
Blue crabs	————————— not surveyed —————————				104	39	429	160
TOTAL FINFISH PLUS BLUE CRABS					1,686	1,748	1,498	3,012
Number of parties	25,500		33,200					
Fishermen trips	—				225,000		290,400	
Number of man hours	338,000		452,800		1,462,500		1,539,122	

(a) English units used as in original sources (Ref. 11 for Potomac estuary; Refs. 3, 22 for Chesapeake Bay); multiply by 0.45359 to get kilograms.
(b) Includes other species caught in low numbers.

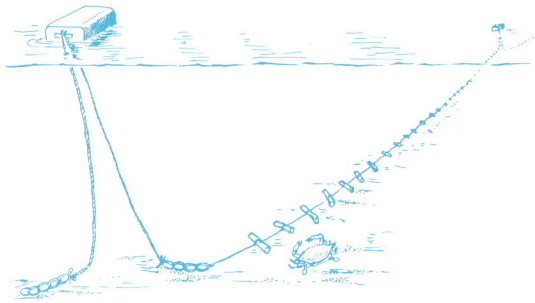

Crab Trot Line

sportfishery since the early 1960s, as can be seen in the results of the 1976 survey. Bluefish overwhelmingly dominate total landings, although striped bass, white perch, and spot continue to be taken in significant numbers.

In spite of shifts in relative importance, the same four species — striped bass, bluefish, spot, and white perch — have dominated the sport catch over the last 17 years in the mesohaline portions of the Chesapeake Bay. The sport catch in the upper oligohaline to tidal freshwater regions is not as well documented. Most sportfish surveys are made in late spring after the herring, shad, and yellow perch fishing seasons, and summer surveys do not usually cover upstream locations where freshwater fishermen may be found. The magnitude of this freshwater fishery is unknown.

In the Potomac, the total sport catch of most species was dwarfed by the commercial harvest of the same species in 1960 and 1961 (Table 10-4). More recent Potomac data are unavailable, but results of the 1976 Chesapeake Bay survey, mentioned earlier,[3] show that the sport catches of spot, Atlantic croaker, and bluefish exceeded commercial catches of these species in the survey area, while catches of white perch and striped bass equaled the commercial harvest. In an area of more intensive commercial fishing such as the Potomac estuary, magnitudes of sport and commercial landings would probably not be as similar.

Table 10-4 — **Comparison of Sport Fish to Commercial Fish Catches in the Potomac Estuary, 1960-61 (Thousands of Pounds Landed)** [a]

	1960		1961	
	Sport	Commercial	Sport	Commercial
Striped bass	357.0	1,583	191.0	1,322
White perch	63.0	219	62.0	183
Spot	41.0	1,083	321.0	89
American eel	3.0	104	0.2	91
Catfish	0.8	586	0.1	766
Yellow perch	0.7	13	0.2	12
Weakfish (gray trout)	—	11	43.0	39
Bluefish	—	9	18.0	68

(a) English units used as in original sources (Refs. 1, 11); multiply by 0.45359 to get kilograms.

Blue Crabs

The largest recorded landings of blue crabs to date, not only in the Potomac estuary but throughout the Chesapeake Bay, were made in the mid 1960s.[23] Peak years occurred in 1965 and 1966 (Fig. 10-5).[1] Apparently, a low survival of the 1967 and 1968 yearclasses caused the low catch in 1969. After several years of greater catches, post-Hurricane Agnes catches also dipped; and although a small degree of recovery has occurred since then, the catches have remained moderate.

211

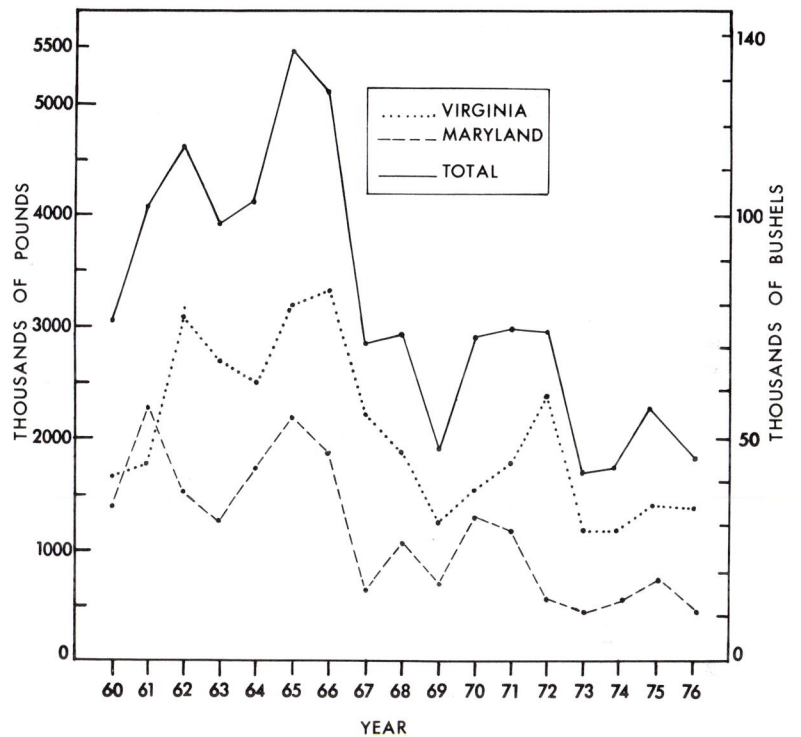

Figure 10-5. *Annual Maryland, Virginia, and total commercial landings of hardshell blue crabs from the Potomac main stem and tributaries, 1960-76 (conversion: 40 lb = 1 bushel). (Source: Ref. 1)*

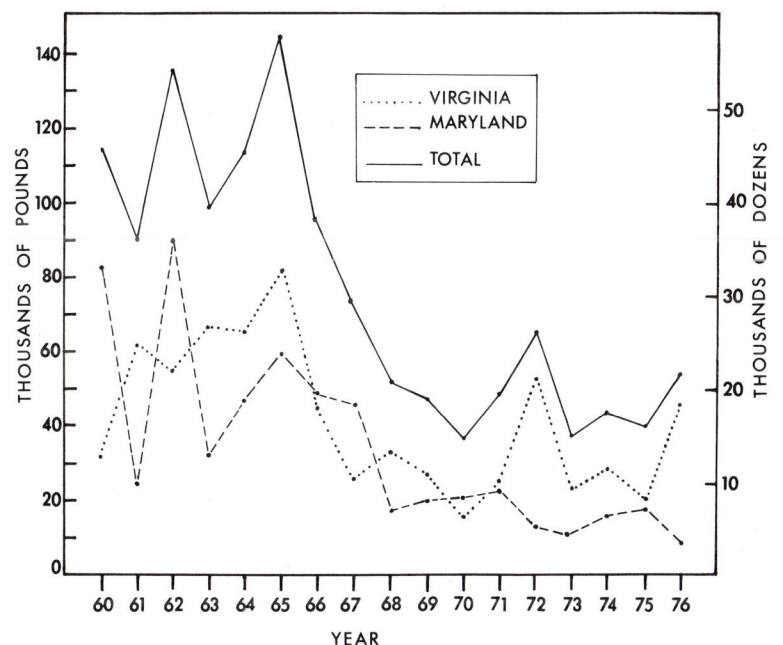

Figure 10-6. *Annual Maryland, Virginia, and total commercial landings of soft and peeler blue crabs from the Potomac main stem and tributaries, 1960-76 (conversion: 2.5 lb = 1 dozen). (Source: Ref. 1)*

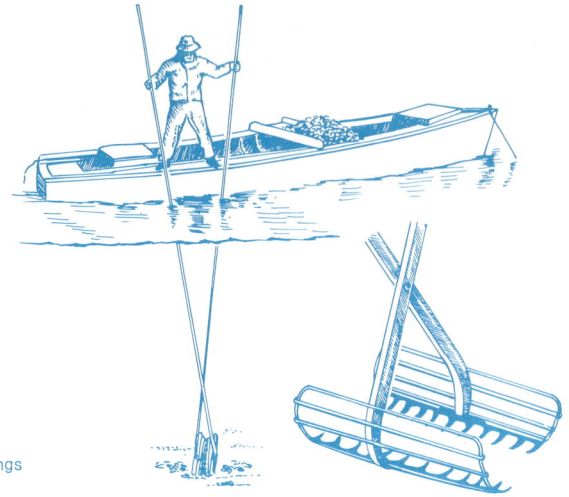

Hand Tongs

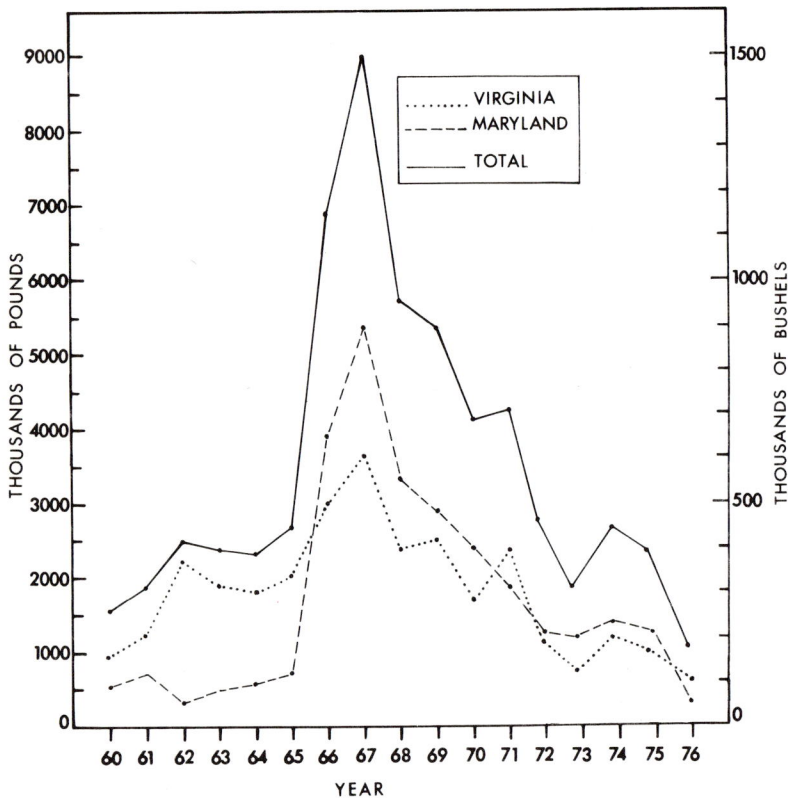

Figure 10-7. *Annual Maryland, Virginia, and total commercial landings of oysters from the Potomac main stem and tributaries, 1960-76 (conversion: 6 lb = 1 bushel). (Source: Ref. 1)*

One interesting aspect of blue crab landings is that total dollar values often remain the same despite fluctuations in the weights landed.[1] Landings of soft-shell and peeler crabs do not follow the same year-to-year dips and peaks as hard crabs, but show similar long-term trends (Fig. 10-6).

Oysters

Figure 10-7 shows the wide variation in oyster landings from 1960 to 1976, with low harvests prevailing in the early sixties and recent seventies, and high catches occurring from 1966 to 1969. The recent decline in landings began prior to Hurricane Agnes, and the hurricane apparently accelerated a change which started several years previously.[24] There has also been a significant shift in the ratio of harvests from tributaries to those from the main stem. Tributary landings, which have been reported separately from mainstem landings since 1963, made up around 40 to 60 percent of the total Potomac catch from 1963 to 1970 (Fig. 10-8). Beginning in 1971, the percentage jumped to 70, and, since 1973, an average of 87 percent of the total oyster landings has come from the tributaries.[1] Factors influencing these variations in oyster catch are not well understood, although years of exceptional sets are responsible for large harvests 3 to 4 years later; and freshwater kills, such as the one caused by Hurricane Agnes, create declines in catch.[24]

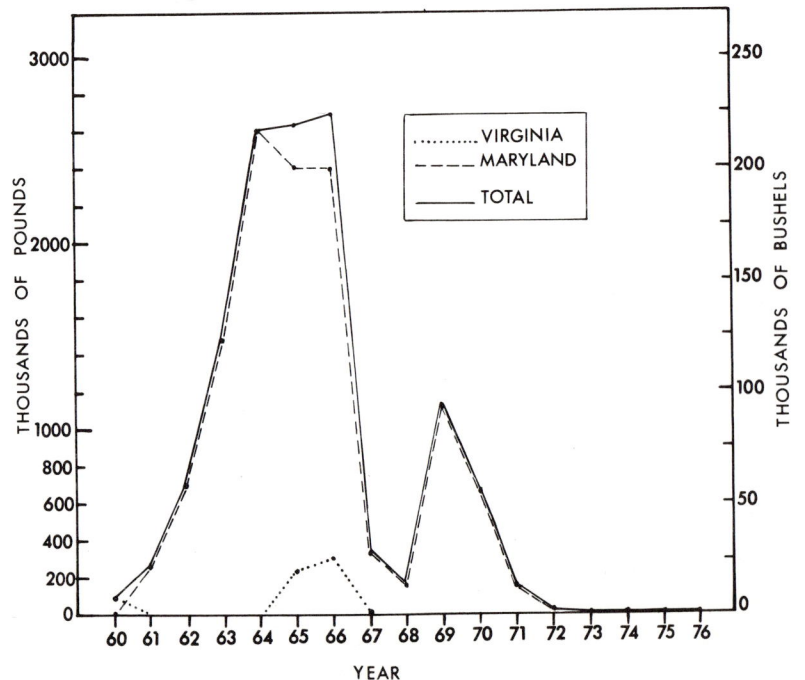

Figure 10-8. *Oyster catch from Maryland and Virginia tributaries as a percentage of total Potomac landings (main stem and tributaries), 1963-76. (Source: Ref. 1)*

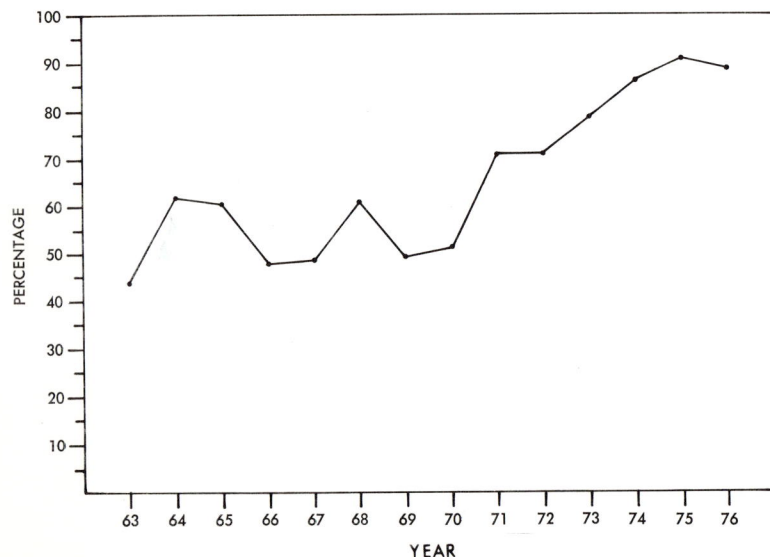

Figure 10-9. *Annual Maryland, Virginia, and total commercial landings of soft-shell clams from the Potomac main stem and tributaries 1960-76 (conversion: 12 lb = 1 bushel). (Source: Ref. 1)*

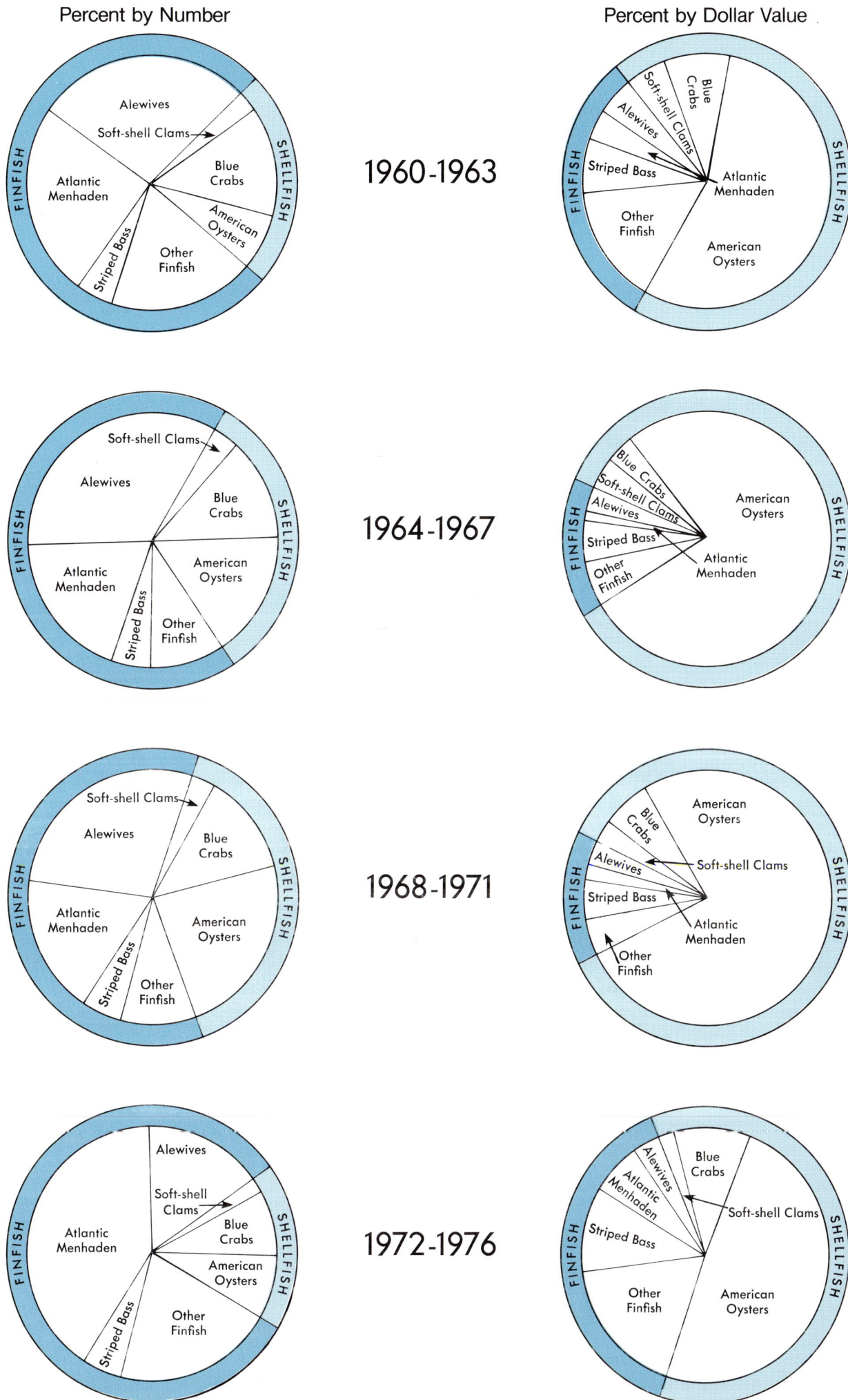

Percent by Number

Percent by Dollar Value

1960-1963

1964-1967

1968-1971

1972-1976

Figure 10-10. *Contribution of three dominant finfish species, other finfish, and three shellfish species to average annual landings in the Potomac estuary main stem and tributaries (percentages given by weight and dollar value). Values are averaged over the periods indicated and are from data in Ref. 1.*

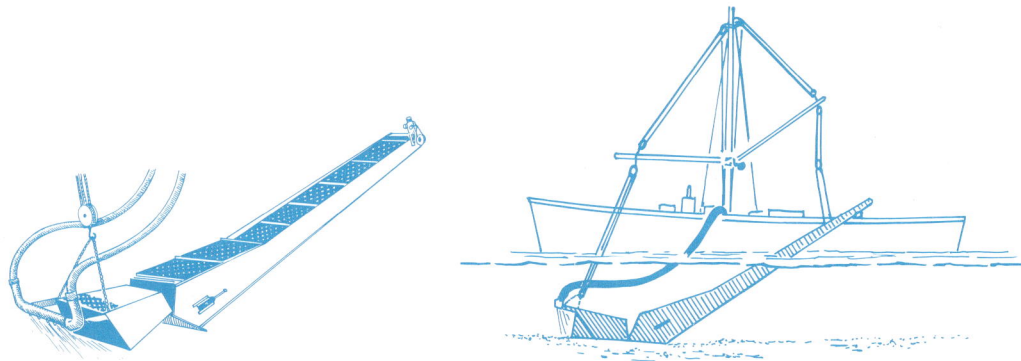

Hydraulic Clam Dredge

Table 10-5 — Average Annual Landings (in Thousands of Pounds) and Average Dollar Value of Landings of Eight Dominant Finfish Species, Other Finfish, and Three Shellfish Species in the Potomac Estuary Main Stem and Tributaries; Values are Averaged Over the Periods Indicated [a]

| | 1960 - 63 | | 1964 - 67 | | 1968 - 71 | | 1972 - 76 | |
	Pounds (Thousands)	Dollar Value	Pounds (Thousands)	Dollar Value	Pounds (Thousands)	Dollar Value	Pounds (Thousands)	Dollar Value
FINFISH								
Alewives	7,896	130,847	11,427	171,630	5,740	111,234	4,020	128,388
Catfishes	766	48,173	389	33,332	109	12,621	112	15,611
American eel	155	21,759	324	38,376	444	72,822	541	151,796
Atlantic menhaden	7,101	105,481	6,365	107,421	3,594	69,411	10,112	233,415
American shad	184	29,715	378	35,064	358	37,770	205	36,472
Spot	359	39,032	446	50,614	313	36,334	162	22,332
Striped bass	1,357	192,538	1,313	252,674	1,056	222,351	1,148	338,650
White perch	246	17,949	274	25,977	376	53,535	107	18,460
Other finfish	3,944	302,528	543	18,572	375	21,530	3,686	342,595
TOTAL FINFISH	22,008	888,022	21,458	733,660	12,365	637,608	20,093	1,287,719
SHELLFISH								
Blue crabs	3,985	249,380	4,471	228,187	2,730	301,621	2,289	321,258
Soft-shell clams	598	113,507	2,048	416,763	513	187,379	4	2,181
Oysters	2,082	1,596,181	5,236	3,694,890	4,788	3,198,873	2,050	1,562,042
TOTAL SHELLFISH	6,665	1,959,068	11,755	4,339,840	8,031	3,687,873	4,343	1,885,481
TOTAL FINFISH AND SHELLFISH	28,673	2,847,090	33,213	5,073,500	20,396	4,325,481	24,436	3,173,200

(a) English units used as in original source (Ref. 1); multiply by 0.45359 to get kilograms.

Soft-shell Clams

From 1960 to 1966, the soft-shell clam industry grew in the Potomac estuary, as it did elsewhere in Maryland. In 1967, unexplained mortalities began in the Potomac and other Chesapeake Bay areas, [17] [25] and clam production fell suddenly. There was a short-lived increase in 1969, followed by another rapid decline (Fig. 10-9). While they were still in low abundance, the soft-shell clams apparently succumbed to the additional stresses brought on by Hurricane Agnes in 1972.[13] [17] Potomac beds have produced virtually no clams since that period, although the clam fisheries in other areas of Maryland (such as the Eastern Shore) have started to revive. Surveys in 1974 and 1975 showed that beds throughout the Potomac did not contain any potentially harvestable soft-shell clams. However, in 1976 and 1977, the beds had numerous young clams, presaging a possible renewal of soft-shell clamming in the near future.[13]

CHANGES IN THE COMMERCIAL FISHERY

The overall nature of the commercial fishery has varied over the period for which records are available, 1960 to 1976. To characterize these variations, annual average landings of the eight dominant finfish species and three shellfish species and their average dollar values were calculated for the periods 1960-63, 1964-67, 1968-71, and 1972-76 (Table 10-5). Figure 10-10 illustrates the relative contribution of the dominant finfish and shellfish species to the average landings and their contributions to the average total dollar value.

In terms of economic value, it is evident that the oyster is the most important species in the Potomac estuary. It consistently accounts for nearly one half or more of the total dollar value of the seafood harvest, even during the post-Hurricane Agnes period. During periods of high production (1964-67 and 1968-71), they accounted for almost three quarters of the total dollar value (Table 10-5). In terms of pounds landed, the emerging dominance of the Atlantic menhaden is evident. Over the 1972-76 period, menhaden accounted for 41 percent of the total landings, while prior to this period, they had fluctuated but stayed close to 20 percent of the total (Fig. 10-10).

Striped bass landings have remained remarkably constant over all four periods, while their dollar value has increased by about 75 percent between the 1960-63 and 1972-76 periods (Table 10-5). An even more dramatic rise in value is shown for the American eels, whose landings rose by less than a factor of four, while their dollar value rose by more than a factor of seven.

The prominence of the category listed as "Other Finfish" in the 1972-76 period (15 percent of pounds landed and 11 percent of dollar value) reflects increases in abundance of such species as summer flounder and bluefish. The fluctuations in landings and dollar values for all the species demonstrate that the commercial fishery of the Potomac estuary is a dynamic entity changing constantly in both magnitude and composition.

1. U.S. Department of Commerce, 1960-1976
2. Murawski and Pacheco, 1977
3. Speir, Weinrich, and Early, 1976
4. Potomac River Fisheries Commission, 1978
5. Carpenter, 1978
6. Roy Mann Associates, 1976
7. Alexandria Drafting Company, 1973
8. Maryland Department of Natural Resources, 1977b
9. deKok, 1975
10. Academy of Natural Sciences of Philadelphia, 1971b
11. Frisbie and Ritchie, 1963
12. Virginia Marine Resources Commission, 1977
13. Norris, 1977
14. Glude, 1955
15. Hanks, 1966
16. Merriner, 1976
17. Lynch, 1977
18. Koo, 1970
19. Naiman, Hixson, and Capizzi, 1978
20. Academy of Natural Sciences of Philadelphia, 1977
21. Boone, 1978
22. Elser, 1965
23. Frisbie, 1971
24. Haven, 1976
25. Shaw and Hamons, 1974

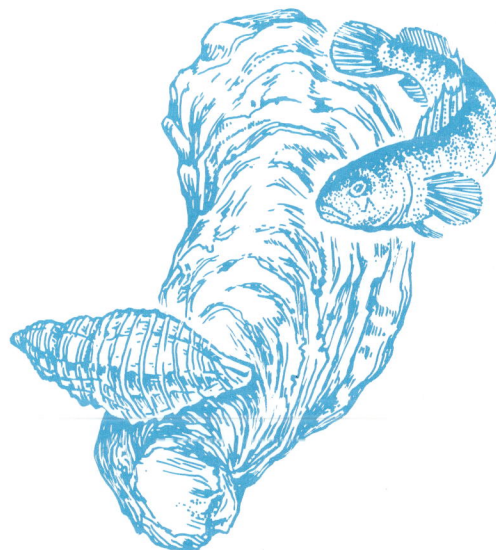

Appendix Tables

Species of phytoplankton and benthic algae recorded from the Potomac estuary are listed. Phytoplankton taxonomy is constantly under revision, and it is difficult to rely on a single source for taxonomic order. The taxonomic references used are numbered below under sources of taxonomy. Except where indicated by footnoting, Refs. 1 and 2 were used. In some instances, where families or species did not appear in the primary reference, they were placed according to the secondary reference or according to the original source of record. Diatoms, yellow-green algae, and golden-brown algae were formerly in a single division (Chrysophyta) and have recently been subdivided into three separate divisions (Chapman and Chapman, 1977). The species list and manuscript reflect these changes.

Sources of Taxonomy: 1) Chapman and Chapman, 1977; 2) Fritsch, 1956; 3) Abbott and Hollenburg, 1976; 4) Taylor, W.R., 1957; 5) Drouet and Daily, 1956; 6) Drouet, 1968; 7) Drouet, 1978; 8) Drouet, 1973; 9) Smith, 1950; 10) Prescott, 1951.

Sources of Record: Academy of Natural Sciences of Philadelphia, 1967, 1968, 1969b, 1971b, 1972, 1973; Dahlberg, 1973; Ecological Analysts, Inc., 1974, 1978; Morrill, 1975; Mountford, K., 1977; Mulford, 1972; Reimer, 1977; Sherman, 1977; Spoon, 1975.

• Found in the Potomac estuary as part of the benthic or aufwuchs flora.
(a) Members of this family are sometimes found in the benthic flora.
(b) Sources of taxonomy for this division: Refs. 3,4.
(c) Source of taxonomy for these two families: Ref. 5.
(d) Source of taxonomy for this family: Ref. 6.
(e) Sources of taxonomy for this family: Refs. 7,8.

DIVISION CHLOROPHYTA (green algae)

CLASS CHLOROPHYCEAE

ORDER VOLVOCALES
FAMILY CHLAMYDOMONADACEAE
Brachiomonas westiana
Carteria cordiformis
Carteria ellipsoidalis
• *Chlamydomonas globosa*
• *Chlamydomonas gracilis*
• *Chlamydomonas monadina*
• *Chlamydomonas* sp.
Lobomonas rostrata
Phacotus lenticularis
Polytomella agilis
Pteromonas aculeata var. *lemmermanni*
Pteromonas angulosa
Pteromonas sp.

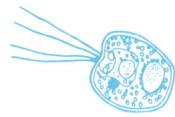

Carteria cordiformis

FAMILY VOLVOCACEAE
Eudorina unicocca
Eudorina sp.
• *Gonium* sp.
• *Pandorina charkowiensis*
Pandorina morum
Platydorina caudata
• *Pleodorina* sp.
• *Volvox* sp.

Phacotus lenticularis

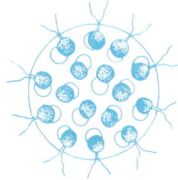

FAMILY HAEMATOCOCCACEAE
Haematococcus lacustris

FAMILY PALMELLACEAE
Gloeocystis planktonica
Sphaerocystis schroeteri

Eudorina

ORDER CHLOROCOCCALES
FAMILY CHLOROCOCCACEAE
Golenkinia paucispina
Golenkinia pusillum
Golenkinia radiata

FAMILY CHLORELLACEAE
Chlorella sp.
Micractinium pusillum

FAMILY OOCYSTACEAE
Ankistrodesmus falcatus
Ankistrodesmus sp.
Chodatella chodati
Chodatella longiseta
Chodatella subsalsa
Chodatella wratislawiensis
Chodatella sp.
Closteriopsis longissima
Echinosphaerella limnetica
Franceia tuberculata
Franceia ovalis
Gloeoactinium limneticum

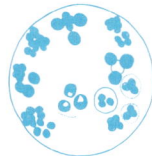

Sphaerocystis schroeteri

217

Kirchneriella obesa
Kirchneriella sp.
Oocystis parva
Oocystis pusilla
Oocystis solitaris
Oocystis submarina
Oocystis sp.
Planktosphaeria gelatinosa
Polyedriopsis quadrispina
Polyedriopsis spinulosa
Quadrigula lacustris
Schroederia setigera
Selenastrum gracile
Tetraëdron caudatum
Tetraëdron constrictum
Tetraëdron limneticum
Tetraëdron minimum
Tetraëdron regulare
Tetraëdron trigonum
Tetraëdron victoriae
Tetraëdron sp.
• *Westella botryoides*

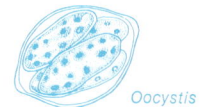

Oocystis

FAMILY HYDRODICTYACEAE
Pediastrum biradiatum
Pediastrum boryanum
Pediastrum duplex
Pediastrum duplex var. *gracilimum*
Pediastrum simplex
Pediastrum simplex var. *duodenarium*
Pediastrum tetras
Pediastrum tetras var. *tetraodon*
• *Pediastrum* sp.

Pediastrum duplex

FAMILY COELASTRACEAE
Actinastrum gracilimum
Actinastrum hantzschii
Coelastrum microporum
Coelastrum reticulatum
Coelastrum sp.
Crucigenia cruciata
Crucigenia fenestrata
Crucigenia quadrata
Crucigenia rectangularis
Crucigenia tetrapedia
Errerella bornhemiensis
Scenedesmus abundans
Scenedesmus abundans var. *brevicauda*
Scenedesmus acuminatus
Scenedesmus anomalous
Scenedesmus armatus
Scenedesmus armatus var. *bicaudatus*
Scenedesmus bicaudatus
Scenedesmus bijuga
Scenedesmus bijuga var. *alternans*
Scenedesmus denticulatus
Scenedesmus denticulatus var. *recurvatus*
Scenedesmus dimorphus
Scenedesmus ecornis
Scenedesmus hystrix

Coelastrum reticulatum

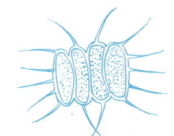

Scenedesmus abundans

Scenedesmus obliquus
Scenedesmus quadricauda
Scenedesmus quadricauda var. maximus
Scenedesmus spinosus
● Scenedesmus sp.
Tetradesmus wisconsinensis
Tetrastrum elegans
Tetrastrum heterocanthum
Tetrastrum staurogeniaeforme

ORDER ULOTRICHALES
FAMILY ULOTRICHACEAE
● Stichococcus subtilis
● Ulothrix aequalis
● Ulothrix implexa
● Ulothrix variabilis
● Ulothrix sp.

FAMILY ULVACEAE
● Enteromorpha intestinalis
● Enteromorpha linza
● Enteromorpha prolifera
● Enteromorpha sp.
● Ulva lactuca (sea lettuce)

FAMILY MONOSTROMATACEAE
● Monostroma grevillei
● Monostroma sp.

ORDER SIPHONOCLADALES
FAMILY CLADOPHORACEAE
● Cladophora glomerata
● Cladophora gracilis
● Cladophora sp.
● Pithophora oedogonia
● Rhizoclonium heiroglyphicum
● Rhizoclonium riparium

ORDER CHAETOPHORALES
FAMILY CHAETOPHORACEAE
● Pseudendoclonium submarinum
● Stigeoclonium lubricum

ORDER OEDOGONIALES
FAMILY OEDOGONIACEAE
● Oedogonium sp.

ORDER ZYGNEMATALES
FAMILY ZYGNEMATACEAE
● Spirogyra sp.

FAMILY DESMIDIACEAE[a]
Closterium dianae
Closterium moniliforme
Closterium parvulum
Closterium pseudodianae
. Cosmarium circulare
Cosmarium formulosum
Cosmarium gracile
Cosmarium margaritaceum
Cosmarium pulchellum
Cosmarium punctulatum
Cosmarium pyramidatum
Cosmarium sp.
Cosmocladium pulchellum
Desmidium sp.
Euastrum sp.
Hyalotheca dissiliens
Staurastrum chaetocerus
Staurastrum natator var. crassum
Staurastrum paradoxum
Staurastrum polymorphum
Staurastrum sp.

FAMILY SELENASTRACEAE
Arthrodesmus phimus

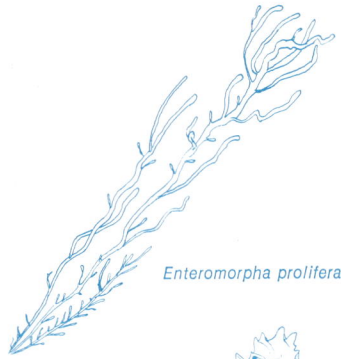
Enteromorpha prolifera

Ulva lactuca

Cladophora gracilis

Spirogyra

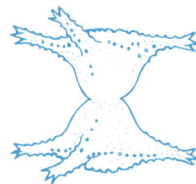
Staurastrum

CLASS PRASINOPHYCEAE
ORDER HALOSPHAERALES
FAMILY HALOSPHAERACEAE
Pyramimonas tetrarhynchus
Pyramimonas sp.

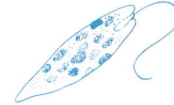
Pyramimonas tetrarhynchus

DIVISION EUGLENOPHYTA (euglenoids)
CLASS EUGLENOPHYCEAE
ORDER EUGLENALES
FAMILY EUGLENACEAE
Euglena acus
Euglena deses
Euglena gracilis
Euglena klebsii
Euglena viridis
● Euglena sp.
Eutreptia marina
Eutreptia sp.
Eutreptiella sp.
Lepocinclis texta
Phacus longicauda
Phacus orbicularis
Phacus sp.
Trachelomonas euchlorum
Trachelomonas hispida
Trachelomonas sp.

Euglena

DIVISION XANTHOPHYTA (yellow-green algae)

CLASS XANTHOPHYCEAE
ORDER VAUCHERIALES
FAMILY VAUCHERIACEAE
● Vaucheria sp.

ORDER TRIBONEMATALES
FAMILY TRIBONEMATACEAE
● Tribonema sp.

Tribonema

DIVISION CHRYSOPHYTA (golden-brown algae)

CLASS CHRYSOPHYCEAE
ORDER CHRYSOMONADALES
FAMILY CHROMULINACEAE
● Chromulina pascheri
Chrysapsis sagene

FAMILY OCHROMONADACEAE
● Ochromonas mutabilis

FAMILY MONADACEAE
Monas guttula

FAMILY PEDINELLACEAE
Apedinella radians
Apedinella sp.
Pseudopedinella pyriforme

FAMILY DINOBRYACEAE
Dinobryon cylindricum
Dinobryon sertularia
Dinobryon sp.

FAMILY SYNURACEAE
● Mallomonas acaroides
● Mallomonas sp.
Synura sp.

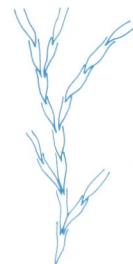
Ochromonas mutabilis

Dinobryon cylindricum

CLASS HAPTOPHYCEAE
ORDER PRYMNESIALES
FAMILY PRYMNESIACEAE
Chrysochromulina sp.

ORDER ISCOCHRYSIDALES
FAMILY ISOCHRYSIDACEAE
Stylochrysalis parasitica

ORDER DICTYOCHALES
FAMILY DICTYOCHACEAE
Dictyocha epiodon
Dictyocha fibula
Dictyocha tripartita (Ebria tripartita)
Distephanum speculum

DIVISION BACILLARIOPHYTA (diatoms)

CLASS BACILLARIOPHYCEAE

ORDER CENTRALES
FAMILY COSCINODISCACEAE

Coscinodiscus apiculatus
Coscinodiscus asteromphalus
Coscinodiscus centralis
Coscinodiscus concinnus
Coscinodiscus curvatulus
Coscinodiscus excentricus
Coscinodiscus granii
Coscinodiscus lacustris
Coscinodiscus lineatus
Coscinodiscus marginatus
Coscinodiscus oculus-iridis
Coscinodiscus perforatus
- *Coscinodiscus pygmaeus* var. *micropunctata*
Coscinodiscus radiatus
Coscinodiscus subtilis
Coscinodiscus wailesii
Coscinodiscus sp.
Coscinosira polychorda
- *Cyclotella aliquantula*
- *Cyclotella atomus*
- *Cyclotella caspia*
Cyclotella florida
Cyclotella glomerata
- *Cyclotella kützingiana*
- *Cyclotella meneghiniana*
- *Cylotella pseudostelligera*
Cylotella stelligera
- *Cyclotella striata*
Cyclotella sp.
Melosira ambigua
- *Melosira borreri*
- *Melosira distans*
- *Melosira granulata*
Melosira granulata var. *angustissima*
Melosira granulata var. *muzzanensis*
Melosira islandica
Melosira juergensii
- *Melosira nummuloides*
Melosira sulcata
- *Melosira varians*
- *Melosira* sp.
Microsiphonia potamos
Skeletonema costatum
Skeletonema potamos
- *Stephanodiscus astraea*
Stephanodiscus hantzschii
Stephanodiscus sp.
Thalassiosira fluviatilis
Thalassiosira nordenskioldii
Thalassiosira rotula
Thalassiosira sp.

FAMILY ACTINODISCACEAE
Actinoptychus undulatus

FAMILY EUPODISCACEAE
Actinocyclus ehrenbergii (Actinocyclus octonarius)

Coscinodiscus

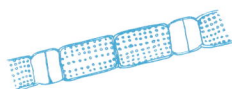
Melosira

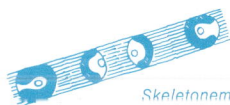
Skeletonema costatum

Thalassiosira

FAMILY BIDDULPHIACEAE
- *Biddulphia laevis*
Biddulphia mobiliensis
Cerataulina bergonii
Ditylum brightwellii

FAMILY ANAULACEAE
Anaulus balticus
Eunotogramma laevis
Eunotogramma sp.

FAMILY CHAETOCERACEAE
Chaetoceros danicus
Chaetoceros decipiens
Chaetoceros densus
Chaetoceros holsaticus
Chaetoceros mitra
Chaetoceros peruvianus (Chaetoceros convexicorne)
Chaetoceros subtilis
Chaetoceros sp.

FAMILY RHIZOSOLENIACEAE
Rhizosolenia calcar avis
Rhizosolenia fragilissima
Rhizosolenia setigera

FAMILY LEPTOCYLINDRACEAE
Leptocylindrus danicus

FAMILY CORETHRONACEAE
Corethron hystrix

ORDER PENNALES
FAMILY FRAGILLARIACEAE
Asterionella formosa
Asterionella gracilis
Asterionella japonica
- *Asterionella* sp.
- *Diatoma anceps*
- *Diatoma hiemale*
- *Diatoma hiemale* var. *mesodon*
- *Diatoma tenue* var. *elongatum*
- *Diatoma vulgare*
- *Dimerogramma lanceolata*
Fragilaria capucina
Fragilaria construens var. *subsalina*
Fragilaria construens var. *venter*
Fragilaria crotonensis
Fragilaria gessneri
Fragilaria intermedia
Fragilaria pinnata
- *Fragilaria vaucheriae*
- *Fragilaria virescens* var. *clavata*
- *Fragilaria* sp.
Grammatophora marina
Licmophora abbreviata
Licmophora ehrenbergii
- *Licmophora ehrenbergii* var. *angustata*
Licmophora paradoxa
- *Meridion circulare*
Opephora marina
Opephora martyi
Opephora pacifica
Opephora schulzi
Plagiogramma sp.
Rhabdonema sp.
Rhaphoneis sp.
Striatella delicatula
Straitella unipunctata
- *Synedra acus*
- *Synedra affinis*
- *Synedra affinis* var. *gracilis*
- *Synedra affinis* var. *obtusa*
Synedra affinis var. *parva*
- *Synedra fasiculata*

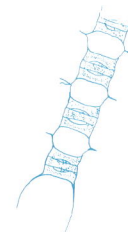
Chaetoceros holsaticus

219

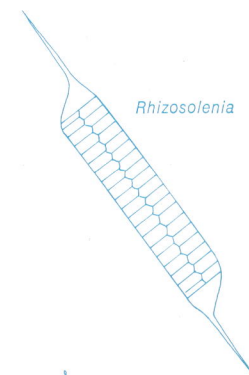
Rhizosolenia

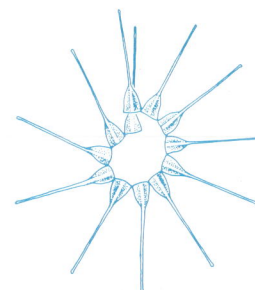
Asterionella japonica

Fragilaria

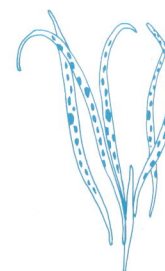
Synedra affinis

Synedra fasiculata var. truncata
Synedra pulchella
● Synedra ulna
● Synedra sp.
Tabellaria fenestrata
Thalassionema nitzschioides
Thalassiothrix sp.

FAMILY EUNOTIACEAE
● Eunotia curvata
● Eunotia valida

FAMILY ACHNANTHACEAE
Achnanthes brevipes
● Achnanthes brevipes var. intermedia
Achnanthes chilensis
Achnanthes delicatula
● Achnanthes exigua
Achnanthes flexulosa
● Achnanthes hauckiana
● Achnanthes hauckiana var. rostrata
● Achnanthes lanceolata
● Achnanthes lemmermannii var. obtusa
Achnanthes longipes
● Achnanthes minutissima
Achnanthes punctatula
● Achnanthes punctifera
● Achnanthes saxonica
Achnanthes temperei
Achnanthes tenera
Achnanthes wellsiae
● Achnanthes spp.
Cocconeis curvatulus
● Cocconeis diminuta
Cocconeis disculus
● Cocconeis pediculus
● Cocconeis placentula
● Cocconeis placentula var. euglypta
● Cocconeis placentula var. lineata
Cocconeis scutellum
Cocconeis sp.
● Rhoicosphenia curvata

Cocconeis placentula

FAMILY NAVICULACEAE
● Amphipleura rutilans
● Amphipleura rutilans var. dillwynii
Amphiprora alata
● Amphiprora paludosa
Amphiprora similis
Amphiprora sp.
● Caloneis bacillum
● Capartogramma crucicula
Diploneis bombus
Diploneis finnica
Diploneis littoralis
Diploneis oblongella
● Diploneis ovalis
● Diploneis puella
Diploneis smithii
● Diploneis smithii var. dilata
Diploneis smithii var. pumila
● Frustulia rhomboides
Frustulia vulgaris
Frustulia sp.
Gyrosigma acuminatum
Gyrosigma balticum
Gyrosigma beaufortianum
Gyrosigma distortum
Gyrosigma fasciola
Gyrosigma macrum
Gyrosigma nodiferum
● Gyrosigma scalproides
Gyrosigma spenceri
Gyrosigma tennuissima
Gyrosigma wormleyi
Gyrosigma sp.

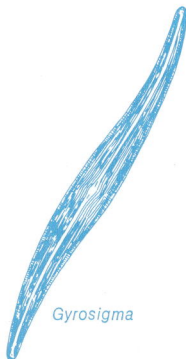

Gyrosigma

Mastogloia dansii
● Mastogloia pumila
Mastogloia pusilla
● Navicula abunda
Navicula adnata
Navicula adrene
Navicula aequoria
● Navicula agnita
Navicula amphipleuroides
● Navicula anglica var. subsalsa
Navicula arenaria
Navicula arenicola
● Navicula bahusiensis
Navicula cancellata
● Navicula capitata
● Navicula capitata var. hungarica
Navicula carminata
● Navicula cincta
Navicula circumtexta
● Navicula clementis
● Navicula complanata
● Navicula confervacea
Navicula contenta
Navicula creuzbergensis
● Navicula cryptocephala
● Navicula cryptocephala var. veneta
Navicula cryptocephaloides
Navicula cuspidata
Navicula decussis
● Navicula diserta
Navicula diversistriata
Navicula exigua
Navicula expansa
Navicula faceta
Navicula fenestrella
Navicula forcipata
Navicula germainii
● Navicula graciloides
● Navicula gravistriata
● Navicula gysingensis
Navicula heufleri var. leptocephala
Navicula hungarica
● Navicula hustedtii
● Navicula incomposita
● Navicula integra
● Navicula lanceolata
Navicula luzonensis
● Navicula menisculus var. upsaliensis
● Navicula minima
Navicula mollis
Navicula muralis
● Navicula mutica
● Navicula mutica var. cohnii
Navicula mutica var. nivalis
● Navicula neoventricosa
Navicula notha
Navicula nyella
Navicula occulta
Navicula oculiformis
Navicula pavillardi
● Navicula peregrina
● Navicula pupula
● Navicula pupula var. mutata
● Navicula pygmaea
Navicula radiosa
Navicula rhynchocephala
● Navicula rhynchocephala var. amphiceros
● Navicula rhynchocephala var. germainii
Navicula rostellata
● Navicula salinarum
● Navicula salinicola
● Navicula sanctaecrucis
Navicula schroeteri
● Navicula secreta
● Navicula secreta var. apilculata
● Navicula seminulum
Navicula spicula
Navicula symmetrica

Navicula

220

- Navicula tenelloides
- Navicula tripunctata
- Navicula tripunctata var. schizonemoides
- Navicula viridula
- Navicula viridula var. avenacea
  Navicula zosteretti
  Navicula spp.
- Neidium affinis var. amphirhynchus
- Neidium apiculatum
- Neidium dubium
- Neidium spp.
- Pinnularia acuminata var. bielawskii
- Pinnularia biceps
- Pinnularia biceps f. petersenii
- Pinnularia braunii var. amphicephala
- Pinnularia brebissonii var. diminuta
- Pinnularia stomatophora
- Pinnularia spp.
  Pleurosigma angulatum
  Pleurosigma australe
  Pleurosigma delicatulum
  Pleurosigma elongatum
  Pleurosigma formosum
  Pleurosigma salinarum
  Pleurosigma sp.
- Stauroneis phoenicentron var. gracilis
  Stauroneis salina
- Stauroneis spp.
  Tropidoneis lepidoptera

Pleurosigma

FAMILY GOMPHONEMATACEAE
- Gomphonema angustatum
- Gomphonema angustatum var. sarcophagus
  Gomphonema constrictum
  Gomphonema gracile
- Gomphonema olivaceum
- Gomphonema parvulum
  Gomphonema sphaerophorum
- Gomphonema subclavatum var. commutatum
- Gomphonema sp.

FAMILY CYMBELLACEAE
- Amphora acutiuscula
- Amphora angusta
  Amphora coffeiformis
- Amphora cymbelloides
  Amphora delicatissima
- Amphora exigua
  Amphora laevissima var. perminuta
- Amphora ovalis
- Amphora ovalis var. affinis
- Amphora ovalis var. pediculus
- Amphora perpusila
  Amphora sublaevis
  Amphora subtilissima
- Amphora tenuissima
  Amphora tumida
- Amphora turgida
- Amphora spp.
  Cymbella amphicephala
  Cymbella cistula
  Cymbella gracilis
  Cymbella microcephala
- Cymbella minuta (ventricosa)
- Cymbella minuta f. latens
- Cymbella sinuata
  Cymbella tumida
- Cymbella turgida
  Cymbella sp.

Cymbella cistula

FAMILY EPITHEMIACEAE
  Denticula tenuis
  Epithemia turgida
  Rhopalodia gibberula
  Rhopalodia gibberula var. musculus

FAMILY NITZSCHIACEAE
- Bacillaria paradoxa
  Gomphonitzschia indica
  Hantzschia amphioxys
- Hantzschia amphioxys f. capitata
  Hantzschia virgata
- Nitzschia acicularis
- Nitzschia amphibia
  Nitzschia angustata
- Nitzschia apiculata
  Nitzschia brevissima
- Nitzschia brittoni
  Nitzschia capitata
  Nitzschia clausii
  Nitzschia closterium
- Nitzschia communis var. hyalina
- Nitzschia confinis
  Nitzschia cursoria
  Nitzschia denticulata
- Nitzschia diserta
- Nitzschia diserta var. 1
- Nitzschia dissipata
  Nitzschia dissipata var. media
  Nitzschia dubioides
  Nitzschia elliptica
  Nitzschia epithemoides
- Nitzschia fasciculata
- Nitzschia filiformis
- Nitzschia fonticola
- Nitzschia frustulum
  Nitzschia frustulum var. indica
- Nitzschia frustulum var. perminuta
- Nitzschia frustulum var. perpusilla
- Nitzschia frustulum var. subsalina
  Nitzschia gracilis var. minor
  Nitzschia holsatica
- Nitzschia hungarica
  Nitzschia hustedtiana
  Nitzschia hybridaeformis
  Nitzschia ignorata
  Nitzschia janischii
- Nitzschia kützingiana
  Nitzschia liebethruthii
- Nitzschia linearis
  Nitzschia longissima
  Nitzschia longissima var. reversa
  Nitzcshia macilenta var. abbreviata
  Nitzschia media
- Nitzschia microcephala
  Nitzschia obtusa var. scalpelliformis
  Nitzschia ovalis
- Nitzschia palea
- Nitzschia paleacea
  Nitzschia paleaformis
  Nitzschia panduriformis var. minor
- Nitzschia paradoxa
- Nitzschia parvula
  Nitzschia punctata
  Nitzschia pungens var. atlantica
- Nitzschia recta
  Nitzschia rhopalodioides
  Nitzschia rigida
- Nitzschia romana
  Nitzschia seriata
  Nitzschia sicula
- Nitzschia sigma
  Nitzschia sigma var. clausii
  Nitzschia sigmaformis
  Nitzschia sigmoidea
  Nitzschia silicula
  Nitzschia silicula var. rigidula
  Nitzschia spathulata
  Nitzschia subcapitellata
  Nitzschia subcohaerens
- Nitzschia sublinearis
  Nitzschia suecica

Bacillaria paradoxa

Nitzschia closterium

221

Nitzschia tarda
Nitzschia thermalis
Nitzschia tropica
Nitzschia tryblionella
• Nitzschia tryblionella var. debilis
Nitzschia tryblionella var. levidensis
• Nitzschia tryblionella var. victoriae
Nitzschia vitrea
Nitzschia sp.

FAMILY SURIRELLACEAE
Cymatopleura solea
• Surirella angustata
Surirella brightwelli
Surirella elegans
Surirella gemma
Surirella littoralis
• Surirella ovalis
• Surirella ovata
Surirella ovata var. salina
Surirella patella
Surirella salina
Surirella striatula
Surirella suecica
Surirella sp.

222

DIVISION CRYPTOPHYTA (blue and red flagellates)

CLASS CRYPTOPHYCEAE
ORDER CRYPTOMONADALES
FAMILY CRYPTOMONADACEAE
Chilomonas paramecium
Cryptomonas acuta
Cryptomonas compressa
Cryptomonas erosa
Cryptomonas ovata
Cryptomonas sp.

Cryptomonas ovata

FAMILY CRYPTOCHRYSIDACEAE
Chroomonas baltica
Chroomonas sp.

Chroomonas

DIVISION PYRROPHYTA (dinoflagellates)

CLASS DESMOPHYCEAE
ORDER PROROCENTRALES
FAMILY PROROCENTRACEAE
Exuviella apora
Exuviella compressa
Exuviella lima
Exuviella marina
Prorocentrum dentatum
Prorocentrum micans
Prorocentrum minimum
Prorocentrum scutellum
Prorocentrum triangulatum

Exuviella marina

CLASS DINOPHYCEAE
ORDER GYMNODINIALES
FAMILY GYMNODINIACEAE
Amphidinium crassa
Amphidinium fusiforme
Amphidinium klebsi
Amphidinium lacustre
Amphidinium operculatum
Amphidinium sp.
Cochlodinium helicoides
Cochlodinium helix
Gymnodinium agilis
Gymnodinium danicum
Gymnodinium estuariale
Gymnodinium marina
Gymnodinium marylandicum
Gymnodinium nelsoni
Gymnodinium punctatum

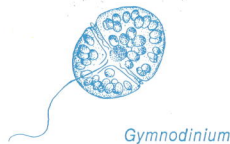
Gymnodinium

Gymnodinium roseostigma
Gymnodinium rotundatum
Gymnodinium simplex
Gymnodinium splendens
Gymnodinium stellatum
Gymnodinium valdecompressum
Gymnodinium sp.
Gyrodinium biconicum
Gyrodinium estuariale
Gyrodinium pellucidum
• Gyrodinium sp.
Katodinium assymetricum
Katodinium rotundatum
Katodinium sp.
Massartia rotundata
Massartia sp.
Nematodinium partitum
Noctiluca miliaris
Oxyrrhis marina
Polykrikos barnegatensis
Polykrikos kofoidi
Polykrikos sp.
Trochodinium prismaticum

ORDER PERIDINIALES
FAMILY PERIDINIACEAE
Diplopsalis lenticula
Peridiniopsis aciculiferum
Peridiniopsis rotunda
Peridinium achromaticum
Peridinium bipes
Peridinium brevipes
Peridinium cerasus
Peridinium cinctum
Peridinium claudicans
Peridinium conicoides
Peridinium crassipes
Peridinium depressum
Peridinium digitale
Peridinium divergens
Peridinium excavatum
Peridinium inconspicuum
Peridinium leonis
Peridinium ovatum
Peridinium rotunda
Peridinium triquetrum
Peridinium trochoideum
Peridinium wisconsinensis
Peridinium sp.

Peridinium

FAMILY GLENODINIACEAE
Glenodinium cinctum
Glenodinium danicum
Glenodinium neglectum
Glenodinium rotundum
Glenodinium ugliginosum
Heterocapsa triquetra

Glenodinium cinctum

FAMILY GONYAULACACEAE
Gonyaulax diacantha
Gonyaulax digitale
Gonyaulax longicornus
Gonyaulax longispina
Gonyaulax monacantha
Gonyaulax polyedra
Gonyaulax polygramma
Gonyaulax scrippsae
Gonyaulax spinifera
Gonyaulax sp.

Gonyaulax polyedra

FAMILY CERATIACEAE
Ceratium furca
Ceratium hirudinella
Ceratium lineatum
Ceratium minutum
Ceratium tripos
Ceratium sp.
Oxytoxum obliquum

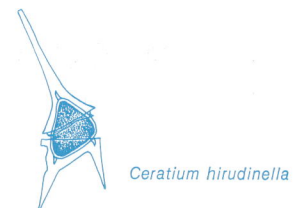
Ceratium hirudinella

ORDER DINOPHYSIALES
  FAMILY DINOPHYSIACEAE
  Dinophysis acuminata

DIVISION PHAEOPHYTA (brown seaweeds)

CLASS PHAEOPHYCEAE

ORDER ECTOCARPALES
  FAMILY ECTOCARPACEAE
- *Ectocarpus confervoides*
- *Ectocarpus tomentosus*
- *Ectocarpus* sp.
- *Plyaiella littoralis*

Ectocarpus tomentosus

DIVISION RHODOPHYTA (red seaweeds)[b]

CLASS BANGIOPHYCEAE

ORDER COMPSOPOGONALES
- *Compsopogon coeruleus*

CLASS FLORIDEOPHYCEAE

ORDER CERAMIALES
  FAMILY RHODOMELACEAE
- *Bostrychia montagnei*
- *Ceramium* sp.
- *Polysiphonia notarisii*
- *Polysiphonia subtilissima*

  FAMILY DELESSERIACEAE
- *Caloglossa leprieurii*

ORDER GIGARTINALES
  FAMILY GRACILARIACEAE
- *Gracilaria verrucosa*

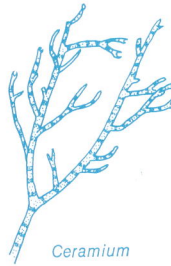

Ceramium

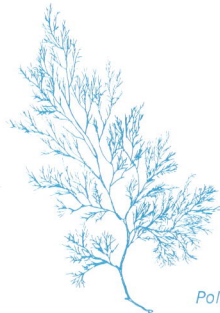

Polysiphonia

DIVISION CYANOPHYTA (blue-green algae)

CLASS MYXOPHYCEAE

ORDER CHROOCOCCALES
  FAMILY CHROOCOCCACEAE[c]
  Agmenellum quadruplicatum
  Agmenellum thermale
  Anacystis cyanea (Anacystis aeruginosa)
  Anacystis dimidiata

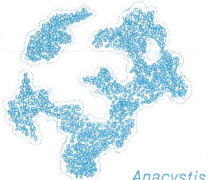

Anacystis

- *Anacystis inserta*
  Anacystis marina
- *Anacystis montana*
  Anacystis thermalis (Chroococcus limneticus)
  Gomphosphaeria lacustris
  Microcroccus geminata

FAMILY ENTOPHYSALIDACEAE[c]
- *Entophysalis deusta*
- *Entophysalis lameniae*
- *Entophysalis rivularis*

ORDER OSCILLATORIALES[d]
  FAMILY OSCILLATORIACEAE
- *Arthrospira neapolitana (Arthrospira brevis)*
- *Microcoleus lyngbyaceus*
- *Microcoleus vaginatus*
  Microcoleus sp. *(Phormidium* sp.)
- *Oscillatoria acuminatum*
- *Oscillatoria lutea*
- *Oscillatoria retzii*
- *Oscillatoria submembranacea*
- *Oscillatoria tenuis*
- *Oscillatoria* sp.
- *Porphyrosiphon notarisii*
  Porphyrosiphon splendidus
- *Schizothrix arenaria*
- *Schizothrix calcicola*
- *Schizothrix mexicana*
- *Schizothrix tenerrima (Microcoleus tenerrimus)*
- *Schizothrix* sp. *(Lyngbya* sp.)
- *Spirulina subsalsa*

FAMILY NOSTOCACEAE[e]
- *Anabaena oscillarioides (Anabaena torulosa)*
- *Anabaena* sp.
- *Calothrix crustacea*
- *Calothrix parietina*
  Nostoc commune (Anabaena circinalis,
    Anabaena flos-aquae, Anabaena
    planktonica, Anabaena spiroides)
- *Nostoc muscorum*
- *Nostoc* sp.

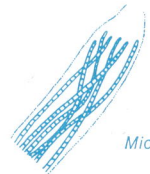

Microcoleus

223

Appendix Table 2 — Wetland Plants of the Potomac Estuarine Drainage

Only species recorded from the Potomac estuarine drainage are listed here. For more extensive listings of Chesapeake Bay wetland plants, see Wass, 1972; for a detailed survey of wetland plants from the Patuxent River (a drainage contiguous with the Potomac drainage) see Anderson, 1972; and for a comprehensive list of Maryland coastal wetland plants, see Jack McCormick & Associates, Inc., 1977.

Sources of record: Maryland Department of Natural Resources, 1967-1968; Stewart, 1962; Wass, 1972; Uhler, 1977; Moore, K. A., 1975a, 1975b, 1975c; Doumlele, 1976; Silberhorn, 1975; and Silberhorn, in press.

Sources of taxonomy and ecological information: Stewart, 1962; Fassett, 1957; Fernald, 1950; Anderson, 1972; Wass, 1972; Tatnall, 1946; Jack McCormick & Associates, Inc., 1977; and Uhler, 1977.

Key

▆ = Major distribution	
▒ = Marginal distribution	
S —	Submerged rooted aquatic plant
E —	Emergent rooted aquatic plant
U —	Upland wetland plant
SH—	Shrub
T —	Tree

224

Type of Vegetation	Salinity Regime (Tidal Fresh 0 ppt — Oligohaline 0.5 — Mesohaline 5 — Polyhaline 18 — 35)	Wetland Plants
		DIVISION ANTHOPHYTA
		ORDER PANDANALES
		FAMILY TYPHACEAE
E		*Typha angustifolia* (narrow leaf cattail)
E		*Typha latifolia* (common cattail)
		FAMILY SPARGANIACEAE
E		*Sparganium* sp. (burreed)
		ORDER NAJADALES
		FAMILY ZOSTERACEAE
S		*Zostera marina* (eelgrass)
		FAMILY POTAMOGETONACEAE
S		*Potamogeton crispus* (curly pondweed, curlyleaf pondweed)
S		*Potamogeton nodosus* (floating pondweed, longleaf pondweed)
S		*Potamogeton perfoliatus* var. *bupleuroides* (redhead grass, redhead pondweed)
S		*Potamogeton pectinatus* (sago pondweed)
S		*Potamogeton richardsonii* (clasping-leaf pondweed)
		FAMILY RUPPIACEAE
S		*Ruppia maritima* (widgeon grass)
		FAMILY ZANNICHELLIACEAE
S		*Zannichellia palustris* (horned pondweed)
		FAMILY NAJADACEAE
S		*Najas flexilis* (bushy pondweed, slender water nymph)
S		*Najas guadalupensis* (southern naiad, common naiad, common water nymph)
		ORDER ALISMALES
		FAMILY ALISMATACEAE
E		*Sagittaria* spp. (arrowheads)
		ORDER HYDROCHARITALES
		FAMILY HYDROCHARITACEAE
S		*Elodea canadensis* (common waterweed)
S		*Vallisneria americana* (wild celery, tapegrass)
		ORDER GRAMINALES
		FAMILY POACEAE
U, E		*Distichlis spicata* (saltgrass, spikegrass)
U, E		*Echinochloa walteri* (Walter millet)
U, E		*Leersia oryzoides* (rice cutgrass)
U		*Panicum agrostoides* (panic grass)
E		*Panicum hemitomon* (maidencane)
U, E		*Panicum virgatum* (switchgrass, panic grass)
U, E		*Phragmites australis (communis)* (giant reedgrass, common reed)
E		*Spartina alterniflora* (saltmarsh cordgrass, smooth cordgrass)
E		*Spartina cynosuroides* (big cordgrass)
E		*Spartina patens* (saltmeadow cordgrass, meadow cordgrass, saltmeadow hay)
U, E		*Zizania aquatica* (wild rice)
		FAMILY CYPERACEAE
U, E		*Carex* spp. (sedges, umbrella sedges)
U, E		*Cyperus engelmanni* (nutgrass)
U, E		*Eleocharis palustris* (common spikerush, creeping spikerush)

Type of Vegetation	Salinity Regime				Wetland Plants
	Tidal Fresh 0 ppt — 0.5	**Oligo-haline** 0.5 — 5	**Meso-haline** 5 — 18	**Poly-haline** 18 — 35	
U, E					*Scirpus americanus* (common threesquare)
E					*Scirpus cyperinus* (woolgrass)
E					*Scirpus fluviatilis* (river bulrush)
U, E					*Scirpus lineatus* (bulrush)
U, E					*Scirpus olneyi* (Olney threesquare)
E					*Scirpus validus* (saltrush, great bulrush)
					FAMILY ARACEAE
E					*Acorus calamus* (sweetflag)
E					*Peltandra virginica* (arrow arum)
					FAMILY LEMNACEAE
E, S					*Lemna minor (perpusilla)* (lesser duckweed, small duckweed)
					FAMILY PONTEDERIACEAE
E					*Heteranthera reniformis* (mud plantain)
E					*Heteranthera dubia* (waterstargrass)
E					*Pontederia cordata* (pickerel weed)
					FAMILY JUNCACEAE
E, U					*Juncus effusus* (soft-rush)
E, U					*Juncus roemerianus* (black needlerush, needlerush)
E					*Juncus* spp. (rushes)
					FAMILY LILIACEAE
SH					*Smilax* spp. (greenbriers)
					FAMILY SALICACEAE
T					*Salix nigra* (black willow)
T					*Salix* spp. (willows)
					FAMILY MYRICACEAE
SH					*Myrica cerifera* (wax myrtle)
SH					*Myrica pensylvanica* (bayberry)
					FAMILY BETULACEAE
T					*Alnus serrulata* (tag alder, smooth alder)
T					*Betula nigra* (river birch)
					FAMILY FAGACEAE
T					*Quercus* spp. (oaks)
					FAMILY POLYGONACEAE
U, E					*Polygonum arifolium* (halberd-leaved tearthumb, halberdleaf tearthumb)
U, E					*Polygonum* spp. (smartweeds, tearthumbs)
E					*Rumex verticillatus* (waterdock, swampdock)
					FAMILY CHENOPODIACEAE
U					*Salicornia* spp. (glassworts)
					FAMILY AMARANTHACEAE
U, E					*Acnida cannabina* (water hemp)
					FAMILY CERATOPHYLLACEAE
S					*Ceratophyllum demersum* (coontail)
					FAMILY NYMPHAEACEAE
S, E					*Nuphar luteum (N. advena)* (spatterdock, yellow pond lily)
					FAMILY NELUMBONACEAE
E					*Nelumbo lutea* (lotus lily)
					FAMILY HAMAMELIDACEAE
T					*Liquidambar styraciflua* (sweet gum)
					FAMILY PLATANACEAE
T					*Platanus occidentalis* (sycamore)
					FAMILY ROSACEAE
SH					*Rosa palustris* (swamp rose)
SH					*Rubus* spp. (blackberries)
					FAMILY ANACARDIACEAE
SH					*Rhus* spp. (sumacs and poison ivy)
					FAMILY AQUIFOLIACEAE
T					*Ilex opaca* (American holly)
					FAMILY ACERACEAE
T					*Acer rubrum* (red maple, swamp maple)
					FAMILY BALSAMINACEAE
E					*Impatiens capensis* (jewelweed, spotted touch-me-not)
					FAMILY MALVACEAE
SH					*Hibiscus moscheutos* (including *palustris* sp.) (rosemallow, marsh hibiscus)
SH					*Kosteletzkya virginica* (seashore mallow, marshmallow)

Type of Vegetation	Salinity Regime (Tidal Fresh: 0 ppt–0.5 / Oligohaline: 0.5–5 / Mesohaline: 5–18 / Polyhaline: 18–35)	Wetland Plants
		FAMILY LYTHRACEAE
E	Tidal Fresh	*Decodon venticillatus* (water willow)
E	Oligohaline–Mesohaline	*Lythrum lineare* (loosestrife)
		FAMILY ONAGRACEAE
S	Tidal Fresh–Oligohaline	*Trapa natans* (water chestnut)
		FAMILY HALORAGACEAE
S	Tidal Fresh–Mesohaline	*Myriophyllum spicatum* (Eurasian water milfoil)
		FAMILY CORNACEAE
SH	Tidal Fresh	*Cornus* spp. (dogwoods)
		FAMILY NYSSACEAE
T	Tidal Fresh	*Nyssa sylvatica* (black gum)
		FAMILY OLEACEAE
T	Tidal Fresh	*Fraxinus pennsylvanica* (red ash)
T	Tidal Fresh	*Fraxinus* spp. (ashes)
		FAMILY GENTIANACEAE
E	Oligohaline–Polyhaline	*Sabatia stellaris* (sea pink, marshpink)
		FAMILY ASCLEPIADACEA
U, E	Tidal Fresh–Mesohaline	*Asclepias* spp. (milkweeds)
		FAMILY CONVOLVULACEAE
U	Tidal Fresh	*Cuscuta coryli* (hazel dodder)
U	Tidal Fresh	*Cuscuta gronovii* (dodder)
		FAMILY ACANTHACEAE
E	Tidal Fresh	*Justicia americana* (water willow)
		FAMILY RUBIACEAE
SH	Tidal Fresh–Mesohaline	*Cephalanthus occidentalis* (button bush)
		FAMILY CAPRIFOLIACEAE
SH	Tidal Fresh–Oligohaline	*Viburnum* spp. (arrowwoods)
		FAMILY CAMPANULACEAE
U	Tidal Fresh	*Lobelia cardinalis* (cardinal flower)
		FAMILY ASTERACEAE
U	Tidal Fresh–Polyhaline	*Aster tenuifolius* (saltmarsh aster)
E	Tidal Fresh–Mesohaline	*Aster* spp. (asters)
E	Tidal Fresh–Mesohaline	*Baccharis halimifolia* (groundselbush, sea myrtle)
E	Tidal Fresh–Oligohaline	*Bidens coronata* (tickseed sunflower)
E	Tidal Fresh–Oligohaline	*Bidens laevis* (smooth burmarigold)
SH	Oligohaline–Mesohaline	*Iva frutescens* (hightide bush, marsh elder)
E	Tidal Fresh–Polyhaline	*Pluchea foetida* (marsh fleabane, stinking fleabane)
U	Tidal Fresh	*Vernonia noveboracensis* (ironweed)
		FAMILY PINACEAE
T	Tidal Fresh–Polyhaline	*Pinus taeda* (loblolly pine)
		FAMILY LEGUMINOSAE
U	Tidal Fresh	*Amorpha fruticosa* (false indigo)
		FAMILY LENTIBULARIACEAE
E	Tidal Fresh	*Utricularia* spp. (bladderworts)

Appendix Table 3 — **Common Plants in Maryland Wetlands Bordering the Potomac Estuarine Drainage**

This matrix of vegetation in Maryland wetlands along the Potomac estuarine drainage includes those species most frequently listed in the Maryland Wetland Habitat Inventory (Maryland Department of Natural Resources, 1967-1968). The wetlands are arranged approximately from upstream to downstream (wetland numbers 1 to 170 on Folio Map 4 and in second column of this table) to suggest a transition from fresh to saline species associations along the estuary. (MDNR numbered wetlands by county in its inventory; see third column for corresponding numbers.)

The wetland types listed in the fourth column are the classifications assigned in the 1967-68 survey. When two types are listed, the type that dominates the acreage is first, and the secondary type follows. The numbers are in parentheses if the wetland no longer exists.

Two common species, a wetland grass (giant reedgrass, *Phragmites australis*) and a submerged rooted aquatic plant (eelgrass, *Zostera marina*), did not appear on the Maryland wetland inventory list. However, giant reedgrass was found in coastal wetlands along the entire Potomac estuary in an inventory made of Virginia wetlands (see Doumlele, 1976; Moore, K.A., 1975a, 1975b, 1975c; Silberhorn, 1975; and Silberhorn, in press). Eelgrass, formerly the dominant aquatic

vegetation in mesohaline and polyhaline waters of the Chesapeake Bay region, began to decline in the fall of 1973 and was virtually gone by 1974. Some viable populations have returned since that time, but the future status of eelgrass populations in the Chesapeake region remains uncertain (Stevenson, 1977).

A number of species that were less frequently listed in the Maryland inventory have been omitted from this appendix. These are the upland trees, shrubs, and weeds: dogwood, holly, pine, sycamore, milkweed, and ragweed; and the wetland plants: common spikerush, lotus lily, and marsh fleabane. Appendix Table 2 includes a more comprehensive listing of species recorded from the Potomac estuarine drainage.

227

Key

■ = species found in dominant wetland type
● = species found in secondary wetland type
▲ = species found in dominant and secondary wetland type

(a) Numbers correspond to wetland numbers marked on State of Maryland County Road Commission maps held by Maryland Department of Natural Resources, Water Resources Administration.

(b) Includes bushy pondweed, *Najas flexilis*, and southern naiad, *Najas guadalupensis*.

County	Wetland No. on Folio Map 4	County Wetland No. (a)	Wetland Types	Birch	Oaks	Gums	Ashes	Maples	Willows	Tag Alder	Button Bush	Greenbriers	Panic Grasses	Wild Rice	Arrowwoods	Najas Species (b)	Pondweeds	Eurasian Water Milfoil	Lesser Duckweed	Arrowheads	Common Waterweed	Coontail	Rose Mallow	Arrow Arum	Wax Myrtle	Sedges	Cattails	Swamp Rose	Smartweeds	Asters	Threesquares & Bulrushes	Saltmarsh Cordgrass	Big Cordgrass	Saltmeadow Cordgrass	Hightide Bush	Groundselbush	Rice Cutgrass	Saltgrass	Black Needlerush
Charles County (Continued)	33	10	12, 7	▲	▲	▲	▲			▲			■							■				■		■	■		■		■								
	34	11	7		■	■	■	■		■	■																		■										
	35	12	12, 7	▲		▲	▲			▲										■				■			■	▲	■		■								
	36	13	12										■											■					■		■								
	37	14	7, 12		■		■	■		■			○						■	▲				▲	▲				▲		▲								
	38	15	7		■		■	■							■									■							■								
	39	16	7, 12	■	■		■	■																■							■								
	40	18	7, 12	■	■		■	■		■					■									▲	▲	▲			▲		▲								
	41	19	7, 12	■	■														■	○				▲	▲	▲		▲	▲		▲								
	42	20	12, 7	▲	▲		▲	▲		▲			■							■					■			▲	■		■								
	43	21	12																	■					■				■		■								
	44	17	12, 7	○	○	○	○			▲										■				■	■			▲	■		▲								
	45	23	12, 7	▲	▲	▲	▲	▲	▲											■				▲	▲	■			■		■	■	■	■					
	46	22	12																	■				■		■	■		■				■	■					
	47	24	12																	■				■		■			■				■	■					
	48	25	7		■	■	■					■			■									■		■							■	■					
	49	28	12												■									■		■	■	■	■				■	■					
	50	27	7	■	■	■	■	■	■						■																								
	51	26	7		■	■	■	■	■																														
	52	29	(12)	Wetland Destroyed																																			
	53	30	12																				■	■									■	■					
	54	31	12						■											■									■				■	■					
	55	33	12										■											■									■	■					
	56	32	12																	■				■									■	■					
	57	34	12																																				
	58	36	12, 7	▲	▲	▲	▲	▲		■	Wetland Partially Destroyed									■		■		▲	■			▲	■		■								
	59	35	12																	■				■	■				■										
	60	38	12																■	■				■					■										
	61	61	(16)	Wetland Destroyed																																			
	62	60	(16)	Wetland Destroyed																																			
	63	59	16	Wetland Partially Destroyed																				■							■	■	■	■					■
	64	58	16							■																					■	■	■	■					
	65	57	16																								■				■	■	■	■					
	66	55	16																	■							■				■	■	■	■					
	67	53	16						■															■			■				■	■							
	68	52	16																						■						■	■	■	■					
	69	54	16																							■				■	■	■	■					
	70	51	16																											■								■	
	71	50	16																					■						■				■					■
	72	49	16																											■	■	■						■	
	73	48	16																											■	■	■	■				■		
	74	47	16																							■				■	■	■					■		
	75	56	16																						■					■	■	■							
	76	46	16																						■					■	■	■					■		
	77	45	16						■																				■	■	■		■				■		
	78	62	16							■																	■	■		■	■	■		■				■	
	79	42	12, 7	▲	▲	▲	▲	▲	▲	▲										■				■	■		■		■	■									
	80	39	12, 7	▲	▲	▲	▲	■	▲											■				■	■														
	81	37	7	■	■	■	■	■	■										■																				
	82	41	7		■	■	■	■	■											■				■															
	83	40	7, 6	■	■	▲	▲		▲						■									▲					▲										
	84	78	5													■	■	■				■																	
	85	43	12																	■				■			■				■		■	■					
	86	44	7		■	■	■	■	■	■																													
	87	75	5													■		■			■																		
	88	74	5													■	■					■																	

228

County	Wetland No. on Folio Map 4	County Wetland No. (a)	Wetland Types	Birch	Oaks	Gums	Ashes	Maples	Willows	Tag Alder	Button Bush	Greenbriers	Panic Grasses	Wild Rice	Arrowwoods	Najas Species (b)	Pondweeds	Eurasian Water Milfoil	Lesser Duckweed	Arrowheads	Common Waterweed	Coontail	Rose Mallow	Arrow Arum	Wax Myrtle	Sedges	Cattails	Swamp Rose	Smartweeds	Asters	Threesquares & Bulrushes	Saltmarsh Cordgrass	Big Cordgrass	Saltmeadow Cordgrass	Highide Bush	Groundselbush	Rice Cutgrass	Saltgrass	Black Needlerush
	89	63	7				■	■		■					■											■				■										
	90	83	12, 7			▲	▲	▲	▲	▲										■				■			■		■		■	■		■					
St. Marys County	91	1	12, 7			▲	▲	▲		▲										■				■			■	▲	■		■								
	92	2	16																						■		■		■		■				■				
	93	3	16																						■		■	■											
	94	4	16																						■		■	■											
	95	5	16																								■	■									■		
	96	6	16																						■		■	■											
	97	7	16																								■	■											
	98	8	16																						■			■											
	99	9	16																						■		■										■		
	100	11	16																						■			■											
	101	12	16																								■	■									■		
	102	13	16																						■		■	■											
	103	14	16																						■		■	■											
	104	16	16																						■		■												
	105	15	16																						■		■	■											
	106	17	16																						■		■												
	107	18	16																						■		■	■											
	108	10	7			■	■	■	■	■															■		■												
	109	24	16																						■		■	■			■	■	■	■	■				
	110	23	12																■				■								■	■	■	■	■				
	111	22	16																								■	■			■	■	■	■		■			
	112	21	16																						■		■	■			■	■	■	■	■				
	113	20	16																						■		■	■			■	■	■	■	■				
	114	19	16, 5													▲	▲	▲						■			■	■			■	■	■	■	■				■
	115	25	12, 7		▲			▲		▲													■			■	■			■	■	■	■	■				■
	116	132	12																						■		■	■			■	■	■	■	■				
	117	30	16																						■		■		■		■	■	■	■					■
	118	27	16																				■		■		■	■			■	■	■	■	■				■
	119	28	16																				■		■		■	■			■	■	■	■	■				
	120	29	12, 16							■										■			■	■		■		■			■	▲	▲	▲	▲	▲				
	121	31	16, 7			▲	▲	▲	▲	▲					▲								○				■	▲	■		■	■	■		■			■	■
	122	32	16																								■	■			■	■	■	■	■				■
	123	33	(16)												*Wetland Destroyed*																									
	124	34	16																								■	■			■	■				■			■
	125	35	7			■	■	■		■																													
	126	37	16																								■	■			■	■	■	■	■		■		
	127	36	16																						■		■	■			■	■	■	■	■				
	128	38	16																								■	■			■	■	■	■	■				■
	129	39	16																								■				■	■	■	■	■				■
	130	40	16																								■	■			■	■	■	■			■		
	131	42	16																								■				■	■	■	■	■				■
	132	129	7				■	■	■	■		■																			■	■	■	■	■				
	133	41	16																								■	■			■	■	■	■					■
	134	43	16																								■	■			■	■	■	■	■				
	135	44	16																						■		■				■	■	■	■	■				
	136	45	16																								■	■			■	■	■	■				■	■
	137	46	16																								■				■	■	■	■					■
	138	47	16																								■	■			■	■	■	■	■				
	139	48	16																	■							■	■			■	■	■	■	■				
	140	49	12																■						■		■		■		■	■	■					
	141	101	7					■	■	■					■												■	■			■		■	■			■		
	142	102	7			■	■	■		■		■			■												■	■			■	■	■	■	■				
	143	103	7			■	■			■		■															■	■			■	■							
	144	128	16																								■				■	■	■	■	■				

County	Wetland No. on Folio Map 4	County Wetland No. (a)	Wetland Types	Birch	Oaks	Gums	Ashes	Maples	Willows	Tag Alder	Button Bush	Greenbriers	Panic Grasses	Wild Rice	Arrowwoods	Najas Species (b)	Pondweeds	Eurasian Water Milfoil	Lesser Duckweed	Arrowheads	Common Waterweed	Coontail	Rose Mallow	Arrow Arum	Wax Myrtle	Sedges	Cattails	Swamp Rose	Smartweeds	Asters	Threesquares & Bulrushes	Saltmarsh Cordgrass	Big Cordgrass	Saltmeadow Cordgrass	Hightide Bush	Groundselbush	Rice Cutgrass	Saltgrass	Black Needlerush
St. Marys County (Continued)	145	104	7			●	●	●	●	●		●			●																									
	146	100	7			●	●	●	●	●					●															●										
	147	50	12																						●			●		●		●		●	●					
	148	51	12																	●					●			●		●		●		●	●					
	149	52	16																									●				●		●	●	●				
	150	53	16																							●		●				●		●	●					
	151	54	16																				●							●		●	●	●	●					
	152	55	16																											●		●	●	●	●		●			
	153	56	16																										●			●	●	●	●					
	154	57	16																													●	●	●	●	●				
	155	58	16																									●				●	●	●	●			●		
	156	60	16																									●				●	●	●	●				●	●
	157	61	14																																					
	158	62	16										●																											
	159	59	16																						●				●			●	●	●	●				●	
	160	63	16																													●	●	●	●			●		●
	161	64	16																						●							●	●	●	●					
	162	65	16																													●	●	●	●		●	●		
	163	67	16																				●		●							●	●	●	●					
	164	66	16																						●				●			●	●	●	●					●
	165	68	16																				●						●			●	●	●	●					
	166	69	16																			●										●	●	●	●					
	167	70	16																			●										●	●	●	●		●			
	168	71	16																										●			●	●	●	●				●	●
	169	72	16																						●							●		●	●	●	●			
	170	74	16																			●			●							●		●	●	●	●			

With few exceptions, only species recorded from the Potomac estuary are listed; there is a good possibility that many more species than those recorded are distributed in the Potomac. Species presumed to occur in the Potomac are noted as such. Protozoans are excluded from this list. Identifications made only to genera may or may not be the same species already listed under the generic name. For a comprehensive list of free-living invertebrates in tidal fresh, oligohaline, and meso-haline waters of the Chesapeake Bay, see Wass, 1972.

Sources of record: Boynton et al., 1977; Burrell and Zwerner, 1972; Calder, 1972a; Cargo, 1971; Dahlberg, 1973; Ecological Analysts, Inc., 1974; Kreuger and Fuller, 1977; Pfitzen-meyer, 1976; Sage and Olson, 1977a; Sage, Summerfield, and Olson, 1976; Spoon; 1976; Wass, 1972; Maryland Department of Natural Resources, 1974-1976.

Sources of taxonomy: Wass, 1972 (general); Coull, 1977 (harpacticoid copepods); Gosner, 1971 (cnidarians, annelids); Pennak, 1953 (rotifers, cladocerans); Usinger, 1963 (insects).

■ Common species in the Potomac estuary.
(a) Presumed occurrence in the Potomac estuary.
(b) Found only as part of the aufwuchs community in tidal fresh regions (Spoon, 1976).
(c) With few exceptions, parasitic arthropods are not included in this list. See Zwerner and Lawler, 1972 for list of Chesapeake Bay species.
(d) Name revisions in this suborder from Coull, 1977. Former names in parentheses.
(e) Larval stages of most decapod crustaceans are meroplanktonic. Only species pelagically oriented as adults are listed here. See Fig. 6-11 for planktonic decapod larvae of the Potomac estuary.
(f) Includes forms found both in the benthos and the water column in the Potomac estuary. See Chapter 6 for those species found most often in the water column.

PHYLUM CNIDARIA
CLASS HYDROZOA (hydromedusae)
Only species with medusoid stages are listed. See Appendix Table 5 for a more complete list of Potomac hydrozoans.
ORDER ATHECATA (anthomedusae)
FAMILY TUBULARIIDAE
Ectopleura dumortieri
FAMILY CORYNIDAE
Dipurena strangulata
Sarsia tubulosa
FAMILY ZANCLEIDAE
Zanclea sp.
FAMILY CLAVIDAE
Turritopsis nutricola
FAMILY RATHKEIDAE
Rathkea octopunctata
FAMILY BOUGAINVILLIIDAE
Bougainvillia rugosa
Nemopsis bachei
ORDER LIMNOMEDUSAE
FAMILY PROBOSCIDACTYLIDAE
Proboscidactyla ornata
ORDER THECATA (leptomedusae)
FAMILY CAMPANULARIDAE
Clytia edwardsi
Clytia hemisphaerica
Clytia longicyatha
Clytia paulensis
Obelia sp.
FAMILY CAMPANULINIDAE (LOVENELLIDAE)
Lovenella gracilis

CLASS SCYPHOZOA (jellyfishes)
ORDER SEMAEOSTOMEAE
FAMILY PELAGIDAE
■ *Chrysaora quinquecirrha* (sea nettle)
FAMILY CYANIDAE
■ *Cyanea capillata* (winter jellyfish)
FAMILY ULMARIDAE
Aurelia aurita (moon jellyfish)
ORDER RHIZOSTOMATIDAE
Rhopiloma verrilli (a)

PHYLUM CTENOPHORA (ctenophores, comb jellies)
CLASS TENTACULATA
ORDER LOBATA
FAMILY MNEMIIDAE
■ *Mnemiopsis leidyi* (sea walnut)

CLASS NUDA
ORDER BEROIDA
FAMILY BEROIDAE
Beroë ovata (mermaids purse)

PHYLUM ANNELIDA (segmented worms)
Trochophore larvae of many benthic annelids are plank-tonic. See Appendix Table 5 for complete listing of species in the Potomac estuary. Species listed here are those with adults that are also often planktonic.

CLASS POLYCHAETA (polychaetes)
ORDER PHYLLODOCIDA
FAMILY NEREIDAE
■ *Nereis succinea*
ORDER SPIONIDA
FAMILY SPIONIDAE
■ *Scolecolepides viridis*

CLASS HIRUDINEA (leeches)
ORDER RHYNCHOBDELLAE
FAMILY PISCICOLIDAE (fish leeches)
Piscicola sp.

PHYLUM MOLLUSCA (molluscs)
Veliger and some trochophore larval stages of molluscs are planktonic. See Appendix Table 5 for species in the Potomac estuary.

PHYLUM ROTIFERA (rotifers)
CLASS DIGONONTA
ORDER BDELLOIDEA
FAMILY PHILODINIDAE
Philodina roseola
Rotaria citrinus
Rotaria sp.
CLASS MONOGONONTA
ORDER FLOSCULARIACEA
FAMILY FLOSCULARIIDAE
Limnias melicerta (b)
Ptygura sp.
FAMILY CONOCHILIDAE
Conochilus unicornis
Conochilus sp. (b)
FAMILY FILINIIDAE
■ *Filinia longiseta*
Filinia sp.
FAMILY TESTUDINELLIDAE
Testudinella patina
Testudinella sp. (b)
FAMILY HEXARTHRIDAE
Hexarthra mira
ORDER COLLOTHECACEA
FAMILY COLLOTHECIDAE
Collotheca pelagica
Collotheca sp. (b)
ORDER PLOIMA
FAMILY BRACHIONIDAE
Subfamily Brachioninae
Brachionus angularis
■ *Brachionus bidentata*
■ *Brachionus calyciflorus*
Brachionus caudatus

Brachionus diversicornis
Brachionus havanensis
■ *Brachionus plicatilis*
Brachionus plerodinoides (b)
■ *Brachionus quadridentata*
Brachionus urcelaris
Brachionus sp.
Erignatha sp. (b)
Euclanis dilatata (b)
Kellicottia bostoniensis
Kellicottia longispina
Keratella americana (b)
■ *Keratella cochlearis*
Keratella quadrata
Keratella valga
Keratella sp.
Lacinularia flosculosa (b)
Lophocharis salpina
Macrochaetus sp.
Notholca acuminata
Notholca labis
Notholca squamula
■ *Notholca* sp.
Platyias patulus
Platyias quadriocornis
Trichotria tetractis
Subfamily Colurinae
Colurella obtusa
Colurella sp.
Lepadella patella
Subfamily Lecaninae
Lecane elasma
Lecane luna
Lecane ohioensis (b)
Lecane sp.
Monostyla bulba
Monostyla clostocerca
Monostyla quadridentata
■ *Monostyla* sp.
FAMILY NOTOMMATIDAE
Cephalodella auriculata (b)
Cephalodella gibba
Cephalodella sp.
Eosphora sp.
Notommata sp.
FAMILY PLOESOMATIDAE
Ploesoma sp.
FAMILY TRICHOCERCIDAE
Trichocerca longispina
Trichocerca multicrinis
Trichocerca similis
Trichocerca sp. (b)
FAMILY GASTROPODIDAE
Chromogaster ovalis
Gastropus sp.
FAMILY ASPLANCHNIDAE
■ *Asplanchna priodonta*
Asplanchna sp.
Asplanchnopus myrmeleo

231

FAMILY SYNCHAETIDAE
Polyarthra dissimulans
Polyarthra euryptera
Polyarthra major
■ *Polyarthra vulgaris*
Synchaeta oblonga
Synchaeta pectinata
Synchaeta stylata
■ *Synchaeta* sp.

PHYLUM ARTHROPODA (arthropods) [c]
CLASS CRUSTACEA (crustaceans)
Subclass Branchiopoda (branchiopods)
ORDER CLADOCERA (water fleas)
FAMILY SIDIDAE
Diaphanosoma brachyurum
Latona setifera
FAMILY LEPTODORIDAE
■ *Leptodora kindtii*
FAMILY DAPHNIDAE
Ceriodaphnia megalops
Ceriodaphnia quadrangula
Ceriodaphnia reticulata
Ceriodaphnia rotunda
Ceriodaphnia sp.
■ *Daphnia catawba*
■ *Daphnia dubia*
Daphnia laevis
Daphnia pulex [b]
Daphnia retrocura
■ *Daphnia* sp.
Moina affinis
Moina brachiata
Moina macrocopa
Moina micrura
FAMILY BOSMINIDAE
■ *Bosmina longirostris*
Bosmina sp.
FAMILY MACROTHRICIDAE
Macrothrix laticornis
Ilyocryptus sp.
FAMILY CHYDORIDAE
■ *Alona affinis*
Alona costata
Alona monacantha
Alona sp.
Chydorus bicornutus
Chydorus piger
Chydorus sphaericus
■ *Chydorus* sp.
Eurycercus lamellatus
Leydigia acanthocercoides
■ *Leydigia quadrangularis*
Leydigia leydidi
Pleuroxus denticulatus
Pleuroxus hamulatus
FAMILY POLYPHEMIDAE
Evadne nordmanni
■ *Podon polyphemoides*
Polyphemus pediculus
FAMILY HOLOPEDIDAE
Holopedium gibberum
Subclass Ostracoda (ostracods)
ORDER PODOCOPA (seed shrimps)
FAMILY CYPRIDAE
Cyclocypris sp.
Cypria sp.
Cypridopsis sp. [b]
Cyprinotus incongruens
Subclass Copepoda (copepods)
ORDER EUCOPEPODA
Suborder Calanoida
FAMILY TEMORIDAE
■ *Eurytemora affinis*
Eurytemora americana
Eurytemora composita

FAMILY CENTROPAGIDAE
Centropages typicus
Limnocalanus macrurus
Limnocalanus sp.
FAMILY DIAPTOMIDAE
Diaptomus minutus
Diaptomus pallidus
Diaptomus sicilis
Diaptomus siciloides
Pseudodiaptomus coronatus
FAMILY ACARTIIDAE
■ *Acartia clausi*
■ *Acartia tonsa*
Suborder Harpacticoida [d]
FAMILY CANUELLIDAE
■ *Scottolana canadensis (Canuella canadensis)*
FAMILY ECTINOSOMIDAE
■ *Halectinosoma curticorne*
    (Ectinosoma curticorne)
Ectinosoma sp.
FAMILY TACHIDIIDAE
Microarthridion littorale (Tachidius littoralis)
Tachidius discipes (Tachidius brevicornis)
FAMILY PELTIDIIDAE
Alteutha depressa
FAMILY DIOSACCIDAE
Amphiascus parvus
Pseudomesocra sp.
FAMILY METIDAE
Metis holothuriae (Metis jousseaumei)
FAMILY AMEIRIDAE
Nitocra sp.
FAMILY CANTHOCAMPTIDAE
Canthocamptus sp.
FAMILY CLETODIDAE
Cletodes longicaudatus
FAMILY LAOPHONTIDAE
Paralaophonte brevirostris
Paralaophonte congenera
Paronychocamptus nanus (Laophonte nana)
Pseudonychocamptus proximus
    (Laophonte proxima)
Suborder Cyclopoida
FAMILY OITHOINIDAE
■ *Oithona colcarva (Oithona brevicornis)*
Oithona similis
FAMILY CYCLOPIDAE
Cyclops bicuspidatus
Cyclops scutifer
Cyclops varicans rubellus
■ *Cyclops vernalis*
Cyclops sp.
Eucyclops agilis
Eucyclops speratus
Halicyclops fosteri
Halicyclops magniceps
Halicyclops sp.
Macrocyclops albidus
Macrocyclops fuscus
■ *Mesocyclops edax*
Paracyclops fimbriatus poppei
Tropocyclops prasinus
FAMILY ERGASILIDAE
■ *Ergasilus* sp. (fish parasite)
ORDER BRACHIURA (fish lice)
Suborder Arguloida
FAMILY ARGULIDAE
Argulus sp.
Subclass Cirripedia (barnacles)
ORDER THORACICA
Suborder Balanomorpha
Nauplii and cypris larval stages
are planktonic.

FAMILY BALANIDAE
Balanus amphitrite
Balanus balanoides [a]
■ *Balanus eburneus*

■ *Balanus improvisus*
FAMILY CHTHAMALIDAE
Chthamalus fragilis [a]
ORDER RHIZOCEPHALA (parasitic barnacles)
FAMILY SACCULINIDAE
Sacculina sp. [a]
Subclass Malacostraca
ORDER MYSIDACEA (mysid shrimps)
FAMILY MYSIDAE
■ *Neomysis americana* (opossum shrimp)
ORDER ISOPODA (isopods)
Suborder Flabellifera
FAMILY CYMOTHOIDAE (fish parasites)
■ *Lironeca ovalis*
Olencira praegustator
Aegathoa medialis
Suborder Valvifera
FAMILY IDOTEIDAE
Chiridotea almyra
Idotea balthica
Edotea triloba
ORDER AMPHIPODA (amphipods)
Suborder Gammaridea
FAMILY AMPITHOIDAE
Cymadusa compta
FAMILY OEDICEROTIDAE
Monoculodes edwardsi
FAMILY PHOTIDAE
■ *Leptocheirus plumulosus*
FAMILY COROPHIIDAE
■ *Corophium lacustre*
FAMILY GAMMARIDAE
■ *Gammarus fasciatus*
■ *Gammarus mucronatus*
Gammarus sp.
■ *Melita nitida*
Suborder Hyperiidea
FAMILY HYPERIIDAE
Hyperia galba [a]
Hyperoche medusarum [a]
ORDER DECAPODA (decapods) [e]
Suborder Natantia (true shrimps)
FAMILY PENAEIDAE
Penaeus aztecus aztecus [a]
Penaeus duorarum duorarum [a]
FAMILY PALAEMONIDAE (grass shrimp)
■ *Palaemonetes intermedius*
■ *Palaemonetes pugio*
Palaemonetes vulgaris [a]
FAMILY CRANGONIDAE
■ *Crangon septemspinosa* (sand shrimp)

CLASS INSECTA [f] (insects)
ORDER COLLEMBOLA (springtails)
Unidentified collembolan
ORDER EPHEMEROPTERA (mayflies)
FAMILY BAETIDAE
Baetidae sp.
Caenis sp.
ORDER ODONATA (dragonflies and damselflies)
FAMILY LIBELLULIDAE
Libellula spp.
FAMILY COENAGRIONIDAE
Enallagma civile
Enallagma divigans
Ischnura verticalis
ORDER HEMIPTERA (true bugs)
FAMILY CORIXIDAE (backswimmers)
Ochteridae sp.
Sigara sp.
FAMILY BELOSTOMATIDAE
Belostoma sp.
ORDER MEGALOPTERA
    (alderflies, dobsonflies, fishflies)

FAMILY SIALIDAE (alderflies)
Sialis sp.

ORDER TRICHOPTERA (caddisflies)
  FAMILY HYDROPSYCHIDAE
    Cheumatopsyche sp.
  FAMILY LEPTOCERIDAE
    Oecetis sp.
  FAMILY PSYCHOMYIIDAE
    Cyrnellus sp.
  FAMILY LIMNEPHILIDAE
    Unidentified limnephilid
ORDER LEPIDOPTERA (aquatic caterpillars)
    Unidentified lepidopteran
    Pyralidae sp.
ORDER COLEOPTERA (beetles)
  FAMILY GYRINIDAE (whirligig beetles)
    Agabus sp.
    Gyrinus sp.
  FAMILY DYTISCIDAE
    Hydroporus sp.
  FAMILY HYDROPHILIDAE
    Helschares sp.
  FAMILY HALIPLIDAE
    Brychius spp.
  FAMILY DRYOPIDAE
    Helichus sp.
  FAMILY ELMIDAE
    Dubiraphia sp.
    Stenelmis sp.
ORDER DIPTERA (flies, gnats, mosquitoes, midges)
  FAMILY LIRIOPEIDAE (PTYCHOPTERIDAE)
    (false crane flies)
    Bittacomorpha sp.
  FAMILY PSYCHODIDAE (moth flies)
    Psychoda sp.
  FAMILY CULICIDAE (mosquitoes)
  ■ *Chaoborus* sp.
  FAMILY TENDIPEDIDAE (CHIRONOMIDAE) (midges)
    Atanytarsus sp.
    Brillia sp.
    Chironomus attenuatus
    Chironomus staegeri
    Chironomus spp.
    Clinotanypus spp.
    Coelotanypus concinnus
    Coelotanypus spp.
    Conchapelopia sp.
    Cricotopus bicinctus
    Cricotopus sp. (sylvestris group)
    Cricotopus spp. (roback group)
    Cryptochironomus fulvus
    Cryptochironomus spp.

Dicrotendipes fumidus
Dicrotendipes modesta
Dicrotendipes neomodestus
Dicrotendipes nervosus
Dicrotendipes sp.
Glyptotendipes lobiferus
Glyptotendipes spp.
Harnischia abortiva
Harnischia nais
Microspecta spp.
Microtendipes sp.
Orthocladius nivoriundus
Parachironomus sp.
Paracladopelma spp.
Polypedilum near *halterale*
Polypedilum scalaenum
Polypedilum spp.
Procladius bellus
Procladius culiciformis
Procladius spp.
Psectrocladius sp.
Pseudochironomus spp.
Pseudosmittia sp.
Rheorthocladius carlatus
Rheotanytarsus exiquus
Rheotanytarsus sp.
Smittia sp.
Stempellina bausei
Stenochironomus sp.
Tanypus sp.
Xenochironomus xenolabis
FAMILY HELEIDAE (CERATOPOGONIDAE)
  (biting midges)
  Unidentified heleids
FAMILY EPHYDRINAE
  Ephydra sp.

PHYLUM CHAETOGNATHA (arrow worms)
    Sagitta elegans

Appendix Table 5 — Benthic Macroinvertebrates of the Potomac Estuary

All species listed are benthic macroinvertebrates recorded from the Potomac estuary or presumed to occur there because of their known distributions in environmentally similar regions of the Chesapeake Bay. This list does not include species of organisms that are smaller than 0.5 mm. There is a good possibility that species other than those listed may occur in the Potomac. Identifications made only to genera may or may not be the same species already listed under the generic name. For a comprehensive list of free-living invertebrates in tidal fresh, oligohaline, mesohaline, and near polyhaline waters of the Chesapeake Bay, see Wass, 1972.

Sources of taxonomy, record, and ecological information:

1. Spoon, 1975
2. Gosner, 1971
3. Wass, 1972
4. Frey, 1946
5. Lippson, 1969-1971
6. Pfitzenmeyer, 1976
7. Hopkins, S.H., 1962
8. Calder, 1971
9. Calder, 1972a
10. Calder and Burrell, 1967
11. Dahlberg, 1973
12. Krueger and Fuller, 1977
13. Abbe, 1977a
14. Cory, 1967
15. Cargo and Schultz, 1967
16. Mountford, N. K., Holland, and Mihursky, 1977
17. Pfitzenmeyer, 1974
18. Polgar, Krainak, and Pfitzenmeyer, 1975
19. Virnstein and Boesch, 1975
20. Pfitzenmeyer and Drobeck, 1963
21. Osburn, 1944
22. Ecological Analysts, Inc., 1974
23. Mariscal, 1975
24. Pennak, 1953
25. Pettibone, 1963
26. Simpson, 1962
27. Haire, 1978
28. Rasmussen, 1973
29. Grassle and Grassle, 1974
30. Mangum, 1964
31. Orth, 1971
32. Muus, 1967
33. Holland, Mountford, N.K., and Mihursky, 1977
34. George, 1966
35. Dunnington et al., 1974
36. Galloway, 1911
37. Chanley and Andrews, 1971
38. Shaw, 1965
39. Castagna and Chanley, 1973
40. Pfitzenmeyer and Drobeck, 1964
41. Holland and Dean, 1977a
42. Holland and Dean, 1977b
43. Gordon, 1969
44. Sage and Olson, 1977a
45. Branscomb, 1976
46. Shaw, 1967
47. Van Engle, 1972b
48. Burbanck, 1967
49. Burbanck, 1963
50. Lindsay and Moran, 1976
51. Bousfield, 1973
52. Feeley, 1967
53. Enequist, 1949
54. Cory, 1978
55. Holland, 1978
56. Bousfield, 1969
57. Hargrave, 1970
58. McCain, 1968

Key

Salinity Range
- █ General distribution
- ▒ Marginal distribution

Season of Reproduction
- █ General
- ▒ Marginal

Column groups: Salinity Range | Habitat Type | Feeding Mode | Reproduction | Season of Reproduction | Sources

Legend for cells: █ = general distribution / general season; ▒ = marginal distribution / marginal season; ■ = black square (presence).

Taxon	Tidal Fresh (0-0.5 ppt)	Oligohaline (0.5-5 ppt)	Mesohaline (5-18 ppt)	Polyhaline (18-30 ppt)	Sand	Mud	Rooted Aquatics	Submerged Objects (Fouling)	Oyster Beds	Intertidal Flats and Marshes	Filter Feeder	Deposit Feeder	Predator	Scavenger	Planktonic Phases	Brooded Young	Other (e.g., budding)	Winter	Spring	Summer	Fall	Sources of Taxonomy, Record, and Ecological Information
PHYLUM PORIFERA (sponges)																						
CLASS DEMOSPONGIAE																						
ORDER HAPLOSCLERIDA																						
FAMILY SPONGILLIDAE																						
Spongilla lacustris	█	▒						■		■	■						■					1, 2
Spongilla spp.	█	█						■		■	■						■					1, 2
FAMILY HALICLONIDAE																						
Haliclona loosanoffi			▒█	█	■		■	■		■	■					■			█	█		2-6
Haliclona permollis			▒	█			■	■		■	■					■		▒	█	█		2-5
ORDER POECILOSCLERIDA																						
FAMILY TEDANIIDAE																						
Lissodendoryx carolinensis			▒					■	■	■	■					■			█			2, 3
FAMILY MICROCIONIDAE																						
Microciona prolifera (redbeard sponge)								■	■		■								█			2-6
ORDER HALICHONDRIDA																						
FAMILY HALICHONDRIDAE																						
Halichondria bowerbanki			█					■			■					■			█			2, 3
ORDER HADROMERIDA																						
FAMILY SUBERITIDAE																						
Prosuberites microsclerus			▒				■	■			■					■						2, 3
FAMILY CLIONIDAE (boring sponges)																						
Cliona truitti		█	█						■		■					■						2-7
PHYLUM CNIDARIA (COELENTERATA) (sea anemones, hydroids, true jellyfishes)																						
CLASS ANTHOZOA (sea anemones)																						
Subclass Zoantharia																						
ORDER ACTINIARIA																						
FAMILY EDWARDSIIDAE																						
Edwardsia elegans		█	■	■				■		■			■		■	■						2, 5, 6, 9
FAMILY AIPTASIOMORPHIDAE																						
Haliplanella (Aiptasiomorpha) luciae		▒	█	█	■	■	■	■	■	■			■		■	■	■		█	█		2, 6, 9
FAMILY DIADUMENIDAE																						
Diadumene leucolena		█	█		■	■	■	■	■	■			■		■	■	■		█	█		2, 4, 5, 9, 12, 13, 16

234

	Salinity Range				Habitat Type						Feeding Mode				Reproduction			Season of Reproduction				Sources of Taxonomy, Record, and Ecological Information
	Tidal Fresh (0–0.5 ppt)	Oligohaline (0.5–5 ppt)	Mesohaline (5–18 ppt)	Polyhaline (18–30 ppt)	Sand	Mud	Rooted Aquatics	Submerged Objects (Fouling)	Oyster Beds	Intertidal Flats and Marshes	Filter Feeder	Deposit Feeder	Predator	Scavenger	Planktonic Phases	Brooded Young	Other (e.g., budding)	Winter	Spring	Summer	Fall	
CLASS HYDROZOA (hydroids)																						
ORDER ATHECATA																						
FAMILY HYDRIDAE																						
Hydra americana	▬	▬					■			■			■		■							1, 6
Protohydra spp.	▬	▬					■			■			■		■							1, 2
FAMILY MOERISIIDAE																						
Moerisia lyonsia			▬				■			■	■				■		■			▬		8–10
FAMILY TUBULARIIDAE																						
Ectopleura dumortieri			▬	▬			■	■		■			■	■	■			▬	▬			2, 8, 9
FAMILY HALOCORDYLIDAE																						
Halocordyle disticha			▬	▬			■	■	■				■	■	■				▬	▬		8, 9
FAMILY CORYNIDAE																						
Dipurena strangulata			▬	▬			■	■					■		■					▬		2, 8, 9
Linvillea agassizi			▬	▬			■	■					■		■					▬		2, 8, 9
Sarsia tubulosa			▬	▬			■	■					■		■				▬			2, 6, 8, 9
FAMILY ZANCLEIDAE																						
Zanclea sp.				▬			■						■		■					▬		2, 6, 8, 9
FAMILY CLAVIDAE																						
Cordylophora lacustris	▬	▬						■	■				■		■		■					1, 2, 6, 8, 9, 11
Turritopsis nutricola				▬			■	■	■				■		■							2, 8, 9
FAMILY HYDRACTINIIDAE																						
Hydractinia arge				▬			■	■					■	■	■					▬		8, 9
FAMILY BOUGAINVILLIIDAE																						
Aselomaris michaeli			▬	▬				■					■		■			▬	▬			8, 9
Bimeria tunicata			▬	▬			■	■	■				■		■		■		▬	▬		2, 4–6, 8, 9, 14
Bougainvillia rugosa			▬	▬			■	■	■				■		■		■		▬			2, 5, 8, 9
Garveia franciscana			▬	▬			■	■	■				■		■		■		▬			2, 4–6, 8, 9, 12, 13
FAMILY RATHKEIDAE																						
Rathkea octopunctata			▬	▬			■						■		■		■		▬			2, 8, 9
FAMILY EUDENDRIIDAE																						
Eudendrium album			▬	▬			■	■	■				■		■		■		▬			2, 8, 9
ORDER LIMNOMEDUSAE																						
FAMILY PROBOSCIDACTYLIDAE																						
Proboscidactyla ornata					■	■							■		■							2, 8, 9
FAMILY OLINDIIDAE																						
Maeotias inexpectata		▬											■		■					▬		8, 9
ORDER THECATA (leptomedusae)																						
FAMILY CAMPANULARIDAE																						
Clytia edwardsi			▬	▬			■	■		■			■		■	■	■			▬		2, 8, 9
Clytia hemisphaerica		▬	▬	▬			■	■		■			■		■	■	■					8, 9
Clytia longicyatha		▬	▬	▬			■	■		■			■		■	■	■		▬			2, 4, 5, 8, 9
Clytia paulensis			▬	▬			■	■		■			■		■	■	■	▬	▬			8, 9
Gonothyraea loveni			▬	▬			■	■		■			■		■	■	■	▬			▬	2, 8, 9
Hartlaubella gelantinosa			▬	▬			■	■		■			■		■	■	■	▬			▬	8, 9
Obelia comissuralis			▬	▬			■	■		■			■		■	■	■		▬	▬		2, 8, 9
Obelia dichotoma			▬	▬			■	■		■			■		■	■	■		▬	▬		2, 8, 9
FAMILY CAMPANULINIDAE																						
"campanulina" spp. 2		▬	▬				■	■		■			■		■	■	■		▬			8, 9
Lovenella gracilis		▬	▬				■			■			■		■					▬		2, 8, 9
FAMILY PHIALELLIDAE																						
Opercularella pumila			▬	▬						■			■		■					▬		8, 9
FAMILY SERTULARIDAE																						
Dynamena cornicina			▬	▬			■	■					■		■				▬			8, 9
Thuiaria argentea			▬	▬		■	■	■					■		■				▬			2, 6, 8, 9
FAMILY PLUMULARIDAE																						
Schizotricha tenella			▬	▬			■	■	■				■				■		▬	▬		8, 9
CLASS SCYPHOZOA (true jellyfishes)																						
ORDER SEMAEOSTOMEAE																						
FAMILY PELAGIDAE																						
Chrysaora quinquecirrha (sea nettle)			▬				■	■		■			■		■	■			▬	▬		8, 9, 15
FAMILY ULMARIDAE																						
Aurelia aurita (moon jellyfish)							■			■			■		■	■			▬			6, 9

Taxon	Tidal Fresh (0-0.5 ppt)	Oligohaline (0.5-5 ppt)	Mesohaline (5-18 ppt)	Polyhaline (18-30 ppt)	Sand	Mud	Rooted Aquatics	Submerged Objects (Fouling)	Oyster Beds	Intertidal Flats and Marshes	Filter Feeder	Deposit Feeder	Predator	Scavenger	Planktonic Phases	Brooded Young	Other (e.g. budding)	Winter	Spring	Summer	Fall	Sources
PHYLUM PLATYHELMINTHES (flat worms)																						
CLASS TURBELLARIA																						
ORDER RHABDOCOELA																						
FAMILY MICROSTOMIDAE																						
Microstomum sp.	█							■					■	■		■	■					1
FAMILY MACROSTOMIDAE																						
Macrostomum sp.	█							■					■	■		■	■					1
FAMILY CALENULIDAE																						
Stenostomum sp.	█												■	■		■	■					1
ORDER TRICHLADIDA																						
Dugesia tigrina	█							■					■	■		■	■					1
ORDER POLYCLADIDA																						
FAMILY STYLOCHIDAE																						
Stylochus ellipticus			█				■	■	■				■	■			■					3-5, 16-19
FAMILY LEPTOPLANIDAE																						
Euplana gracilis			█	█		■	■	■	■	■			■	■			■					3, 12, 20
CLASS TREMATODA																						
Bucephalus sp.			█						■						parasitic	■						4
PHYLUM RHYNCHOCOELA (proboscis worms)																						
CLASS ANOPLA																						
ORDER PALEONEMERTEA																						
FAMILY CARINOMIDAE																						
Carinoma tremaphoros			█		■	■																2, 3, 12
ORDER HETERONEMERTEA																						
FAMILY LINEIDAE																						
Cerebratulus lacteus			█		■								■	■		■						2, 3, 6, 20
Micrura leidyi		█	█		■	■	■	■		■			■	■		■						2-4, 6, 12, 16-18
CLASS ENOPLA																						
ORDER HOPLONEMERTEA																						
FAMILY CARCINONEMERTIDAE																						
Carcinonemertes carcinophila			█		commensal on blue crabs									■		■						2, 3, 5
FAMILY AMPHIPORIDAE																						
Amphiporus sp.			█		■		■															2, 3
FAMILY TETRASTEMMATIDAE																						
Tetrastemma sp.			█				■									■						2, 3
Zygonemertes virescens			█				■									■						2, 3
ORDER BDELLONEMERTEA																						
FAMILY MALACOBDELLIDAE																						
Malacobdella grossa			█		commensal on oysters and clams																	2, 3
PHYLUM BRYOZOA (bryozoans)																						
FAMILY PEDICELLINIDAE																						
Barentsia gracilis	█	█						■			■				■	■	■					6, 21-23
Pedicellina cernua			█					■			■				■	■	■			█		6, 11, 21-23
FAMILY LOPHOPODIDAE																						
Lophopodella sp.	█	█						■			■					■	■					6, 11, 23, 24
Pectinatella magnifica	█	█						■			■					■	■					6, 11, 23, 24
CLASS GYMNOLAEMATA																						
ORDER CTENOSTOMATA																						
FAMILY ALCYONIDIIDAE																						
Alcyonidium verrilli		█	█					■	■		■					■	■	█				2, 3, 21
FAMILY VICTORELLIDAE																						
Victorella pavida		█	█					■	■	■	■					■	■			█		2, 3, 6, 12-14, 17, 18, 21
FAMILY NOLELLIDAE																						
Anguinella palmata			█					■	■		■					■	■					2, 3, 21

	Salinity Range				Habitat Type						Feeding Mode				Reproduction			Season of Reproduction				Sources of Taxonomy, Record, and Ecological Information
	Tidal Fresh (0–0.5 ppt)	Oligohaline (0.5–5 ppt)	Mesohaline (5–18 ppt)	Polyhaline (18–30 ppt)	Sand	Mud	Rooted Aquatics	Submerged Objects (Fouling)	Oyster Beds	Intertidal Flats and Marshes	Filter Feeder	Deposit Feeder	Predator	Scavenger	Planktonic Phases	Brooded Young	Other (e.g., budding)	Winter	Spring	Summer	Fall	
FAMILY VESICULARIDAE																						
Amathia vidovici			▬	▬					■		■				■	■						2, 3, 21
Bowerbankia gracilis			▬	▬				■	■		■				■	■			▬	▬		2–5, 14, 21
FAMILY WALKERIIDAE																						
Aeverrillia armata			▬	▬				■			■				■	■						2, 3, 21
ORDER CHEILOSTOMATA																						
FAMILY MEMBRANIPORIDAE																						
Membranipora membranacea		▬	▬	▬			■	■	■		■				■	■						2, 3, 14
Membranipora tenuis			▬	▬			■	■	■	■	■				■	■						2, 3, 6, 12–14, 21
FAMILY ELECTRIDAE																						
Electra crustulenta			▬	▬				■	■		■				■	■						2, 3, 5, 6, 14, 21
FAMILY HIPPOTHOIDAE																						
Hippothoa hyalina			▬	▬				■			■				■	■						2, 3, 21
CLASS PHYLACTOLAEMATA																						
ORDER PLUMATELLINA																						
FAMILY PLUMATELLINAE																						
Hyalinella punctata	▬							■							■	■			▬	▬		1, 22
Plumatella repens	▬							■							■	■			▬	▬		1
PHYLUM PHORONIDA																						
(phoronid worms)																						
Phoronis architecta			▬		■	■					■				■							2, 3
PHYLUM ANNELIDA																						
(segmented worms)																						
CLASS POLYCHAETA (bristle worms, polychaetes)																						
ORDER PHYLLODOCIDA																						
FAMILY PHYLLODOCIDAE																						
Eteone heteropoda		▬	▬	▬	■	■	■	■	■		■				■			▬	▬			2, 3, 16–19, 25
Eteone lactea		▬	▬	▬	■	■	■	■		■	■											2, 3, 16, 25
Nereiphylla fragilis			▬	▬	■	■			■						■							2–4, 25
FAMILY POLYNOIDAE																						
Lepidonotus sublevis			▬	▬	■	■		■	■				■		■							2, 3, 25, 54
FAMILY GLYCERIDAE																						
Glycera americana			▬	▬	■	■		■					■		■				▬	▬		2, 3, 25, 26
Glycera dibranchiata (common bloodworm)		▬	▬	▬	■	■							■		■				▬	▬		2, 3, 16, 19, 20, 25, 26
FAMILY GONIADIDAE																						
Glycinde solitaria			▬	▬	■	■							■		■							2, 3, 16, 19, 25
FAMILY NEPHTYIDAE																						
Nephtys incisa				▬	■	■							■		■							2, 3, 16, 25
FAMILY HESIONIDAE																						
Gyptis vittata			▬	▬	■	■							■	■	■							2, 3, 25, 54
Micropthalmus sczelkowii			▬	▬	■	■														▬		2, 19, 25
FAMILY NEREIDAE																						
Laeonereis culveri		▬	▬							■				■								2–4, 6, 12, 16, 19, 25
Nereis succinea	▬	▬	▬	▬				■	■	■				■	■				▬			2–4, 6, 12, 13, 16–19, 25, 27
Platynereis dumerilii			▬	▬				■						■	■				▬	▬		2, 3, 25, 28
ORDER CAPITELLIDA																						
FAMILY CAPITELLIDAE																						
Capitella capitata		▬	▬	▬	■	■	■			■		■			■	■			▬	▬	▬	2, 3, 12, 26, 28, 29
Heteromastus filiformis		▬	▬	▬	■	■				■		■			■				▬	▬		2, 5, 6, 12, 16–19, 22, 26–28
Notomastus latericeus			▬	▬	■	■						■										2, 3, 12, 26, 28
FAMILY ARENICOLIDAE																						
Arenicola cristata			▬	▬	■					■		■										2, 3
FAMILY MALDANIDAE																						
Clymenella torquata			▬	▬	■	■					■	■			■							2, 3, 30

Column groups: **Salinity Range** · **Habitat Type** · **Feeding Mode** · **Reproduction** · **Season of Reproduction** · **Sources of Taxonomy, Record, and Ecological Information**

Taxon	Tidal Fresh (0-0.5 ppt)	Oligohaline (0.5-5 ppt)	Mesohaline (5-18 ppt)	Polyhaline (18-30 ppt)	Sand	Mud	Rooted Aquatics	Submerged Objects (Fouling)	Oyster Beds	Intertidal Flats and Marshes	Filter Feeder	Deposit Feeder	Predator	Scavenger	Planktonic Phases	Brooded Young	Other (e.g., budding)	Winter	Spring	Summer	Fall	Sources
ORDER SPIONIDA																						
FAMILY SPIONIDAE																						
Polydora ligni		▓	▓	▓	■	■	■	■	■			■			■	■			▓	▓		2, 3, 6, 12, 13, 16-19, 22, 31
Polydora websteri			▓	▓					■		■				■	■						2-4, 6, 32
Prionospio heterobranchia				▓		■						■			■	■						2, 3
Paraprionospio pinnata			▓	▓	■	■				■	■	■			■					▓		2, 3, 16, 19, 33
Scolecolepides viridis		▓	▓	▓		■		■	■			■			■	■			▓			2, 3, 6, 12, 16-19, 22, 26, 27, 34
Spiophanes bombyx			░	▓	■							■			■					▓		2, 3, 12
Streblospio benedicti		▓	▓	▓	■	■					■	■			■	■			▓			2, 3, 16, 19
FAMILY CHAETOPTERIDAE																						
Chaetopterus variopedatus			░	▓	■	■				■					■							2, 3
Spiochaetopterus oculatus			░	▓	■	■									■							2, 3
FAMILY SABELLARIIDAE																						
Sabellaria vulgaris				▓	■				■	■					■							2, 3, 16, 19, 35
ORDER EUNICIDA																						
FAMILY ONUPHIDAE																						
Diopatra cuprea				▓	■	■						■			■	■			░	▓		2, 3, 25
FAMILY EUNICIDAE																						
Marphysa sanguinea				▓	■	■							■		■				░	▓		2, 3, 25
FAMILY ARABELLIDAE																						
Drilonereis longa				▓	■								■		■							2, 3, 25
FAMILY DORVILLEIDAE																						
Stauronereis rudolphi				▓	■	■						■	■		■				▓			2, 3, 25
ORDER ARICIIDA																						
FAMILY ORBINIIDAE																						
Scoloplos fragilis			▓	▓	■	■				■		■			■	■		▓	░			2, 3, 16, 26, 33
Scoloplos robustus			▓	▓	■	■	■			■		■			■	■		▓	░			2, 3, 6, 12, 16, 19, 26
ORDER CIRRATULIDA																						
FAMILY CIRRATULIDAE																						
Cirratulus grandis			▓	▓								■										2, 3
Tharyx setigera		▓	▓									■										2, 3
ORDER TEREBELLIDA																						
FAMILY PECTINARIIDAE																						
Pectinaria gouldii			▓	▓	■	■		■				■			■							2, 3, 16, 19, 20, 33
FAMILY AMPHARETIDAE																						
Asabellides oculata		░	▓	▓	■	■	■					■			■							2, 3, 19
Hypaniola grayi		▓	▓	░	■	■					■											3, 6, 11, 12, 17, 18
FAMILY TEREBELLIDAE																						
Amphitrite ornata			▓	▓	■		■			■		■			■							2, 3, 6, 20
Loimia medusa			▓	▓	■	■						■			■							2, 3, 16, 19
Pista sp.				▓	■	■						■			■							2, 3
CLASS OLIGOCHAETA (aquatic earthworms, oligochaetes)																						
ORDER PLEISLOPORA																						
FAMILY AEOLOSOMATIDAE																						
Aeolosoma hemprichi	▓						■					■					■					1, 36
Aeolosoma spp.	▓						■										■					1, 36
Aulophorus furcatus	▓						■										■					1, 36
FAMILY NADIDAE																						
Chaetogaster diaphanus	▓						■				■	■					■					1, 11, 36
Chaetogaster diastrophus	▓										■						■					1, 36
Dero sp.	▓	░		■		■					■	■					■					1, 6, 11, 36
Nais communis	▓			■		■	■				■	■			■		■					1, 11, 22, 36
Nais elinguis	▓			■		■	■				■	■			■		■					6, 11, 22, 36
Nais spp.	▓	░		■		■	■				■	■			■		■					11, 22, 36
Ophidonais spp.	▓					■						■					■					1, 36
Paranais simplex	▓	░									■	■					■					6, 11, 36
Pristina leidyi	▓	░		■		■											■					6, 11, 36
Pristina longiseta	▓			■		■											■					6, 11, 36
Pristina menoni	▓	░		■		■											■					6, 11, 36
Pristina sp.	▓						■					■					■					1, 36

Legend: █ = salinity/season range (blue bar); ▒ = light blue (extended range); ■ = black presence mark

Species	Tidal Fresh (0–0.5 ppt)	Oligohaline (0.5–5 ppt)	Mesohaline (5–18 ppt)	Polyhaline (18–30 ppt)	Sand	Mud	Rooted Aquatics	Submerged Objects (Fouling)	Oyster Beds	Intertidal Flats and Marshes	Filter Feeder	Deposit Feeder	Predator	Scavenger	Planktonic Phases	Brooded Young	Other (e.g., budding)	Winter	Spring	Summer	Fall	Sources of Taxonomy, Record, and Ecological Information
Specaria josinae	█	▒				■					■					■						11, 36
Stylaria proboscidea	█					■		■			■	■				■						1, 36
FAMILY TUBIFICIDAE																						
Aulodrilus pigueti	█	▒				■					■				■							6, 11, 22, 36
Limnodrilus hoffmeisteri	█	▒				■					■				■							6, 11, 17, 18, 22, 36
Limnodrilus profundicola	█	▒				■					■				■							6, 22, 36
Limnodrilus spiralis	█					■					■				■							11, 36
Peloscolex freyi	█				■						■				■							6, 22, 36
Peloscolex gabriellae			█	▒	■										■							3, 16, 19, 36
Peloscolex multisetosus	█					■					■				■							6, 22, 36
Protamothrix hammoniensis	█					■					■				■							6, 11, 36
Protamothrix vejdovskyi	█					■					■				■							6, 11, 22, 36
Psammorcytides curvisetosus	█					■					■				■							6, 22, 36
Tubifex tubifex	█					■					■				■							6, 11, 36
CLASS HIRUDINEA (leeches)																						
FAMILY GLOSSIPHONIIDAE																						
Glossophonia complanata	█							■							■	■						1
Hemobdella sp.	█							■							■	■						1, 4
Placobdella montifera	█	▒				■									■	■						22
Placobdella parasitica	█	▒													■	■						22
FAMILY PISCICOLIDAE																						
Illinobdella richardsoni	█							■							■	■						1
FAMILY ICHTHYOBDELLIDAE																						
Ichthyobdella rapax	█	▒				■					*commensal on grass shrimp*				■							3, 11
PHYLUM MOLLUSCA (molluscs)																						
CLASS GASTROPODA																						
Subclass Prosobranchia																						
ORDER MESOGASTROPODA																						
FAMILY LITTORINIDAE																						
Littorina irrorata			▒	█			■	■	■	■					■	■						2–4
FAMILY HYDROBIIDAE																						
Hydrobia jacksoni		▒	█									■			■	■						3
Hydrobia truncata		█	█				■								■	■						3
Littoridinops tenuipes																						12
FAMILY RISSOIDAE																						
Sayella chesapeakea		█	█		■	■									■	■						3
FAMILY ASSIMINEIDAE																						
Assiminea succinea			█	█				■	■							■						3
FAMILY CERITHIIDAE																						
Bittium varium			█				■								■	■				█	█	2, 3
Cerithiopsis greeni			█												■	■						2, 3
FAMILY EPITONIIDAE																						
Epitonium rupicolum			█		■	■									■	■						2, 3, 16
FAMILY CALYPTRAEIDAE																						
Crepidula acuta			█					■	■		■				■	■						3
Crepidula convexa			█	█	■			■	■		■				■	■						2–4
Crepidula fornicata			█	█	■			■	■		■				■	■						2, 3, 5, 20
ORDER NEOGASTROPODA																						
FAMILY MURICIDAE (oyster drills)																						
Urosalpinx cinereus			█	█	■	■	■	■	■	■			■			■						2, 3, 5
FAMILY COLUMBELLIDAE																						
Mitrella lunata			█				■								■	■						2, 3
FAMILY NASSARIIDAE (mud snails)																						
Nassarius obsoleta															■	■						2, 3, 6
Nassarius vibex															■	■						2–4, 6, 16
Subclass Opisthobranchia																						
ORDER PYRAMIDELLACEA																						
FAMILY PYRAMIDELLIDAE																						
Odostomia impressa		▒	█	█	■	■		■			■				■	■						3, 4, 22
Pyramidella sp.			█	█	■	■									■	■						3, 19

Taxon	Tidal Fresh (0–0.5 ppt)	Oligohaline (0.5–5 ppt)	Mesohaline (5–18 ppt)	Polyhaline (18–30 ppt)	Sand	Mud	Rooted Aquatics	Submerged Objects (Fouling)	Oyster Beds	Intertidal Flats and Marshes	Filter Feeder	Deposit Feeder	Predator	Scavenger	Planktonic Phases	Brooded Young	Other (e.g., budding)	Winter	Spring	Summer	Fall	Sources of Taxonomy, Record, and Ecological Information
ORDER CEPHALASPIDEA																						
FAMILY ACTEONIDAE																						
Acteon punctostriatus			░	▓	■	■							■									2, 3, 16, 19
FAMILY RETUSIDAE																						
Acteocina canaliculata			░	▓	■	■						■										2, 3, 16, 19
FAMILY ATYIDAE																						
Haminoea solitaria					■	■						■	■									2, 3, 16, 19
ORDER SACOGLOSSA																						
FAMILY ELYSIIDAE																						
Elysia chlorotica			▓	░				■					■	■								2, 3
ORDER NUDIBRANCHIA (sea slugs)																						
FAMILY CORAMBIDAE																						
Doridella (Corambe) obscura			▓	▓			■	■					■	■			■					2, 3, 12, 16, 19
FAMILY POLYCERIDAE																						
Polycerella conyma			▓	▓				■					■	■								3
FAMILY CRATENIDAE																						
Cratena pilata			▓	▓				■					■	■								2, 3, 16, 19
Subclass Pulmonata																						
ORDER BASOMMATOPHORA																						
FAMILY ELLOBIIDAE																						
Phytia myosotis			▓	▓	■	■																3
FAMILY MELAMPIDAE																						
Melampus bidentatus			░	▓						■		■										3, 6, 12
FAMILY PLANORBIDAE																						
Goniobasis virginica	▓						■							■	■	■						1
Gyraulus sp.	▓						■							■	■	■						1
Lanx sp.	▓						■							■	■	■						1
Planorbula sp.	▓	░					■							■	■	■						1, 22
FAMILY LYMNAEIDAE																						
Bulumid sp.	▓						■							■	■	■						1
Lymnaea sp.	▓	░					■							■	■	■						1, 22
FAMILY ANCYLIDAE																						
Ferrissia sp.	▓	░					■							■	■	■						6, 22
FAMILY PHYSIDAE																						
Physa gyrina	▓	░					■							■	■	■						1, 12, 22
FAMILY AMNICOLIDAE																						
Bithynia spp.	▓						■							■		■						1
CLASS BIVALVIA (bivalves)																						
Subclass Pteriomorphia																						
ORDER PTEROCONCHIDA																						
FAMILY MYTILIDAE																						
Amygdalum papyria			▓	▓				■	■	■	■											2–4
Brachidontes recurvus (curved mussel)		░	▓		■			■	■	■	■								▓	▓	▓	2–4, 6, 12–14, 16–19, 27, 37–39
Modiolus demissus		░	▓		■			■	■	■	■								▓	▓	▓	2–5, 12, 37, 39
FAMILY OSTREIDAE																						
Crassostrea virginica (American oyster)			░	▓	■	■		■	■		■									▓		2–6, 12, 13, 16–20, 27, 37, 39
ORDER HETERODONTIDA																						
FAMILY CORBICULIIDAE																						
Corbicula manilensis (Asian clam)	▓	░			■	■					■											24, 54
Polymesoda caroliniana		▓	▓		■	■				■	■											2, 3
FAMILY SPHAERIIDAE																						
Musculinium spp.	▓	░			■	■					■					■	■					24, 54
Pisidium dibium	▓				■	■					■											1
Sphaerium spp.	▓				■	■				■						■	■					5, 22, 24
FAMILY DREISSENIDAE																						
Congeria leucopheata (Conrad's false mussel)		░	▓					■	■		■											2–4, 6, 12, 17, 18, 20, 22, 27, 37, 39

Legend: ▬ = blue bar (salinity range / season of reproduction); ■ = filled (black) square.

Taxon	Tidal Fresh (0–0.5 ppt)	Oligohaline (0.5–5 ppt)	Mesohaline (5–18 ppt)	Polyhaline (18–30 ppt)	Sand	Mud	Rooted Aquatics	Submerged Objects (Fouling)	Oyster Beds	Intertidal Flats and Marshes	Filter Feeder	Deposit Feeder	Predator	Scavenger	Planktonic Phases	Brooded Young	Other (e.g., budding)	Winter	Spring	Summer	Fall	Sources of Taxonomy, Record, and Ecological Information
FAMILY CARDIIDAE																						
Laevicardium mortoni				▬	■				■	■	■				■				▬	▬		2, 3, 6, 37, 39
FAMILY VENERIDAE																						
Gemma gemma (gem clam)			▬	▬	■					■						■				▬		2-4, 16, 19, 37, 39
FAMILY PETRICOLIDAE																						
Petricola pholadiformis				▬	■	■				■	■				■				▬	▬		2, 3, 6, 16, 20, 37, 39
FAMILY MACTRIDAE																						
Mulinia lateralis (coot clam)			▬	▬	■	■				■	■				■			▬	▬	▬	▬	2-4, 6, 12, 16, 19, 20, 37-39
Rangia cuneata (brackish-water clam)	▬	▬	▬		■	■				■	■				■				▬	▬		2-4, 6, 12, 16-20, 27, 37, 39, 40
FAMILY TELLINIDAE (tellin clams)																						
Macoma balthica		▬	▬	▬	■	■				■	■	■			■				▬	▬	▬	2, 3, 6, 11, 12, 16-19, 27, 37, 39
Macoma phenax			▬	▬	■	■				■	■	■			■				▬	▬		2-6, 12, 16-19, 37-39
Macoma tenta				▬	■	■				■	■	■			■							2-4, 37, 39
Tellina agilis			▬	▬	■					■	■	■			■				▬	▬		2, 3, 6, 12, 37, 39
FAMILY SOLECURTIDAE																						
Tagelus plebeius (stout razor clam)				▬	■	■				■	■	■			■					▬		2, 3, 16, 20, 37, 39, 41, 42
FAMILY SOLENIDAE																						
Ensis directus (Atlantic jackknife clam)				▬	■	■				■	■				■			▬	▬			2, 3, 6, 16, 19, 37, 39
FAMILY MYIDAE																						
Mya arenaria (soft-shell clam)			▬	▬	■	■				■	■				■			▬	▬	▬		2-6, 12, 16, 19, 20, 27, 33, 37, 39
FAMILY PHOLADIDAE																						
Barnea truncata				▬		■					■				■				▬	▬		2, 3, 19, 37
FAMILY TEREDINIDAE																						
Bankia gouldi			▬	▬				■		■	■				■				▬	▬		2, 3, 12, 39
Subclass Anomalodesmata																						
ORDER EUDESMODONTIDA																						
FAMILY LYONSIIDAE																						
Lyonsia hyalina			▬	▬	■	■				■	■				■			▬				2, 3, 19, 37, 39
ORDER EULAMELLIBRANCHIA																						
FAMILY UNIONIDAE																						
Anodonta spp.	▬	▬				■				■						■						5, 22
Lampsilis spp.	▬	▬				■				■						■						5, 22

PHYLUM ARTHROPODA

(joint-legged animals, arthropods)

Subphylum Chelicerata

CLASS MEROSTOMATA

ORDER XIPHOSURIDA

FAMILY LIMULIDAE

Taxon	Tidal Fresh	Oligohaline	Mesohaline	Polyhaline	Sand	Mud	Rooted Aquatics	Submerged Objects	Oyster Beds	Intertidal Flats	Filter Feeder	Deposit Feeder	Predator	Scavenger	Planktonic Phases	Brooded Young	Other	Winter	Spring	Summer	Fall	Sources
Limulus polyphemus (horseshoe crab)			▬	▬	■	■	■						■	■	■				▬			2, 3, 16

Subphylum Mandibulata

CLASS CRUSTACEA (crustaceans)

Subclass Branchiura

ORDER ARGULOIDA (fish lice)

FAMILY ARGULIDAE

Taxon	Tidal Fresh	Oligohaline	Mesohaline	Polyhaline	Habitat						Filter Feeder	Deposit Feeder	Predator	Scavenger	Planktonic Phases	Brooded Young	Other	Winter	Spring	Summer	Fall	Sources
Argulus alosae		▬			parasitic on gills of fish											■	■					2, 12, 13

Subclass Cirripedia (barnacles)

ORDER THORACICA

Suborder Lepadomorpha

FAMILY LEPADIDAE

Taxon	Tidal Fresh	Oligohaline	Mesohaline	Polyhaline	Habitat			Submerged Objects	Oyster Beds		Filter Feeder	Deposit Feeder	Predator	Scavenger	Planktonic	Brooded	Other	Winter	Spring	Summer	Fall	Sources
Octolasmis lowei			▬		live attached to gills of adult female blue crabs			■			■											2, 3, 5

Suborder Balanomorpha

FAMILY CHTHAMALIDAE

Taxon	Tidal Fresh	Oligohaline	Mesohaline	Polyhaline	Sand	Mud	Rooted	Submerged Objects	Oyster Beds	Intertidal Flats	Filter Feeder	Deposit Feeder	Predator	Scavenger	Planktonic	Brooded	Other	Winter	Spring	Summer	Fall	Sources
Chthamalus fragilis				▬				■	■	■	■				■	■			▬	▬		2, 3, 16, 43-46

	Salinity Range				Habitat Type						Feeding Mode				Reproduction			Season of Reproduction				
	Tidal Fresh (0-0.5 ppt)	Oligohaline (0.5-5 ppt)	Mesohaline (5-18 ppt)	Polyhaline (18-30 ppt)	Sand	Mud	Rooted Aquatics	Submerged Objects (Fouling)	Oyster Beds	Intertidal Flats and Marshes	Filter Feeder	Deposit Feeder	Predator	Scavenger	Planktonic Phases	Brooded Young	Other (e.g., budding)	Winter	Spring	Summer	Fall	Sources of Taxonomy, Record, and Ecological Information
FAMILY BALANIDAE																						
Balanus amphitrite			▒	█				■	■	■	■				■	■		█	█			2, 3, 13, 17, 18, 43-46
Balanus balanoides				▒				■	■		■				■	■		█	█	█		2, 3, 43-46
Balanus eburneus				█				■	■		■				■	■		█	█	█	█	2, 3, 43-46
Balanus improvisus		█	█	█				■	■	■	■				■	■		█	█	█	█	2, 3, 19, 43-46
Subclass Malacostraca																						
Superorder Hoplocarida																						
ORDER STOMATOPODA (mantid shrimp)																						
FAMILY SQUILLIDAE																						
Squilla empusa				█	■	■	■						■		■	■						2, 3, 5
Superorder Peracarida																						
ORDER CUMACEA (cumaceans)																						
FAMILY BODOTRIIDAE																						
Cyclaspis varians			▒	█	■	■	■	■		■						■						2, 19, 47
FAMILY LEUCONIDAE																						
Leucon americanus			▒	█	■	■			■	■						■						2, 6, 16, 19, 47
FAMILY DIASTYLIDAE																						
Oxyurostylis smithi			▒	█	■	■										■						2, 6, 16, 19, 47
ORDER TANAIDACEA (tanaids)																						
FAMILY PARATANAIDAE																						
Leptochelia rapax			▒	█	■			■		■						■						2, 6, 12, 19
ORDER ISOPODA (isopods)																						
Suborder Gnathiidea																						
FAMILY GNATHIIDAE																						
Gnathia sp.		▒	█	█	■	■	■						■	■		■						2, 22
Suborder Anthuridea																						
FAMILY ANTHURIDAE																						
Cyathura polita		█	█	█	■	■	■	■					■	■		■			█	█		2, 3, 6, 12, 16-19, 22, 27, 33, 48, 49
Suborder Flabellifera																						
FAMILY CYMOTHOIDAE																						
Lironeca ovalis			█	█	parasitic on fish											■						2, 12, 50
Olencira praegustator			█	█	parasitic on fish											■						2, 12, 50
FAMILY SPHAEROMIDAE																						
Cassidinisca lunifrons			▒	█				■	■				■	■		■						2-4, 6, 12, 17, 18
Paracerceis caudata				█		■							■	■		■						3
Sphaeroma quadridentatum			█	█	■	■		■	■	■				■		■						2, 3, 19
Suborder Valvifera																						
FAMILY IDOTEIDAE																						
Chiridotea almyra		▒	█	█	■	■	■						■	■		■				█		2, 3, 6, 17, 18, 22, 27
Chiridotea coeca			█	█	■	■	■						■	■		■						2, 3, 11, 16
Edotea triloba		▒	█	█	■	■	■						■	■		■				█		2, 3, 6, 16-19, 27
Erichsonella attenuata			█	█	■	■	■						■	■		■						2-4
Suborder Asellota																						
FAMILY ASELLIDAE																						
Asellus attenuatus	█	█			■	■	■						■	■		■						1, 2, 22
Suborder Onoscoidea																						
FAMILY LIGIDIDAE																						
Ligia exotica				█				■	■				■	■		■						2, 3, 19
ORDER AMPHIPODA (amphipods)																						
Suborder Gammaridea																						
FAMILY AMPELISCIDAE																						
Ampelisca spp.			█	█	■	■	■				■	■				■						2, 3, 19, 51
FAMILY AMPITHOIDAE																						
Ampithoe longimana		█	█	█			■	■	■			■				■		█	█			2, 3, 51
Ampithoe valida		█	█	█			■					■				■		█	█	█		2, 3, 51
Cymadusa compta			▒	█	■		■	■				■				■		█	█	█	█	2-4, 12, 51, 52

242

Benthic macroinvertebrate ecological chart — Amphipoda, Mysidacea, and Decapoda.

Taxon	Salinity Range: Tidal Fresh (0–0.5 ppt)	Oligohaline (0.5–5 ppt)	Mesohaline (5–18 ppt)	Polyhaline (18–30 ppt)	Habitat: Sand	Mud	Rooted Aquatics	Submerged Objects (Fouling)	Oyster Beds	Intertidal Flats and Marshes	Feeding: Filter Feeder	Deposit Feeder	Predator	Scavenger	Repro: Planktonic Phases	Brooded Young	Other (e.g., budding)	Season: Winter	Spring	Summer	Fall	Sources of Taxonomy, Record, and Ecological Information
FAMILY BATEIDAE																						
Batea catharinensis			X	X			X	X					X	X	X				X	X		2, 3, 12, 51, 52
FAMILY COROPHIIDAE																						
Corophium lacustre		X	X		X	X	X	X		X	X	X			X	X			X	X		2-4, 6, 11-14, 16-19, 27, 51-53
Corophium simile		X	X		X	X	X	X		X					X	X			X			51, 54
FAMILY GAMMARIDAE																						
Gammarus daiberi		X	X		X	X	X	X		X		X				X			X	X	X	2, 3, 6, 51, 55, 56
Gammarus fasciatus	X	X			X	X	X	X		X		X				X			X	X	X	1, 6, 11, 54
Gammarus mucronatus			X	X	X	X	X	X		X		X				X			X	X	X	2, 3, 6, 12, 13, 16, 19, 51, 55, 56
Gammarus palustris		X	X		X	X	X	X		X		X				X			X	X	X	2, 3, 6, 11, 51, 52, 56
Gammarus tigrinus		X	X		X	X	X	X		X		X				X			X	X	X	2, 3, 6, 51, 52, 56
Gammarus spp.		X	X	X	X	X	X	X		X		X				X		X	X	X	X	2, 3, 12, 22, 27, 52, 56
Hyalella azteca	X	X					X			X		X				X			X	X	X	22, 51, 57
Melita nitida			X	X	X	X	X	X		X		X				X			X	X	X	2, 3, 6, 12, 13, 16-19, 27, 51, 52
FAMILY HAUSTORIIDAE																						
Lepidactylus dytiscus			X	X	X					X		X				X			X	X		2, 3, 6, 12, 27, 53, 55
Neohaustorius schmitzi		X	X	X	X					X		X				X			X	X		2, 3, 51, 53, 55
FAMILY OEDICEROTIDAE																						
Monoculodes edwardsi		X	X	X	X	X		X				X				X		X	X			2, 3, 6, 11, 12, 16, 19, 27, 51-53
FAMILY PHOTIDAE																						
Leptocheirus plumulosus		X	X		X	X				X	X	X				X				X		2, 3, 6, 12, 16-19, 27, 51-53
FAMILY TALITRIDAE																						
Orchestia platensis			X							X		X		X		X			X	X		2, 3, 12, 51, 55
Talorchestia longicornis			X							X				X		X			X	X		2, 3, 12, 51, 55
ORDER CAPRELLIDEA																						
FAMILY CAPRELLIDAE																						
Caprella penantis		X	X				X	X		X	X	X			X			X	X			2-4, 58
ORDER MYSIDACEA (mysid shrimp)																						
FAMILY MYSIDAE																						
Neomysis americana (opossum shrimp)		X	X		X	X	X	X	X	X	X		X			X			X	X		2, 3, 6, 12, 16, 19
ORDER DECAPODA (decapods)																						
Intraorder Penaeidea (penaeid shrimp)																						
FAMILY PENAEIDAE																						
Penaeus spp.			X	X	X	X	X	X				X	X		X							2, 3
Intraorder Caridea (caridean shrimp)																						
FAMILY PALAEMONIDAE (grass shrimp)																						
Palaemonetes intermedius		X	X	X	X	X	X	X	X				X	X	X							2, 3, 12, 13
Palaemonetes pugio		X	X	X	X	X	X	X	X				X	X	X							2, 3, 12, 13
FAMILY CRANGONIDAE																						
Crangon septemspinosa (sand shrimp)		X	X	X	X	X			X				X		X							2-4, 6, 12, 16, 19
Intraorder Brachyura (true crabs)																						
Section Brachyryhncha																						
FAMILY PORTUNIDAE																						
Callinectes sapidus (blue crab)	X	X	X	X	X	X	X	X	X				X	X	X				X	X		2, 3, 5, 6
Ovalipes ocellatus			X	X	X	X	X	X					X	X	X							2, 3, 19
FAMILY XANTHIDAE (mud crabs)																						
Eurypanopeus depressus			X	X				X	X				X	X	X							2-6
Neopanope texana sayi		X	X	X		X		X	X				X	X	X							2, 3
Panopeus herbstii			X	X	X			X	X				X	X	X					X		2, 3, 19
Rhithropanopeus harrisii		X	X	X		X	X	X	X				X	X	X					X	X	2, 3, 5, 6, 11-13, 16-18, 27
FAMILY PINNOTHERIDAE																						
Pinnotheres ostreum (pea crab)		X	X		live in oysters					X			X									2-4
FAMILY GRAPSIDAE																						
Sesarma cinereum		X	X				X	X	X				X	X	X							2-4, 6
FAMILY OCYPODIDAE (fiddler crabs)																						
Uca minax	X	X	X							X		X	X	X	X							2, 3
Uca pugilator			X	X						X		X										2, 3
Uca pugnax			X							X		X										2, 3

PHYLUM ECHINODERMATA

CLASS HOLOTHUROIDEA

ORDER APODIDA

FAMILY SYNAPTIDAE

Leptosynapta inhaerens

PHYLUM CHORDATA (tunicates, sea squirts)

Subphylum Urochordata

CLASS ASCIDIACEA

ORDER PLEUROGONA

Suborder Stolidobranchiata

FAMILY MOLGULIDAE

Molgula manhattensis (sea squirt)

Species	Salinity Range: Tidal Fresh (0–0.5 ppt)	Oligohaline (0.5–5 ppt)	Mesohaline (5–18 ppt)	Polyhaline (18–30 ppt)	Habitat Type: Sand	Mud	Rooted Aquatics	Submerged Objects (Fouling)	Oyster Beds	Intertidal Flats and Marshes	Feeding Mode: Filter Feeder	Deposit Feeder	Predator	Scavenger	Reproduction: Planktonic Phases	Brooded Young	Other (e.g., budding)	Season of Reproduction: Winter	Spring	Summer	Fall	Sources of Taxonomy, Record, and Ecological Information
Leptosynapta inhaerens			■	■	■	■									■							2, 3, 20
Molgula manhattensis (sea squirt)			■	■			■	■	■		■				■				■	■		2, 3, 5, 6, 12, 16, 19

244

Appendix Table 6 — **Fishes of the Potomac Estuary**

Fishes listed are recorded from the Potomac estuary, including those from fresh waters above the tide line of the tributaries (excluding the Potomac River). Some additional species are also included because their known distributional range overlaps the Potomac.

Each fish is categorized according to its primary habitat as either freshwater, estuarine, marine, anadromous or semianadromous, or catadromous.

The occurrence of each species in the Potomac estuary is designated as:

A — Abundant. A common species found in large numbers.

C — Common. A species often encountered and recorded in most surveys of the estuary.

MS — Marine straggler. A marine species, rare to estuaries, but recorded as captured within the Potomac estuary or the Chesapeake Bay, most often as a single individual.

OC — Occasional. A freshwater or marine species found sporadically within the Potomac estuary, sometimes in high numbers.

OR — Old record. A species recorded in the past, but probably not currently present in the Potomac estuary.

P — Presumed occurrence. Known to occur in the Chesapeake Bay in habitats similar to those found in the Potomac estuary.

R — Rare. A seldom encountered indigenous or introduced species.

Sources of record in estuarine portion: Wiley, 1970; Musick, 1972; Scott, R. F., and Boone, 1973; Loos, 1975; O'Dell, King, and Gabor, 1973; Hildebrand and Schroeder, 1928; Smith, 1892; Smith and Beane, 1899; Ecological Analysts, Inc., 1974; McAtee and Weed, 1915; Dintaman, 1975; Schwartz, 1968 (marsh killifish); Schwartz, 1971 (gobies); and Jenkins and Zorach, 1970 (bridle shiner).

Sources of record in fresh waters above the tide line: Allee, Kelso, and Rosenblatt, 1976 (bluntnose minnow); Breder and Crawford, 1922; Dintaman, 1975; Elser, 1950 (warmouth); Giraldi and Dietemann, 1974; Howden and R. J. Mansueti, 1951; Mansueti, R. J., 1955c (rockbass, white crappie); McAtee and Weed, 1915; Musick, 1972; Schwartz, 1963 (minnows); Smith and Beane, 1899.

Sources of taxonomy and nomenclature: American Fisheries Society, 1970; Hubbs and Potter, 1971 (least brook lamprey); and Chao and Musick, 1977 (silver perch and Atlantic croaker).

245

● Recorded from Potomac estuarine drainage in fresh waters above the tide line; designation of occurrence applies to nontidal fresh waters; all species would seldom be encountered in tidal waters.

Freshwater	Estuarine	Marine	Anadromous or Semianadromous	Catadromous		Occurrence in the Potomac Estuary and Its Drainage
					PHYLUM CHORDATA	
					Subphylum Vertebrata	
					CLASS AGNATHA (jawless fishes)	
					ORDER PETROMYZONTIFORMES	
					FAMILY PETROMYZONTIDAE (lampreys)	
■	■				*Okkelbergia aepyptera* (least brook lamprey)	C
■	■				*Lampetra lamottei* (American brook lamprey)	R
			■		*Petromyzon marinus* (sea lamprey)	C
					CLASS CHONDRICHTHYES (cartilaginous fishes)	
					ORDER SQUALIFORMES	
					FAMILY CARCHARHINIDAE (requiem sharks)	
		■			*Carcharhinus leucas* (bull shark)	MS
		■			*Carcharhinus milberti* (sandbar shark)	MS
		■			*Mustelus canis* (smooth dogfish)	MS
					ORDER RAJIFORMES	
					FAMILY DASYATIDAE (stingrays)	
		■			*Dasyatis americana* (southern stingray)	MS
		■			*Dasyatis sabina* (Atlantic stingray)	MS
					FAMILY MYLIOBATIDAE (eagle rays)	
		■			*Rhinoptera bonasus* (cownose ray)	C
					CLASS OSTEICHTHYES (bony fishes)	
					ORDER ACIPENSIFORMES	
					FAMILY ACIPENSERIDAE (sturgeons)	
			■		*Acipenser brevirostrum* (shortnose sturgeon)	R, endangered
			■		*Acipenser oxyrhynchus* (Atlantic sturgeon)	R
					ORDER SEMIONOTIFORMES	
					FAMILY LEPISOSTEIDAE (gars)	
■					*Lepisosteus osseus* (longnose gar)	C
					ORDER ANGUILLIFORMES	
					FAMILY ANGUILLIDAE (freshwater eels)	
				■	*Anguilla rostrata* (American eel)	A
					ORDER CLUPEIFORMES	
					FAMILY CLUPEIDAE (herrings)	
			■		*Alosa aestivalis* (blueback herring)	A
			■		*Alosa mediocris* (hickory shad)	C
			■		*Alosa pseudoharengus* (alewife)	A

246

Freshwater	Estuarine	Marine	Anadromous or Semianadromous	Catadromous		Occurrence in the Potomac Estuary and Its Drainage
			●		*Alosa sapidissima* (American shad)	A
		●			*Brevoortia tyrannus* (Atlantic menhaden)	A
		●			*Clupea harengus harengus* (Atlantic herring)	P, MS
			●		*Dorosoma cepedianum* (gizzard shad)	A
●					*Dorosoma petenense* (threadfin shad)	OC
		●			*Opisthonema oglinum* (Atlantic thread herring)	P, MS
					FAMILY ENGRAULIDAE (anchovies)	
		●			*Anchoa hepsetus* (striped anchovy)	OC
	●				*Anchoa mitchilli* (bay anchovy)	A
					ORDER SALMONIFORMES	
					FAMILY UMBRIDAE (mudminnows)	
●					*Umbra pygmaea* (eastern mudminnow)	R
					FAMILY ESOCIDAE (pikes)	
●					*Esox americanus americanus* (redfin pickerel)	OC
●					*Esox niger* (chain pickerel)	C
					ORDER MYCTOPHIFORMES	
					FAMILY SYNODONTIDAE (lizardfishes)	
		●			*Synodus foetens* (inshore lizardfish)	OC
					ORDER CYPRINIFORMES	
					FAMILY CYPRINIDAE (minnows and carps)	
●					*Carassius auratus* (goldfish)	R
●					●*Clinostomus funduloides* (rosyside dace)	C
●					*Cyprinus carpio* (carp)	C
●					●*Ericymba buccata* (silverjaw minnow)	OR
●					●*Exoglossum maxillingua* (cutlips minnow)	C
●					*Hybognathus nuchalis* (silvery minnow)	C
●					●*Nocomis micropogon* (river chub)	R
●					*Notemigonus crysoleucas* (golden shiner)	C
●					●*Notropis amoenus* (comely shiner)	R
●					*Notropis analostanus* (satinfin shiner)	C
●					*Notropis bifrenatus* (bridle shiner)	R
●					●*Notropis chalybaeus* (ironcolor shiner)	R
●					●*Notropis cornutus* (common shiner)	C
●					*Notropis hudsonius* (spottail shiner)	C
●					●*Notropis procne* (swallowtail shiner)	C
●					●*Notropis rubellus* (rosyface shiner)	R
●					●*Notropis spilopterus* (spotfin shiner)	R
●					●*Pimephales notatus* (bluntnose minnow)	R
●					*Rhinichthys atratulus* (blacknose dace)	C
●					*Semotilus atromaculatus* (creek chub)	OC-C
●					*Semotilus corporalis* (fallfish)	C
●					●*Tinca tinca* (tench)	OR
					FAMILY CATOSTOMIDAE (suckers)	
●					*Carpiodes cyprinus* (quillback)	R
●					*Catostomus commersoni* (white sucker)	C
●					*Erimyzon oblongus* (creek chubsucker)	C
●					*Hypentelium nigricans* (northern hog sucker)	OC
●					*Moxostoma macrolepidotum* (shorthead redhorse)	OC
					ORDER SILURIFORMES	
					FAMILY ICTALURIDAE (freshwater catfishes)	
●					*Ictalurus catus* (white catfish)	A
●					●*Ictalurus furcatus* (blue catfish)	OR
●					*Ictalurus natalis* (yellow bullhead)	OC
●					*Ictalurus nebulosus* (brown bullhead)	A
●					*Ictalurus punctatus* (channel catfish)	A
●					*Noturus gyrinus* (tadpole madtom)	R
●					*Noturus insignis* (margined madtom)	OC
					FAMILY ARIIDAE (sea catfishes)	
		●			*Bagre marinus* (gafftopsail catfish)	OR
					ORDER PERCOPSIFORMES	
					FAMILY APHREDODERIDAE (pirate perches)	
●					●*Aphredoderus sayanus* (pirate perch)	OC
					FAMILY PERCOPSIDAE (trout-perches)	
●					●*Percopsis omiscomaycus* (trout-perch)	OR
					ORDER BATRACHOIDIFORMES	
					FAMILY BATRACHOIDIDAE (toadfishes)	
	●				*Opsanus tau* (oyster toadfish)	C

Freshwater	Estuarine	Marine	Anadromous or Semianadromous	Catadromous		Occurrence in the Potomac Estuary and Its Drainage
					ORDER GOBIESOCIFORMES	
					FAMILY GOBIESOCIDAE (clingfishes)	
	X				*Gobiesox strumosus* (skilletfish)	C
					ORDER GADIFORMES	
					FAMILY GADIDAE (codfishes)	
		X			*Urophycis chuss* (red hake)	P, MS
		X			*Urophycis regius* (spotted hake)	OC
					ORDER ATHERINIFORMES	
					FAMILY EXOCOETIDAE (flying fishes and halfbeaks)	
		X			*Hemiramphus* balao (balao)	P, MS
		X			*Hemiramphus brasiliensis* (ballyhoo)	P, MS
		X			*Hyporhamphus unifasciatus* (halfbeak)	OC
					FAMILY BELONIDAE (needlefishes)	
		X			*Strongylura marina* (Atlantic needlefish)	C
					FAMILY CYPRINODONTIDAE (killifishes)	
	X				*Cyprinodon variegatus* (sheepshead minnow)	C
	X				*Fundulus confluentus* (marsh killifish)	R
	X				*Fundulus diaphanus* (banded killifish)	A
	X				*Fundulus heteroclitus* (mummichog)	A
	X				*Fundulus luciae* (spotfin killifish)	R
	X				*Fundulus majalis* (striped killifish)	A
	X				*Lucania parva* (rainwater killifish)	C
					FAMILY POECILIIDAE (livebearers)	
X					*Gambusia affinis* (mosquitofish)	C
					FAMILY ATHERINIDAE (silversides)	
	X				*Membras martinica* (rough silverside)	A
	X				*Menidia beryllina* (tidewater silverside)	A
	X				*Menidia menidia* (Atlantic silverside)	A
					ORDER GASTEROSTEIFORMES	
					FAMILY GASTEROSTEIDAE (sticklebacks)	
	X				*Apeltes quadracus* (fourspine stickleback)	C
	X				*Gasterosteus aculeatus* (threespine stickleback)	P
					FAMILY SYNGNATHIDAE (pipefishes and seahorses)	
		X			*Hippocampus erectus* (lined seahorse)	OC
		X			*Syngnathus floridae* (dusky pipefish)	OC
	X				*Syngnathus fuscus* (northern pipefish)	C
					ORDER PERCIFORMES	
					FAMILY PERCICHTHYIDAE (temperate basses)	
			X		*Morone americana* (white perch)	A
			X		*Morone saxatilis* (striped bass)	A
					FAMILY SERRANIDAE (sea basses)	
		X			*Centropristes striata* (black sea bass)	OC
					FAMILY CENTRARCHIDAE (sunfishes)	
X					●*Ambloplites rupestris* (rock bass)	OR
X					*Enneacanthus gloriosus* (bluespotted sunfish)	OC
X					*Lepomis auritus* (redbreast sunfish)	C
X					*Lepomis cyanellus* (green sunfish)	OC
X					*Lepomis gibbosus* (pumpkinseed)	C
X					●*Lepomis gulosus* (warmouth)	R
X					*Lepomis macrochirus* (bluegill)	C
X					*Lepomis megalotis* (longear sunfish)	R
X					●*Micropterus dolomieui* (smallmouth bass)	R
X					*Micropterus salmoides* (largemouth bass)	C
X					●*Pomoxis annularis* (white crappie)	R
X					*Pomoxis nigromaculatus* (black crappie)	C
					FAMILY PERCIDAE (perches)	
X					●*Etheostoma flabellare* (fantail darter)	R
X					●*Etheostoma fusiforme* (swamp darter)	R
X					*Etheostoma olmstedi* (tessellated darter)	C
X					●*Etheostoma vitreum* (glassy darter)	R
			X		*Perca flavescens* (yellow perch)	A
X					●*Percina caprodes* (logperch)	R
X					●*Percina notogramma* (stripeback darter)	R
X					●*Percina peltata* (shield darter)	OR
X					●*Stizostedion vitreum vitreum* (walleye)	OR
					FAMILY POMATOMIDAE (bluefishes)	
		X			*Pomatomus saltatrix* (bluefish)	A

Freshwater	Estuarine	Marine	Anadromous or Semianadromous	Catadromous		Occurrence in the Potomac Estuary and Its Drainage
					FAMILY RACHYCENTRIDAE (cobias)	
		▆			*Rachycentron canadum* (cobia)	OC
					FAMILY ECHENEIDAE (remoras)	
		▆			*Echeneis naucrates* (sharksucker)	P, MS
					FAMILY CARANGIDAE (jacks and pompanos)	
		▆			*Alectis crinitus* (African pompano)	P, MS
		▆			*Caranx crysos* (blue runner)	OC
		▆			*Caranx hippos* (crevalle jack)	P, MS
		▆			*Selar crumenophthalmus* (bigeyed scad)	P, MS
		▆			*Selene vomer* (lookdown)	MS
		▆			*Seriola dumerili* (greater amberjack)	MS
		▆			*Trachurus lathami* (rough scad)	P, MS
		▆			*Vomer setapinnis* (Atlantic moonfish)	P, MS
					FAMILY POMADASYIDAE (grunts)	
		▆			*Orthopristis chrysoptera* (pigfish)	MS
					FAMILY SPARIDAE (porgies)	
		▆			*Archosargus probatocephalus* (sheepshead)	MS
					FAMILY SCIAENIDAE (drums)	
		▆			*Bairdiella chrysura* (silver perch)	C
		▆			*Cynoscion nebulosus* (spotted seatrout)	OC-C
		▆			*Cynoscion regalis* (weakfish)	C
		▆			*Leiostomus xanthurus* (spot)	A
		▆			*Menticirrhus americanus* (southern kingfish)	OC
		▆			*Menticirrhus saxatilis* (northern kingfish)	OC
		▆			*Micropogonias undulatus* (Atlantic croaker)	C
		▆			*Pogonias cromis* (black drum)	OC
		▆			*Scianops ocellata* (red drum)	OC
					FAMILY EPHIPPIDAE (spadefishes)	
		▆			*Chaetodipterus faber* (spadefish)	OR
					FAMILY LABRIDAE (wrasses)	
		▆			*Tautoga onitis* (tautog)	OC
					FAMILY MUGILIDAE (mullets)	
		▆			*Mugil cephalus* (striped mullet)	OC
		▆			*Mugil curema* (white mullet)	OC
					FAMILY URANOSCOPIDAE (stargazers)	
		▆			*Astroscopus guttatus* (northern stargazer)	OC
					FAMILY BLENNIIDAE (combtooth blennies)	
	▆				*Chasmodes bosquianus* (striped blenny)	C
	▆				*Hypsoblennius hentzi* (feather blenny)	P
					FAMILY GOBIIDAE (gobies)	
	▆				*Gobiosoma bosci* (naked goby)	A
	▆				*Gobiosoma ginsburgi* (seaboard goby)	P
	▆				*Gobiosoma robustum* (code goby)	P, R
	▆				*Microgobius gulosus* (clown goby)	P, R
	▆				*Microgobius thalassinus* (green goby)	OC
					FAMILY TRICHIURIDAE (cutlassfishes)	
		▆			*Trichiurus lepturus* (Atlantic cutlassfish)	MS
					FAMILY SCOMBRIDAE (mackerels and tunas)	
		▆			*Euthynnus alletteratus* (little tunny)	MS
		▆			*Sarda sarda* (Atlantic bonito)	MS
		▆			*Scomberomorus cavalla* (king mackerel)	P, MS
		▆			*Scomberomorus maculatus* (Spanish mackerel)	MS
					FAMILY STROMATEIDAE (butterfishes)	
		▆			*Peprilus alepidotus* (harvestfish)	C
		▆			*Peprilus triacanthus* (butterfish)	OC
					FAMILY TRIGLIDAE (searobins)	
		▆			*Prionotus carolinus* (northern searobin)	C
		▆			*Prionotus evolans* (striped searobin)	OC
					ORDER PLEURONECTIFORMES	
					FAMILY BOTHIDAE (lefteye flounders)	
		▆			*Paralichthys dentatus* (summer flounder)	C
		▆			*Scopthalmus aquosus* (windowpane)	P
					FAMILY PLEURONECTIDAE (righteye flounders)	
		▆			*Pseudopleuronectes americanus* (winter flounder)	C
					FAMILY SOLEIDAE (soles)	
	▆				*Trinectes maculatus* (hogchoker)	A
					FAMILY CYNOGLOSSIDAE (tonguefishes)	
		▆			*Symphurus plagiusa* (blackcheek tonguefish)	P

Freshwater	Estuarine	Marine	Anadromous or Semianadromous	Catadromous		Occurrence in the Potomac Estuary and Its Drainage
					ORDER TETRODONTIFORMES	
					FAMILY BALISTIDAE (triggerfishes and filefishes)	
		▬			*Aluterus schoepfi* (orange filefish)	OC
		▬			*Balistes vetula* (queen triggerfish)	MS
					FAMILY TETRAODONTIDAE (puffers)	
		▬			*Lagocephalus laevigatus* (smooth puffer)	P, MS
		▬			*Sphoeroides maculatus* (northern puffer)	C
					FAMILY DIODONTIDAE (porcupinefishes)	
		▬			*Chilomycterus schoepfi* (striped burrfish)	OC

Appendix Table 7 — **Birds of the Potomac Estuary**

Species listed have been recorded in various inventories and studies of birds that are not only year-round residents of the Potomac estuarine region, but that may also frequent the area during seasonal migrations along the Atlantic flyway. There is a good possibility that species other than those listed may occur in the Potomac estuarine region. For a comprehensive list of birds dependent on open waters or wetlands of the Chesapeake Bay area, see Wass, 1972.

Sources of record: Studholme et al., 1965; Maryland Department of Natural Resources, 1956-1976; Maryland Department of Natural Resources, 1967-1968; Stewart, 1962; Maryland Department of Natural Resources, 1960-1975; Wass, 1972; and Stotts, 1978.

Sources of taxonomy: Wass, 1972; Peterson, 1947; Johnsgard, 1975; and Orr, 1976.

PHYLUM CHORDATA
Subphylum Vertebrata
CLASS AVES (birds)
ORDER ANSERIFORMES (waterfowl)

FAMILY ANATIDAE (swans, geese, ducks)

Subfamily Cygninae (swans)

Olor columbianus (whistling swan)

Subfamily Anserinae (geese)

Branta bernicla (brant)

Branta canadensis (Canada goose)

Subfamily Anatinae (dabbling ducks)

Aix sponsa (wood duck)

Anas acuta (pintail)

Anas americana (baldpate, American widgeon)

Anas carolinensis (green-winged teal)

Anas discors (blue-winged teal)

Anas platyrhynchos (mallard)

Anas rubripes (black duck)

Anas strepera (gadwall)

Spatula clypeata (shoveler)

Subfamily Aythyinae (diving ducks)

Aythya affinis (lesser scaup)

Aythya americana (redhead)

Aythya collaris (ring-necked duck)

Aythya marila (greater scaup)

Aythya valisineria (canvasback)

Bucephala albeola (bufflehead)

Bucephala clangula (common goldeneye)

Clangula hyemalis (oldsquaw)

Melanitta deglandi (white-winged scoter)

Melanitta perspicillata (surf scoter)

Oidemia nigra (common scoter)

Subfamily Oxyurinae (stiff-tailed ducks)

Oxyura jamaicensis (ruddy duck)

Subfamily Merginae (mergansers)

Lophodytes cucullatus (hooded merganser)

Mergus merganser (common merganser)

Mergus serrator (red-breasted merganser)

ORDER CICONIIFORMES (herons, storks, and their allies)

FAMILY ARDEIDAE (herons, egrets, bitterns)

Subfamily Ardeinae (herons, egrets)

Ardea herodias (great blue heron)

Casmerodius albus (common egret)

Florida caerulea (little blue heron)

Subfamily Botaurinae (bitterns)

Botaurus lentiginosus (American bittern)

ORDER FALCONIFORMES (diurnal birds of prey)

FAMILY ACCIPITRIDAE (hawks, falcons, vultures, eagles)

Subfamily Haliaeetinae (eagles)

Haliaeetus leucocephalus (bald eagle)

FAMILY PANDIONIDAE (ospreys)

Subfamily Pandioninae (ospreys)

Pandion haliaetus (osprey)

ORDER GRUIFORMES (cranes, rails, and their allies)

FAMILY RALLIDAE (rails and coots)

Subfamily Rallinae (rails)

Rallus elegans (king rail)

Rallus longirostris (clapper rail)

Subfamily Fulicinae (coots)

Fulica americana (American coot)

All species listed are mammals recorded from the Potomac estuarine region or that are presumed to occur there because of their known distributions in environmentally similar regions. There is a good possibility that species other than those listed occur in the Potomac estuarine region. For a comprehensive list of mammals found in the waters, in the wetlands, and on the barrier islands of the Chesapeake Bay, see Wass, 1972.

Sources of record: Maryland Department of Natural Resources, 1967-1968; Wass, 1972; and U.S. Department of the Army, 1976.

Sources of taxonomy: Wass, 1972; Walker, 1964; and Orr, 1976.

● Rare occurrence

PHYLUM CHORDATA
Subphylum Vertebrata
  CLASS MAMMALIA (mammals)
    ORDER INSECTIVORA (insectivores)
      FAMILY TALPIDAE (desmans, moles, and shrew-moles)
        Condylura cristata (star-nosed mole)
      FAMILY SORICIDAE (shrews)
        Cryptotis parva (least shrew)
    ORDER RODENTIA (rodents)
      FAMILY CASTORIDAE (beaver)
        Castor canadensis (beaver)
      FAMILY CRICETIDAE (cricetids)
        Microtus pennsylvanicus (meadow vole)
        Ondatra zibethicus (muskrat)
        Oryzomys palustris (marsh rice rat)
        Peromyscus leucopus (white-footed mouse)
        ●*Synaptomys cooperi* (southern bog lemming)
      FAMILY MURIDAE (old world rats and mice)
        Mus musculus (house mouse)
      FAMILY CAPROMYIDAE (coypu and hutias)
        Myocastor coypu (nutria)
    ORDER CETACEA (whales, porpoises, dolphins)
      FAMILY STENIDAE (rough-toothed, river, and coastal dolphins)
        ●*Tursiops truncatus* (Atlantic bottle-nosed dolphin)

      FAMILY BALAENOPTERIDAE (rorquals)
        ●*Balaenoptera borealis* (sei whale)
    ORDER CARNIVORA (carnivores)
      FAMILY CANIDAE (dog-like carnivores)
        Urocyon cinereoargenteus (gray fox)
        Vulpes fulva (red fox)
      FAMILY PROCYONIDAE (raccoon-like animals)
        Procyon lotor (raccoon)
      FAMILY MUSTELIDAE (mustelids)
        Lutra canadensis (river otter)
        Mustela vison (eastern mink)
    ORDER PINNIPEDIA
      FAMILY PHOCIDAE (hair or earless seals)
        ●*Phoca vitulina* (harbor seal)
    ORDER MARSUPIALA (marsupials)
      FAMILY DIDELPHIDAE (American opossums)
        Didelphis virginiana (opossum)
    ORDER ARTIODACTYLA (even-toed, hoofed mammals)
      FAMILY CERVIDAE (deer-like animals)
        Odocoileus virginianus (white-tailed deer)

250

Appendix Table 9 — **Endangered Species of the Potomac Estuary**

The following species have been found or are presumed to occur in the Potomac estuary and have been declared endangered by state and/or federal agencies. A publication of the Maryland Department of Natural Resources (1977a) lists these species as well as other species that are considered endangered in Maryland.

● Not officially recorded in the Potomac estuary; however, documented sightings in other areas with similar habitats indicate that these two species may be found in the Potomac estuary.

Common Name	Scientific Name	Agency
●Atlantic hawksbill turtle	*Eretmochelys imbricata imbricata*	State and Federal
Atlantic Ridley turtle	*Lepidochelys kempi*	State and Federal
Eastern narrow-mouthed toad	*Gastrophryne carolinensis*	State
●Loggerhead turtle	*Caretta caretta caretta*	State
Rainbow snake	*Farancia erytrogramma erytrogramma*	State
Sei whale	*Balaenoptera borealis*	Federal
Shortnose sturgeon	*Acipenser brevirostrum*	Federal
Southern bald eagle	*Haliaeetus leucocephalus leucocephalus*	Federal

Appendix Table 10 — **Inventory of Pertinent Environmental Monitoring Programs and Scientific Investigations in the Potomac Estuary**

Most data and information used in the Atlas were obtained from the programs and studies listed below. The individual reports, data sheets, or data banks resulting from those investigations that were cited in the Atlas are included in the Reference section (following the conversion table) and are listed under the name of the program, sponsor, institution, and/or investigator mentioned in this table.

• Programs that contributed substantially to the information base for the synthesis of material in the Atlas.

Program: • **Anadromous Fish Stream Survey Program, Potomac River Basin**

Sponsor: U.S. Department of Commerce, National Oceanic and Atmospheric Administration, National Marine Fisheries Service

Institution and Investigators: Maryland Department of Natural Resources, Fisheries Administration (J. O'Dell, H.J. King, and J.P. Gabor)

Purpose: To inventory streams supporting runs of anadromous fish and to document species and their spawning and nursery areas in streams entering the Maryland portion of the Potomac River.

Study Period: 1970-1971

Subjects Investigated: Adults of all fishes present, eggs and larvae of anadromous fish species, water quality, stream sediments, and stream barriers

Data Collection Procedures or Sampling Design: Potomac main stem and 110 tributaries: haul seining for young-of-the-year fish, trap sampling for adults, plankton sampling for eggs and larvae

Presentation of Study Results and Data: Reports and data base stored at Maryland Department of Natural Resources

Program: • **Annual Shore Zone Seining Survey**

Sponsor: Maryland Department of Natural Resources

Institution: Maryland Department of Natural Resources, Fisheries Administration

Purpose: To determine changes in abundances and distributions of tidewater fish species in Maryland, with major emphasis on juvenile striped bass

Study Period: 1961-present

Subjects Investigated: Fish abundances and sizes, salinity, and water temperature

Data Collection Procedures or Sampling Design: Two seine hauls at each of 56 locations concentrated at the head of the Chesapeake Bay, and in the Potomac, Choptank, and Nanticoke rivers; sampling conducted during July, August, and September of each year

Presentation of Study Results and Data: Data report MFA 73-1 available from Maryland Fisheries Administration and unpublished data sheets

Program: • **Blue Crab Study**

Sponsor: U.S. Department of the Interior, Bureau of Commercial Fisheries

Institution and Investigator: University of Maryland, Center for Environmental and Estuarine Studies, Chesapeake Biological Laboratory (R.L. Lippson)

Purpose: To monitor distribution of blue crab populations throughout the Chesapeake Bay and its tributaries

Study Period: 1969-1971

Subjects Investigated: Blue crab abundance and distribution by size and sex; salinity, water temperature, and bottom sediments

Data Collection Procedures or Sampling Design: In the Potomac, monthly trawl surveys throughout the year, at 15 transects with 3 sampling stations each from Hains Point to the mouth

Presentation of Study Results and Data: Unpublished, raw data sheets available from University of Maryland, Center for Environmental and Estuarine Studies

Program: • **Commercial Fisheries Landing Statistics**

Sponsor: U.S. Department of Commerce, National Oceanic and Atmospheric Administration, National Marine Fisheries Service (NMFS)

Institution: Potomac River Fisheries Commission

Purpose: To record commercial fish harvests in the Potomac River

Study Period: 1960-present

Subjects Investigated: Finfish and shellfish harvests

Data Collection Procedures or Sampling Design: Monthly harvest reports from commercial fishermen for entire estuarine portion in four statistical reporting zones

Presentation of Study Results and Data: Maryland and Virginia landings in Current Fisheries Statistics published annually and monthly by the Department of Commerce, National Oceanic and Atmospheric Administration, National Marine Fisheries Service and data stored at NMFS

Program: • **Douglas Point Site Evaluation Studies**

Sponsor: Maryland Department of Natural Resources, Power Plant Siting Program

Institutions: Ecological Analysts, Inc.; The Johns Hopkins University, Applied Physics Laboratory and Department of Geography and Environmental Engineering

Purpose: To provide a scientific basis for the prediction of the environmental impact of a proposed nuclear power plant at Douglas Point

Study Period: 1972-1974

Subjects Investigated: Phytoplankton, zooplankton, ichthyoplankton, benthic invertebrates, adult and juvenile fish, and hydrography

Data Collection Procedures or Sampling Design: Monthly sampling with standard gear at stations along 15 transects from Maryland Point to Indian Head

Presentation of Study Results and Data: Reports to Maryland Department of Natural Resources, Power Plant Siting Program by the Johns Hopkins University, Applied Physics Laboratory and Department of Geography and Environmental Engineering and by Ecological Analysts, Inc., and data in the Johns Hopkins University, Applied Physics Laboratory data bank

Program: • **Douglas Point Site Evaluation Studies**

Sponsor: Potomac Electric Power Company

Institution: NUS Corporation, Ecological Sciences Department

Purpose: To provide a scientific basis for the prediction of the environmental impact of a proposed nuclear power plant at Douglas Point

Study Period: 1972-1973

Subjects Investigated: Phytoplankton, zooplankton, ichthyoplankton, benthic macroinvertebrates, fish, and physical and chemical variables

Data Collection Procedures or Sampling Design: Monthly sampling with standard gear along transects from Maryland Point to Indian Head

Presentation of Study Results and Data: Reports by NUS Corporation to Potomac Electric Power Company

Program: **Evaluation of the Coastal Wetlands of Maryland**

Sponsor: Maryland Department of Natural Resources, Coastal Zone Management Program

Institution: Jack McCormick & Associates, Inc.

Purpose: To refine and expand existing information on tidal coastal wetlands of Maryland

Study Period: 1975-1978

Subjects Investigated: Wetland vegetation, wildlife, water quality, and sedimentation

Data Collection Procedures or Sampling Design: Compilation of information through aerial photographs and literature review

Presentation of Study Results and Data: Report by Jack McCormick & Associates, Inc. to Maryland Department of Natural Resources, Coastal Zone Management Program

Program: **Furbearers Program**

Sponsor: Maryland Department of Natural Resources

Institution and Investigators: Maryland Department of Natural Resources, Wildlife Administration (D. Pursley and J. DiStephana)

Purpose: To preserve, enhance, and manage the furbearer population

Study Period: ongoing

Subjects Investigated: Furbearer population sizes and fur harvest

Data Collection Procedures or Sampling Design: Collection of fur dealer reports on sale and transport of pelts; tagging of beaver and otter pelts; life history studies of furbearers

Presentation of Study Results and Data: Special publications and data reports available from Maryland Wildlife Administration

Program: **Lower Potomac River (Piney Point, Maryland) Study**

Sponsor: Steuart Petroleum Company

Institution and Investigators: Virginia Institute of Marine Science (R. W. Virnstein and D. F. Boesch)

Purpose: To gather baseline data on benthic organisms of the lower Potomac River to assess the potential impact of expanding pier facilities at Steuart Petroleum Company

Study Period: 1975

Subjects Investigated: Benthic invertebrates and bottom sediments

Data Collection Procedures or Sampling Design: Triplicate bottom samples at 15 stations were obtained in February 1975

Presentation of Study Results and Data: Report by Virginia Institute of Marine Science to Steuart Petroleum Company

Program: • **Marsh and Waterfowl Investigations**

Sponsor: Maryland Department of Natural Resources, Wildlife Administration

Institutions and Investigators: Maryland Wildlife Administration (V. Stotts and L. J. Hindman); U.S. Department of the Interior, Fish and Wildlife Service

Purpose: To characterize waterfowl populations by species and habitat and to inventory abundances of resident and transient waterfowl species

Study Period: 1956-present

Subjects Investigated: Waterfowl and waterfowl habitat types

Data Collection Procedures or Sampling Design: Aerial surveys made annually during winter months

Presentation of Study Results and Data: Unpublished data sheets available from U.S. Fish and Wildlife Service

Program: • **Morgantown Steam Electric Station Studies**

Sponsor: Maryland Department of Natural Resources, Power Plant Siting Program

Institutions: Martin Marietta Corporation, Environmental Center; University of Maryland, Chesapeake Biological Laboratory

Purpose: To measure the effects of power plant operations on the aquatic ecosystem of the Potomac River

Study Period: 1972-1975

Subjects Investigated: Entrainment of planktonic organisms; nearfield studies of phytoplankton, zooplankton, macroplankton, benthic invertebrates, and finfish; and monitoring of physical and chemical variables

Data Collection Procedures or Sampling Design: Intensive spring and summer sampling with standard gears over 4 periods: fall 1972, winter 1972, spring 1973, summer 1973

Presentation of Study Results and Data: Reports by Martin Marietta Corporation, Environmental Center and University of Maryland, Chesapeake Biological Laboratory to Maryland Department of Natural Resources, Power Plant Siting Program

Program: • **Morgantown Steam Electric Station Studies**

Sponsor: Potomac Electric Power Company

Institution: Academy of Natural Sciences of Philadelphia

Purpose: To provide legally required pre- and post-operational monitoring of the aquatic biota and environment in the Potomac estuary near and at the Morgantown Steam Electric Station site

Study Period: 1966-present

Subjects Investigated: Phytoplankton, protozoa, bacteria, zooplankton, benthic invertebrates, fishes, and physical and chemical variables

Data Collection Procedures or Sampling Design: From Mathias Point to Lower Cedar Point: semiannual nearshore surveys of littoral and benthic plants and animals; biweekly studies of changes in diatom communities; monthly studies of blue crab, oyster, and fish populations

Presentation of Study Results and Data: Reports by Academy of Natural Sciences of Philadelphia to Potomac Electric Power Company

252

Program: **Non-Game and Endangered Species Program**

Sponsor: Maryland Department of Natural Resources

Institution: Maryand Department of Natural Resources, Wildlife Administration

Purpose: To determine the distributions and status of all species listed as endangered, and to determine means for restoring endangered populations

Study Period: 1976-present

Subject Investigated: Endangered animal species

Data Collection Procedures or Sampling Design: State-wide surveys: daily or seasonal observations, depending on species

Presentation of Study Results and Data: Unpublished, raw data available from Maryland Wildlife Administration

Program: **Oyster Bar Surveys**

Sponsor: Potomac River Fisheries Commission

Institutions: Potomac River Fisheries Commission; Maryland Department of Natural Resources, Fisheries Administration; University of Maryland, Chesapeake Biological Laboratory; Virginia Institute of Marine Science

Purpose: To determine the extent and general productivity of oyster bars

Study Period: 1963-present

Subjects Investigated: Status of oyster bars; size distributions and abundance of oyster adults, juveniles, and spat

Data Collection Procedures or Sampling Design: Semiannual sampling at known oyster bars, by dredge and patent tong

Presentation of Study Results and Data: Reports available from Potomac River Fisheries Commission

Program: **Piscataway Creek and Potomac River Wetland Studies**

Sponsor: Washington Suburban Sanitary Commission

Institution: Ecological Analysts, Inc.

Purpose: To gather baseline data to assess the influence of the Piscataway Wastewater Treatment Plant on the Piscataway Creek and nearby Potomac River ecosystem

Study Period: Baseline study: October 1976 - October 1977; Monitoring Program: October 1977 - present

Subjects Investigated: Wetland vegetation, phytoplankton, periphyton, zooplankton, benthic invertebrates, fish, water quality, and hydrology

Data Collection Procedures or Sampling Design: Annual quantitative sampling, April to August: wetland vegetation - June; phytoplankton and periphyton - April, June, August; zooplankton - April, August; benthic invertebrates - April, August; fish - April, May, June, July; water quality - April, June, August; and a one-time hydrological study

Presentation of Study Results and Data: Reports by Ecological Analysts, Inc. to Washington Suburban Sanitary Commission

Program: **Possum Point Steam Electric Station Studies**

Sponsor: Virginia Electric and Power Company

Institution and Investigators: Virginia Electric and Power Company, Environmental Control Department (G. M. Simmons and B. J. Armitage)

Purpose: To evaluate the ecological effects of the heated discharge from the Possum Point Power Station, specifically in relation to phytoplankton blooms in the Potomac River estuary

Study Period: April 1971-December 1971

Subjects Investigated: Phytoplankton density and taxonomy; levels of nitrates, nitrites, ammonia, and phosphates; and pH values

Data Collection Procedures or Sampling Design: Biweekly sampling at 15 stations from Occoquan Bay downstream as far as Aquia Creek

Presentation of Study Results and Data: Reports by G. M. Simmons and B. J. Armitage, Virginia Polytechnic Institute, and Virginia State University to Virginia Electric and Power Company

Program: **Possum Point Steam Electric Station Studies**

Sponsor: Virginia Electric and Power Company

Institution: Ecological Analysts, Inc.

Purpose: To characterize the fish fauna and invertebrate and benthic communities in the vicinity of Possum Point Steam Electric Station

Study Period: November 1977-December 1978

Subjects Investigated: Microzooplankton and macrozooplankton, all fish life stages

Data Collection Procedures or Sampling Design: Monthly in summer, semimonthly in spring, bimonthly in winter

Presentation of Study Results and Data: Reports by Ecological Analysts, Inc. to Virginia Electric and Power Company

Program: **Potomac Estuary Study**

Sponsor: U.S. Department of the Interior, U.S. Geological Survey

Institution: U.S. Geological Survey

Purpose: To monitor water quality, to characterize the downstream transport of nutrients and suspended sediments in the Potomac estuary, and to study biota in relation to nutrient cycling

Study Period: October 1977-present

Subjects Investigated: Levels of chlorophyll *a*, organic carbon, organic and inorganic nitrogen species, and phosphate; oxygen demand; physical variables; biota, including benthos, rooted aquatics, and plankton

Data Collection Procedures or Sampling Design: Weekly at four stations; and also before, during, and after severe climatic events such as flooding; samples are also taken over three or four successive tidal cycles

Presentation of Study Results and Data: Unpublished data available from U.S. Geological Survey

Program: • **Potomac River Fisheries Program**

Sponsor: Maryland Department of Natural Resources, Power Plant Siting Program

Institutions: Academy of Natural Sciences of Philadelphia; Martin Marietta Corporation, Environmental Center; University of Maryland, Chesapeake Biological Laboratory

Purpose: To assess the potential effects of power plant cooling systems on the fisheries of the Potomac

Study Period: 1974-1977

Subjects Investigated: Striped bass spawning stock assessment; ichthyoplankton survey; shore and tributary ichthyoplankton, zooplankton, juvenile fish, and their food habits; current velocity, hydrography, and water quality parameters

Data Collection Procedures or Sampling Design: From Morgantown to Washington, D.C.: ichthyoplankton - March to June; juveniles - June to September; adults - February to May; hydrography - April

Presentation of Study Results and Data: Reports by Academy of Natural Sciences of Philadelphia; Martin Marietta Corporation, Environmental Center; and University of Maryland, Chesapeake Biological Laboratory to Maryland Department of Natural Resources, Power Plant Siting Program, and data in Potomac River Fisheries Program data bank

Program: **Tidal Marsh Inventories**

Sponsor: U.S. Department of Commerce, National Oceanic and Atmospheric Administration, Office of Coastal Zone Management

Institution: Virginia Institute of Marine Science, Department of Wetlands Research

Purpose: To inventory tidal wetlands in Virginia

Study Period: 1975-present

Subjects Investigated: Plants and physical characteristics of marsh areas

Data Collection Procedures or Sampling Design: Aerial photographs and topographic maps used to obtain wetland locations, boundaries, and patterns of vegetation. Acreages and wetland boundaries substantiated by observations on foot, by boat, and by low-level overflights

Presentation of Study Results and Data: Reports by Virginia Institute of Marine Science to National Oceanic and Atmospheric Administration, Office of Coastal Zone Management

Program: • **Water Quality Survey, Potomac River**

Sponsor: D.C. Department of Environmental Services, Washington, D.C.

Institution: D.C. Department of Environmental Services

Purpose: To monitor water quality in the Potomac River system

Study Period: 1969-present

Subjects Investigated: Nitrates, nitrites, TKN, phosphates, BOD, suspended solids, fecal coliform, chlorophyll *a,* and physical variables

Data Collection Procedures or Sampling Design: Weekly sampling in the upper Potomac estuary from Little Falls to approximately 10 miles below the Blue Plains Sewage Treatment Plant

Presentation of Study Results and Data: Data available from U.S. Environmental Protection Agency, STORET system, Philadelphia, Pa.

Program: • **Water Quality Survey, Potomac River**

Sponsor: U.S. Environmental Protection Agency, Region III, Annapolis Field Office

Institution: U.S. Environmental Protection Agency, Annapolis Field Office

Purpose: To monitor water quality in the Potomac estuary

Study Period: 1965-present

Subjects Investigated: Nitrites, nitrates, TKN, phosphates, BOD, other chemical variables, and physical variables

Data Collection Procedures or Sampling Design: Monthly sampling through 1975; and since 1977, intensive summer sampling (July through September)

Presentation of Study Results and Data: Reports and data available from U.S. Environmental Protection Agency STORET system, Philadelphia, Pa.

Program: • **Wetland Habitat Inventory of Maryland**

Sponsor: Maryland Department of Natural Resources

Institution: Maryland Department of Natural Resources, Department of State Planning

Purpose: To inventory tidal and nontidal wetlands in Maryland

Study Period: 1967-1968

Subjects Investigated: Wetland areas, wetland plants, birds, mammals, and other wildlife

Data Collection Procedures or Sampling Design: Statewide aerial reconnaissance and field surveys; interviews with natural resources management personnel

Presentation of Study Results and Data: Unpublished data sheets and data bank output available from Maryland Department of Natural Resources

Program: • **Winter Inventory of Waterfowl and Their Distribution in the Upper Chesapeake Region**

Sponsor: U.S. Department of the Interior

Institutions and Investigator: Maryland Department of Natural Resources, Wildlife Administration (V. Stotts) and U.S. Department of the Interior, Bureau of Sport Fisheries and Wildlife

Purpose: To take annual censuses of overwintering migratory waterfowl in the Chesapeake Bay region

Study Period: 1954-present

Subject Investigated: Waterfowl

Data Collection Procedures or Sampling Design: Annual aerial surveys

Presentation of Study Results and Data: Unpublished data sheets and data bank output available from Maryland Department of Natural Resources

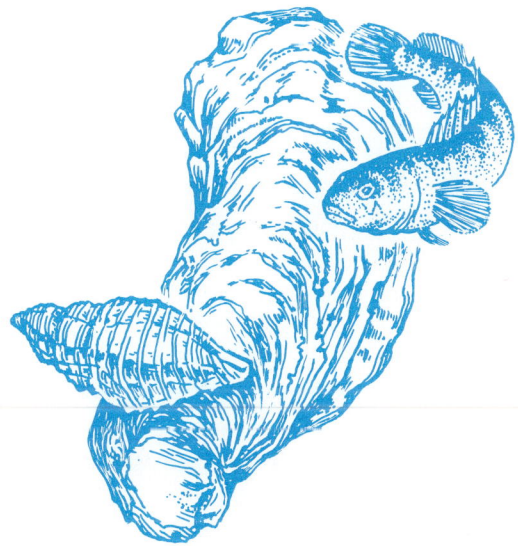

Glossary*

abiotic — not involving living organisms

accretion — the gradual build-up of sediments by the settling out of water-borne particles, sometimes resulting in land build-up

aeration — the process of dissolving gases from the air into water by turbulent mixing and molecular diffusion

algae — any of a group of chiefly aquatic nonvascular plants; most have chlorophyll

ammocoete — worm-like larval stage of the lampreys

amphipods — invertebrate animals of the crustacean order Amphipoda that are generally characterized by laterally flattened bodies; comprising the sand fleas and related forms

anadromous — fishes that ascend from their primary habitat in the ocean to fresh waters to spawn

anchor gill net — gill net that has the ends secured by anchors and the top kept at the water surface by floats

annelid worms — segmented worms of the phylum Annelida; including polychaetes, oligochaetes, and the leeches

arthropods — invertebrate animals of the phylum Arthropoda that are characterized by articulate bodies and limbs and chitinous exoskeletons; including insects and crustaceans

aufwuchs — organisms that are attached to or move about on a submerged substrate, but do not penetrate it; also called periphyton

barbel — fleshy, elongated projection found below the lower jaw, under the snout, or around the mouth in catfishes, cods, sturgeons, and other fishes

barnacles — invertebrate animals of the crustacean order Cirripedia with feathery appendages for gathering food; free-swimming as larvae, but fixed to substrates and having a hard outer covering as adults

bathymetry — measurement of the depth of a body of water, also the mapping of its floor

benthic organisms — organisms living in or on bottom substrates in aquatic habitats

biochemical oxygen demand — the oxygen necessary for the metabolic functioning of organisms and for the decomposition of organic material in the water

biodeposition — the process by which organisms deposit suspended materials onto the sediment

biogeochemical cycle — paths by which elements that are essential to life circulate from the nonliving environment to living organisms and back again to the environment

biomass — the quantity of living matter; frequently used as a measure of standing crop

biotic — involving living organisms

bioturbation — the process by which bottom sediments are disturbed by infaunal deposit feeders, primarily by their continual ingestion and reingestion of the sediments

bivalves — molluscs with paired shells; including clams, oysters, and mussels

bloodworms — segmented worms of the family Glyceridae that have a dark red respiratory fluid, are detritus feeders, and are often used as bait by sportfishermen

bloom — high concentrations of algae or phytoplankton that occur when sufficient light and nutrients are available

bottom trawl — any of a variety of cone or wedged-shaped nets towed along the bottom to catch demersal fish

brackish — having a salt content between 0.5 and 18 parts per thousand

brooding — reproductive mode in which adults retain developing eggs or embryonic stages until they are sufficiently developed to be released into the environment

carapace — hard protective outer covering such as the shell of the blue crab

carnivores — organisms that feed on animal matter

cartilaginous fish — fish of the class Chondrichthyes, having a skeleton that is mostly cartilage rather than bone, such as sharks and rays

catadromous — fishes that descend from their primary habitat in fresh water to the ocean to spawn

caudal — pertaining to the tail, or posterior end

cfs — cubic feet per second

channel habitat — term used in this book for a region where waters are 30 feet or more in depth

Chlorophyll a — a group of green pigments that occur in plant cells, chiefly in bodies called chloroplasts, and that are active in photosynthesis

chlorophyll — the most important of the principal photosynthetic pigments, often used as a measure of plant biomass

cilia — hair-like structures of cells, which beat rhythmically and cause locomotion in single cells and multicellular organisms, and which are used to create water currents and/or movement of particles

cladocerans — invertebrate animals of the crustacean order Cladocera, each having a single sessile compound eye; often called water fleas

clam bed — an area of bottom supporting a dense population of clams

class — see phylogenetic group

coastal plain — a lowland area extending in a gentle slope inland from the shoreline of an ocean

coelenterates — invertebrate animals of the phylum Coelenterata (Cnidaria) that are basically radially symmetrical; including jellyfishes, sea anemones, and hydroids

comb jellies — invertebrate animals of the phylum Ctenophora that superficially resemble jellyfish (but have few or no tentacles and no stinging nematocysts) and that have cilia, which they use for swimming, arranged in long combs around their bodies; also called ctenophores

community — a natural assemblage of animals and plants interacting with each other and with abiotic components of the environment

competition — interaction between organisms for limited resources needed for survival, including food, shelter, and space

---

* The glossary terms and abbreviations included are defined as they are used in this book, especially for those words that may have several definitions.

continental shelf — the submerged edge of a continent extending from the low-water line to a region with a distinct change in slope

copepods — minute shrimp-like crustaceans that are often the most common zooplankters in estuarine and oceanic waters

copepodite — a juvenile stage of copepods

crab pot — a baited cage with conical-shaped entrance ways, used to capture crabs

crustacean — any of the large class Crustacea of mainly aquatic arthopods that characteristically possess chitinous exoskeletons and jointed appendages; including crabs and shrimps

ctenophores — see comb jellies

cultch — shells or other material spread over oyster grounds by man, on which oyster larvae can attach and develop

cypris — a planktonic larval stage of a barnacle that eventually attaches itself to a hard substrate and then metamorphoses into an adult

DDT — (dichloro-diphenyl-trichloro-ethane) a colorless, odorless, water-insoluable crystalline insecticide ($C_{14}H_9Cl_5$) that may concentrate in certain organisms and whose effects may be toxic

decapods — invertebrate animals of the crustacean order Decapoda that are characterized by five pairs of legs; including crabs and shrimps

decomposers — organisms (chiefly bacteria and fungi) that break down nonliving matter, absorb some of the products of decomposition, and release compounds that become usable by producers

demersal — living close to or on the bottom

deposit feeders — infaunal organisms that meet their nutritional needs by either selectively or indiscriminately ingesting the sediments in which they live

detritus — particulate matter, especially that of organic origin, floating in the water or settling to the bottom

diatoms — unicellular algae of the division Bacillariophyta that have siliceous cell walls; a major division of primary producers found in aquatic ecosystems

diatomaceous earth — soil composed of the siliceous cell walls of dead diatoms that have settled and accumulated on bottoms of bodies of water

dinoflagellates — aquatic, unicellular algae of the division Pyrrophyta that are major primary producers in temperate estuaries

dip net — any of a variety of nets that are held by hand and used to catch fish or aquatic invertebrates; generally used in small streams

diurnal vertical migration — the daily vertical movement of organisms up and down in the water column

diversity — a measure of community complexity that is generally a function of both the number of species present and the relative proportions of their numbers

division — see phylogenetic group

dominant yearclass — a segment of a population composed of a single age group resulting from a spawn with a particularly high survival; this segment may make up a significant portion of the total population for a period of several years

dorsal — pertaining to the back

drainage — see watershed

dredged spoil — the sediment removed by man from the bottoms of aquatic habitats

drift net — gill net that is held upright by floats and counterweights and that drifts freely in the water column

ecosystem — an interactive system which includes the organisms of a natural community together with their environment

ectoparasite — a parasitic organism that lives on, rather than in, its host and uses the host for energy and habitat

entrainment — the process of being drawn in and transported by the flow of water; frequently refers to the process whereby organisms are drawn into the cooling systems of electrical power plants and into water intake systems of other industrial facilities

ephyrae — flattened juvenile medusoid jellyfish released from the top of the polyp form when strobilation is completed

epibenthic — term used for organisms that live on the surfaces of bottom substrates

epifauna — organisms that live on substrates in aquatic habitats; including crabs, shrimps, snails, oysters, and mussels

estuary — a semienclosed, tidal, coastal body of saline water with a free connection to the sea and within which sea water is measurably diluted with fresh water derived from land drainage; commonly the lower end of a river

euglenoids — unicellular algae of the division Euglenophyta, which have one or two flagella

euphotic zone — the zone in a water column through which there is sufficient penetration by sunlight for photosynthesis to occur

euryhaline — physiologically adapted for survival in aquatic environments over a broad range of salinities

eutrophication — a process in which the nutrient levels and productivity within a water body increase, often resulting in depletion of dissolved oxygen

eyespot — light-sensitive organelle found on some unicellular organisms, used for orientation to light

f. — form; a taxonomic category for phytoplankton below the subspecies and variety level, used to make the finest distinction of grouping within a species classification

family — see phylogenetic group

filter feeders — organisms that obtain food particles from the water column by filtering large quantities of water via a wide variety of mechanisms; mostly invertebrates, plus certain fishes

finfish — term used to refer to true fishes; excluding shellfish species such as clams and oysters

fishing effort — a quantifiable measure of the allotment of time, energy, money, and/or equipment for fishing

fishery management — the application of the principles of population dynamics, fish culturing and stocking, marketing, and conservation for the maintenance and sustained use of fish resources

fission — asexual reproduction by cell division

flagella — long hair-like appendages projecting from a cell; primarily used as a means of locomotion

flatfish — any of a number of asymmetrical fishes that compose the order Pleuronectiformes and live on the bottom; having a laterally compressed body and both eyes on the same side of the head

flatworms — unsegmented worms of the phylum Platyhelminthes, which are usually flattened, have soft bodies, and have no appendages

flyway — the principal migratory pathway by which birds travel between their summer breeding grounds and overwintering areas

food chain — the sequence in which energy as food is transferred from one group of organisms to another (from a lower to a higher trophic level)

food web — the complex interaction of food chains in a biological community, including the processes of production, consumption, and decomposition

forage species — a species that is actively sought by another as a major food source

Foraminifera — order of benthic or planktonic protozoans that generally possess shells, which are usually composed of calcium carbonate

fouling organisms — aquatic organisms that encrust submerged surfaces such as piers, boats, water intake ducts, etc.

fusiform — tapering toward both ends, as in many diatoms

fyke net — see hoop net

gastropods — molluscs of the class Gastropoda, including shell-less as well as frequently coiled, univalve species; usually have distinct heads that bear sensory organs

genus — see phylogenetic group; *pl.*-genera

geochemical cycle — the process of introduction, storage, transformation, and output of a particular chemical component in the nonliving environment

gill net — a type of gear that captures fish by entangling their gill covers in the meshes of the net

grass shrimp — shrimp of the genus *Palaemonetes,* which are most abundant in submerged aquatic vegetation

hand tongs — a hand-operated, hinged, grasping device used to collect oysters from shallow water

harpacticoids — copepods of the suborder Harpacticoida; most genera are meiofaunal

haul seine — a net operated in the shallows, usually by hand, which is used to encircle fish and is then hauled to shore

herbivores — organisms that feed on plants (primary producers) and are considered primary consumers

holdfasts — structures by which seaweeds attach themselves to a solid substrate

holoplankton — group of organisms that are planktonic throughout their lives

hoop net — a tubular net consisting of a series of conical shaped components into which fish enter and become trapped

hydraulic clam dredge — apparatus used in clam harvesting, in which clams are collected on a towed conveyor-dredge

hydroids — invertebrate animals of the jellyfish class Hydrozoa (phylum Cnidaria)

ichthyoplankton — planktonic stages of fish (i.e., eggs and larvae)

ICPRB — Interstate Commission on the Potomac River Basin

impingement — process of being forced and held against structures; frequently refers to the process whereby organisms are forced by water flow against screens in the water intake systems of electrical power plants and industrial facilities

infauna — animals that live in or burrow through bottom sediments

interface region — region in which saltwater and freshwater masses come into contact

intertidal habitat — a shoreline area that is alternately covered and uncovered by tidal waters

invertebrate — animal that has no backbone (i.e., lacking a notochord or a dorsal vertebral column)

isohalines — lines drawn on a map connecting points of equal salinity

isopods — invertebrate animals of the crustacean order Isopoda that are characterized by dorso-ventrally flattened bodies

jellyfish — invertebrate animals of the phyla Ctenophora (comb jellies) and Cnidaria (Coelenterata)

kinorhynchs — minute, superficially segmented, worm-like marine animals that feed on detritus and microscopic algae

landings — the quantity of fish and/or shellfish harvested

leeches — segmented worms of the class Hirudinea, each having a distinct sucker at each end of a cylindrically-shaped body

leptocephali — the transparent, leaf-shaped marine larvae of various eel species, including *Anguilla rostrata*

livebearers — organisms that give birth to live young, as opposed to those that deposit eggs

lithophytes — algae (seaweeds) that grow attached to submerged rocks or stones

low brackish — having a salt content between 0.5 and 5 parts per thousand

macrofauna — benthic animals larger than 0.5 millimeters

macroinvertebrates — see macrofauna

macroplankton — see macrozooplankton

macrozooplankton — group of generally mobile zooplankters, that are usually retained by a plankton net with a mesh size of 505 micrometers

main stem — the main body of a river or estuary, excluding its tributaries

marsh — soft wetland area, usually with grasses and other low vegetation

mass transport — a volume of water transported per unit time across a given plane in a body of water

MDNR — Maryland Department of Natural Resources

mean low water — (MLW) the average height of all low tides at a given location

medusa — free-swimming life stage of coelenterate jellyfishes, having a disc- or bell-shaped body of jelly-like consistency; *pl.*-medusae, *adj.*- medusoid

megalops — last larval stage of decapods before they metamorphose into juveniles

meiofauna — benthic animals between 0.5 and 63 micrometers in size

meroplankton — group of organisms that are planktonic for only a part of their life cycles

mesohaline — having a salt content between 5 and 18 parts per thousand

mesozooplankton — group of zooplankters that are retained by a plankton net with a mesh size of 202 micrometers

MGD — million gallons per day

microbiota — organisms that are less than 63 micrometers in size, e.g., bacteria

microzooplankton — group of zooplankters that can pass through a plankton net with a mesh size of 202 micrometers

mid-depth habitat — term used in this book for a region where waters are 3 to 30 feet in depth; also called shoal habitat

mid-water trawl — any of a variety of nets towed through the water column to capture pelagic fish

MLW — see mean low water

moderately brackish — having a salt content between 5 and 18 parts per thousand; mesohaline

molluscs — invertebrate animals of the phylum Mollusca, each having a soft, unsegmented body that is usually enclosed in a calcareous shell; including shellfish such as snails and clams

morphology — the physical form and structure of plants or animals

mud flat — an unvegetated muddy region that is alternately exposed and covered by the tide

mysid shrimp — invertebrate animals of the crustacean order Mysidacea that are shrimp-like in appearance and make up a significant component of the diet of many juvenile fish species

nannoplankton — group of phytoplankters that are between 5 and 60 micrometers in size

nauplii — the earliest larval stages of zooplanktonic groups such as copepods and barnacles; *sing.* - nauplius; *adj.* - nauplioid

nautical river mile — see nautical mile in conversion table

nekton — actively-swimming aquatic organisms such as fish and mammals

nematocysts — stinging cells of many coelenterates

nematodes — cylindrical worms of the group Nematoda that have no appendages and are parasitic or free-living, commonly called roundworms

net plankton — group of phytoplankters that are larger than 60 micrometers

NOAA — National Oceanographic and Atmospheric Administration

nudibranch — gastropods in which the shell is entirely absent in the adult form

nuisance species — those species that interfere with normal recreational or occupational uses of a body of water

nursery grounds — areas utilized by juvenile fish or shellfish during their development

nutrients — chemicals, primarily nitrogen and phosphorus, that are necessary for growth, development, and reproduction of plants

nutrient loading — the input of nutrients into a body of water, which leads to increased primary productivity (usually implies excess input of nutrients)

oligochaetes — segmented worms of the class Oligochaeta (phylum Annelida) having segments that are similar from head to tail

oligohaline — having a salt content between 0.5 and 5 parts per thousand

omnivores — organisms that consume both animal and plant matter

opossum shrimp — the mysid shrimp, *Neomysis americana,* which is common in the Potomac estuary

order — see phylogenetic group

organic compounds — chemical compounds containing carbon, usually of living origin

ostracods — organisms of the crustacean order Ostracoda, each characterized by a carapace which completely encloses the body

oyster bar — an area of sea bottom covered by oyster shells, frequently supporting dense populations of oysters

oyster drill — the snail, *Urosalpinx cinereus,* which preys on oysters by boring small holes through their shells

oyster seed — young newly settled oysters that are transplanted by man to new habitats for maturation

oyster spat — oyster juveniles that have settled and attached to a substrate

parapodia — appendages that are frequently paddle-like and that are found on each body segment of polychaete worms; used in locomotion, feeding, and respiration

parthenogenetic reproduction — reproduction by development of an unfertilized egg

parts per thousand — (ppt) term generally used for salinity measurements as parts salt per thousand parts water

patent tongs — hydraulically operated grab-like apparatus operated from a boom on a boat to harvest oysters

PCB — (polychlorinated biphenyls) a class of industrial compounds that are toxic environmental pollutants and tend to accumulate in animal tissue

pectoral fins — the front pair of fins on fishes, corresponding to the front legs of a four-limbed animal

peeler crab — a blue crab 1 to 3 days before it sheds its shell

pelagic — pertaining to organisms living in the water column

photosynthesis — the synthesis of organic compounds from water and carbon dioxide using light energy (photons) in the presence of chlorophyll

phylogenetic group — a group of organisms defined by evolutionary origin; also called taxonomic category or taxon; categories from least to most specific include phylum (for animals) or division (for plants), class, order, family, genus, species, subspecies, variety, and form

phylum — see phylogenetic group; *pl.* - phyla

phytoplankton — group of generally unicellular plants freely drifting in the water column that are primary producers; *sing.* - phytoplankter

plankton — group of passively drifting or weakly swimming organisms (animal or plant); *sing.* - plankter

planula larva — a freely-swimming larval stage of jellyfish, produced by sexual reproduction

Pleistocene glacial age — a geological epoch of the Quaternary period of the Cenozoic Era, lasting from 2.5 million years to about 10,000 years ago, during which there were four glacial and three interglacial periods

polychaetes — estuarine and marine segmented worms of the class Polychaeta (phylum Annelida) that have foot-like appendages called parapodia

polyhaline — having a high salt content, between 18 and 30 parts per thousand

polyp — the attached stage of a coelenterate that develops from a planula larva

population — in general, a group of organisms of the same species occupying a particular habitat

pound net — a type of stationary fishing gear with a long net attached to stakes (a wall), which directs the fish through a maze and into a trap or pocket

ppt — see parts per thousand

predation — process of feeding by which one animal preys on another animal

PRFC — Potomac River Fisheries Commission

primary consumers — plant-eating organisms that make up the second trophic level

primary producers — those organisms capable of synthesizing complex organic compounds from simple inorganic substances by photosynthesis

(green plants) or by chemosynthesis; first trophic level

proboscis worm — invertebrate animal of the phylum Rhynchocoela that is carnivorous and characterized by a tubular tongue-like structure (proboscis) which it extends to capture prey

productivity, biological —in a general sense, the rate at which biomass is produced at a particular trophic level

productivity, primary — the rate at which radiant energy is converted by means of photosynthetic processes of producer organisms, into the form of organic substances that become usable as food for higher trophic levels

protozoans — microscopic, single-celled organisms characterized by typical cellular structures

Radiolaria — order of single-celled planktonic organisms that have skeletons of siliceous spicules

radula — a rasping tongue-like organ used in food-gathering by snails

rooted aquatics — vascular plants that are not free-floating, but rooted to the bottom of a water body

rotifers — minute aquatic invertebrates of the phylum Rotifera; also known as wheel animalcules, the name being suggested by the rotating motion of their ciliated crowns, which they use in feeding and locomotion

salt balance — phenomenon of salt discharge equaling salt input in a given body of water

sand shrimp — shrimp of the genus *Crangon*

scavengers — animals that eat dead plant or animal material

sea anemone — invertebrate animal of the class Anthozoa; coelenterate that has tentacles and somewhat resembles a flower

sea nettle — the stinging jellyfish *Chrysaora quinquecirrha*

sea walnut — a species of the phylum Ctenophora (*Mnemiopsis leidyi*) commonly occurring in the Potomac estuary

seaweeds — macroscopic, multicellular species of benthic algae

secondary consumers — carnivores that make up the third trophic level

sediment — particulate organic and inorganic matter that accumulates in a loose, unconsolidated form on the bottom of a body of water

semianadromous — fishes that ascend from their primary habitat in an estuary to fresh waters to spawn

sessile — permanently attached; not free to move about

setting — process by which pelagic larvae attach to substrates

shallow habitat — term used in this book for a region where waters are less than 3 feet in depth

shellfish — a general term used to describe commercially harvested molluscs or crustaceans

shoal habitat — see mid-depth habitat

siphons — tube-like structures used by some benthic organisms during feeding and respiration

soft crab — a blue crab after shedding and before its new shell hardens

sp. — used to refer to one species in a genus when the species name is not known

spp. — used to refer to more than one species in a genus when the species names are not known

spawning — the release of eggs and sperm or the release of brooded young

species — a reproductively isolated group of interbreeding organisms; see phylogenetic group

stake gill net — gill net held in position by long stakes or poles spaced along the length of the net

standing crop — the amount of organic material in a trophic level; usually expressed in terms of the number per unit area or in terms of biomass (i.e., the quantity of living matter); also called standing stock

still fishing — the act of fishing from a stationary boat, other structure, or shoreline

strobilation — asexual body division in which jellyfish polyps produce ephyrae and eventually the medusoid sexual stage (see Fig. 6-12)

swamp — wetland saturated with water, sometimes inundated with water; vegetation usually dominated by shrubs and trees

tardigrades — animals of the phylum Tardigrada; characterized by short, cylindrical bodies, each with four pairs of stubby legs; commonly called water bears

taxon — see phylogenetic group; *pl.*-taxa

taxonomic category — see phylogenetic group

tidal excursion — the displacement of a parcel of water (or a float) during one half-tidal cycle

tidal flats — marshy or muddy areas that are covered and uncovered by the rise and fall of the tide; the vegetated parts are called tidal marshes

tidal marshes — vegetated areas that are covered and uncovered by the rise and fall of the tide

Tinntinnidae — family of microscopic planktonic protozoans having tubular or vase-shaped outer shells

TKN — total Kjeldahl nitrogen; analytical determination of the sum of organic nitrogen and ammonia nitrogen

topography — the physical features of a geographical area, particularly land elevations

trace elements — elements or compounds that are necessary for the operation of living systems, but that are present in the environment only in minute quantities

transport processes — mechanisms that result either in displacement, advection, or mixing (turbulence) of water masses

trochopore — the free-swimming early life stage of some segmented worms and molluscs

trolling — method of fishing in which the bait or lure is towed through the water

trophic level — collection of organisms that occupy the same stratum of a food web, and that are the same number of steps from the primary producers. Primary producers (phytoplankton) constitute the first trophic level, herbivorous zooplankton the second, and carnivorous organisms the third and higher trophic levels in many estuarine ecosystems.

trot line — long line with bait attached at intervals. Fish trot lines have bait attached to hooks, and hookless lines are used for crabbing.

tunicates — organisms of the phylum Chordata that are invertebrate-like as adults but have larval stages with distinct chordate characteristics; commonly called sea squirts

turbidity — decreased clarity of water caused by the presence of dissolved or suspended matter

tychoplankton — organisms of the benthic community occurring accidently in the plankton

ultraplankton — planktonic organisms less than 5 micrometers in length or diameter

var. — variety; a taxonomic category below the subspecies level, used to separate members of a species that have special, but similar, differences from other members of their species group

veliger — free-swimming early life stage of molluscs.

ventral — pertaining to the lower surface

vertebrate — animal that has a backbone (i.e., possessing a notochord or a dorsal vertebral column)

water column — term used to refer to a water body in its vertical extent

waterfowl — aquatic or semiaquatic birds of the order Anseriformes (ducks, geese, and swans)

watershed — the region drained by a stream, a river with its tributaries, a lake, or other bodies of water

wetlands — marshes, swamps, and other land-water interface areas that receive enough moisture and sunlight to support extensive growths of specially adapted vegetation

winter jellyfish — the medusoid stage of *Cyanea capillata,* which is abundant during winter months in the Chesapeake Bay region

yearclass — part of a species population consisting of all individuals produced during a given year (see also dominant yearclass)

zoea — early larval planktonic stage of many decapod crustaceans; including crabs and shrimp

zooplankton — group of animals whose distributions in the water column are governed primarily by currents; *sing.* - zooplankter

Conversion Table for Metric and English Units Used in Atlas

Metric Unit	Symbol	Multiply by	To Find	Symbol
Length				
micrometer	μm	10^{-6}	meter	m
millimeter	mm	10^{-3}	meter	m
millimeter	mm	0.0394	inch	in
centimeter	cm	10^{-2}	meter	m
centimeter	cm	0.3937	inch	in
meter	m	3.281	feet	ft
meter	m	39.37	inch	in
kilometer	km	10^3	meter	m
kilometer	km	3,281.0	foot	ft
kilometer	km	0.6214	statute mile	mi
kilometer	km	0.54	nautical mile	
Area				
square centimeter	cm²	10^{-4}	square meter	m²
square centimeter	cm²	0.155	square inch	in²
square meter	m²	10^{-6}	square kilometer	km²
square meter	m²	10.765	square foot	ft²
square meter	m²	0.000247	acre	
square kilometer	km²	10^6	square meter	m²
square kilometer	km²	0.386	square mile	mi²
square kilometer	km²	247.1	acre	
Volume				
cubic centimeter	cm³	10^{-6}	cubic meter	m³
cubic centimeter	cm³	10^{-3}	liter	l
cubic centimeter	cm³	0.061	cubic inch	in³
milliliter	ml	10^{-3}	liter	l
milliliter	ml	1.0	cubic centimeter	cm³
milliliter	ml	0.0338	fluid ounce	oz
liter	l	10^3	cubic centimeter	cm³
liter	l	10^{-3}	cubic meter	m³
liter	l	61.02	cubic inch	in³
liter	l	0.035	cubic foot	ft³
liter	l	0.908	dry quart	
liter	l	1.06	liquid quart	qt
cubic meter	m³	10^3	liter	l
cubic meter	m³	35.31	cubic foot	ft³
cubic meter	m³	264.2	gallon	gal
Velocity				
centimeters per second	cm/sec	0.0328	foot per second	ft/sec
centimeters per second	cm/sec	0.0224	mile per hour	mph
centimeters per second	cm/sec	0.0194	knot	
meters per second	m/sec	2.236	mile per hour	mph
Mass — Weight				
microgram	μg	10^{-9}	kilogram	kg
milligram	mg	10^{-6}	kilogram	kg
milligram	mg	0.000035	ounce	oz
gram	g	10^{-3}	kilogram	kg
gram	g	0.0353	ounce	oz
kilogram	kg	10^3	gram	g
kilogram	kg	2.205	pound	lb
Flow Rates				
cubic meters per second	m³/sec	35.31	cubic feet per second	cfs
cubic meters per second	m³/sec	15,852.0	gallons per minute	gpm
cubic meters per second	m³/sec	22.83	millions of gallons per day	MGD

English Unit	Symbol	Multiply by	To Find	Symbol
Length				
inch	in	25.40	millimeter	mm
inch	in	2.54	centimeter	cm
inch	in	0.0833	foot	ft
foot	ft	30.480	centimeter	cm
foot	ft	0.3048	meter	m
statute mile	mi	1.609	kilometer	km
statute mile	mi	0.869	nautical mile	
nautical mile		1.852	kilometer	km
nautical mile		1.151	statute mile	mi
Area				
square foot	ft²	0.0929	square meter	m²
square foot	ft²	144.0	square inch	in²
acre		4,047.0	square meter	m²
acre		0.4047	hectare	ha
acre		0.00156	square mile	mi²
square mile	mi²	2.59	square kilometer	km²
square mile	mi²	259.0	hectare	ha
square mile	mi²	640.0	acre	
Volume				
cubic foot	ft³	28.32	liter	l
cubic foot	ft³	0.0283	cubic meter	m³
cubic foot	ft³	29.92	quart	qt
cubic foot	ft³	7.48	gallon	gal
gallon	gal	3.785	liter	l
gallon	gal	0.003785	cubic meter	m³
gallon	gal	0.1337	cubic foot	ft³
Velocity				
knot		1.0	nautical mile per hour	
knot		1.151	mile per hour	mph
knot		51.44	centimeter per second	cm/sec
Mass — Weight				
pound	lb	0.454	kilogram	kg
ton		2,000.0	pound	lb
ton		907.0	kilogram	kg
Flow Rates				
cubic feet per second	cfs	0.0283	cubic meters per second	m³/sec
cubic feet per second	cfs	448.8	gallons per minute	gpm
cubic feet per second	cfs	0.6463	millions of gallons per day	MGD
gallons per minute	gpm	0.00006	cubic meters per second	m³/sec
gallons per minute	gpm	0.00223	cubic feet per second	cfs
gallons per minute	gpm	0.00144	millions of gallons per day	MGD
millions of gallons per day	MGD	0.0438	cubic meters per second	m³/sec
millions of gallons per day	MGD	1.547	cubic feet per second	cfs
millions of gallons per day	MGD	694.4	gallons per minute	gpm

Temperature Scale Conversion

degree Celsius	°C	($°C \times 9/5$) + 32	degree Fahrenheit	°F
degree Fahrenheit	°F	($°F - 32$) 5/9	degree Celsius	°C

261

References

Abbe, G. R. 1977a. Blue crab studies. Pages 10-101 to 10-113 in Morgantown Station and the Potomac estuary: A 316 environmental demonstration, Vol. II. Prepared for Potomac Electric Power Co. by Academy of Natural Sciences of Philadelphia, Pa.

Abbe, G. R. 1977b. Substrate studies. Pages 10-63 to 10-76 in Morgantown Station and the Potomac estuary: A 316 environmental demonstration, Vol. II. Prepared for Potomac Electric Power Co. by Academy of Natural Sciences of Philadelphia, Pa.

Abbott, I. A., and G. J. Hollenberg. 1976. Marine Algae of California. Stanford Univ. Press, Stanford, Calif. 827 pp.

Academy of Natural Sciences of Philadelphia. 1967. Potomac River surveys (Morgantown): 1966 river survey report. Prepared for Potomac Electric Power Co. ANSP, Philadelphia, Pa.

Academy of Natural Sciences of Philadelphia. 1968. Potomac River surveys (Morgantown): 1967 river survey report. Prepared for Potomac Electric Power Co. ANSP, Philadelphia, Pa.

Academy of Natural Sciences of Philadelphia. 1969a. Potomac River, Maryland: Studies on the blue crab, I, June-October, 1968. Prepared for Potomac Electric Power Co. ANSP, Philadelphia, Pa.

Academy of Natural Sciences of Philadelphia. 1969b. Potomac River surveys (Morgantown): 1968 river survey report. Prepared for Potomac Electric Power Co. ANSP, Philadelphia, Pa.

Academy of Natural Sciences of Philadelphia. 1970a. Blue crab studies on the Potomac River at Morgantown, Maryland, 1969. Prepared for Potomac Electric Power Co. ANSP, Philadelphia, Pa.

Academy of Natural Sciences of Philadelphia. 1970b. Potomac River, Maryland: Fish trawl survey, progress report II, October 1968 - September 1969. Prepared for Potomac Electric Power Co. ANSP, Philadelphia, Pa.

Academy of Natural Sciences of Philadelphia. 1971a. Blue crab studies on the Potomac River at Morgantown, Maryland, III, 1970. Prepared for Potomac Electric Power Co. ANSP, Philadelphia, Pa.

Academy of Natural Sciences of Philadelphia. 1971b. Potomac River surveys (Morgantown): 1969 river survey report. Prepared for Potomac Electric Power Co. ANSP, Philadelphia, Pa.

Academy of Natural Sciences of Philadelphia. 1972. Potomac River surveys, Morgantown: 1970. Prepared for Potomac Electric Power Co. ANSP, Philadelphia, Pa.

Academy of Natural Sciences of Philadelphia. 1977. Morgantown Station and the Potomac estuary: A 316 environmental demonstration, Vol. II. Prepared for Potomac Electric Power Co. ANSP, Philadelphia, Pa.

Alexandria Drafting Company. 1973. Salt Water Sport Fishing and Boating in Maryland. Alexandria, Va. 37 pp.

Allan, J. D. 1976. Life history patterns in zooplankton. Amer. Naturalist 110:165-180.

Allee, D. C., D. Kelso, and P. L. Rosenblatt. 1976. Impact of construction and urban development on the aquatic environment. Pages 38 to 45 in The Potomac estuary: Biological resources, trends and options, W. T. Mason and K. C. Flynn, eds. Proc. of a symposium sponsored by Interstate Commission on the Potomac River Basin and Power Plant Siting Program, Md. Dept. of Natural Resources. ICPRB, Bethesda, Md. Tech. Publ. 76-2.

American Fisheries Society. 1970. A list of common and scientific names of fishes from the United States and Canada. 3rd ed., Special Publ. No. 6. American Fisheries Society, Washington, D.C. 150 pp.

Andersen, A. M., W. J. Davis, M. P. Lynch, and J. R. Schubel, eds. 1973. Effects of Hurricane Agnes on the environment and organisms of Chesapeake Bay: Early findings and recommendations. Prepared for U.S. Army Corps of Engineers, Philadelphia District, by the Chesapeake Bay Research Council, Natural Resources Institute, Chesapeake Biological Laboratory, Solomons, Md. Contr. No. 529.

Anderson, R. R. 1972. Submerged vascular plants of the Chesapeake Bay and tributaries. Pages S87 to S89 in Biota of the Chesapeake Bay, A. J. McErlean, C. Kerby, and M. L. Wass, eds. Chesapeake Sci., Vol. 13, Supplement.

Anjard, C. A. 1974. Centrarchidae — sunfishes. Pages 178 to 195 in Manual for identification of early developmental stages of fishes of the Potomac River estuary, A. J. Lippson and R. L. Moran, eds. Prepared for Power Plant Siting Program, Md. Dept. of Natural Resources by Martin Marietta Corp., Environmental Technology Center, Baltimore, Md. PPSP-MP-13.

Barnes, H. 1962. Note on variations in the release of nauplii of Balanus balanoides with special reference to the spring diatom outburst. Crustaceana 4:118-122.

Barnes, R. D. 1968. Invertebrate Zoology. W. B. Saunders Co., Philadelphia, Pa 743 pp.

Bayley, S., H. Rabin, and C. H. Southwick. 1968. Recent decline in the distribution and abundance of Eurasian milfoil in Chesapeake Bay. Chesapeake Sci. 9(3):173-181.

Beaven, G. F. 1954. Various aspects of oyster setting in Maryland. Proc. Natl. Shellfisheries Assoc. 45:29-37.

Bigelow, H. B., and W. C. Schroeder. 1948. Cyclostomes. Pages 29 to 58 in Fishes of the Western North Atlantic. Yale Univ., Sears Foundation for Marine Research, New Haven, Conn. Mem. 1(1).

Bigelow, H. B., and W. C. Schroeder. 1953. Sawfishes, guitarfishes, skates, and rays. Pages 1 to 514 in Fishes of the Western North Atlantic. Yale Univ., Sears Foundation for Marine Research, New Haven, Conn. Mem. 1(2).

Biological Methods Panel Committee on Oceanography. 1969. Recommended procedures for measuring the productivity of plankton standing stock and related oceanic properties. Natl. Acad. of Sciences, Washington, D.C. 59 pp.

Bishop, J. W. 1967. Feeding rates of the ctenophore, Mnemiopsis leidyi. Chesapeake Sci. 8(4): 259-261.

Boesch, D. F. 1973. Classification and community structure of macrobenthos in the Hampton Roads area, Virginia. Mar. Biol. 21:226-244.

Boesch, D. F. 1977. A new look at the distribution of benthos along the estuarine gradient. Pages 245 to 266 in Ecology of Marine Benthos, No. 6, Belle W. Baruch Library in Marine Science, B. C. Coull, ed. Univ. of South Carolina Press, Columbia, S.C.

Boesch, D. F. 1978. Personal communication. Virginia Institute of Marine Science, Gloucester Point, Va.

Bold, H. C. 1973. Morphology of plants. Harper and Row Publs. Inc., New York, N.Y. 668 pp.

Bongers, L. H., T. T. Polgar, A. J. Lippson, G. M. Krainak, L. R. Moran, A. F. Holland, and W. A. Richkus. 1975. The impact of the Morgantown power plant on the Potomac estuary: An interpretive summary of the 1972-1973 investigations. Prepared for Power Plant Siting Program, Md. Dept. of Natural Resources by Martin Marietta Corp., Environmental Technology Center, Baltimore, Md. PPSP-MP-15.

Boone, J. G. 1978. Chesapeake Bay enjoying bluefish bonanza. Md. Dept. of Natural Resources, Commerical Fish. News 11(4):1,3.

Bosch, H. F., and W. R. Taylor. 1967. Marine cladocerans in the Chesapeake Bay estuary. Crustaceana 15(2):161-164.

Bosch, H. F., and W. R. Taylor. 1970. Ecology of Podon polyphemoides (Crustacea: Branchipoda) in the Chesapeake Bay. The Johns Hopkins Univ., Chesapeake Bay Institute, Baltimore, Md. Tech. Rept. 66, Ref. 70-5.

Bosch, H. F., and W. R. Taylor. 1973. Distribution of the cladoceran Podon polyphemoides in the Chesapeake Bay. Mar. Biol. 19:161-171.

Bousfield, E. L. 1969. New records of Gammarus (Crustacea: Amphipoda) from the Middle Atlantic region. Chesapeake Sci. 10(1):1-17.

Bousfield, E. L. 1973. Shallow-water Gammaridean Amphipoda of New England. Cornell Univ. Press, Ithaca, N.Y. 312 pp.

Boyce Thompson Institute for Plant Research. 1977. An Atlas of the Biological Resources of the Hudson Estuary, L. H. Weinstein, ed. Boyce Thompson Institute for Plant Research, Inc., Yonkers, N.Y. 104 pp.

Boynton, W. R., E. M. Setzler, K. V. Wood, H. H. Zion, M. Homer, and J. A. Mihursky. 1977. Potomac river fisheries program, ichthyoplankton and juvenile investigations, 1976, draft report. Prepared for Power Plant Siting Program, Md. Dept. of Natural Resources by Univ. of Md., Center for Environmental and Estuarine Studies, Chesapeake Biological Laboratory, Solomons, Md. Ref. No. 77-169 CBL.

Branscomb, E. S. 1976. Proximate causes of mortality determining the distribution and abundance of the barnacle Balanus improvisus Darwin in Chesapeake Bay. Chesapeake Sci. 17(4):281-288.

Bratina, I. C. 1978. Personal communication. Piscataway Wastewater Treatment Plant, Piscataway, Md.

Breder, C. M., Jr. 1948. Field Book of Marine Fishes of the Atlantic Coast from Labrador to Texas. G. P. Putnam's Sons, New York, N.Y. 332 pp.

Breder, C. M., Jr., and D. R. Crawford. 1922. The food of certain minnows. *Zoologica* 2(14):287-327.

Browne, M. E., L. B. DeLancey, D. A. Randle, and W. H. Whitmore. 1976. A study of zooplankton in the Delaware River in the vicinity of Artificial Island in 1974. Pages 1 to 151 *in* An ecological study of the Delaware River in the vicinity of Artificial Island, Vol. II. Prepared for Public Service Electric and Gas Co. by Ichthyological Associates, Inc., Ithaca, N.Y.

Burbanck, W. D. 1963. Some observations on the isopod, *Cyathura polita* in Chesapeake Bay. *Chesapeake Sci.* 4(2):104-105.

Burbanck, W. D. 1967. Evolutionary and ecological implications of the zoogeography, physiology, and morphology of *Cyathura* (Isopoda). Pages 564 to 573 *in* Estuaries, G. H. Lauff, ed. American Association for the Advancement of Science, Washington, D. C. Publ. No. 83.

Burchard, R. P. 1971. Chesapeake Bay bacteria able to cycle carbon, nitrogen, sulfur, and phosphorus. *Chesapeake Sci.* 12(3):179-180.

Burrell, V. G. 1968. The ecological significance of a ctenophore, *Mnemiopsis leidyi* (A. Agassiz) in a fish nursery ground. M.A. Thesis. College of William and Mary, Williamsburg, Va. 61 pp.

Burrell, V. G., and D. E. Zwerner. 1972. Subclass Copepoda. Pages 138 to 142 *in* A check list of the biota of lower Chesapeake Bay, M. L. Wass, ed. Virginia Institute of Marine Science, Gloucester Point, Va. Special Scientific Rept. No. 65.

Cain, T. D. 1973. The combined effects of temperature and salinity on embryos and larvae of the clam, *Rangia cuneata*. *Mar. Biol.* 21:1-6.

Calabrese, A., and H. C. Davis. 1970. Tolerances and requirements of embryos and larvae of bivalve molluscs. *Helgol. Wiss. Meeresunters.* 20:553-564.

Calder, D. R. 1971. Hydroids and hydromedusae of southern Chesapeake Bay. Virginia Institute of Marine Science, Gloucester Point, Va. Special Papers in Marine Science 1. 125 pp.

Calder, D. R. 1972a. Phylum Cnidaria. Pages 97 to 102 *in* A check list of the biota of lower Chesapeake Bay, M. L. Wass, ed. Virginia Institute of Marine Science, Gloucester Point, Va. Special Scientific Rept. No. 65.

Calder, D. R. 1972b. Tentative outline for inventory of planktonic Cnidaria: *Chrysaora quinquecirrha* (stinging nettle). Pages S179 to S181 *in* Biota of the Chesapeake Bay, A. J. McErlean, C. Kerby, and M. L. Wass, eds. *Chesapeake Sci.,* Vol. 13, Supplement.

Calder, D. R. 1972c. Cnidaria of the Chesapeake Bay. Pages S100 to S102 *in* Biota of the Chesapeake Bay, A. J. McErlean, C. Kerby, and M. L. Wass, eds. *Chesapeake Sci.,* Vol. 13, Supplement.

Calder, D. R., and V. G. Burrell, Jr. 1967. Occurrence of *Moerisia lyonsi* (Limnomedusae, Moerisiidae) in North America. *Amer. Midland Naturalist* 78:540-541.

Cargo, D. G. 1971. The sessile stages of a scyphozoan identified as *Rhopilema verrilli. Tulane Studies in Zoology and Botany* 17(2):31-34.

Cargo, D. G., and L. P. Schultz. 1966. Notes on the biology of the sea nettle, *Chrysaora quinquecirrha* in Chesapeake Bay. *Chesapeake Sci.* 7(2):95-100.

Cargo, D. G., and L. P. Schultz. 1967. Further observations on the biology of the sea nettle and jellyfishes in Chesapeake Bay. *Chesapeake Sci.* 8(4):209-220.

Carpenter, K. A. 1978. Personal communication. Potomac River Fisheries Commission, Colonial Beach, Va.

Carpenter, J. H., D. W. Pritchard, and R. C. Whaley. 1969. Observations of eutrophication and nutrient cycles in some coastal plain estuaries. Pages 210 to 221 *in* Proc. of the International Symposium on Eutrophication: Causes, Consequences, Correctives. Natl. Acad. of Sciences, Washington, D.C.

Carriker, M. R. 1967. Ecology of estuarine benthic invertebrates: A perspective. Pages 442 to 487 *in* Estuaries, G. H. Lauff, ed. American Association for the Advancement of Science, Washington, D.C. Publ. No. 83.

Carriker, M. R., and L. G. Williams. 1978. Chemical mechanisms of shell penetration by *Urosalpinx*. An hypothesis. *Malacologia* 17:142-156.

Castagna, M., and P. Chanley. 1973. Salinity tolerance of some marine bivalves from inshore and estuarine environments in Virginia waters of the western mid-Atlantic Coast. *Malacologia* 12:47-96.

Caulk, E. T. 1965. Some common birds of Maryland. Md. Dept. of Game and Inland Fish, Annapolis, Md. 64 pp.

Chambers, J. R., R. G. Burbidge, and W. A. Van Engel. 1970. The occurrence of *Leptodora kindtii* (Focke) (Cladocera) in tributaries of Chesapeake Bay. *Chesapeake Sci.* 11(4):255-258.

Chanley, P., and J. D. Andrews. 1971. Aids for identification of bivalve larvae of Virginia. *Malacologia* 11(1):45-119.

Chao, L. N., and J. A. Musick. 1977. Life history, feeding habits, and functional morphology of juvenile sciaenid fishes in the York River estuary, Virginia. *Fishery Bull.* 75(4):657-702.

Chapman, V. J. 1964. The Algae. MacMillan and Company, Ltd., London, 472 pp.

Chapman, V. J., and D. J. Chapman. 1977. The Algae, 2nd ed. The MacMillan Press, Ltd., London. 497 pp.

Churchill, E. P. 1919. Life history of the blue crab. *Bull. U.S. Bureau of Fisheries* 36:91-128.

Cole, C. F. 1967. A study of the eastern johnny darter, *Etheostoma olmstedi* Storer (Teleostei, Percidae). *Chesapeake Sci.* 8(1):28-51.

Colwell, R. R. 1972. Bacteria, yeasts, viruses, and related microorganisms of the Chesapeake Bay. Pages S67 to S70 *in* Biota of the Chesapeake Bay, A. J. McErlean, C. Kerby, and M. L. Wass, eds. *Chesapeake Sci.*, Vol. 13, Supplement.

Cones, H. N., Jr., and D. S. Haven. 1969. Distribution of *Chrysaora quinquecirrha* in the York River. *Chesapeake Sci.* 10(2):75-84.

Connell, J. H. 1961. The influence of interspecific competition and other factors on the distribution of the barnacle *Chthamalus stellatus*. *Ecology* 42(4):710-723.

Conover, R. J. 1956. Oceanography of Long Island Sound, 1952-1954, VI: Biology of *Acartia clausi* and *Acartia tonsa*. *Bull. Bingham Oceanogr. College* 15:156-233.

Corddry, M. 1978. Baltimore Gas and Electric helps save heron haven. Page C1 *in* the *Baltimore Sun,* January 24, Baltimore, Md.

Cory, R. L. 1967. Epifauna of the Patuxent River estuary, Maryland, for 1963 and 1964. *Chesapeake Sci.* 8(2):71-89.

Cory, R. L. 1978. Personal communication. Chesapeake Bay Center for Environmental Studies, Smithsonian Institution, Edgewater, Md.

Costlow, J. D. 1967. The effect of salinity and temperature on survival and metamorphosis of megalops of the blue crab, *Callinectes sapidus. Helgol. Wiss. Meeresunters.* 15:84-97.

Costlow, J. D., and C. G. Bookout. 1959. The larval development of *Callinectes sapidus* Rathbun reared in the laboratory. *Biol. Bull.* 116(3):373-396.

Coull, B. C. 1977. Marine flora and fauna of the northeastern United States: Copepoda: Harpacticoida. U.S. Dept. of Commerce, National Oceanic and Atmospheric Adm., National Marine Fisheries Service. NOAA Tech. Rept., NMFS Circular 397.

Cowardin, L. M., F. C. Golet, and E. T. LaRoe. 1977. Classification of wetlands and deepwater habitats of the United States (an operational draft). U.S. Department of the Interior, Fish and Wildlife Service, Washington, D.C.

Cronin, L. E. 1973. The Chesapeake Bay. Pages 2 to 3 *in* The Chesapeake Bay in Maryland: An Atlas of Natural Resources, A. J. Lippson, ed. The Johns Hopkins Univ. Press, Baltimore, Md.

Cronin, L. E., J. C. Daiber, and E. M. Hulbert. 1962. Quantitative seasonal aspects of zooplankton in the Delaware River estuary. *Chesapeake Sci.* 3(2):63-93.

Cronin, L. E., and A. J. Mansueti. 1971. The biology of an estuary. Pages 14 to 39 *in* A Symposium on the Biological Significance of Estuaries. Sport Fishing Institute, Washington, D.C.

Cronin, W. B. 1971. Volumetric, areal, and tidal statistics of the Chesapeake Bay estuary and its tributaries. Chesapeake Bay Institute, The Johns Hopkins Univ., Baltimore, Md. Special Rept. 20, Ref. 71-2.

Cronin, W. B., and A. W. Pritchard. 1975. Additional statistics on the dimensions of the Chesapeake Bay and its tributaries: Cross-section widths and segment volumes per meter depth. Chesapeake Bay Institute, The Johns Hopkins Univ., Baltimore, Md. Special Rept. 42, Ref. 75-3.

Dahlberg, M. D. 1973. An ecological survey of the Potomac River at Douglas Point, Maryland. Prepared for Potomac Electric Power Co. by NUS Corp., Cyrus Wm. Rice Division, Ecological Sciences Dept., Pittsburgh, Pa. 365 pp.

Dahlberg, M. D. 1975. Guide to Coastal Fishes of Georgia and Nearby States. Univ. of Georgia Press, Athens, Ga. 186 pp.

Darnell, R. M. 1959. Studies of the life history of the blue crab (*Callinectes sapidus* Rathbun) in Louisiana waters. *Trans. Amer. Fish. Soc.* 88:294-304.

Dawson, E. Y. 1966. Marine Botany: An Introduction. Holt, Rinehart and Winston, Inc., New York, N.Y. 371 pp.

deKok, A. 1975. Personal communication. Chesapeake Biological Laboratory, Solomons, Md.

Delaware Coastal Management Program. 1976. An atlas of Delaware's wetlands and estuarine resources. Univ. of Delaware, College of Marine Studies, Newark, Del. Tech. Rept. No. 2.

Diaz, R. J. 1976. Personal communication. Virginia Institute of Marine Science, Gloucester Point, Va.

Dintaman, R. C. 1975. A preliminary study of the occurrence of the American eel and other finfish species in Maryland, Vol. I, 1974: Upper Chesapeake Bay drainage, Potomac River drainage. Md. Dept. of Natural Resources, Fisheries Adm., Anadromous Fish Stream Survey Program, Annapolis, Md.

Doroshev, S. I. 1970. Biological features of the eggs, larvae and young of the striped bass, *Roccus saxatilis* (Walbaum), in connection with the problem of its acclimatization in the USSR. *J. Ichthyol.* 10(2):235-248.

Doumlele, D. G. 1976. Fairfax County tidal marsh inventory including City of Alexandria and Arlington County. Virginia Institute of Marine Science, Applied Marine Science and Ocean Engineering, Gloucester Point, Va. Special Rept. No. 108.

Dovel, W. L. 1967. Fish eggs and larvae of the Magothy River, Maryland. *Chesapeake Sci.* 8(2):125-129.

Dovel, W. L. 1968. Predation by striped bass as a possible influence on population size of Atlantic croaker. *Trans. Amer. Fish. Soc.* 97(4):313-319.

Dovel, W. L. 1971. Fish eggs and larvae of the upper Chesapeake Bay. Univ. of Md., Natural Resources Institute, Solomons, Md. Special Rept. No. 4, Contribution No. 460.

Dovel, W. L., and A. J. Lippson. 1973. Spot, croaker, and weakfish. Pages 42 to 43 *in* The Chesapeake Bay in Maryland: An Atlas of Natural Resources, A. J. Lippson, ed. The Johns Hopkins Univ. Press, Baltimore, Md.

Dovel, W. L, J. A. Mihursky, and A. J. McErlean. 1969. Life history aspects of the hogchoker, *Trinectes maculatus,* in the Patuxent River estuary, Maryland. *Chesapeake Sci.* 10(2):104-119.

Drouet, F. 1968. Revision of the classification of the Oscillatoriaceae. Academy of Natural Sciences of Philadelphia, Pa. Monograph 15. 370 pp.

Drouet, F. 1973. Revision of the Nostocaceae with Cylindrical Trichomes. Hafner Press, New York, N.Y. 292 pp.

Drouet, F. 1978. Revision of the Nostocaceae with Constricted Trichomes. *Beiheste zur Noza hedwigia,* Vol. 57. Lubrecht and Cramer, Monticello, N.Y. 258 pp.

Drouet, F., and W. A. Daily. 1956. Revision of the Coccoid Myxophyceae. *Butler Univ. Botanical Studies,* Vol. 12. Lubrecht and Cramer, Monticello, N.Y. 218 pp.

Dunnington, E. A., D. S. Haven, and K. G. Drobeck. 1974. Summer 1974 survey of Potomac oyster bars. Rept. to Potomac River Fisheries Commission by Chesapeake Biological Laboratory, Solomons, Md. CBL Ref. No. 74-119.

DuPaul, W. D., and J. D. McEachran. 1973. Age and growth of the butterfish, *Peprilus tricanthus,* in the lower York River. *Chesapeake Sci.* 14(3):205-207.

Ecological Analysts, Inc. 1974. Douglas Point site, power plant site evaluation, aquatic biology, final report, Vol. II. Prepared for Power Plant Siting Program, Md. Dept. of Natural Resources, EAI, Baltimore, Md. PPSE 4-2.

Ecological Analysts, Inc. 1978. Baseline inventory of wetlands of Piscataway Creek and Potomac River. Prepared for Washington Suburban Sanitary Commission. EAI, Baltimore, Md.

Elser, H. J. 1950. The common fishes of Maryland and how to tell them apart. Md. Dept. of Research and Education, Chesapeake Biological Laboratory, Solomons, Md. Publ. No. 88.

Elser, H. J. 1965. Chesapeake Bay creel census, 1962. Univ. of Md.,Natural Resources Institute, Solomons, Md. Ref. No. 65-17.

Elser, H. J., and R. J. Mansueti. 1961. Notes on the chain pickerel in Maryland. Md. Dept. of Research and Education, Chesapeake Biological Laboratory and Inland Resources Division, Solomons, Md. Ref. No. 61-14.

Enequist, P. 1949. Studies on the soft-bottom amphipods of the Skagerak. *Zoologiska bidrag fraan Uppsala* 28:297-492.

Fassett, N. C. 1957. A Manual of Aquatic Plants. The Univ. of Wisconsin Press, Madison, Wisc. 405 pp.

Feeley, J. B. 1967. The distribution and ecology of the Gammaridea (Crustacea: Amphipoda) of the lower Chesapeake estuaries. M.S. Thesis. The College of William and Mary, Williamsburg, Va. 75 pp.

Fell, P. E. 1974. Porifera. Pages 51 to 125 *in* Reproduction in Marine Invertebrates, Vol. I: Acoelomate and Pseudocoelomate Metazoans, A. C. Giese and J. S. Pearse, eds. Academic Press, Inc., New York, N.Y.

Fenchel, T. 1972. Aspects of decomposer food chains in marine benthos. *Verh. Dtsch. Zool. Ges.* 65:14-22.

Fenchel, T., L. H. Kofoed, and A. Lappalainen. 1975. Particle size selection of two deposit feeders: The amphipod, *Corophium volutator,* and the prosobranch, *Hydrobia ulvae.* *Mar. Biol.* 30:119-128.

Fernald, M. L. 1950. Gray's Manual of Botany, 8th ed. American Book Company, New York, N.Y. 1,632 pp.

Flemer, D. A. 1973. Phytoplankton: Microscopic plants. Pages 14 to 15 *in* The Chesapeake Bay in Maryland: An Atlas of Natural Resources, A. J. Lippson, ed. The Johns Hopkins Univ. Press, Baltimore, Md.

Fogg, G. E. 1965. Algal Cultures and Phytoplankton Ecology. Univ. of Wisconsin Press, Madison, Wisc. 126 pp.

Foster, N. R. 1967. Comparative studies on the biology of killifishes (Pisces, Cyprinodontidae). Ph.D. Dissertation. Cornell Univ., N.Y. Univ. Microfilms, Ann Arbor, Mich. No. 67-08759, 369 pp.

Foster, N. R. 1974. Cyprinodontidae — Killifishes. Pages 127 to 141 *in* Manual for identification of early developmental stages of fishes of the Potomac River estuary, A. J. Lippson and R. L. Moran, eds. Prepared for Power Plant Siting Program, Md. Dept. of Natural Resources by Martin Marietta Corp., Environmental Technology Center, Baltimore, Md. PPSP-MP-13.

Frey, D. G. 1946. Oyster bars of the Potomac River. U.S. Dept. of the Interior, Fish and Wildlife Service, Washington, D. C. Special Scientific Rept. No. 32.

Friedrich, H. 1969. Marine Biology. Univ. of Washington Press, Seattle, Wash. 474 pp.

Frisbie, C. M. 1971. Recent trends and importance of Maryland's commercial fisheries, 1960-1970. Md. Dept. of Natural Resources, Fish and Wildlife Adm., Finfish Div., Annapolis, Md. Ref. No. 71-11-16.

Frisbie, C. M., and D. E. Ritchie, Jr. 1963. Sport fishery survey of the lower Potomac estuary, 1959-1961. *Chesapeake Sci.* 4(4):175-191.

Fritsch, F. E. 1952. Structure and Reproduction of the Algae, Vol. II. Cambridge Univ. Press, London, 939 pp.

Fritsch, F. E. 1956. Structure and Reproduction of the Algae, Vol. I. Cambridge Univ. Press, London. 791 pp.

Fritz, E. S., W. H. Meredith, and V. A. Lotrich. 1975. Fall and winter movements and activity level of the mummichog, *Fundulus heteroclitus,* in a tidal creek. *Chesapeake Sci.* 16(3):211-214.

Galloway, J. W. 1911. The common freshwater Oligochaeta of the United States. *Trans. of Amer. Microscop. Soc.* 30:285-317.

Galtsoff, P. S. 1964. The American oyster, *Crassostrea virginica* (Gmelin). *Fishery Bull.* 64:1-480.

George, J. D. 1966. Reproduction and early development of the Spionid polychaete, *Scolecolepides viridis* (Verrill). *Biol. Bull.* 130(1):76-93.

Giraldi, A., and A. Dietemann. 1974. Fishes of the upper Anacostia River system of Maryland. *Atlantic Naturalist* 29(2):61-67.

Glude, J. B. 1955. The effects of temperature and predators on the abundance of the soft-shell clam, *Mya arenaria,* in New England. *Trans. Amer. Fish. Soc.* 84:13-26.

Goldsberry, J. R. 1971. Nutria — What to do with it? *Maryland Fish and Wildlife News* (October), Md. Dept. of Natural Resources, Annapolis, Md.

Goldsberry, J. R. 1973. Personal communication. Md. Dept. of Natural Resources, Wildlife Adm., Annapolis, Md.

Gordon, C. M. 1969. The apparent influence of salinity on the distribution of barnacle species in Chesapeake Bay (Cirripedia). *Crustaceana* 16(2):139-142.

Gosner, K. L. 1971. Guide to Identification of Marine and Estuarine Invertebrates. John Wiley & Sons, Inc., New York, N.Y. 693 pp.

Grassle, J. F., and J. P. Grassle. 1974. Opportunistic life histories and genetic systems in marine benthic polychaetes. *J. Mar. Research* 32(2):253-284.

Gross, M. G. 1972. Oceanography: A View of the Earth. Prentice-Hall, Inc., Englewood Cliffs, N.J. 581 pp.

Haefner, P. A., Jr. 1976. Seasonal distribution and abundance of sand shrimp, *Crangon septemspinosa,* in the York River-Chesapeake Bay estuary. *Chesapeake Sci.* 17(2):131-134.

Haire, M. S. 1978. An investigation of the macrobenthic communities in the vicinity of the Morgantown power plant. M.S. Thesis. Towson State Univ., Towson, Md. 79 pp.

Hanks, R. W. 1966. The soft-shell clam. U.S. Dept. of the Interior, Fish and Wildlife Service, Bureau of Commercial Fisheries, National Oceanic and Atmospheric Adm., Washington, D.C. Circular 162.

Hardy, A. 1965. The Open Sea: Its Natural History, Part I: The World of Plankton. Houghton Mifflin Co., Boston, Mass. 322 pp.

Hardy, J. D. 1978. Development of Fishes of the Mid-Atlantic Bight: An Atlas of Egg, Larval and Juvenile Stages, Vol. II: Anguillidae through Syngnathidae. U.S. Dept. of the Interior, Fish and Wildlife Service, Biological Services Program, Washington, D.C. FWS/ OBS - 78/12. 458 pp.

Hargrave, B. T. 1970. Distribution, growth, and seasonal abundance of *Hyalella azteca* (Amphipoda) in relation to sediment microflora. *J. Fish. Res. Bd. Canada* 27:685-699.

Haven, D. S. 1976. The shellfish fisheries of the Potomac River. Pages 88 to 94 *in* The Potomac estuary: Biological resources, trends and options, W. T. Mason and K. C. Flynn, eds. Proc. of a symposium sponsored by Interstate Commission on the Potomac River Basin and Power Plant Siting Program, Md. Dept. of Natural Resources. ICPRB, Bethesda, Md. Tech. Publ. 76-2.

Haven, D. S., and J. Davis. 1973. Survey for deep water oysters in the Potomac River. Report to Potomac River Fisheries Commission by the Virginia Institute of Marine Science, Gloucester Point, Va.

Heinle, D. R. 1966. Production of a calanoid copepod, *Acartia tonsa,* in the Patuxent River estuary. *Chesapeake Sci.* 7(2):59-74.

Heinle, D. R. 1973. Copepods:Microscopic animals.Pages 16 to 17 *in* The Chesapeake Bay in Maryland: An Atlas of Natural Resources, A. J. Lippson, ed. The Johns Hopkins Univ. Press, Baltimore, Md.

Heinle, D. R., D. A. Flemer, and J. F. Ustach. 1977. Contribution of tidal marshlands to mid-Atlantic estuarine food chains. Pages 309 to 319 *in* Estuarine Processes, Vol. II: Circulation, Sediments, and Transfer of Material in the Estuary. Academic Press, Inc., New York, N.Y.

Heinle, D. R., H. S. Millsaps, and J. K. Lawson. 1973. Zooplankton investigations at the Morgantown power plant: Phase III report. Prepared for Power Plant Siting Program, Md. Dept. of Natural Resources by Univ. of Md., Natural Resources Institute, Chesapeake Biological Laboratory, Solomons, Md. Ref. No. 73-132.

Heinle, D. R., H. S. Millsaps, and J. K. Lawson. 1974. Zooplankton investigations at the Morgantown power plant: Phase V, final report. Prepared for Power Plant Siting Program, Md. Dept. of Natural Resources by Univ. of Md., Natural Resources Institute, Chesapeake Biological Laboratory, Solomons, Md. Ref. No. 74-10.

Heinle, D. R., H. S. Millsaps, J. K. Lawson, and K. V. Millsaps. 1973a. Sampling and methodology studies at the Morgantown power plant: Phase I report — zooplankton. Prepared for Power Plant Siting Program, Md. Dept. of Natural

Resources by Univ. of Md., Natural Resources Institute, Chesapeake Biological Laboratory, Solomons, Md. Ref. No. 73-49.

Heinle, D. R., H. S. Millsaps, J. K. Lawson, and K. V. Millsaps. 1973b. Zooplankton investigations at the Morgantown Power Plant: Phase II report. Prepared for Power Plant Siting Program, Md. Dept. of Natural Resources by Univ. of Md., Natural Resources Institute, Chesapeake Biological Laboratory, Solomons, Md. Ref. No. 73-118.

Heinle, D. R., and K. V. Millsaps. 1973. Zooplankton studies. Pages 128 to 151 in The effects of Morgantown steam electric station operations on organisms pumped through the cooling water system, July 1972 - December 1972, final report. Prepared for Power Plant Siting Program, Md. Dept. of Natural Resources by Univ. of Md., Natural Resources Institute, Chesapeake Biological Laboratory, Solomons, Md. Ref. No. 73-20.

Henny, C. J., M. M. Smith, and V. D. Stotts. 1974. The 1973 distribution and abundance of breeding ospreys in the Chesapeake Bay. Chesapeake Sci. 15(3):125-133.

Henry, K. A., E. B. Joseph, C. M. Bearden, and J. W. Reintjes. 1965. Atlantic menhaden. Atlantic States Marine Fisheries Comm., Tallahassee, Fla. Marine Resources of the Atlantic Coast. Leaflet No. 2.

Herman, S. S., J. A. Mihursky, and A. J. McErlean. 1968. Zooplankton and environmental characteristics of the Patuxent River estuary, 1963-1965. Chesapeake Sci. 9(2):67-82.

Hidu, H., N. H. Roosenburg, K. G. Drobeck, A. J. McErlean, and J. A. Mihursky. 1974. Thermal tolerance of oyster larvae, Crassostrea virginica (Gmalin), as related to power plant operations. Proc. Natl. Shellfisheries Assoc. 64:102-110.

Higgins, R. P. 1972a. Kinorhyncha of the Chesapeake Bay. Pages S105 to S106 in Biota of the Chesapeake Bay, A. J. McErlean, C. Kerby, and M. L. Wass, eds. Chesapeake Sci., Vol. 13, Supplement.

Higgins, R. P. 1972b. Tardigrada of the Chesapeake Bay. Pages S103 to S104 in Biota of the Chesapeake Bay, A. J. McErlean, C. Kerby, and M. L. Wass, eds. Chesapeake Sci., Vol. 13, Supplement.

Higgins, R. P. 1972c. Priapulida of the Chesapeake Bay. Pages S102 to S103 in Biota of the Chesapeake Bay, A. J. McErlean, C. Kerby, and M. L. Wass, eds. Chesapeake Sci., Vol. 13, Supplement.

Hildebrand, S. F. 1963. Family Clupeidae. Pages 257 to 385, 397 to 442, 452 to 454 in Fishes of the Western North Atlantic. Sears Foundation for Marine Research, Yale Univ., New Haven, Conn. Mem. 1(3).

Hildebrand, S. F., and W. C. Schroeder. 1928. Fishes of the Chesapeake Bay. Reprinted from Bull. U.S. Bureau Fisheries, Vol. 53, Part I. Smithsonian Institution Press, Washington, D.C. 366 pp.

Hjulstrom, F. 1939. Transportation of detritus by moving water. Pages 5 to 31 in Recent Marine Sediments, P. D. Trask, ed. American Association of Petroleum Geology, Tulsa, Okla.

Hochachka, P. W. 1975. An exploration of metabolic and enzyme mechanisms underlying animal life without oxygen. Pages 107 to 137 in Biochemical and Biophysical Perspectives in Marine Biology, Vol. 2, D.C. Malins and J. R. Sargent, eds. Academic Press, Inc, New York, N.Y.

Holland, A. F. 1978. Unpublished data from personal collections of amphipods from the Potomac estuary. Martin Marietta Corp., Environmental Center, Baltimore, Md.

Holland, A. F., and J. M. Dean. 1977a. The biology of the stout razor clam, Tagelus plebeius: I. Animal-sediment relations, feeding mechanism and community biology. Chesapeake Sci. 18(1):58-66.

Holland, A. F., and J. M. Dean. 1977b. The biology of the stout razor clam, Tagelus plebeius: II. Some aspects of the population dynamics. Chesapeake Sci. 18(2):188-196.

Holland, A. F., K. Kaumeyer, N. K. Mountford, M. Hiegel, and J. A. Mihursky. 1977. Results of benthic studies at Calvert Cliffs: Interim report on August and December 1976 community data and March 1977 oyster spat and shell survey. Prepared for Power Plant Siting Program, Md. Dept. of Natural Resources by Martin Marietta Corp., Environmental Center, Baltimore, Md. and Chesapeake Biological Laboratory, Solomons, Md. MMC Ref. No. CC-77-1.

Holland, A. F., N. K. Mountford, M. Hiegel, K. Kaumeyer, D. Cargo, and J. A. Mihursky. 1978. Results of benthic studies at Calvert Cliffs: Annual report on the August 1976 through May 1977 data. Prepared for Power Plant Siting Program, Md. Dept. of Natural Resources by Martin Marietta Corp., Environmental Center, Baltimore, Md. and Univ. of Md., Center for Environmental and Estuarine Studies, Chesapeake Biological Laboratory, Solomons, Md. MMC Ref. No. CC-78-2.

Holland, A. F., N. K. Mountford, and J. A. Mihursky. 1977. Temporal variation in upper bay mesohaline benthic communities, I: The 9-m mud habitat. Chesapeake Sci. 18(4): 370-378.

Hopkins, S. H. 1962. Distribution of species of Cliona (boring sponge) on the eastern shore of Virginia in relation to salinity. Chesapeake Sci. 3(2):121-124.

Hopkins, S. H., J. N. Anderson, and K. Horvath. 1973. The brackish water clam, Rangia cuneata, an indicator of ecological effects of salinity changes in coastal waters. Report to the U.S. Army Engineer Waterways Experimental Station, Vicksburg, Miss. 250 pp.

Hopkins, T. L. 1965. Mysid shrimp abundance in surface waters of Indian River Inlet, Delaware. Chesapeake Sci. 6(2):86-91.

Howden, H. F., and R. J. Mansueti. 1951. Fishes of the tributaries of the Anacostia River, Maryland. Proc. Biol. Soc. Wash. 64:93-96.

Hubbs, C. O., and I. C. Potter. 1971. Distribution, phylogeny, and taxonomy. Pages 1 to 65 in The Biology of Lampreys, M. W. Hardesty and I. C. Potter, eds. Academic Press, London.

Jack McCormick & Associates, Inc. 1977. Evaluation of the coastal wetlands of Maryland, draft. Prepared for Md. Dept. of Natural Resources, Coastal Zone Management Program. JMA, Devon, Pa. 319 pp.

Jacobs, F. 1978. Zooplankton distribution, biomass, biochemical composition and seasonal community structure in lower Chesapeake Bay. Ph.D. Dissertation. Univ. of Virginia, Charlottesville, Va. 104 pp.

Jaworski, N. A., D. W. Lear, and J. A. Aalto. 1969. A technical assessment of current water quality conditions and factors affecting water quality in the upper Potomac estuary. U.S. Dept. of the Interior, Federal Water Pollution Control Adm., Chesapeake Tech. Support Laboratory, Middle Atlantic Region, Annapolis, Md. Tech. Rept. No. 5.

Jaworski, N. A., D. W. Lear, and O. Villa. 1971. Nutrient management in the Potomac estuary. U. S. Environmental Protection Agency, Chesapeake Tech. Support Laboratory, Annapolis, Md. Tech. Rept. 45.

Jeffries, H. P. 1962. Succession of two Acartia species in estuaries. Limnol. Oceanogr. 7:354-364.

Jenkins, R. E., and T. Zorach. 1970. Zoogeography and characters of the American cyprinid fish, Notropis bifrenatus. Chesapeake Sci. 11(3):174-182.

The Johns Hopkins University. 1949-1972. Data Inventory Bank of Chesapeake Bay Institute, Baltimore, Md.

Johnsgard, P. A. 1975. Waterfowl of North America. Indiana Univ. Press, Bloomington, Ind. 575 pp.

Jones, P. W., J. S. Wilson, R. P. Morgan, H. R. Lunsford, and J. Lawson. 1977. Potomac River fisheries study: Striped bass spawning stock assessment, interpretive report 1974-1976. Submitted to Power Plant Siting Program, Md. Dept. of Natural Resources by Univ. of Md., Center for Environmental and Estuarine Studies, Chesapeake Biological Laboratory, Solomons, Md. UMCEES Ref. No. 77-56 CBL.

Jørgensen, C. B. 1955. Quantitative aspects of filter-feeding in invertebrates. Biol. Rev. 30:391-454.

Jørgensen, C. B., and T. Fenchel. 1974. The sulfur cycle of a marine sediment model system. Mar. Biol. 24(13):189-201.

Joseph, E. B. 1972. The status of the sciaenid stocks of the middle Atlantic coast. Chesapeake Sci. 13(2):87-100.

June, F. C., and W. R. Nicholson. 1964. Age and size composition of the menhaden catch along the Atlantic Coast of the United States, 1958. U. S. Dept. of the Interior, Fish and Wildlife Service, Washington, D.C. Spec. Sci. Rept., Fisheries, No. 446.

Kachur, M. E. 1977. Phytoplankton. Pages 7-1 to 7-33 in Non-radiological environmental monitoring report, Calvert Cliffs Nuclear Power Plant, July to December 1976. Prepared for Baltimore Gas and Electric Co. by Academy of Natural Sciences of Philadelphia, Pa.

Kachur, M. E. 1978. Phytoplankton. Pages 7-1 to 7-39 in Non-radiological environmental monitoring report, Calvert Cliffs Nuclear Power Plant, January to December 1977. Prepared by Baltimore Gas and Electric Co. and Academy of Natural Sciences of Philadelphia, Pa.

Kendall, A. W., Jr., and F. J. Schwartz. 1968. Lethal temperature and salinity tolerances for white catfish, Ictalurus catus, from the Patuxent River, Maryland. Chesapeake Sci. 9(2):103-108.

Kinne, O. 1970. Temperature. Pages 407 to 415 in Marine Ecology, Vol. I, Part 1, O. Kinne, ed. John Wiley and Sons, Inc., New York, N.Y.

Kinne, O. 1971. Salinity. Pages 821 to 997 in Marine Ecology, Vol. I, Part 2, O. Kinne, ed. John Wiley and Sons, Inc., New York, N.Y.

Knowlten, R. E. 1970. Effects of environmental factors on the larval development of Alpheus heterochaelis (Say) and Palaemonetes vulgaris (Say), (Crustacea, Decapoda, Caridea), with ecological notes on larval and adult Alpheidae and Palaemonidae. Ph.D. Dissertation. Univ. of North Carolina, Chapel Hill, N.C. 544 pp.

Kohlenstein, L. C. In Press. On the proportion of the Chesapeake Bay stock of striped bass that migrates into the coastal fishery. Proc. of Session on Advances in Striped Bass Life History and Population Dynamics, 108th Annual Meeting, American Fisheries Society, August 22, 1978, at Univ. of Rhode Island.

Koo, T. S. Y. 1967. Commerical Fisheries Research and Development Act: Final report for contract year ending August 31, 1967. Univ. of Md., Natural Resources Institute. Ref. No. 67-6-C.

Koo, T. S. Y. 1970. The striped bass fishery in the Atlantic states. Chesapeake Sci. 11(2):73-93.

Koo, T. S. Y. 1973. Winter flounder. Pages 46 to 47 in The Chesapeake Bay in Maryland: An Atlas of Natural Resources, A. J. Lippson, ed. The Johns Hopkins Univ. Press, Baltimore, Md.

Korringa, P. 1941. Experiments and observation on swarming, pelagic life, and setting in the European flat oyster, Ostreas edulis (L.). Arch. Neerl. Zool. 5:1-254.

Krauss, R. W., and P. Orris. 1972. Benthic macroalgae of the Maryland portion of the

Chesapeake Bay. Pages S81 to S83 *in* Biota of the Chesapeake Bay, A. J. McErlean, C. Kerby, and M. L. Wass, eds. *Chesapeake Sci.,* Vol. 13, Supplement.

Krueger, F. E. 1977. Macroinvertebrates: Benthic grid study. Pages 10-39 to 10-62 *in* Morgantown Station and the Potomac estuary: A 316 environmental demonstration, Vol. II. Prepared for Potomac Electric Power Co. by Academy of Natural Sciences of Philadelphia, Pa.

Krueger, F. E., and S. L. Fuller. 1977. Macroinvertebrates: Benthic and shoreline macroinvertebrates. Pages 10-1 to 10-38 *in* Morgantown Station and the Potomac estuary: A 316 environmental demonstration, Vol. II. Prepared for Potomac Electric Power Co. by Academy of Natural Sciences of Philadelphia, Pa.

Kumar, K. D., and W. Van Winkle. 1978. Estimates of relative stock composition of the Atlantic coast striped bass population based on the 1975 Texas Instruments data set. Paper presented February 27, 1978 at Northeast Fish and Wildlife Conference, The Greenbrier, W. Va. Sponsored by American Fisheries Society, N.E. Division.

Lackey, J. B. 1967. The microbiota of estuaries and their roles. Pages 291 to 302 *in* Estuaries, G. H. Lauff, ed. American Association for the Advancement of Science, Washington, D.C. Publ. 83.

Lear, D. W., and S. K. Smith. 1976. Phytoplankton of the Potomac estuary. Pages 70 to 74 *in* The Potomac estuary: Biological resources, trends and options, W. T. Mason and K. C. Flynn, eds. Proc. of a symposium sponsored by the Interstate Commission on the Potomac River Basin and Power Plant Siting Program, Md. Dept. of Natural Resources. ICPRB, Bethesda, Md. Tech. Publ. 76-2.

Lebour, M. V. 1933. The importance of larval molluscs in the plankton. *J. Cons. Cons. Int. Explor. Mer.* 8:335-343.

Lee, D. S., A. Norden, C. R. Gilbert, and R. Franz. 1976. A list of the freshwater fishes of Maryland and Delaware. *Chesapeake Sci.* 17(3):205-211.

Leggett, W. C., and R. R. Whitney. 1972. Water temperature and the migration of American shad. *Fishery Bull.* 70(3):659-670.

Lindsay, J. A., and R. L. Moran. 1976. Relationships of parasitic isopods, *Lironeca ovalis* and *Olencira praegustator,* to marine fish hosts in Delaware Bay. *Trans. Amer. Fish. Soc.* 105:327-332.

Lippson, A. J., ed. 1973. The Chesapeake Bay in Maryland: An Atlas of Natural Resources. The Johns Hopkins Univ. Press, Baltimore, Md. 55 pp.

Lippson, A. J., and R. L. Moran. 1974. Manual for identification of early developmental stages of fishes of the Potomac River estuary. Prepared for Power Plant Siting Program, Md. Dept. of Natural Resources by Martin Marietta Corp., Environmental Technology Center, Baltimore, Md. PPSP-MP-13.

Lippson, R. L. 1969-1971. Unpublished data sheets, Chesapeake Biological Laboratory Blue Crab Survey, Potomac River. Univ. of Md., Center for Environmental and Estuarine Studies, CBL, Solomons, Md.

Lippson, R. L. 1971. Blue crab study in Chesapeake Bay, Maryland: Annual progress report. Univ. of Md., Natural Resources Institute, Chesapeake Biological Laboratory, Solomons, Md. Ref. No. 71-9.

Lippson, R. L. 1973. American eel. Pages 30 to 31 *in* The Chesapeake Bay in Maryland: An Atlas of Natural Resources, A. J. Lippson, ed. The Johns Hopkins Univ. Press, Baltimore, Md.

Lippson, R. L. 1978. Personal communication. National Oceanographic and Atmospheric Adm., Environmental Assessment Division, Oxford, Md.

Littleford, R. A. 1939. The life cycle of *Dactylometra quinquecirrha,* L. Agassiz, in the Chesapeake Bay. *Biol. Bull.* 77:368-381.

Loos, J. 1975. Shore and tributary distribution of ichthyoplankton and juvenile fish with a study of their food habits. Prepared for Power Plant Siting Program, Md. Dept. of Natural Resources by Academy of Natural Sciences of Philadelphia, Pa.

Loosanoff, V. L. 1966. Time and intensity of setting of the oyster, *Crassostrea virginica,* in Long Island Sound. *Biol. Bull.* 130(2):211-227.

Lux, F. E., P. E. Hamer, and J. C. Poole. 1966. Summer flounder. Marine Resources of the Atlantic Coast. Atlantic States Marine Fisheries Comm., Tallahassee, Fla. Leaflet No. 6.

Lynch, M. P. 1977. Workshop report on fisheries and wildlife. Proc. of the Bi-State Conference on the Chesapeake Bay, Chesapeake Research Consortium, Annapolis, Md. CRC Publ. 61.

MacKenzie, C. L., Jr. 1974. Maryland's oyster seed areas are in excellent condition for set. *Commercial Fisheries News* 7(5):1.

Mangum, C. P. 1964. Studies on speciation in maldanid polychaetes of the North American Atlantic Coast, IV: Distribution and competitive interactions of five sympatric species. *Limnol. Oceanogr.* 9:12-26.

Mangum, C. P., S. L. Santos, and W. R. Rhodes, Jr. 1968. Distribution and feeding in the omphid polychaete, *Diopatra cuprea* (Bosc). *Mar. Biol.* 2:33-40.

Mangum, C. P., and W. Van Winkle. 1973. Responses of aquatic invertebrates to declining oxygen conditions. *Amer. Zool.* 13:529-541.

Mansueti, A. J. 1964. Early development of the yellow perch, *Perca flavescens. Chesapeake Sci.* 5(1-2):46-66.

Mansueti, A. J., and J. D. Hardy, Jr. 1967. Development of Fishes of the Chesapeake Bay Region, Part I: An Atlas of Egg, Larval and Juvenile Stages. Univ. of Md., Natural Resources Institute, Chesapeake Biological Laboratory, Solomons, Md. 202 pp.

Mansueti, R. J. 1955a. Natural history of the American shad in Maryland waters. Reprinted from *Maryland Tidewater News,* Vol. 11, No. 11., Suppl. No. 4. Univ. of Md., Natural Resources Institute, Chesapeake Biological Laboratory, Solomons, Md.

Mansueti, R. J. 1955b. Life history of the striped bass in Maryland waters. Reprinted from *Maryland Tidewater News,* Vol. 11, No. 9., Suppl. No. 3. Md. Dept. of Research and Education, Chesapeake Biological Laboratory, Solomons, Md.

Mansueti, R. J. 1955c. Important Potomac River fishes recorded from marine and fresh waters between Point Lookout, St. Mary's County, and Little Falls, Montgomery County, Maryland, with a bibliography to Potomac fisheries. Md. Dept. of Research and Education, Chesapeake Biological Laboratory, Solomons, Md.

Mansueti, R. J. 1957. Shades of Moby Dick. *The Skipper Magazine* 17(8):28-46.

Mansueti, R. J. 1960a. Restriction of very young red drum, *Sciaenops ocellata,* to shallow estuarine waters of Chesapeake Bay during late autumn. *Chesapeake Sci.* 1(3-4):207-210.

Mansueti, R. J. 1960b. Comparison of the movements of stocked and resident yellow perch, *Perca flavescens,* in tributaries of Chesapeake Bay, Maryland. *Chesapeake Sci.* 1(1):21-35.

Mansueti, R. J. 1961a. Age, growth, and movements of the striped bass, *Roccus saxatilis,* taken in size selective fishing gear in Maryland. *Chesapeake Sci.* 2(1-2):9-36.

Mansueti, R. J. 1961b. Movements, reproduction and mortality of the white perch, *Roccus americanus,* in the Patuxent estuary, Maryland. *Chesapeake Sci.* 2(3-4):142-205.

Mansueti, R. J. 1963. Symbiotic behavior between small fishes and jellyfish, with new data on that between the stromateid, *Peprelus alepidotus,* and the scyphomedusa, *Chrysaora quinquecirrha. Copeia* 1:40-80.

Mansueti, R. J., and H. Kolb. 1953. A historical review of the shad fisheries of North America. State of Md., Board of Natural Resources, Dept. of Research and Education, Chesapeake Biological Laboratory, Solomons, Md. Publ. No. 97.

Mansueti, R. J., and R. S. Scheltema. 1953. Summary of fish collections made in the Chesapeake Bay area of Maryland and Virginia during October 1953. Md. Dept. of Research and Education, Chesapeake Biological Laboratory, Solomons, Md. Field Summary #1.

Marcy, B. 1969. Age determination from scales of *Alosa pseudoharengus* (Wilson) and *Alosa aestivalis* (Mitchill) in Connecticut waters. *Trans. Amer. Fish. Soc.* 98:622-630.

Mare, M. F. 1942. A study of marine benthic communities with special reference to the microorganisms. *J. Marine Biol. Assoc. U.K.* 25:517-554.

Marsical, R. N. 1975. Entoprocta. Pages 1 to 36 *in* Reproduction of Marine Invertebrates, Vol. II: Entroprocts and Lesser Coelomates, A. C. Giese, and J. S. Pearse, eds. Academic Press, Inc., New York, N.Y.

Maryland Department of Natural Resources. 1956-1976. Marsh and waterfowl investigations. Pittman-Robertson W-30-R and W-45-R data sheets (3-year period averages). Md. Wildlife Adm., Annapolis, Md.

Maryland Department of Natural Resources. 1960-1975. Marsh and waterfowl investigations. Pittman-Robertson W-30-R and W-45-R data sheets (5-year period averages). Md. Wildlife Adm., Annapolis, Md.

Maryland Department of Natural Resources. 1967-1968. Unpublished data from Wetland Habitat Inventory. Water Resources Adm., Annapolis, Md.

Maryland Department of Natural Resources. 1969-1971. Anadromous Fish Stream Survey Program: Data base. Md. Fisheries Adm., Annapolis, Md.

Maryland Department of Natural Resources. 1969-1974. Power Plant Siting Program: Chesapeake Bay oceanographic data base. Annapolis, Md.

Maryland Department of Natural Resources. 1974-1976. Potomac River Fisheries Program data base. Md. Fisheries Adm. for Power Plant Siting Program. Stored at Martin Marietta Corp., Environmental Center, Baltimore, Md.

Maryland Department of Natural Resources. 1977a. Maryland endangered species, mimeo. Md. Wildlife Adm., Non-Game and Endangered Species Program, Annapolis, Md.

Maryland Department of Natural Resources. 1977b. Rules and Regulations, Annotated Code of Maryland. Annapolis, Md.

Maryland Department of Natural Resources. 1978a. Maryland Register, Vol. 5, Issue 10.

Maryland Department of Natural Resources. 1978b. Maryland fur harvest summary for 1975-77 seasons (internal communciation). Md. Wildlife Adm., Annapolis, Md.

Maryland Department of State Planning and Chesapeake Bay Interagency Planning Committee. 1972. Maryland Chesapeake Bay Study. Prepared by Wallace, McHarg, Roberts, and Todd, Inc., Philadelphia, Pa. 403 pp.

Maryland Geological Survey. 1975. Historical shorelines and erosion rates: Lower western shore (Calvert, Charles, Prince Georges, and St. Marys counties). Md. Dept. of Natural Resources, Md. Coastal Zone Management Program, Annapolis, Md.

Mason, W. T., ed. 1974. Potomac River basin baseline water quality monitoring network. Interstate Commission on the Potomac River Basin, Bethesda, Md. Tech. Publ.

Mason, W. T., ed. 1977. Potomac River basin baseline water quality monitoring network. Interstate Commission on the Potomac River Basin, Bethesda, Md. Tech. Publ. 77-1.

Mason, W. T., and K. C. Flynn, eds. 1976. The Potomac estuary: Biological resources, trends and options. Proc. of a symposium sponsored by Interstate Commission on the Potomac River Basin and Power Plant Siting Program, Md. Dept. of Natural Resources. ICPRB, Bethesda, Md. Tech. Publ. 76-2.

Mason, W. T., V. J. Rasin, W. J. McCaw and K. C. Flynn. 1976. Potomac River basin water quality. Interstate Commission on the Potomac River Basin, Bethesda, Md. Tech. Publ. 76-3.

Massmann, W. H. 1954. Marine fishes in fresh and brackish waters of Virginia rivers. *Ecology* 35(1):75-78.

Massmann, W. H., E. C. Ladd, and H. N. McCutcheon. 1952. A biological survey of the Rappahannock River, Virginia. Virginia Fisheries Laboratory, Gloucester Point, Va. Special Scientific Rept. 6, Part I.

Massmann, W. H., and A. L. Pacheco. 1961. Movements of striped bass tagged in Virginia waters of Chesapeake Bay. *Chesapeake Sci.* 2(1-2):37-44.

Maurer, D., L. Watling, P. Kinner, W. Leathem, and C. Wethe. 1978. Benthic invertebrate assemblages of Delaware Bay. *Mar. Biol.* 45(1):65-78.

McAtee, W. L., and A. C. Weed. 1915. First list of the fishes of the vicinity of Plummers Island, Maryland. *Proc. Biol. Soc. of Washington* 28:1-14.

McCain, J. C. 1968. The Caprellidae (Crustacea: Amphipoda) of the western North Atlantic United States. National Museum Bulletin 278. Smithsonian Institution Press, Washington, D.C. 147 pp.

McCarthy, J. J., W. R. Taylor, and J. L. Taft. 1977. Nitrogenous nutrition of the plankton in the Chesapeake Bay, 1: Nutrient availability and phytoplankton preferences. *Limnol. Oceanogr.* 22(6):996-1011.

McCaw, W. J. 1978. Personal communciation. Planning and Environmental Analysis, Interstate Commission on the Potomac River Basin, Rockville, Md.

Meadows, P. S., and J. I. Campbell. 1972. Habitat selection by aquatic invertebrates. *Advances in Mar. Biol.* 10:271-382.

Meglitsch, P. A. 1972. Invertebrate Zoology. Oxford Univ. Press, New York, N.Y. 834 pp.

Meritt, D. W. 1977. Oyster spat set on natural cultch in the Maryland portion of the Chesapeake Bay (1939-1975). Univ. of Md., Center for Environmental and Estuarine Studies, Horn Point Environmental Laboratories, Cambridge, Md. CEES Special Rept. No. 7.

Merriner, J. V. 1976. Anadromous fishes of the Potomac estuary. Pages 105 to 109 *in* The Potomac estuary: Biological resources, trends and options, W. T. Mason and K. C. Flynn, eds. Proc. of a symposium sponsored by Interstate Commission on the Potomac River Basin and Power Plant Siting Program, Md. Dept. of Natural Resources. ICPRB, Bethesda, Md. Tech. Publ. 76-2.

Metzgar, R. G. 1973. Wetlands in Maryland. Md. Dept. of State Planning, Baltimore, Md. Publ. No. 157.

Mihursky, J. A. 1973. The effects of steam electric station operations on organisms pumped through the cooling water system: Macroplankton studies, Morgantown S.E.S. site, May-June 1973, progress report. Prepared for Power Plant Siting Program, Md. Dept. of Natural Resources by Univ. of Md., Natural Resources Institute, Chesapeake Biological Laboratory, Solomons, Md. Ref. No. 73-76f.

Mihursky, J. A., W. R. Boynton, E. M. Setzler, K. V. Wood, H. H. Zion, E. W. Gordon, P. Pulles, and J. Leo. 1976. Potomac estuary fisheries study: Ichthyoplankton and juvenile investigations, final report. Submitted to Power Plant Siting Program, Md. Dept. of Natural Resources by Univ. of Md., Center for Environmental and Estuarine Studies, Chesapeake Biological Laboratory, Solomons, Md. CEES Ref. No. 76-12 CBL.

Mileikovsky, S. A. 1971. Types of larval development in marine bottom invertebrates, their distribution and ecological significance: A re-evaluation. *Mar. Biol.* 10:193-213.

Miller, P. E. 1976. Experimental study and modeling of striped bass egg and larval mortality. Ph.D. Dissertation. The Johns Hopkins Univ., Baltimore, Md. 99 pp.

Mills, E. L. 1969. The community concept in marine zoology, with comments on continua and instability in some marine communities. *J. Fish. Res. Bd. Canada* 26:1415-1428.

Miner, R. W. 1950. Field Book of Seashore Life. G. P. Putnam's Sons, New York, N.Y. 888 pp.

Moore, G. N. 1978. Personal communication. Virginia State Water Control Board, Northern Regional Office, Alexandria, Va.

Moore, K. A. 1975a. Stafford County tidal marsh inventory. Virginia Institute of Marine Science, Applied Marine Science and Ocean Engineering, Gloucester Point, Va. Special Rept. No. 62.

Moore, K. A. 1975b. Prince William County tidal marsh inventory. Virginia Institute of Marine Science, Applied Marine Science and Ocean Engineering, Gloucester Point, Va. Special Rept. No. 78.

Moore, K. A. 1975c. King George County tidal marsh inventory. Virginia Institute of Marine Science, Applied Marine Science and Ocean Engineering, Gloucester Point, Va. Special Rept. No. 63.

Morgan, R. P. 1973. Marking fish eggs with biological stains. *Chesapeake Sci.* 14(4):303-305.

Morgan, R. P., T. S. Koo, and G. E. Krantz. 1973. Electrophoretic determination of populations of the striped bass, *Morone saxatilis*, in the upper Chesapeake Bay. *Trans. Amer. Fish. Soc.* 102(1):21-32.

Morgan, R. P., and J. Rasin. 1973. Hydrographic and ecological effects of enlargement of the Chesapeake and Delaware canal, Appendix X: Effects of salinity and temperature on the development of eggs and larvae of the striped bass and white perch. Univ. of Md., Center for Environmental and Estuarine Studies, Natural Resources Institute, Chesapeake Biological Laboratory, Solomons, Md. NRI Ref. No. 73-109.

Morrill, J. 1975. Phytoplankton studies. Pages II.1-107 to II.1-136 *in* Semi-annual report, 1974: Biological and chemical baseline investigations in the vicinity of Calvert Cliffs Nuclear Power Plant. Prepared for Baltimore Gas and Electric Co. by Academy of Natural Sciences of Philadelphia, Pa.

Morrill, J., and M. E. Kachur. 1976. Phytoplankton. Pages 7-1 to 7-47 *in* Semi-annual environmental monitoring report, Calvert Cliffs Nuclear Power Plant, March 1976. Prepared by Academy of Natural Sciences of Philadelphia for Baltimore Gas and Electric Co., Baltimore, Md.

Mountford, K. 1977. Phytoplankton and primary productivity. Pages 8-75 to 8-122 *in* Morgantown Station and the Potomac estuary: A 316 environmental demonstration, Vol. II. Prepared for Potomac Electric Power Co. by Academy of Natural Sciences of Philadelphia, Pa.

Mountford, K., G. Chisholm, R. Donahoe, and A. Bacheler. 1976. Phytoplankton: Productivity and biomass. Pages 6-1 to 6-39 *in* Semi-annual environmental monitoring report, Calvert Cliffs Nuclear Power Plant, March 1976. Prepared by Academy of Natural Sciences of Philadelphia for Baltimore Gas and Electric Co., Baltimore, Md.

Mountford, N. K., A. F. Holland, and J. A. Mihursky. 1977. Identification and description of macrobenthic communities in the Calvert Cliffs region of the Chesapeake Bay. *Chesapeake Sci.* 18(4):360-369.

Mowbray, E. E., J. A. Chapman, and J. R. Goldsberry. 1976. Preliminary observations on otter distribution and habitat preferences in Maryland with descriptions of otter field sign. Pages 125 to 131 *in* Proc. of the 33rd Northeast Fish and Wildlife Conference, April 26-29, 1976, Hershey, Pa. Univ. of Md., Appalachian Environmental Laboratory, Frostburg, Md. Ref. No. 77-114-AEL.

Mulford, R. A. 1972. Phytoplankton of the Chesapeake Bay. Pages S74 to S81 *in* Biota of the Chesapeake Bay, A. J. McErlean, C. Kerby, and M. L. Wass, eds. *Chesapeake Sci.*, Vol. 13, Supplement.

Mulford, R. A., and S. D. Van Valkenberg. 1973. Phytoplankton taxonomic studies. Chapter 4 *in* The effects of Morgantown steam electric station operations on organisms pumped through the cooling water system, July 1972-December 1972, final report, loose-leaf pub. n.p. Submitted to Md. Dept. of Natural Resources by Univ. of Md., Natural Resources Institute, Solomons, Md. Ref. No. 73-20.

Muncy, R. J. 1962. Life history of the yellow perch, *Perca flavescens*, in estuarine waters of Severn River, a tributary of Chesapeake Bay, Maryland. *Chesapeake Sci.* 3(3):143-159.

Munkittrick, G. T. 1976. The effect of agriculture on the biota of the Potomac River basin. Pages 46 to 48 *in* The Potomac estuary: Biological resources, trends and options, W. T. Mason and K. C. Flynn, eds. Proc. of a symposium sponsored by Interstate Commission on the Potomac River Basin and Power Plant Siting Program, Md. Dept. of Natural Resources. ICPRB, Bethesda, Md. Tech. Publ. 76-2.

Murawski, S. A., and A. L. Pacheco. 1977. Biological and fisheries data on Atlantic sturgeon, *Acipenser oxyrhynchus* (Mitchill). U.S. Dept. of Commerce, National Oceanographic and Atmospheric Adm., National Marine Fisheries Service. Northeast Fisheries Center, Sandy Hook Laboratory, Highlands, N.J. Tech. Series Rept. No. 10.

Musick, J. A. 1972. Fishes of Chesapeake Bay and the adjacent coastal plain. Pages 175 to 212 *in* A check list of the biota of lower Chesapeake Bay, M. L. Wass, ed. Virginia Institute of Marine Science, Gloucester Point, Va. Special Scientific Rept. No. 65.

Muus, B. J. 1967. The fauna of Danish estuaries and lagoons. *Meddelelser fra Danmarks Fiskeri-og Havundersøgelser* 5(1):1-316.

Naiman, R. J., H. Hixson, and T. Capizzi. 1978. Fish bottom trawling. Pages 8-1 to 8-37 *in* Non-radiological environmental monitoring report, Calvert Cliffs Nuclear Power Plant, January to December 1977. Prepared by Baltimore Gas and Electric Co. and Academy of Natural Sciences of Philadelphia, Pa.

Newell, R. 1965. The role of detritus in the nutrition of two marine deposit-feeders, the prosobranch *Hydrobia ulvae* and the bivalve *Macoma balthica*. *Proc. Zool. Soc. London* 144:25-45.

Nicholas, P. R., and R. V. Miller. 1967. Seasonal movements of striped bass, *Roccus saxatilis* (Walbaum), tagged and released in the Potomac River, Maryland, 1959-1961. *Chesapeake Sci.* 8(2):102-124.

Nicholson, W. R. 1972. Population structure and movements of Atlantic menhaden, *Brevoortia tyrannus*, as inferred from back-calculated length frequencies. *Chesapeake Sci.* 13(3):161-174.

Norris, R. M. 1975. Personal communication. Potomac River Fisheries Commission, Colonial Beach, Va.

Norris, R. M. 1977. Personal communication, Potomac River Fisheries Commission, Colonial Beach, Va.

O'Dell, J., R. C. Dintaman, and J. Gabor. 1976. Fishes of the Potomac estuary. Pages 100 to 104 in The Potomac estuary: Biological resources, trends and options. Proc. of a symposium sponsored by Interstate Commission on the Potomac River Basin and Power Plant Siting Program, Md. Dept. of Natural Resources. ICPRB, Bethesda, Md. Tech. Publ. 76-2.

O'Dell, J., H. J. King, III, and J. P. Gabor, 1973. Survey of anadromous fish spawning areas. Submitted to U.S. Dept. of Commerce, National Oceanic and Atmospheric Adm., National Marine Fisheries Service by Md. Dept. of Natural Resources, Fisheries Adm., Annapolis, Md. 65 pp.

Odum, E. P. 1959. Fundamentals of Ecology, 2nd ed. W. B. Saunders Co., Philadelphia, Pa. 546 pp.

Old, M. C. 1941. The taxonomy and distribution of the boring sponges (Clionidae) along the Atlantic Coast of North America. State of Md., Board of Natural Resources, Dept. of Research and Education, Chesapeake Biological Laboratory, Solomons, Md. Publ. No. 44, pp. 3-16.

Olson, M., and L. E. Sage. 1978. Nearfield zooplankton studies at the Calvert Cliffs Nuclear Power Plant, May 1974 through December 1976. Prepared for Baltimore Gas and Electric Company by Academy of Natural Sciences of Philadelphia, Pa. Rept. No. 78-18.

Orr, R. T. 1976. Vertebrate Biology. W. B. Saunders Co., Philadelphia, Pa. 472 pp.

Orth, R. J. 1971. Observations on the planktonic larvae of Polydora ligni Webster (Polychaeta: Spionidae) in the York River, Virginia. Chesapeake Sci. 12(3):121-124.

Orth, R. J. 1975. Destruction of eelgrass, Zostera marina, by the cownose ray, Rhinoptera bonasus, in the Chesapeake Bay. Chesapeake Sci. 16(3):205-208.

Osburn, R. C. 1944. A survey of the bryozoa of Chesapeake Bay. State of Md., Board of Natural Resources, Dept. of Research and Education, Chesapeake Biological Laboratory, Solomons, Md. Publ. No. 63.

Ott, F. D. 1972. Macroalgae of the Chesapeake Bay. Pages S83 to S84 in Biota of the Chesapeake Bay, A. J. McErlean, C. Kerby, and M. L. Wass, eds. Chesapeake Sci., Vol. 13, Supplement.

Pacheco, A. L. 1962. Age and growth of spot in lower Chesapeake Bay, with notes on distribution and abundance of juveniles in the York River system. Chesapeake Sci. 3(1):18-28.

Paine, R. T. 1966. Food web complexity and species diversity. Amer. Naturalist 100:65-75.

Palmer, R. N. 1975. Non-point pollution in the Potomac River basin. Interstate Commission on the Potomac River Basin, Bethesda, Md. Tech. Publ. No. 75-2.

Pennak, R. W. 1953. Fresh-Water Invertebrates of the United States. The Ronald Press Co., New York, N.Y. 769 pp.

Perry, M. C., R. Andrews, and P. P. Beaman. 1976. Distribution and abundance of canvasbacks in Chesapeake Bay in relation to food organisms. Paper presented at meeting of Atlantic Estuarine Research Society, October 14-16, 1976, Cape May, N.J.

Petersen, C. G. 1913. Valuation of the Sea, II: The animal communities of the sea bottom and their importance to marine zoogeography. Report of the Danish Biological Station 21:1-44.

Peterson, R. T. 1947. A Field Guide to the Birds. Houghton Mifflin Co., Boston, Mass. 230 pp.

Pettibone, M. H. 1963. Marine polychaete worms of the New England region, I: Aphroditidae through Trochochaetidae. Smithsonian Institute, Washington, D.C. U.S. Natural Museum Bull. 227. 356 pp.

Pfitzenmeyer, H. T. 1962. Periods of spawning and setting of the soft-shelled clam, Mya arenaria, at Solomons, Maryland. Chesapeake Sci. 3(2):114-120.

Pfitzenmeyer, H. T. 1974. The effects of Morgantown steam electric station operations on organisms pumped through the cooling water system, Part IV: Benthic investigations. Prepared for Power Plant Siting Program, Md. Dept. of Natural Resources by Univ. of Md., Natural Resources Institute, Chesapeake Biological Laboratory, Solomons, Md. Ref. No. 74-13.

Pfitzenmeyer, H. T. 1976. Some effects of salinity on the benthic macroinvertebrates of the lower Potomac. Pages 75 to 80 in The Potomac estuary: Biological resources, trends and options, W. T. Mason and K. C. Flynn, eds. Proc. of a symposium sponsored by Interstate Commission on the Potomac River Basin and Power Plant Siting Program, Md. Dept. of Natural Resources. ICPRB, Bethesda, Md. Tech. Publ. 76-2.

Pfitzenmeyer, H. T., and K. G. Drobeck. 1963. Benthic survey for populations of soft-shelled clams, Mya arenaria, in the lower Potomac River, Maryland. Chesapeake Sci. 4(2):67-74.

Pfitzenmeyer, H. T., and K. G. Drobeck. 1964. The occurrence of the brackish water clam, Rangia cuneata, in the Potomac River, Maryland. Chesapeake Sci. 5(4):209-212.

Pheiffer, T. H. 1976. Current nutrient assessment of the upper Potomac estuary. Pages 28 to 37 in The Potomac estuary: Biological resources, trends and options, W. T. Mason and K. C. Flynn, eds. Proc. of a symposium sponsored by Interstate Commission on the Potomac River Basin and Power Plant Siting Program, Md. Dept. of Natural Resources. ICPRB, Bethesda, Md. Tech. Publ. 76-2.

Polgar, T. T. 1975. Impact of Potomac River power plants on early life stages of striped bass: Preliminary results. Record of the Maryland Power Plant Siting Act, Vol. 4, No. 3. Md. Dept. of Natural Resources, Annapolis, Md.

Polgar, T. T., G. M. Krainak, and H. T. Pfitzenmeyer. 1975. A methodology for quantifying the responses of benthic communities to environmental perturbations. Prepared for Power Plant Siting Program, Md. Dept. of Natural Resources by Martin Marietta Corp., Environmental Technology Center, Baltimore, Md. Morgantown Monitoring Program Rept. Series. Ref. No. MT-75-2.

Polgar, T. T., J. A. Mihursky, R. E. Ulanowicz, R. P. Morgan, and J. S. Wilson. 1976. An analysis of 1974 striped bass spawning success in the Potomac estuary. Pages 151 to 165 in Estuarine Processes, Vol. I: Uses, Stresses and Adaptation to the Estuary, M. Wiley, ed. Academic Press, Inc., New York, N.Y.

Polgar, T. T., R. E. Ulanowicz, D. A. Pyne, and G. M. Krainak. 1975. Investigations of the role of physical-transport processes in determining ichthyoplankton distributions in the Potomac River. Prepared for Power Plant Siting Program, Md. Dept. of Natural Resources by Martin Marietta Corp., Environmental Technology Center, Baltimore, Md. PPRP-11, PPMP-14.

Postma, H. 1967. Sediment transport and sedimentation in the estuarine environment. Pages 158 to 179 in Estuaries, G. H. Lauff, ed. American Association for the Advancement of Science, Washington, D.C. Publ. 83.

Potomac Basin Reporter. 1974. Lower Potomac hoping for oyster recovery. Potomac Basin Reporter, Vol. 30, No. 9. Interstate Commission on the Potomac River Basin, Bethesda, Md.

Potomac River Fisheries Commission. 1978. Regulations of the Potomac River Fisheries Commission. Colonial Beach, Va.

Pough, R. H. 1951. Audubon Water Bird Guide: Water, Game, and Large Land Birds of Eastern and Central North America. Doubleday and Co., Inc., Garden City, N.Y. 352 pp.

Prescott, G. W. 1951. Algae of the Western Great Lakes Area. Wm. C. Brown Co. Publishers, Dubuque, Iowa. 977 pp.

Prescott, G. W. 1968. The Algae: A Review. Houghton Mifflin Co., Boston, Mass. 436 pp.

Price, K. S., Jr. 1962. Biology of the sand shrimp, Crangon septemspinosa, in the shore zone of the Delaware Bay region. Chesapeake Sci. 3(4):244-255.

Pritchard, D. W. 1967a. What is an estuary: Physical viewpoint. Pages 3 to 5 in Estuaries, G. H. Lauff, ed. American Association for the Advancement of Science, Washington, D.C. Publ. 83.

Pritchard, D. W. 1967b. Observations of circulation in coastal plain estuaries. Pages 37 to 44 in Estuaries, G. H. Lauff, ed. American Association for the Advancement of Science, Washington, D.C. Publ. 83.

Rasin, V. J. 1977. Potomac River basin water quality, 1975-1976. Interstate Commission on the Potomac River Basin, Bethesda, Md. Tech. Publ. 77-4.

Rasmussen, E. 1973. Systematics and ecology of the Isefjord Marine Fauna (Denmark). The Isefjord Laboratory, Vellerup Vig, Denmark. Reprinted from Ophelia, Vol. II. 495 pp.

Rawls, C. 1964. Aquatic plant nuisances. Univ. of Md., Natural Resources Institute, Solomons, Md. Ref. No. 64-15.

Raymont, J. E. 1963. Plankton and Productivity in the Oceans. Pergamon Press, Inc., Elmsford, N.Y. 660 pp.

Reimer, C. W. 1977. Benthic and littoral algae (including epiphytic forms). Pages 8-1 to 8-34 in Morgantown Station and the Potomac estuary: A 316 environmental demonstration, Vol. II. Prepared for Potomac Electric Power Co. by Academy of Natural Sciences of Philadelphia, Pa.

Rhoads, D. C. 1974. Organism-sediment relations on the muddy sea floor. Oceanogr. Mar. Biol. Ann. Rev. 12:263-300.

Rice, N. E., and W. A. Powell. 1970. Observations on three species of jellyfishes from Chesapeake Bay with special reference to their toxins, I: Chrysaora (Dactylometra) quinquecirrha. Biol. Bull. 139(1):180-187.

Richards, C. E., and R. L. Bailey. 1967. Occurrence of Fundulus luciae, spotfin killifish, on the seaside of Virginia's eastern shore. Chesapeake Sci. 8(3):204-205.

Richards, C. E., and M. Castagna. 1970. Marine fishes of Virginia's eastern shore (inlet, marsh, and seaside waters). Chesapeake Sci. 11(4):235-248.

Richardson, S. L., and E. B. Joseph. 1975. Occurrence of larvae of the green goby, Microgobius thalassinus, in the York River, Virginia. Chesapeake Sci. 16(3):215-218.

Riser, N.W. 1974. Nemertinea. Pages 359 to 389 in Reproduction of Marine Invertebrates, Vol. I: Acoelomate and Pseudocoelomate Metazoans, A. C. Giese and J. S. Pearse, eds. Academic Press, Inc., New York, N.Y.

Ritchie, D. E., Jr. 1977. Groundfish resources assessment of Choptank, Nanticoke and Wicomico rivers from November 1973 to October 1974. Univ. of Md., Center for Environmental and Estuarine Studies, Chesapeake Biological Laboratory, Solomons, Md. Ref. No. 77-06.

Ritchie, D. E, H. J. King, and A. J. Lippson. 1973. White perch. Pages 34 to 35 in The Chesapeake Bay in Maryland: An Atlas of Natural Resources, A. J. Lippson, ed. The Johns Hopkins Univ. Press, Baltimore, Md.

Commission on the Potomac River Basin and Power Plant Siting Program, Md. Dept. of Natural Resources. ICPRB, Bethesda, Md. Tech. Publ. 76-2.

Robertson, P. G. 1977. Back River: An assessment of water quality and related fish mortalities. Md. Dept. of Natural Resources, Water Resources Adm., Annapolis, Md. 132 pp.

Rohde, F. C. 1974. Percidae — perches. Pages 196 to 205 *in* Manual for identification of early developmental stages of fishes of the Potomac River estuary, A. J. Lippson and R. L. Moran, eds. Prepared for Power Plant Siting Program, Md. Dept. of Natural Resources by Martin Marietta Corp., Environmental Technology Center, Baltimore, Md. PPSP-MP-13.

Rosenberg, R. 1973. Succession in benthic macrofauna in a swedish fjord subsequent to the closure of a sulphite paper mill. *Oikos* 24:344-358.

Roy Mann Associates, Inc. 1976. Recreational boating on the tidal waters of Maryland: A management planning study. Prepared for Energy and Coastal Zone Adm., Md. Dept. of Natural Resources. Cambridge, Mass. 177 pp.

Russell-Hunter, W. D. 1970. Aquatic Productivity: An Introduction to Some Basic Aspects of Biological Oceanography and Limnology. MacMillan Publishing Co., Inc., New York, N.Y. 306 pp.

Sage, L. E. 1976. Zooplankton entrainment. Pages 13.2-1 to 13.2-62 *in* Semi-annual environmental monitoring report, Calvert Cliffs Nuclear Power Plant, March 1976. Prepared for Baltimore Gas and Electric Co. by Academy of Natural Sciences of Philadelphia, Pa.

Sage, L. E., and A. G. Bacheler. 1978. Zooplankton entrainment. Pages 12.2-1 to 12.2-54 *in* Non-radiological environmental monitoring report, Calvert Cliffs Nuclear Power Plant, January to December, 1977. Prepared by Baltimore Gas and Electric Co. and Academy of Natural Sciences of Philadelphia. B. G. and E., Baltimore, Md.

Sage, L. E., J. M. Summerfield, and M. M. Olson. 1976. Zooplankton of the Potomac estuary. Pages 81 to 87 *in* The Potomac estuary: Biological resources, trends and options. Proc. of a symposium sponsored by the Interstate Commission on the Potomac River Basin and Power Plant Siting Program, Md. Dept. of Natural Resources. ICPRB, Bethesda, Md. Tech. Publ. 76-2.

Sage, L. E., and M. M. Olson. 1977a. Zooplankton. Pages 9-21 to 9-74 *in* Morgantown Station and the Potomac estuary: A 316 environmental demonstration, Vol. II. Prepared for Potomac Electric Power Co. by Academy of Natural Sciences of Philadelphia, Pa.

Sage, L. E., and M. M. Olson. 1977b. Zooplankton entrainment. Pages 13.2-1 to 13.2-72 *in* Non-radiological environmental monitoring report, Calvert Cliffs Nuclear Power Plant, July-December 1976. Prepared for Baltimore Gas and Electric Co. by Academy of Natural Sciences of Philadelphia, Pa.

Sanders, H. L. 1958. Benthic studies in Buzzards Bay, I: Animal-sediment relationships. *Limnol. Oceanogr.* 3:245-358.

Sandifer, P. A. 1972. Morphology and ecology of Chesapeake Bay decapod crustacean larvae. Ph.D. Dissertation. Univ. of Virginia. 532 pp. Univ. Microfilms, Ann Arbor, Mich. No. 72-23441.

Sandifer, P. A. 1973. Distribution and abundance of decapod crustacean larvae in the York River estuary and adjacent lower Chesapeake Bay, Virginia, 1968-1969. *Chesapeake Sci.* 14(4):235-257.

Sandoz, O., and K. H. Johnston. 1966. Culture of striped bass, *Roccus saxatalis* (Walbaum). Pages 390 to 394 *in* Proc. of the 19th Annual Conference of the Southeastern Assoc. of Game and Fish Commissioners, 1965. Southeastern Assoc. of Game and Fish Commissioners, Columbia, S.C.

Scherk, J. A. 1973. Sediments. Pages 8 to 9 *in* The Chesapeake Bay in Maryland: An Atlas of Natural Resources, A. J. Lippson, ed. The Johns Hopkins Univ. Press, Baltimore, Md.

Schubel, J. R. 1972. Suspended sediment discharge up the Susquehanna River at Conowingo, Maryland during 1969. *Chesapeake Sci.* 13(1):53-58.

Schubel, J. R., and R. H. Meade. 1977. Man's impact on estuarine sedimentation. Pages 193 to 209 *in* Estuarine pollution control and assessment, proceedings of a conference, Vol. I. U.S. Environmental Protection Agency, Office of Water Planning and Standards, Washington, D.C.

Schultz, L. P., and D. G. Cargo. 1971. The sea nettle of Chesapeake Bay. Univ. of Md., Natural Resources Institute, Solomons, Md. Educational Series No. 93.

Schwartz, F. J. 1960. The pickerels. *Md. Conservationist* 37(4):21-26.

Schwartz, F. J. 1961a. Lampreys and eels. *Md. Conservationist* 38(2):18-27.

Schwartz, F. J. 1961b. Catfishes. *Md. Conservationist* 38(5):21-26.

Schwartz, F. J. 1962. The beaked fishes of Maryland. *Md. Conservationist* 39(2):21-25.

Schwartz, F. J. 1963. The fresh-water minnows of Maryland. *Md. Conservationist* 40(2):19-29.

Schwartz, F. J. 1968. Bull minnows. *Md. Conservationist* 4(3):2-5.

Schwartz, F. J. 1971. Biology of *Microgobius thalassinus* (Pisces: Gobiidae), a sponge-inhabiting goby of Chesapeake Bay, with range extension of two goby associates. *Chesapeake Sci.* 12(3):156-166.

Schwartz, F. J. 1974. Movements of the oyster toadfish (Pisces: Batrachoididae) about Solomons, Maryland. *Chesapeake Sci.* 15(3):155-159.

Scott, R. F., and J. G. Boone. 1973. Fish distribution in various areas of Maryland tidewater as derived from shore-zone seining 1956-1972. Md. Dept. of Natural Resources, Fisheries Adm., Annapolis, Md. Data Report MFA-73-1.

Scott, W. B., and E. J. Crossman. 1973. Freshwater Fishes of Canada. Fisheries Research Board of Canada, Ottawa. Bull. 184. 966 pp.

Segal, E. 1970. Light. Pages 159 to 212 *in* Marine Ecology, Vol. I, Part 1, 0. Kinne, ed. John Wiley and Sons, Inc., New York, N.Y.

Shaw, W. N. 1965. Seasonal setting patterns of five species of bivalves in the Tred Avon River, Md. *Chesapeake Sci.* 6(1):33-37.

Shaw, W. N. 1967. Seasonal fouling and oyster setting on asbestos plates in Broad Creek, Talbot County, Maryland, 1963-65. *Chesapeake Sci.* 8(4):228-236.

Shaw, W. N., and F. Hamons. 1974. The present status of the soft-shell clam in Maryland. *Proc. Natl. Shellfisheries Assoc.* 64:38-44.

Sherman, J. S. 1977. Diatom studies using artificial substrates. Pages 8-39 to 8-74 *in* Morgantown Station and the Potomac estuary: A 316 environmental demonstration, Vol. II. Prepared for Potomac Electric Power Co. by Academy of Natural Sciences of Philadelphia, Pa.

Sikora, J. P., W. B. Sikora, C. W. Erkenbrecher, and B. C. Coull. 1977. Significance of ATP, carbon, and caloric content of meiobenthic nematodes in partitioning benthic biomass. *Mar. Biol.* 44(1):7-14.

Silberhorn, G. M. 1975. Northumberland County tidal marsh inventory. Virginia Institute of Marine Science, Applied Marine Science and Ocean Engineering, Gloucester Point, Va. Special Rept. No. 58.

Silberhorn, G. M. In Press. Westmoreland County tidal marsh inventory. Virginia Institute of Marine Science, Applied Marine Science and Ocean Engineering, Gloucester Point, Va. Special Rept. No. 59.

Simmons, G. M., and B. J. Armitage. 1972. An ecological evaluation of the heated discharge from the Possum Point power station on phytoplankton blooms in the Potomac River estuary. Virginia Electric and Power Co., Environmental Control Dept., Richmond, Va.

Simpson, M. 1962. Reproduction of the polychaete *Glycera dibranchiata* at Solomons, Maryland. *Biol. Bull.* 123(2):396-411.

Sipple, W. S. 1978. Personal communication. Water Resources Adm., Annapolis, Md.

Smith, H. M. 1892. Notes on a collection of fishes from the lower Potomac River, Maryland. *Bull. U.S. Fish. Comm.* 10(1890):63-72.

Smith, H. M., and B. A. Bean. 1899. List of fishes known to inhabit the waters of the District of Columbia and vicinity. *Bull. U.S. Fish. Comm.* 18(1898):179-187.

Speir, H. J., D. R. Weinrich, and R. S. Early. 1976. Maryland Chesapeake Bay sport fishing survey. Md. Dept. of Natural Resources, Fisheries Adm., Annapolis, Md. 99 pp.

Spoon, D. M. 1975. Survey, ecology, and systematics of the upper Potomac estuary biota: Aufwuchs microfauna, Phase I, final report. Water Resources Research Center, Washington Tech. Institute, Washington, D.C. Rept. No. 6.

Spoon, D. M. 1976. Microbial communities of the upper Potomac estuary: The aufwuchs. Pages 63 to 69 *in* The Potomac estuary: Biological resources, trends and options, W. T. Mason and K. C. Flynn, eds. Proc. of a symposium sponsored by Interstate Commission on the Potomac River Basin and Power Plant Siting Program, Md. Dept. of Natural Resources. ICPRB, Bethesda, Md. Tech. Publ. 76-2.

Stevenson, J. C. 1977. Summary of available information on Chesapeake Bay submerged vegetation, draft report. Univ. of Md., Center for Environmental and Estuarine Studies, Solomons, Md.

Stewart, R. E. 1962. Waterfowl populations in the upper Chesapeake region. U.S. Dept. of the Interior, Fish and Wildlife Service, Bureau of Sport Fisheries. Special Scientific Rept. — Wildlife No. 65.

Stotts, V. D. 1978. Personal communication on unpublished data from waterfowl habitat inventory. Md. Dept. of Natural Resources, Wildlife Adm., Annapolis, Md.

Strealy, L. A. 1978. Personal communication. Engineering and Construction Adm., Blue Plains Wastewater Treatment Plant, Dept. of Environmental Services, Washington, D.C.

Studholme, A. T., J. W. Aldrich, C. P. Gilchrist, F. G. Gillett, V. D. Stotts, and F. M. Uhler. 1965. Unpublished data. Patuxent Wildlife Research Center, Laurel, Md.

Sulkin, S. D. 1973. Blue crab study in Chesapeake Bay, Maryland. Univ. of Md., Natural Resources Institute, Chesapeake Biological Laboratory, Solomons, Md. Ref. No. 73-94.

Taft, J. L., and W. R. Taylor. 1976. Phosphorus distribution in the Chesapeake Bay. *Chesapeake Sci.* 17(2):67-73.

Tagatz, M. E. 1968. Biology of the blue crab, *Callinectes sapidus* Rathbun, in the St. Johns River, Florida. *Fishery Bull.* 67:17-33.

Tatnall, R. R. 1946. Flora of Delaware and the Eastern Shore: An annotated list of the ferns and flowering plants of the peninsula of Delaware, Maryland, and Virginia. The Society of Natural History of Delaware, Wilmington, Del.

Tatum, B. L., J. D. Bayless, B. G. McCoy, and W. B. Smith. 1966. Preliminary experiments in the artificial propogation of striped bass, *Roccus saxatilis*. Pages 374 to 389 *in* Proc. of the 19th Annual Conference of the Southeastern Assoc. of Game and Fish Commissioners, 1965. Southeastern Assoc. of Game and Fish Commissioners, Columbia, S.C.

Taylor, G. J. 1977. Personal communication. Endangered Species Project Leader, Maryland Department of Natural Resources, Wildlife Adm., Cambridge, Md.

Taylor, W. R. 1957. Marine Algae of the Northeastern Coast of North America, 2nd ed. The Univ. of Michigan Press, Ann Arbor, Mich. 509 pp.

Thorson, G. 1946. Reproduction and larval development of Danish marine bottom invertebrates, with special reference to planktonic larvae in the sound (Øresund). *Meddelelser fra Danmarks Fiskeri-og Havundersøgelser.* (Ser. Plankton) 4:1-523.

Thorson, G. 1950. Reproduction and larval ecology of marine bottom invertebrates. *Biol. Rev. Cambridge Philos. Soc.* 25(1):1-45.

Thorson, G. 1957. Bottom communities (sublittoral or shallow shelf). Pages 461 to 534 *in* Treatise on Marine Ecology and Paleo-ecology, Vol. I, J. W. Hedgpeth, ed. Geological Society of America, Washington, D.C. Mem. 67.

Todd, D. K., ed. 1970. The Water Encyclopedia. Water Information Center, Port Washington, N.Y. 559 pp.

Trippensee, R. E. 1953. Wildlife Management, Vol. II. McGraw-Hill Book Co., New York, N.Y. 572 pp.

Truitt, R. V. 1939. Our water resources and their conservation. Chesapeake Biological Laboratory, Solomons, Md. Contribution No. 27. 103 pp.

Tyler, A. V. 1963. A cleaning symbiosis between the rainwater fish, *Lucania parva*, and the stickleback, *Apeltes quadracus. Chesapeake Sci.* 4(2):105-106.

Tyrrell, W. B. 1936. The ospreys of Smith's Point, Virginia. *The Auk* 53(3):261-271.

Uhler, F. M. 1977. Personal communication. U.S. Dept. of the Interior, U.S. Fish and Wildlife Service, Patuxent Wildlife Research Center, Laurel, Md.

U.S. Department of the Army. 1973. Chesapeake Bay Existing Conditions Report. Baltimore District Corps of Engineers, Baltimore, Md. Appendices A, B, C, and D.

U.S. Department of the Army. 1974. Tropical Storm Agnes, June 1972, post flood report, Vol. I: Meteorology and hydrology. Baltimore District Corps of Engineers, Baltimore, Md.

U.S. Department of the Army. 1976. Chesapeake Bay future conditions report, Appendix 12: Fish and wildlife. Prepared for Baltimore District Army Corps of Engineers by U.S. Dept. of the Interior, U.S. Fish and Wildlife Service, Washington, D.C.

U.S. Department of Commerce. 1960-1976. Current fisheries statistics, Maryland and Virginia landings, annual summaries. National Oceanic and Atmospheric Adm., National Marine Fisheries Service, Washington, D.C.

U.S. Department of Commerce. 1974-1975. Nautical charts. Catalog No. 1, Panel D, Chart Nos. 12233, 12286, 12288, 12289. National Oceanic and Atmospheric Adm., National Ocean Survey, Washington, D.C.

U.S. Department of Commerce. 1977a. Tidal current tables 1978, Atlantic Coast of North America. National Oceanic and Atmospheric Adm., National Ocean Survey, Washington, D.C.

U.S. Department of Commerce. 1977b. Tide tables 1978: High and low water predictions, East Coast of North and South America including Greenland. National Oceanic and Atmospheric Adm., National Ocean Survey, Washington, D.C.

U.S. Department of the Interior. 1965-1977. Water resources data for Maryland and Delaware, Part I: Surface water records. U.S. Geological Survey, Washington, D.C.

U.S. Environmental Protection Agency. 1965-1975. STORET system, water quality data base, U.S. EPA, Region III, Philadelphia, Pa.

U.S. Environmental Protection Agency. 1969-1972. Water quality data sheets. Annapolis Field Office, Annapolis, Md.

Usinger, R. L., ed. 1963. Aquatic Insects of California with Keys to North American Genera and California Species. Univ. of California Press, Berkeley and Los Angeles, Calif. 508 pp.

Vance, R. R. 1973. On reproductive strategies in marine benthic invertebrates. *Amer. Naturalist* 107:339-352.

Van Engel, W. A. 1958. The blue crab and its fishery in Chesapeake Bay, Part I: Reproduction, early development, growth, and migration. *Commerical Fisheries Review* 20(6):6-17.

Van Engel, W. A. 1972a. Subclass Cirripedia. Page 143 *in* A check list of the biota of lower Chesapeake Bay, M. L. Wass, ed. Virginia Institute of Marine Science, Gloucester Point, Va. Special Scientific Rept. No. 65.

Van Engel, W. A. 1972b. Order Cumacea. Pages 141 to 145 *in* A check list of the biota of lower Chesapeake Bay, M. L. Wass, ed. Virginia Institute of Marine Science, Gloucester Point, Va. Special Scientific Rept. No. 65.

Van Valkenburg, S. 1972. Nannoplankton of the Chesapeake Bay. Pages S72 to S74 *in* Biota of the Chesapeake Bay, A. J. McErlean, C. Kerby, and M. L. Wass, eds. *Chesapeake Sci.,* Vol. 13, Supplement.

Virginia Marine Resources Commission. 1977. Laws of Virginia relating to fisheries of tidal waters. Commonwealth of Va., Newport News, Va.

Virnstein, R. W. 1977. The importance of predation by crabs and fishes on benthic infauna in Chesapeake Bay. *Ecology* 58(6):1199-1217.

Virnstein, R. W., and D. F. Boesch. 1975. Survey of benthic organisms of the lower Potomac River in the vicinity of Piney Point, Maryland, final report. Submitted to Steuart Petroleum Co. by Virginia Institute of Marine Science, Gloucester Point, Va. 9 pp.

Vokes, H. E., and J. Edwards, Jr. 1968. Geography and geology of Maryland. Maryland Geological Survey, Baltimore, Md. Bull. 19.

Von Vaupel-Klein, J. C., and R. E. Weber. 1975. Distribution of *Eurytemora affinis* (Copepoda: Calanoida) in relation to salinity: Field and laboratory observations. *Neth. J. Sea Res.* 9:297-310.

Wakefield, W. W. 1977. Macroplankton. Pages 13.3-1 to 13.3-50 *in* Non-radiological environmental monitoring report, Calvert Cliffs Nuclear Power Plant, July to December 1976. Prepared for Baltimore Gas and Electric Co. by Academy of Natural Sciences of Philadelphia, Pa.

Walford, L., S. Wilk, B. Olla, A. Kendall, B. Freeman, D. Deuel, and M. Silverman. 1978. The bluefish (*Pomatomus saltatrix*): A synoptic review of its biology. U.S. Dept. of Commerce, National Oceanic and Atmospheric Adm., National Marine Fisheries Service, Northeast Fisheries Center, Sandy Hook Laboratory, Highland, N.J. Tech. Series Rept.

Walker, E. P. 1964. Mammals of the World, Vols. I and II. The Johns Hopkins Press, Baltimore, Md. 1,500 pp.

Warinner, J. E., J. P. Miller, and J. Davis. 1970. Distribution of juvenile river herring in the Potomac River. Pages 384 to 387 *in* Proc. of 23rd Annual Conference of the Southeastern Assoc. of Game and Fish Commissioners, 1970. Available from Virginia Institute of Marine Science, Gloucester Point, Va.

Warner, W. W. 1976. Beautiful Swimmers: Watermen, Crabs and the Chesapeake Bay. Little Brown and Co., Boston, Mass. 304 pp.

Washington Post. 1974. Design changes are pushing up plant estimates. Page A8 *in The Washington Post,* December 1.

Wass, M. L., ed. 1972. A check list of the biota of lower Chesapeake Bay. Virginia Insitute of Marine Science, Gloucester Point, Va. Special Scientific Rept. No. 65.

Wiemeyer, S. N. 1971. Reproductive success of Potomac River ospreys-1970. *Chesapeake Sci.* 12(4):278-280.

Wigley, R. C., and B. R. Burns. 1971. Distribution and biology of mysids (Crustacea: Mysidacea) from the Atlantic Coast of the United States in the NMFS Woods Hole collection. *Fish. Bull.* 69:717-746.

Wiley, M. L. 1970. Fishes of the lower Potomac River. *Atlantic Naturalist* 25(4):151-159.

Wiley, M. L., and J. G. Boone. 1973. Bluefish. Pages 40 to 41 *in* The Chesapeake Bay in Maryland: An Atlas of Natural Resources, A. J. Lippson, ed. The Johns Hopkins Univ. Press, Baltimore, Md.

Williams, A. B. 1974. The swimming crabs of the genus *Callinectes* (Decapoda: Portunidae). *Fish. Bull.* 72(3):685-798.

Williams, A. B., and E. E. Deubler. 1968. A ten-year study of meroplankton in North Carolina estuaries: Assessment of environmental factors and sampling success among bothid flounders and penaeid shrimps. *Chesapeake Sci.* 9(1):27-41.

Willner, G. R., J. A. Chapman, and D. Pursley. In press. Reproduction, physiological responses, food habits, and abundance of nutria in Maryland marshes. *J. Wildlife Management.*

Wilson, J. S., R. P. Morgan, G. B. Gray, P. W. Jones, J. Lawson, and R. Lunsford. 1974. Potomac River fisheries study: Striped bass spawning stock assessment, interim data report. Prepared for Power Plant Siting Program, Md. Dept. of Natural Resources by Univ. of Md., Center for Environmental and Estuarine Studies, Natural Resources Institute, Chesapeake Biological Laboratory, Solomons, Md. Ref. No. 74-91.

Wilson, J. S., R. P. Morgan, P. W. Jones, J. Lawson, R. Lunsford, and S. Murphy. 1975. Potomac River fisheries study: Striped bass spawning stock assessment, interim data report. Prepared for Power Plant Siting Program, Md. Dept. of Natural Resources by Univ. of Md., Center for Environmental and Estuarine Studies, Natural Resources Institute, Chesapeake Biological Laboratory, Solomons, Md. Ref. No. 75-91.

Wilson, J. S., R. P. Morgan, P. W. Jones, H. R. Lunsford, J. Lawson, and S. Murphy. 1976. Potomac River fisheries study: Striped bass spawning stock assessment, final report 1975. Prepared for Power Plant Siting Program, Md. Dept. of Natural Resources by Univ. of Md., Center for Environmental and Estuarine Studies, Chesapeake Biological Laboratory, Solomons, Md. UMCEES Ref. No. 76-14 CBL.

Woodin, S. A. 1974. Polychaete abundance patterns in a marine soft-sediment environment: The importance of biological interactions. *Ecological Monographs* 44(2):171-187.

Young, R. H. 1953. An investigation of the inshore populations of the spot (*Leiostomos xanthurus,* Lacepede) with particular reference to seasonal growth and size distribution in Chesapeake Bay. M. S. Thesis. Univ. of Md. 53 pp.

Zaneveld, J. S., and W. D. Barnes. 1965. Reproductive periodicities of some benthic algae in lower Chesapeake Bay. *Chesapeake Sci.* 6(1):17-32.

Zobell, C. E., and C. B. Feltham. 1942. The bacterial flora of a marine mud flat as an ecological factor. *Ecology* 23:69-77.

Zwerner, D. E., and A. R. Lawler. 1972. Some parasites of Chesapeake Bay fauna. Pages 78 to 88 *in* A check list of the biota of lower Chesapeake Bay, M. L. Wass, ed. Virginia Institute of Marine Science, Gloucester Point, Va. Special Scientific Rept. No. 65.

Index

Boldface numbers indicate pages that include information on a topic in a figure or illustration. The letters *fm* precede folio map numbers.

272

273

276

277

Summary Matrix

DC   **Nautical River Mile**   **Mouth**

95 90 85 80 75 70 65 60 55 50 45 40 35 30 25 20 15 10 5 0

PHYSICAL AND CHEMICAL CHARACTERISTICS

AVERAGE SURFACE SALINITY ZONES [a]
- High Flow (Spring)
- Low Flow (Fall)

Legend:
- Tidal Fresh
- Oligohaline
- Mesohaline

SEDIMENTS [b]
- Maryland Shoal
- Channel
- Virginia Shoal

Legend:
- Soft Muds
- Sands
- Firm Muds

WATER QUALITY [c]
- Present Status
- State Classification

Legend:
- Fair to Good
- Good
- Good to Excellent
- D.C. — DEF
- Md. Class I
- Md. Class II

BIOTIC DISTRIBUTIONS

PHYTOPLANKTON [d]
- Greens and Blue-Greens (Summer)
- (Winter)
- Diatoms (Summer)
- (Winter)
- Dinoflagellates (Summer)
- (Winter)

Legend:
- Significant Component of Phytoplankton Community
- Present, but Not a Significant Component of Phytoplankton Community

WETLANDS ALONG MAIN STEM [e]
- Maryland Shore
- Virginia Shore

Legend:
- Locations

ZOOPLANKTON [f]
- Rotifers
- Cladocerans (Freshwater Species)
- (Estuarine Species)
- Copepods
  - *Eurytemora affinis* (Winter-Spring)
  - *Acartia tonsa* (Summer)

Legend:
- Dense Population
- Moderate Population
- Sparse Population

BENTHIC MACROINVERTEBRATES AND SHELLFISHING [g]
- Blue Crabs (Summer)
- (Winter)
- Oysters
- Soft-shell Clams

Legend:
- Commercially Harvestable Areas
- Extent of Beds

FISH SPAWNING AND NURSERY AREAS [h]
- Striped Bass
- White Perch
- Herrings and Shad
- Forage Species [i]
- Estuarine Dependent Species [j]

Legend:
- Major Spawning Area
- Secondary Spawning Area
- Major Nursery Area
- Secondary Nursery Area

RESOURCE USE

FINFISHING [k]
- Recreational
- Commercial Gill Netting
- Commercial Pound Netting

Legend:
- Most Heavily Used Areas

PARKS [l]

MILITARY LANDS [l]

Legend:
- Along Maryland or Virginia Shore

HUMAN POPULATION CENTERS [l]

Legend:
- > 5,000 People per Square Mile

WASTE WATER DISCHARGE AREAS [m]

Legend:
- > 100 Million Gallons/Day
- < 100 Million Gallons/Day

POWER PLANT COOLING WATER DISCHARGE AREAS [m]

Legend:
- Locations

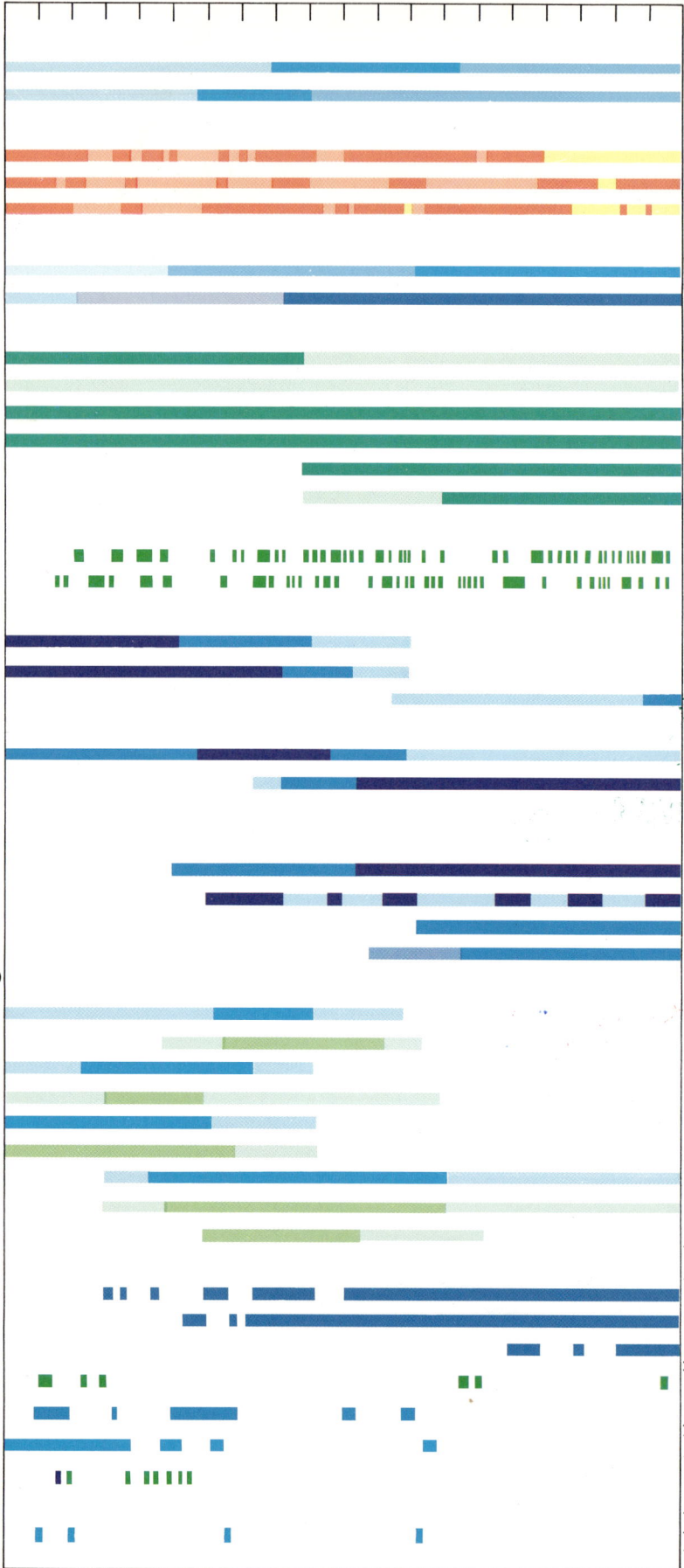

(a) Zone location may vary with river flow and runoff. See Chapter 3.
(b) See Folio Map 3, Chapters 1 and 3.
(c) See Chapter 3.
(d) See Chapter 4.
(e) See Folio Map 4, Chapter 5.
(f) See Chapter 6.
(g) See Folio Maps 5 and 6, Chapters 7 and 10.

(h) See Folio Map 7, Chapter 8.
(i) Bay anchovies and three species of silversides.
(j) Atlantic menhaden, spot, Atlantic croaker- spawning occurs in the Atlantic Ocean.
(k) See Chapter 10.
(l) See Folio Map 1.
(m) Discharge into main stem or into tributary area within 1 nautical mile of main stem. See Folio Map 9.